SLEEP AND AFFECT
Assessment, Theory, and Clinical Implications

SLEEP AND AFFECT

Assessment, Theory, and Clinical Implications

Edited by

KIMBERLY A. BABSON

National Center for PTSD and Center for Innovation to Implementation, VA Palo Alto Health Care System, Menlo Park, CA, USA & Department of Psychiatry and Behavioral Sciences, Stanford School of Medicine, Stanford, CA, USA

MATTHEW T. FELDNER

Department of Psychological Science, University of Arkansas, Fayetteville, AR, USA & Laureate Institute for Brain Research, Tulsa, OK, USA

AMSTERDAM • BOSTON • HEIDELBERG • LONDON
NEW YORK • OXFORD • PARIS • SAN DIEGO
SAN FRANCISCO • SINGAPORE • SYDNEY • TOKYO

Academic Press is an imprint of Elsevier

Academic Press is an imprint of Elsevier
32 Jamestown Road, London NW1 7BY, UK
525 B Street, Suite 1800, San Diego, CA 92101-4495, USA
225 Wyman Street, Waltham, MA 02451, USA
The Boulevard, Langford Lane, Kidlington, Oxford OX5 1GB, UK

© 2015 Elsevier Inc. All rights reserved.

Front cover credit: "Sleep Stage N3" by NascarEd - Own work. Licensed under Creative Commons Attribution-Share Alike 3.0 via Wikimedia Commons -http://commons.wikimedia.org/wiki/File:Sleep_Stage_N3.png#mediaviewer/File:Sleep_Stage_N3.png

No part of this publication may be reproduced or transmitted in any form or by any means, electronic or mechanical, including photocopying, recording, or any information storage and retrieval system, without permission in writing from the publisher. Details on how to seek permission, further information about the Publisher's permissions policies and our arrangements with organizations such as the Copyright Clearance Center and the Copyright Licensing Agency, can be found at our website: www.elsevier.com/permissions.

This book and the individual contributions contained in it are protected under copyright by the Publisher (other than as may be noted herein).

Notices
Knowledge and best practice in this field are constantly changing. As new research and experience broaden our understanding, changes in research methods, professional practices, or medical treatment may become necessary.

Practitioners and researchers must always rely on their own experience and knowledge in evaluating and using any information, methods, compounds, or experiments described herein. In using such information or methods they should be mindful of their own safety and the safety of others, including parties for whom they have a professional responsibility.

To the fullest extent of the law, neither the Publisher nor the authors, contributors, or editors, assume any liability for any injury and/or damage to persons or property as a matter of products liability, negligence or otherwise, or from any use or operation of any methods, products, instructions, or ideas contained in the material herein.

British Library Cataloguing in Publication Data
A catalogue record for this book is available from the British Library

Library of Congress Cataloging-in-Publication Data
A catalog record for this book is available from the Library of Congress

ISBN: 978-0-12-417188-6

For information on all Elsevier publications
visit our website at http://store.elsevier.com/

Contents

Foreword: Interrelationships between Sleep and Affect   xiii
EDWARD PACE-SCHOTT

Preface   xxvii
Acknowledgements   xxix
Contributors   xxxi

Part 1
DEFINITIONS

1. Neurophysiology of Sleep and Circadian Rhythms
JEFF DYCHE, KATHERINE C. COUTURIER, AND M. KATE HALL

Overview	4
Wakefulness	5
NREM Sleep	9
REM Sleep	11
Circadian Rhythms	13
Sleep Deprivation	15

2. Human Emotions: A Conceptual Overview
ANDREAS KEIL AND VLADIMIR MISKOVIC

Foundations of Affect and Emotion	23
Current Definitions and Theoretical Approaches	29
The Data of Emotion: Empirical Approaches	33
The Role of "Regulation" in Emotion Research	37
Conclusions and Elements of a Comprehensive Approach for the Study of Emotion	39

3. Sleep, Emotions, and Emotion Regulation: An Overview
CHRISTOPHER P. FAIRHOLME AND RACHEL MANBER

Sleep, Emotions, and Emotion Regulation: An Overview	45
Emotions and Emotion Regulation	46
Sleep and its Impact on Affect and Affect Regulation	47
Affect and its Impact on Sleep	50
Affect Regulation and its Impact on Sleep	53
Sleep and Affect: Bidirectional Relations	54
Discussion	56

Part 2
METHODS

4. Methodology for the Assessment of Sleep
CHRISTOPHER B. MILLER, SIMON D. KYLE, KERRI L. MELEHAN, AND DELWYN J. BARTLETT

Part 1: Objective Diagnostic Measures of Sleep	69
Part 2: Subjective Diagnostic Measures of Sleep	76
Part 3: Research-Focused Measures of Sleep	80
Summary	84

5. The Elicitation and Assessment of Emotional Responding
SARAH J. BUJARSKI, EMILY MISCHEL, COURTNEY DUTTON, J. SCOTT STEELE, AND JOSHUA CISLER

Primary Perspectives	92
Emotion Elicitation: Laboratory Methods	92
Trajectories of Emotional Experience	103
Elicitation Method and Assessment Strategy Considerations	108
Conclusions	110

6. Methodological Considerations When Integrating Experimental Manipulations of Sleep and Emotion
JARED MINKEL AND SAMANTHA PHILLIPS

Emotion Elicitation	120
Measurement of Emotional Responses	125
Special Considerations when Designing Experimental Studies of Sleep and Affect	132
Conclusion	135

Part 3
EVIDENCE REGARDING SLEEP AND SPECIFIC TYPES OF AFFECT
Section 1
SLEEP AND NEGATIVE AFFECT

7. The Interrelations Between Sleep and Fear/Anxiety: Implications for Behavioral Treatment
KIMBERLY A. BABSON

Background and Definition of Key Concepts	143
Fear/Anxiety Is Associated with Sleep Disturbances	145

Sleep Disturbance Is Associated with Elevations in Fear/Anxiety	147
Fear/Anxiety Is Associated with Sleep Disturbance	149
Theoretical Models	150
Proposed Mechanisms of Action	152
Implications for Intervention	154
Conclusions and Future Directions	157

8. Nightmares and the Mood Regulatory Functions of Sleep

PATRICK J. McNAMARA, UMBERTO PRUNOTTO, SANFORD H. AUERBACH, AND ALINA A. GUSEV

Nightmares and the Mood Regulatory Functions of Sleep	163
REM Properties	164
REM and Nightmares	165
Sleep Disorders with Nightmares	166
Mood Disturbances, Depression, and Nightmares	168
Theories of the Nightmare	170
Theory and Treatment Strategies	172
Future Directions	175

9. Isolated Sleep Paralysis and Affect

BRIAN A. SHARPLESS

Isolated Sleep Paralysis and Affect	181
Vignette	182
Diagnostic Criteria	184
Prevalence Rates	185
Associated Comorbidities	185
Sleep Paralysis and Affect	186
Clinical Impairment as a Result of Sleep Paralysis	194
Conclusions	195

10. Sleep and Repetitive Thought: The Role of Rumination and Worry in Sleep Disturbance

VIVEK PILLAI AND CHRISTOPHER L. DRAKE

Sleep and Repetitive Thought: The Role of Rumination and Worry in Sleep Disturbance	201
Worry: Phenomenology and Assessment	202
Worry and Sleep	203
Rumination: Phenomenology and Assessment	206
Rumination and Sleep	208
Rumination and Worry: Common and Distinguishing Features	210
Rumination, Worry, and Sleep: Theoretical Models	212
Rumination, Worry, and Sleep: Treatment Implications	215
Rumination, Worry, and Sleep: Other Clinical Implications	216
Conclusion	218

11. Sleep, Sadness, and Depression
RACHEL MANBER AND CHRISTOPHER P. FAIRHOLME

Sleep, Sadness, and Depression	227
How Sleep Impacts Sadness or Depression	228
How Poor and Insufficient Sleep Impact a Future Depressive Episode	229
The Impact of Poor Sleep on Next-Day Depressed Mood in Nondepressed Samples	230
The Impact of Sleep Deprivation on Mood During a Depressive Episode	231
The Impact of Poor Sleep on the Severity of Depression	232
The Impact of Poor Sleep on the Course of Depression Treatment	233
The Impact of Sadness or Depression on Sleep	234
Subjective Sleep Symptoms in Depression	234
Objective Sleep Abnormalities in Depression	234
How Sadness and Depression Impact Future Sleep Disorders	236
How Mood Impacts Next-Day Sleep	238
Summary and Implications for Future Research	239

12. The Interrelations Between Sleep, Anger, and Loss of Aggression Control
JEANINE KAMPHUIS AND MARIKE LANCEL

Introduction	247
Aspects of Sleep, Anger, and Aggression	248
Correlational Studies Showing the Relationship Between Sleep and Anger	249
What is the Causal Direction?	252
Possible Neurobiological Mechanisms Underlying the Relationship Between Sleep and Aggression	259
Groups at Risk	263
Clinical Implications	264
Conclusions and Future Directions	265

Section 2
SLEEP AND POSITIVE AFFECT

13. Positive Affect as Resilience and Vulnerability in Sleep
ANTHONY D. ONG, EMILY D. BASTARACHE, AND ANDREW STEPTOE

Positive Affect as Promotive Influence and Resilience	276
Promotive Effects	276
Protective Effects	281
Mechanisms	281
Summary	282
Positive Affect as Risk and Vulnerability	282

Mood Disorders and Sleep	282
Mechanisms	284
Summary	285
Conclusions	285

14. Sleep and Biological Rhythms in Mania
RÉBECCA ROBILLARD AND IAN B. HICKIE

Mania as an Activation Switch Dysfunction	293
Sleep and Circadian Disturbances in Mania-Related Illnesses	298
The Impact of Sleep Disturbances on Affective Regulation, Executive Functions, and Risk-Taking Behavior	303
Therapeutic Implications	305
Conclusion	309

15. Physical Activity, Sleep, and Biobehavioral Synergies for Health
MATTHEW P. BUMAN AND SHAWN D. YOUNGSTEDT

Exercise and Sleep: Overview of Primary Findings	322
Temporal Exercise-Sleep Relationships	323
Exercise Modalities and Sleep	324
Putative Biological Mechanisms of Exercise Effects on Sleep	325
Exercise and Obstructive Sleep Apnea	327
Exercise and Restless Leg Syndrome	328
Sedentary Behavior and Sleep	328
Synergies Among Sleep, Sedentary Behavior, and Exercise	329
A 24-h Approach to Chronic Disease Prevention	330
Summary	332

16. Mindfulness, Affect, and Sleep: Current Perspectives and Future Directions
SHEILA N. GARLAND, WILLOUGHBY BRITTON, NOEMI AGAGIANIAN, ROBERTA E. GOLDMAN, LINDA E. CARLSON, AND JASON C. ONG

Mindfulness, Affect, and Sleep: Current Perspectives and Future Directions	339
Mindfulness-Based Interventions	340
Evidence for the Efficacy of MBIs on Sleep	342
Study Characteristics	342
Outcome Measures for Sleep	354
Comparison Conditions	356
Theoretical Models of the Relationship of Mindfulness To Sleep and Affect?	358
Cognitive Models	359
Neurobiological Models	360
Future Directions and Implications	362

Section 3
EVIDENCE REGARDING SLEEP AND AFFECT AMONG SPECIAL POPULATIONS

17. Pain and Sleep
TIMOTHY ROEHRS AND THOMAS ROTH

Pain and Sleep	377
Methodological Issues	378
Association of Disturbed Sleep and Pain	379
Experimental Sleep Manipulations and Pain Sensitivity	384
Mechanisms Underlying the Sleep-Pain Nexus	386
Modulators of the Sleep-Pain Nexus: Mood and Cognitive Processes	387
Pharmacological and Behavioral Treatment of Sleep and Pain	389
Summary	392

18. The Impact of Sleep on Emotion in Typically Developing Children
REUT GRUBER, SOUKAINA PAQUIN, JAMIE CASSOFF, AND MERRILL S. WISE

Associations Among Key Aspects of Sleep Regulation, Emotion, and Mood	400
Circadian Tendencies and Socioemotional Adjustment	407
Neural Sleep Regulation and its Associations with Affect	408
Developmental Considerations Related to the Interplay Between Sleep and Affect	408
Directionality and the Contributions of Additional Factors	412
Summary	413

19. Sleep and Adolescents
ELEANOR L. McGLINCHEY

Sleep and Adolescents	421
Changes in Normal Sleep	422
Conceptual Models	424
Current State of the Research	426
Discussion of the Impact of Sleep on Adolescent Well-Being and Prevention	432

20. The Relationship Between Sleep and Emotion Among the Elderly
PASCAL HOT, ISABELLA ZSOLDOS, AND JULIE CARRIER

Emotion Regulation in Aging: The "Paradox of Well-Being"	442
Decline in Emotional Processing with Aging	444
Age-Related Sleep Modifications	446
Are Age-Related Sleep Modifications Linked to Emotion Regulation in Aging?	448

Part 4
FUTURE DIRECTIONS

21. Sleep and Affect: An Integrative Synthesis and Future Directions
KIMBERLY A. BABSON AND MATTHEW T. FELDNER

The Complex Interplay of Sleep and Affect	464
Factors Implicated in Links Between Sleep and Affect	467
Predisposing Factors	467
Causal Links	470
Clinical Implications	473
General Conclusions	477

Index   **485**

Foreword

Interrelationships between Sleep and Affect

Edward Pace-Schott
Harvard Medical School and Massachusetts General Hospital, Boston, MA, USA

Sleep and Affect: Assessment, Theory, and Clinical Implications provides the first comprehensive review of the emerging synthesis between the affective neurosciences and sleep psychology and medicine. Researchers frequently hypothesize that normal sleep helps regulate emotions in healthy humans (e.g., Cartwright, Luten, Young, Mercer, & Bears, 1998; Dahl & Lewin, 2002; Deliens, Gilson, & Peigneux, 2014; Germain, Buysse, & Nofzinger, 2008; Goldstein & Walker, 2014; Kramer, 1993; Levin & Nielsen, 2007; Soffer-Dudek, Sadeh, Dahl, & Rosenblat-Stein, 2011; Walker, 2009; Walker & van der Helm, 2009). The bulk of the empirical evidence supporting this assertion has emerged only recently, however, and it represents a dynamic and expanding field of inquiry brought together in this volume. After introducing key concepts and methodologies in each field, the book discusses the interactions between sleep and the full complement of negative and positive emotions, while explaining the clinical implications of those interactions. Some relationships between sleep and emotion have been widely studied, especially in the context of mood and anxiety disorders, whereas researchers are only beginning to examine others, such as the possible link between sleep quality and anger. Once we have considered the more general interactions between sleep and emotion, we explore how these interactions manifest in special populations such as children and adolescents, among whom sleep problems may presage and even contribute to later psychiatric disorders. Importantly, this volume deals with sleep and emotion across the spectrum of mental health from the normal covariance of mood and sleep to the pathological extremes.

People often take for granted the fact that sleep profoundly affects one's emotional state. In most cases, a good night of sleep improves mood, along with other subjective experiences of mind-body health such as restedness and stamina. In fact, the absence of such benefits can be associated with symptoms of sleep or psychiatric disorders, referred to as nonrestorative sleep or anxious awakenings. As a tried and true adage of folk psychology, the individual expects to be a different person after a good night's sleep. So, what takes place across sleep in healthy individuals to produce such a reliable change? In part, improved mood is expected to result from sleep based on Borbely's two-process model (Borbely, 1982) in which sleep propensity results from an interaction between circadian and sleep-homeostatic factors. For example, circadian rhythms, such as the preawakening rise in cortisol (Kalsbeek et al., 2012) and the cortisol awakening response (Federenko et al., 2004), are believed to prepare us for the challenges of the new day, and these processes could promote a sense of vigor upon awakening. Similarly, having slept away homeostatic sleep pressure and reduced levels of endogenous somnogens, such as adenosine (Porkka-Heiskanen & Kalinchuk, 2011), in the central nervous system (CNS), one might expect to experience improved mood, which can be further boosted by the adenosine-receptor antagonist, caffeine (Fisone, Borgkvist, & Usiello, 2004). An abrupt change in the forebrain's neuromodulatory milieu also coincides with abundant REM sleep late in the sleep period, during which monoamines, such as serotonin, are at their nadir, ending when the individual wakes, at which time they return to high levels (Pace-Schott & Hobson, 2002). Such state-dependent changes are undoubtedly important for producing a mild morning euphoria (with notable exceptions being many adolescents and evening chronotypes!). Other nightly changes promote enduring aspects of emotional health regardless of waking mood, however. Such changes have now become long-overdue subjects of many scientific investigations, such as those addressing the sleep-dependent consolidation of emotional memories, and this volume provides the first systematic compendium of their findings.

These new investigations into sleep and emotion are increasingly necessary, given the well-documented trend toward voluntary curtailment of sleep in Western societies, a trend that may be broadly problematic for physical, mental, and, especially, emotional health. From the 1950s to the first decade of this century, average sleep duration in adults has decreased from over 8 to under 7h per night (Van Cauter, Knutson, Leproult, & Spiegel, 2005). Recent research has shown that many physiological systems are negatively influenced by insufficient sleep (Van Cauter et al., 2007). For example, sleep deprivation is associated with endocrine abnormalities such as elevated evening cortisol (Leproult, Copinschi, Buxton, & Van Cauter, 1997), immunological abnormalities such as increased inflammatory markers (Mullington, Haack, Toth, Serrador, & Meier-Ewert, 2009; Pejovic et al., 2013), and heightened risk of cardiovascular disease (Solarz, Mullington, & Meier-Ewert, 2012). Similarly, sleep deprivation is associated with metabolic abnormalities contributing to obesity, insulin resistance, and, ultimately, type II diabetes (Knutson, Spiegel, Penev, & Van Cauter, 2007). Sleep disorders such as obstructive sleep apnea and insomnia are linked to hypertension (Fernandez-Mendoza et al., 2012; Palagini et al., 2013) and elevated sympathetic nervous system activity (Zhong et al., 2005). Notably, many of these physiological abnormalities can be reversed by naps or recovery sleep (Pejovic et al., 2013; Vgontzas et al., 2007). Although emotional state is intimately linked to physiological homeostasis in both the peripheral nervous system and CNS (Craig, 2002; Damasio, 2003), the effect of insufficient sleep on physiological aspects of normal emotional regulation has only begun to attract inquiry.

CNS function and cognition are similarly impacted by insufficient sleep. Loss of vigilance, especially at unfavorable circadian periods for maintenance of wakefulness, leads to a large number of automobile, public transportation, and industrial accidents (Garbarino, Nobili, Beelke, De Carli, & Ferrillo, 2001). In the elderly, sleep disturbances may be a risk factor for incident cognitive impairment (Blackwell et al., 2011). Cognitive skills such as working memory, short-term memory, and logical reasoning are especially disrupted by sleep deprivation (Chee & Chuah, 2008). Such functions rely on the prefrontal regions of the brain and include executive processes such as decision-making and behavioral inhibition (Chee & Chuah, 2008; Drummond, Paulus, & Tapert, 2006; Killgore, Balkin, & Wesensten, 2006). These prefrontal areas include the major loci of emotion regulation (Ochsner & Gross, 2005; Schiller & Delgado, 2010). Thus, sleep loss also impacts cognitive processes for which emotional information is essential, such as moral reasoning (Killgore et al., 2007), emotional intelligence (Killgore et al., 2008), and affect-guided decision-making (Killgore et al., 2006).

The following brief overview of some recent findings on sleep and emotion will whet the reader's appetite for the extensive treatment provided later in the book. Many of these studies have used total sleep deprivation (TSD) and sleep restriction protocols. TSD can reportedly impair recognition of facial emotion (van der Helm, Gujar, & Walker, 2010), and sleep restriction slows the expression of facial emotions (Schwarz et al., 2013). Similarly, even mild sleep restriction can impair emotional regulation in children (Gruber, Cassoff, Frenette, Wiebe, & Carrier, 2012), and normal variation in sleep quality can affect an individual's high-level ability to reappraise negative stimuli in normal adults (Mauss, Troy, & Lebourgeois, 2013). Functional neuroimaging studies have shown distinct effects of TSD on the neural circuits involved in emotion regulation (Gujar, Yoo, Hu, & Walker, 2011; Yoo, Gujar, Hu, Jolesz, & Walker, 2007). Following TSD, activation is reduced in regions, such as the ventromedial prefrontal cortex (vmPFC), that inhibit expression of negative emotion (Thomas et al., 2000; Yoo et al., 2007) but increased in the amygdala in response to emotional stimuli (Yoo et al., 2007). Moreover, sleep deprivation disrupts the functional connectivity between the vmPFC and the amygdala (Yoo et al., 2007). A recent fMRI study has further shown that the sleep debt accrued by prolonged sleep restriction can produce a similar hyperresponsivity of the amygdala, while reducing its functional connectivity with the vmPFC (Motomura et al., 2013).

Unlike normal emotion regulation, psychopathology appears to be strongly linked to the sleep disturbances that are ubiquitous in affective and anxiety disorders (Ford & Cooper-Patrick, 2001; Harvey, 2008, 2011; Kobayashi, Boarts, & Delahanty, 2007; Mellman, 2006, 2008; Peterson & Benca, 2006; Riemann, Berger, & Voderholzer, 2001). Among the anxiety disorders, poor sleep quality (e.g., low efficiency, prolonged onset latency) is common in posttraumatic stress disorder (PTSD), generalized anxiety disorder (GAD), and panic disorder (Mellman, 2006, 2008). Moreover, the comorbidity of insomnia with both GAD (Monti & Monti, 2000; Ohayon, 1997) and major depressive disorder (MDD) (Staner, 2010) is extremely high. Among specific sleep stages, decreased slow-wave sleep (SWS) is seen in both depression (Peterson & Benca, 2006; Steiger & Kimura, 2010) and PTSD (Kobayashi et al., 2007). REM abnormalities, often interpreted as elevated REM pressure, are also widely reported in MDD and include shortened REM latency and increased REM density (Modell & Lauer, 2007; Steiger & Kimura, 2010; Tsuno, Besset, & Ritchie, 2005). Elevated REM density is also seen in PTSD (Kobayashi et al., 2007). In MDD, TSD produces a transient antidepressant effect. In mania, sleep loss can produce a similar, albeit pathological, directionality of mood change by triggering manic episodes (Harvey, 2008, 2011). In striking contrast, TSD has been shown to have a mood-lowering effect in healthy individuals (Haack & Mullington, 2005; Zohar, Tzischinsky, Epstein, & Lavie, 2005).

Sleep may regulate emotion via its influences on emotional memory. Given the increasing interest in the sleep dependency of memory consolidation over the past decade (Diekelmann & Born, 2010; Diekelmann, Wilhelm, & Born, 2009; Stickgold, 2005; Walker & Stickgold, 2006), reports of sleep effects on emotional memory have multiplied. For example, the consolidation of emotional memory is enhanced by nocturnal sleep (Baran, Pace-Schott, Ericson, & Spencer, 2012; Hu, Stylos-Allan, & Walker, 2006; Wagner, Gais, & Born, 2001; Wagner, Hallschmid, Rasch, & Born, 2006) as well as by daytime naps (Nishida, Pearsall, Buckner, & Walker, 2009). Sleep can also promote the offline reorganization of emotional memory: A night of sleep enhances the trade-off in which an emotionally salient foreground image is selectively recalled over a neutral background (Payne, Chambers, & Kensinger, 2012; Payne, Stickgold, Swanberg, & Kensinger, 2008). Functional neuroimaging studies show distinct effects of TSD on the neural circuits involved in emotional memory (Menz et al., 2013; Payne & Kensinger, 2011; Sterpenich et al., 2007). There is differential activation in key components of the brain's emotional circuitry during recall of emotional stimuli depending upon whether or not sleep occurred after encoding. For example, the successful recall of emotional stimuli was associated with greater activation of the amygdala, hippocampus, and vmPFC in a group that slept versus one that was deprived of sleep following learning, and this group difference remained 6 months later (Sterpenich et al., 2007, 2009). Such findings provide a possible brain basis for an earlier behavioral study showing that superior performance on an emotional verbal memory task, measured shortly following postlearning sleep vs. continued wakefulness, was maintained for a full 4 years (Wagner et al., 2006). Another fMRI study has suggested that the retrieval of an emotional memory after a night of sleep engaged a more discrete and integrated set of limbic structures than retrieval of the same memory following a day awake (Payne & Kensinger, 2011).

Although emotional salience clearly interacts with sleep's effects on human declarative memory, evolutionarily ancient learning and memory processes may be equally important for human emotion regulation. These include processes involving simple associative memories, such as classical conditioning, through which an emotionally neutral stimulus is paired in time with a rewarding or punishing stimulus so that, when the emotionally salient stimulus is removed, the formerly neutral stimulus elicits a similar response (Pavlov, 1927). Following fear conditioning, another form of associative learning, fear extinction, allows the subject to learn that the once-feared stimulus is no longer dangerous (Hermans, Craske, Mineka, & Lovibond, 2006; Milad & Quirk, 2012). Rather than erasing fear, extinction forms a new inhibitory memory that coexists and competes with the neural processes of fear expression (Quirk & Mueller, 2008). Crucially, this indicates that extinction learning represents brain plasticity that, like other forms of memory, must consolidate following encoding in order to persist for later retrieval. A similarly primitive, nonassociative form of memory, habituation, involves reduced reactivity to frequently encountered stimuli (Grissom & Bhatnagar, 2009; Leussis & Bolivar, 2006; Thompson & Spencer, 1966). As with its "opposite," sensitization, habituation also involves neuroplastic changes that must consolidate in order to persist. Extinction memory is of particular clinical interest because its deficiency is likely associated with the development and perpetuation of anxiety disorders. For example, the failure to subsequently extinguish or remember the extinction of trauma-related memories and cues may perpetuate PTSD (Pitman et al., 2012), and extinction memory is impaired in PTSD at both the behavioral and neural level (Milad et al., 2008, 2009). Moreover, exposure therapy, which forms therapeutic extinction memories through *in vivo*, virtual reality, or imaginal exposure to fear, is the gold-standard treatment for anxiety disorders such as PTSD, specific phobia, social phobia, and obsessive compulsive disorder (Craske et al., 2008; Foa, Hembree, & Rothbaum, 2007; McNally, 2007).

Sleep promotes consolidation of fear extinction and habituation memory. For example, in one study, researchers used electric shock to condition subjects to fear two color stimuli, but only one color stimulus was then extinguished (Pace-Schott et al., 2009). Some subjects were subsequently allowed to sleep. After this period, both groups showed reduced conditioned responding to the extinguished stimulus (extinction recall), but only the subjects who had slept exhibited a reduced response to the unextinguished stimulus (extinction generalization). Similarly, sleep has been shown to enhance extinction memory consolidation and generalization following simulated exposure therapy for spider phobia (Pace-Schott, Verga, Bennett, & Spencer, 2012), a finding recently replicated using actual exposure therapy (Kleim et al., 2013). Sleep has also reportedly enhanced intersession habituation (Pace-Schott et al., 2011, 2014). Although circadian rhythms may additionally influence memory for extinction and habituation (Pace-Schott et al., 2013, 2014), other studies controlling for this influence confirm the importance of sleep between learning and recall (Kleim et al., 2013; Pace-Schott et al., 2011, 2014, 2012). Notably, despite supporting more uniquely human forms of emotion regulation, such as cognitive reappraisal of negative stimuli, higher-level associational areas, such as the dorsomedial and dorsolateral prefrontal cortices, may not be highly sensitive to normal variations in self-reported sleep quality (Minkel et al., 2012), even though behavioral expression may be affected (Mauss et al., 2013). Therefore, the effects of sleep quality on the more elemental means of achieving emotional homeostasis, such as those that rely on autonomic structures and somatovisceral feedback, may be one mechanism by which poor sleep leads to impairments of mood.

To conclude this introduction to sleep and affect, we introduce an interesting controversy that has already emerged on the bases of early studies—the relative importance of different sleep stages to emotion regulation. Both the occurrence and amount of REM sleep have been associated with enhanced emotional memory (Hu et al., 2006; Nishida et al., 2009; Wagner et al., 2001, 2006). Based upon such evidence, Walker and colleagues have advanced a "Sleep to Remember, Sleep to Forget" (SRSF) model (Goldstein & Walker, 2014; Walker, 2009; Walker & van der Helm, 2009). In this model, REM serves to both consolidate the contents of an emotional memory and to simultaneously depotentiate the emotional charge associated with that memory. As evidence for this model, the authors have shown that a daytime nap containing REM prevents an increase in reactivity to facial expressions of a negative emotion (anger and fear) observed following a day without such a nap (Gujar, McDonald, Nishida, & Walker, 2010). Similarly, following exposure to negative stimuli, the spectral power of high-frequency gamma (>30 Hz) EEG oscillations during subsequent nocturnal REM was negatively associated with the next-day subjective negative ratings of the previously seen stimuli, as well as with amygdala activation measured via fMRI (van der Helm et al., 2011). The authors suggest that higher gamma power indicates higher REM-sleep levels of norepinephrine, a neuromodulator that normally reaches its daily nadir during REM (Pace-Schott & Hobson, 2002), and, hence, a less favorable physiological milieu in REM for the depotentiation of emotion. Deliens and colleagues (Deliens, Gilson, Schmitz, & Peigneux, 2013) have provided some support for the SRSF model, using the mood-dependent memory (MDM) effect, whereby memories encoded in one emotional state are better retrieved in that same emotional state. Using a mood-induction technique, participants learned word pairs in one mood and were tested in another. When one night's TSD followed encoding, the MDM effect remained after two night's recovery sleep, whereas if normal sleep followed encoding, the MDM effect was eliminated. However, in a follow-up study, the MDM effect remained following late-night, REM-rich sleep, as well as following early-night SWS-rich sleep (Deliens, Neu, & Peigneux, 2013). Supporting SRSF, a neuroimaging study contrasting selective REM deprivation (REMD) with selective NREM deprivation (NREMD) showed an increase in behavioral reactivity to emotional images following REMD in comparison to a baseline night, whereas no change was seen following NREMD (Rosales-Lagarde et al., 2012).

fMRI has provided additional evidence for the involvement of REM sleep in emotion regulation. Rosales-Lagarde et al. (2012) studied subjects who performed an emotional reactivity task while in the scanner. The task was repeated with the same stimuli before and after a night of REMD or NREMD. In this task, subjects indicated behaviorally, with a defensive choice, whether or not they felt threatened when imagining themselves "in" an aversive photo. Those having REMD showed more defensive choices at the second presentation of the same images, but those with NREMD made similar choices. In addition, the expected habituation of a large number of cortical areas was see across responses to all stimuli at their second presentation following NREMD, whereas responses following REMD were of the same magnitude as initial responses. Moreover, for the contrast of perceived threatening versus nonthreatening images, activation of visual association areas increased between sessions following REMD but not NREMD.

Among lower-level forms of emotion regulation, Spoormaker et al. (2010) showed that extinction recall and activation of the vmPFC were greater in individuals who achieved REM during a 90-min afternoon nap that followed fear conditioning and extinction learning. In a subsequent study using all-night instrumental REMD compared to a NREMD, REMD resulted in poorer extinction memory, an effect accompanied by differential activation of the left middle temporal gyrus. Pace-Schott et al. (2014) showed that an index of extinction memory correlated with REM percent during the preceding night. In animal models, experimental stressors have been widely shown to reduce and fragment REM (Pawlyk, Morrison, Ross, & Brennan, 2008) and a REM-related brainstem-generated potential, the p-wave, has been shown to be essential to extinction memory (Datta & O'Malley, 2013).

However, other studies have advanced an entirely opposite prediction for the effects of REM on aversive memories, implicating REM in the preservation, rather than removal, of the emotional components of declarative memory (Baran et al., 2012; Groch, Wilhelm, Diekelmann, & Born, 2013; Lara-Carrasco, Nielsen, Solomonova, Levrier, & Popova, 2009; Pace-Schott et al., 2011; Wagner, Fischer, & Born, 2002). Adding to this apparent contradiction are studies showing that TSD following exposure attenuates aversive memories in humans (Kuriyama, Soshi, & Kim, 2010) and rats (Graves, Heller, Pack, & Abel, 2003; Kumar & Jha, 2012). For example, Kuriyama et al. (2010) associated particular pictorial contexts with neutral or aversive outcomes in video clips. These investigators then showed that sleep deprivation on the first subsequent night prevented the generalization of subjective and physiologically expressed fear from the aversive to the neutral context following recovery sleep. They therefore suggested that posttraumatic insomnia may adaptively serve to diminish the generalization of conditioned fear that might follow trauma. A later test of this theory, which added an active memory suppression procedure following encoding, showed that sleep deprivation increased rather than diminished physiological fear expression, however (Kuriyama, Honma, Yoshiike, & Kim, 2013). Similarly, in another study, time spent in REM sleep correlated with consolidation of conditioned fear but not extinction memory (Menz et al., 2013).

In addition, emotional memory can be modulated in NREM as well as REM. For example, an exposure therapy study showed symptom reduction to be associated with NREM, not REM sleep (Kleim et al., 2013), and inter-session habituation to aversive images has been negatively associated with REM and positively with SWS (Pace-Schott et al., 2011). Similarly, when an olfactory cue associated with a stimulus that was fear-conditioned during a prior period of wakefulness was presented during SWS, a diminished response to that cue during subsequent waking retrieval was observed (Hauner, Howard, Zelano, & Gottfried, 2013).

Therefore, at its earliest stages, this emerging field faces an apparent paradox and opposing implications for clinical interventions in traumatized individuals, in other words, to promote sleep and especially REM (e.g., Pace-Schott et al., 2009; Spoormaker & Montgomery, 2008; Walker & van der Helm, 2009) or, alternatively, to reduce sleep and especially REM (e.g., Baran et al., 2012; Kuriyama et al., 2010; Wagner et al., 2006). Interestingly, experimental stressors show negative effects on both REM and NREM sleep in humans (Vandekerckhove et al., 2011) or negative effects on REM and positive ones on SWS (Talamini, Bringmann, de Boer, & Hofman, 2013). Therefore, the specific sleep stage favoring emotion regulation may depend on the specific nature of presleep stress experienced or the type of emotion regulated. A nonexclusive alternate possibility is that certain combinations, sequences, or intensities of sleep stages or phasic events (e.g., sleep spindles) may provide optimal physiological conditions for specific types of emotional regulation.

Such paradoxical findings and consequent scientific controversies are an important sign of the health and vigor of the emerging field of affective neuroscience of sleep. *Sleep and Affect* provides an important introduction to this topic, and it provides a common basis of knowledge for experienced investigators as well as their students.

References

Baran, B., Pace-Schott, E. F., Ericson, C., & Spencer, R. M. (2012). Processing of emotional reactivity and emotional memory over sleep. *Journal of Neuroscience, 32*(3), 1035–1042.

Blackwell, T., Yaffe, K., Ancoli-Israel, S., Redline, S., Ensrud, K. E., Stefanick, M. L., et al. (2011). Association of sleep characteristics and cognition in older community-dwelling men: the MrOS sleep study. *Sleep, 34*(10), 1347–1356.

Borbely, A. A. (1982). A two process model of sleep regulation. *Human Neurobiology, 1*(3), 195–204.

Cartwright, R., Luten, A., Young, M., Mercer, P., & Bears, M. (1998). Role of REM sleep and dream affect in overnight mood regulation: A study of normal volunteers. *Psychiatry Research, 81*(1), 1–8.

Chee, M. W., & Chuah, L. Y. (2008). Functional neuroimaging insights into how sleep and sleep deprivation affect memory and cognition. *Current Opinion in Neurology, 21*(4), 417–423.

Craig, A. D. (2002). How do you feel? Interoception: The sense of the physiological condition of the body. *Nature Reviews Neuroscience, 3*(8), 655–666.

Craske, M. G., Kircanski, K., Zelikowsky, M., Mystkowski, J., Chowdhury, N., & Baker, A. (2008). Optimizing inhibitory learning during exposure therapy. *Behaviour Research and Therapy, 46*(1), 5–27.

Dahl, R. E., & Lewin, D. S. (2002). Pathways to adolescent health sleep regulation and behavior. *Journal of Adolescent Health, 31*(Suppl. 6), 175–184.

Damasio, A. (2003). Feelings of emotion and the self. *Annals of the New York Academy of Sciences, 1001*, 253–261.

Datta, S., & O'Malley, M. W. (2013). Fear extinction memory consolidation requires potentiation of pontine-wave activity during REM sleep. *Journal of Neuroscience, 33*(10), 4561–4569.

Deliens, G., Gilson, M., & Peigneux, P. (2014). Sleep and the processing of emotions. *Experimental Brain Research, 232*, 1403–1414.

Deliens, G., Gilson, M., Schmitz, R., & Peigneux, P. (2013). Sleep unbinds memories from their emotional context. *Cortex, 49*(8), 2221–2228.

Deliens, G., Neu, D., & Peigneux, P. (2013). Rapid eye movement sleep does not seem to unbind memories from their emotional context. *Journal of Sleep Research, 22*(6), 656–662.

Diekelmann, S., & Born, J. (2010). The memory function of sleep. *Nature Reviews Neuroscience, 11*(2), 114–126.

Diekelmann, S., Wilhelm, I., & Born, J. (2009). The whats and whens of sleep-dependent memory consolidation. *Sleep Medicine Reviews, 13*(5), 309–321.

Drummond, S. P., Paulus, M. P., & Tapert, S. F. (2006). Effects of two nights sleep deprivation and two nights recovery sleep on response inhibition. *Journal of Sleep Research, 15*(3), 261–265.

Federenko, I., Wust, S., Hellhammer, D. H., Dechoux, R., Kumsta, R., & Kirschbaum, C. (2004). Free cortisol awakening responses are influenced by awakening time. *Psychoneuroendocrinology, 29*(2), 174–184.

Fernandez-Mendoza, J., Vgontzas, A. N., Liao, D., Shaffer, M. L., Vela-Bueno, A., Basta, M., et al. (2012). Insomnia with objective short sleep duration and incident hypertension: The Penn State Cohort. *Hypertension, 60*(4), 929–935.

Fisone, G., Borgkvist, A., & Usiello, A. (2004). Caffeine as a psychomotor stimulant: Mechanism of action. *Cellular and Molecular Life Sciences, 61*(7–8), 857–872.

Foa, E., Hembree, E., & Rothbaum, B. (2007). *Prolonged exposure therapy for PTSD. Emotional processing of traumatic experiences. Therapist guide.* USA: Oxford University Press.

Ford, D. E., & Cooper-Patrick, L. (2001). Sleep disturbances and mood disorders: An epidemiologic perspective. *Depression and Anxiety, 14*(1), 3–6.

Garbarino, S., Nobili, L., Beelke, M., De Carli, F., & Ferrillo, F. (2001). The contributing role of sleepiness in highway vehicle accidents. *Sleep, 24*(2), 203–206.

Germain, A., Buysse, D. J., & Nofzinger, E. (2008). Sleep-specific mechanisms underlying posttraumatic stress disorder: Integrative review and neurobiological hypotheses. *Sleep Medicine Reviews, 12*(3), 185–195.

Goldstein, A. N., & Walker, M. P. (2014). The role of sleep in emotional brain function. *Annual Reviews of Clinical Psychology, 10,* 679–708.

Graves, L. A., Heller, E. A., Pack, A. I., & Abel, T. (2003). Sleep deprivation selectively impairs memory consolidation for contextual fear conditioning. *Learning & Memory, 10*(3), 168–176.

Grissom, N., & Bhatnagar, S. (2009). Habituation to repeated stress: Get used to it. *Neurobiology of Learning and Memory, 92*(2), 215–224.

Groch, S., Wilhelm, I., Diekelmann, S., & Born, J. (2013). The role of REM sleep in the processing of emotional memories: Evidence from behavior and event-related potentials. *Neurobiology of Learning and Memory, 99,* 1–9.

Gruber, R., Cassoff, J., Frenette, S., Wiebe, S., & Carrier, J. (2012). Impact of sleep extension and restriction on children's emotional lability and impulsivity. *Pediatrics, 130*(5), e1155–e1161.

Gujar, N., McDonald, S. A., Nishida, M., & Walker, M. P. (2010). A role for REM sleep in recalibrating the sensitivity of the human brain to specific emotions. *Cerebral Cortex, 21*(1), 115–123.

Gujar, N., Yoo, S. S., Hu, P., & Walker, M. P. (2011). Sleep deprivation amplifies reactivity of brain reward networks, biasing the appraisal of positive emotional experiences. *The Journal of Neuroscience, 31*(12), 4466–4474.

Haack, M., & Mullington, J. M. (2005). Sustained sleep restriction reduces emotional and physical well-being. *Pain, 119*(1–3), 56–64.

Harvey, A. G. (2008). Sleep and circadian rhythms in bipolar disorder: Seeking synchrony, harmony, and regulation. *The American Journal of Psychiatry, 165*(7), 820–829.

Harvey, A. G. (2011). Sleep and circadian functioning: Critical mechanisms in the mood disorders? *Annual Review of Clinical Psychology, 7,* 297–319.

Hauner, K. K., Howard, J. D., Zelano, C., & Gottfried, J. A. (2013). Stimulus-specific enhancement of fear extinction during slow wave sleep. *Nature Neuroscience, 16,* 1553–1555. http://dx.doi.org/10.1038/nn.3527 (published online 22 September 2013).

Hermans, D., Craske, M. G., Mineka, S., & Lovibond, P. F. (2006). Extinction in human fear conditioning. *Biological Psychiatry, 60*(4), 361–368.

Hu, P., Stylos-Allan, M., & Walker, M. P. (2006). Sleep facilitates consolidation of emotional declarative memory. *Psychological Science, 17*(10), 891–898.

Kalsbeek, A., van der Spek, R., Lei, J., Endert, E., Buijs, R. M., & Fliers, E. (2012). Circadian rhythms in the hypothalamo-pituitary-adrenal (HPA) axis. *Molecular and Cellular Endocrinology, 349*(1), 20–29.

Killgore, W. D., Balkin, T. J., & Wesensten, N. J. (2006). Impaired decision making following 49 h of sleep deprivation. *Journal of Sleep Research, 15*(1), 7–13.

Killgore, W. D., Kahn-Greene, E. T., Lipizzi, E. L., Newman, R. A., Kamimori, G. H., & Balkin, T. J. (2008). Sleep deprivation reduces perceived emotional intelligence and constructive thinking skills. *Sleep Medicine, 9*(5), 517–526.

Killgore, W. D., Killgore, D. B., Day, L. M., Li, C., Kamimori, G. H., & Balkin, T. J. (2007). The effects of 53 hours of sleep deprivation on moral judgment. *Sleep, 30*(3), 345–352.

Kleim, B., Wilhelm, F. H., Temp, L., Margraf, J., Wiederhold, B. K., & Rasch, B. (2013). Sleep enhances exposure therapy. *Psychological Medicine*, 1–9. http://dx.doi.org/10.1017/S0033291713001748.

Knutson, K. L., Spiegel, K., Penev, P., & Van Cauter, E. (2007). The metabolic consequences of sleep deprivation. *Sleep Medicine Reviews*, 11(3), 163–178.

Kobayashi, I., Boarts, J. M., & Delahanty, D. L. (2007). Polysomnographically measured sleep abnormalities in PTSD: A meta-analytic review. *Psychophysiology*, 44(4), 660–669.

Kramer, M. (1993). The selective mood regulatory function of dreaming: An update and revision. In A. Moffitt, M. Kramer, & R. Hoffman (Eds.), *The functions of dreaming*. Albany, NY: State University of New York Press.

Kumar, T., & Jha, S. K. (2012). Sleep deprivation impairs consolidation of cued fear memory in rats. *PLoS ONE*, 7(10), e47042.

Kuriyama, K., Honma, M., Yoshiike, T., & Kim, Y. (2013). Memory suppression trades prolonged fear and sleep-dependent fear plasticity for the avoidance of current fear. *Scientific Reports*, 3, 2227.

Kuriyama, K., Soshi, T., & Kim, Y. (2010). Sleep deprivation facilitates extinction of implicit fear generalization and physiological response to fear. *Biological Psychiatry*, 68(11), 991–998.

Lara-Carrasco, J., Nielsen, T. A., Solomonova, E., Levrier, K., & Popova, A. (2009). Overnight emotional adaptation to negative stimuli is altered by REM sleep deprivation and is correlated with intervening dream emotions. *Journal of Sleep Research*, 18(2), 178–187.

Leproult, R., Copinschi, G., Buxton, O., & Van Cauter, E. (1997). Sleep loss results in an elevation of cortisol levels the next evening. *Sleep*, 20(10), 865–870.

Leussis, M. P., & Bolivar, V. J. (2006). Habituation in rodents: a review of behavior, neurobiology, and genetics. *Neuroscience & Biobehavioral Reviews*, 30(7), 1045–1064.

Levin, R., & Nielsen, T. A. (2007). Disturbed dreaming, posttraumatic stress disorder, and affect distress: A review and neurocognitive model. *Psychological Bulletin*, 133(3), 482–528.

Mauss, I. B., Troy, A. S., & Lebourgeois, M. K. (2013). Poorer sleep quality is associated with lower emotion-regulation ability in a laboratory paradigm. *Cognition and Emotion*, 27(3), 567–576.

McNally, R. J. (2007). Mechanisms of exposure therapy: How neuroscience can improve psychological treatments for anxiety disorders. *Clinical Psychology Review*, 27(6), 750–759.

Mellman, T. A. (2006). Sleep and anxiety disorders. *The Psychiatric Clinics of North America*, 29(4), 1047–1058.

Mellman, T. A. (2008). Sleep and anxiety disorders. *Sleep Medicine Clinics*, 3, 261–268.

Menz, M. M., Rihm, J. S., Salari, N., Born, J., Kalisch, R., Pape, H. C., et al. (2013). The role of sleep and sleep deprivation in consolidating fear memories. *NeuroImage*, 75, 87–96.

Milad, M. R., Orr, S. P., Lasko, N. B., Chang, Y., Rauch, S. L., & Pitman, R. K. (2008). Presence and acquired origin of reduced recall for fear extinction in PTSD: Results of a twin study. *Journal of Psychiatric Research*, 42(7), 515–520.

Milad, M. R., Pitman, R. K., Ellis, C. B., Gold, A. L., Shin, L. M., Lasko, N. B., et al. (2009). Neurobiological basis of failure to recall extinction memory in posttraumatic stress disorder. *Biological Psychiatry*, 66(12), 1075–1082.

Milad, M. R., & Quirk, G. J. (2012). Fear extinction as a model for translational neuroscience: Ten years of progress. *Annual Review of Psychology*, 63, 129–151.

Minkel, J. D., McNealy, K., Gianaros, P. J., Drabant, E. M., Gross, J. J., Manuck, S. B., et al. (2012). Sleep quality and neural circuit function supporting emotion regulation. *Biology of Mood & Anxiety Disorders*, 2(1), 22.

Modell, S., & Lauer, C. J. (2007). Rapid eye movement (REM) sleep: An endophenotype for depression. *Current Psychiatry Reports*, 9(6), 480–485.

Monti, J. M., & Monti, D. (2000). Sleep disturbance in generalized anxiety disorder and its treatment. *Sleep Medicine Reviews*, 4(3), 263–276.

Motomura, Y., Kitamura, S., Oba, K., Terasawa, Y., Enomoto, M., Katayose, Y., et al. (2013). Sleep debt elicits negative emotional reaction through diminished amygdala-anterior cingulate functional connectivity. *PLoS ONE*, 8(2), e56578.

Mullington, J. M., Haack, M., Toth, M., Serrador, J. M., & Meier-Ewert, H. K. (2009). Cardiovascular, inflammatory, and metabolic consequences of sleep deprivation. *Progress in Cardiovascular Diseases, 51*(4), 294–302.

Nishida, M., Pearsall, J., Buckner, R. L., & Walker, M. P. (2009). REM sleep, prefrontal theta, and the consolidation of human emotional memory. *Cerebral Cortex, 19*(5), 1158–1166.

Ochsner, K. N., & Gross, J. J. (2005). The cognitive control of emotion. *Trends in Cognitive Sciences, 9*(5), 242–249.

Ohayon, M. M. (1997). Prevalence of DSM-IV diagnostic criteria of insomnia: Distinguishing insomnia related to mental disorders from sleep disorders. *Journal of Psychiatric Research, 31*(3), 333–346.

Pace-Schott, E. F., & Hobson, J. A. (2002). The neurobiology of sleep: Genetics, cellular physiology and subcortical networks. *Nature Reviews Neuroscience, 3*(8), 591–605.

Pace-Schott, E. F., Milad, M. R., Orr, S. P., Rauch, S. L., Stickgold, R., & Pitman, R. K. (2009). Sleep promotes generalization of extinction of conditioned fear. *Sleep, 32*(1), 19–26.

Pace-Schott, E. F., Shepherd, E., Spencer, R. M., Marcello, M., Tucker, M., Propper, R. E., et al. (2011). Napping promotes inter-session habituation to emotional stimuli. *Neurobiology of Learning and Memory, 95*(1), 24–36.

Pace-Schott, E. F., Spencer, R. M., Vijayakumar, S., Ahmed, N. A., Verga, P. W., Orr, S. P., et al. (2013). Extinction of conditioned fear is better learned and recalled in the morning than in the evening. *Journal of Psychiatric Research, 47*(11), 1776–1784.

Pace-Schott, E. F., Tracy, L. E., Rubin, Z., Mollica, A. G., Ellenbogen, J. M., Bianchi, M. T., et al. (2014). Interactions of time of day and sleep with between-session habituation and extinction memory in young adult males. *Experimental Brain Research*, http://dx.doi.org/10.1007/s00221-014-3829-9.

Pace-Schott, E. F., Verga, P. W., Bennett, T. S., & Spencer, R. M. (2012). Sleep promotes consolidation and generalization of extinction learning in simulated exposure therapy for spider fear. *Journal of Psychiatric Research, 46*(8), 1036–1044.

Palagini, L., Bruno, R. M., Gemignani, A., Baglioni, C., Ghiadoni, L., & Riemann, D. (2013). Sleep loss and hypertension: A systematic review. *Current Pharmaceutical Design, 19*(13), 2409–2419.

Pavlov, I. (1927). *Conditioned reflexes*. London: Oxford University Press.

Pawlyk, A. C., Morrison, A. R., Ross, R. J., & Brennan, F. X. (2008). Stress-induced changes in sleep in rodents: Models and mechanisms. *Neuroscience & Biobehavioral Reviews, 32*(1), 99–117.

Payne, J. D., Chambers, A. M., & Kensinger, E. A. (2012). Sleep promotes lasting changes in selective memory for emotional scenes. *Frontiers in Integrative Neuroscience, 6*, 108.

Payne, J. D., & Kensinger, E. A. (2011). Sleep leads to changes in the emotional memory trace: Evidence from FMRI. *Journal of Cognitive Neuroscience, 23*(6), 1285–1297.

Payne, J. D., Stickgold, R., Swanberg, K., & Kensinger, E. A. (2008). Sleep preferentially enhances memory for emotional components of scenes. *Psychological Science, 19*(8), 781–788.

Pejovic, S., Basta, M., Vgontzas, A. N., Kritikou, I., Shaffer, M. L., Tsaoussoglou, M., et al. (2013). Effects of recovery sleep after one work week of mild sleep restriction on interleukin-6 and cortisol secretion and daytime sleepiness and performance. *American Journal of Physiology Endocrinology and Metabolism, 305*(7), E890–E896.

Peterson, M. J., & Benca, R. M. (2006). Sleep in mood disorders. *Psychiatric Clinics of North America, 29*(4), 1009–1032 abstract ix.

Pitman, R. K., Rasmusson, A. M., Koenen, K. C., Shin, L. M., Orr, S. P., Gilbertson, M. W., et al. (2012). Biological studies of post-traumatic stress disorder. *Nature Reviews Neuroscience, 13*(11), 769–787.

Porkka-Heiskanen, T., & Kalinchuk, A. V. (2011). Adenosine, energy metabolism and sleep homeostasis. *Sleep Medicine Reviews, 15*(2), 123–135.

Quirk, G. J., & Mueller, D. (2008). Neural mechanisms of extinction learning and retrieval. *Neuropsychopharmacology, 33*(1), 56–72.

Riemann, D., Berger, M., & Voderholzer, U. (2001). Sleep and depression—Results from psychobiological studies: An overview. *Biological Psychology, 57*(1–3), 67–103.

Rosales-Lagarde, A., Armony, J. L., Del Rio-Portilla, Y., Trejo-Martinez, D., Conde, R., & Corsi-Cabrera, M. (2012). Enhanced emotional reactivity after selective REM sleep deprivation in humans: An fMRI study. *Frontiers in Behavioral Neuroscience, 6,* 25.

Schiller, D., & Delgado, M. R. (2010). Overlapping neural systems mediating extinction, reversal and regulation of fear. *Trends in Cognitive Sciences, 14*(6), 268–276.

Schwarz, J. F., Popp, R., Haas, J., Zulley, J., Geisler, P., Alpers, G. W., et al. (2013). Shortened night sleep impairs facial responsiveness to emotional stimuli. *Biological Psychology, 93*(1), 41–44.

Soffer-Dudek, N., Sadeh, A., Dahl, R. E., & Rosenblat-Stein, S. (2011). Poor sleep quality predicts deficient emotion information processing over time in early adolescence. *Sleep, 34*(11), 1499–1508.

Solarz, D. E., Mullington, J. M., & Meier-Ewert, H. K. (2012). Sleep, inflammation and cardiovascular disease. *Frontiers in Bioscience (Elite Edition), 4,* 2490–2501.

Spoormaker, V. I., & Montgomery, P. (2008). Disturbed sleep in post-traumatic stress disorder: Secondary symptom or core feature? *Sleep Medicine Reviews, 12*(3), 169–184.

Spoormaker, V. I., Sturm, A., Andrade, K. C., Schroter, M. S., Goya-Maldonado, R., Holsboer, F., et al. (2010). The neural correlates and temporal sequence of the relationship between shock exposure, disturbed sleep and impaired consolidation of fear extinction. *Journal of Psychiatric Research, 44*(16), 1121–1128.

Staner, L. (2010). Comorbidity of insomnia and depression. *Sleep Medicine Reviews, 14*(1), 35–46.

Steiger, A., & Kimura, M. (2010). Wake and sleep EEG provide biomarkers in depression. *Journal of Psychiatric Research, 44*(4), 242–252.

Sterpenich, V., Albouy, G., Boly, M., Vandewalle, G., Darsaud, A., Balteau, E., et al. (2007). Sleep-related hippocampo-cortical interplay during emotional memory recollection. *PLoS Biology, 5*(11), e282.

Sterpenich, V., Albouy, G., Darsaud, A., Schmidt, C., Vandewalle, G., Dang Vu, T. T., et al. (2009). Sleep promotes the neural reorganization of remote emotional memory. *The Journal of Neuroscience, 29*(16), 5143–5152.

Stickgold, R. (2005). Sleep-dependent memory consolidation. *Nature, 437*(7063), 1272–1278.

Talamini, L. M., Bringmann, L. F., de Boer, M., & Hofman, W. F. (2013). Sleeping worries away or worrying away sleep? Physiological evidence on sleep-emotion interactions. *PLoS ONE, 8*(5), e62480.

Thomas, M., Sing, H., Belenky, G., Holcomb, H., Mayberg, H., Dannals, R., et al. (2000). Neural basis of alertness and cognitive performance impairments during sleepiness. I. Effects of 24 h of sleep deprivation on waking human regional brain activity. *Journal of Sleep Research, 9*(4), 335–352.

Thompson, R. F., & Spencer, W. A. (1966). Habituation: A model phenomenon for the study of neuronal substrates of behavior. *Psychological Review, 73*(1), 16–43.

Tsuno, N., Besset, A., & Ritchie, K. (2005). Sleep and depression. *Journal of Clinical Psychiatry, 66*(10), 1254–1269.

Van Cauter, E., Holmback, U., Knutson, K., Leproult, R., Miller, A., Nedeltcheva, A., et al. (2007). Impact of sleep and sleep loss on neuroendocrine and metabolic function. *Hormone Research, 67*(Suppl. 1), 2–9.

Van Cauter, E., Knutson, K., Leproult, R., & Spiegel, K. (2005). The impact of sleep deprivation on hormones and metabolism. *Medscape Neurology & Neurosurgery, 7*(1), http://www.medscape.org/viewarticle/502825.

van der Helm, E., Gujar, N., & Walker, M. P. (2010). Sleep deprivation impairs the accurate recognition of human emotions. *Sleep, 33*(3), 335–342.

van der Helm, E., Yao, J., Dutt, S., Rao, V., Saletin, J. M., & Walker, M. P. (2011). REM sleep depotentiates amygdala activity to previous emotional experiences. *Current Biology, 21*(23), 2029–2032.

Vandekerckhove, M., Weiss, R., Schotte, C., Exadaktylos, V., Haex, B., Verbraecken, J., et al. (2011). The role of presleep negative emotion in sleep physiology. *Psychophysiology*, *48*(12), 1738–1744.

Vgontzas, A. N., Pejovic, S., Zoumakis, E., Lin, H. M., Bixler, E. O., Basta, M., et al. (2007). Daytime napping after a night of sleep loss decreases sleepiness, improves performance, and causes beneficial changes in cortisol and interleukin-6 secretion. *American Journal of Physiology Endocrinology and Metabolism*, *292*(1), E253–E261.

Wagner, U., Fischer, S., & Born, J. (2002). Changes in emotional responses to aversive pictures across periods rich in slow-wave sleep versus rapid eye movement sleep. *Psychosomatic Medicine*, *64*(4), 627–634.

Wagner, U., Gais, S., & Born, J. (2001). Emotional memory formation is enhanced across sleep intervals with high amounts of rapid eye movement sleep. *Learning & Memory*, *8*(2), 112–119.

Wagner, U., Hallschmid, M., Rasch, B., & Born, J. (2006). Brief sleep after learning keeps emotional memories alive for years. *Biological Psychiatry*, *60*(7), 788–790.

Walker, M. P. (2009). The role of sleep in cognition and emotion. *Annals of the New York Academy of Sciences*, *1156*, 168–197.

Walker, M. P., & Stickgold, R. (2006). Sleep, memory, and plasticity. *Annual Review of Psychology*, *57*, 139–166.

Walker, M. P., & van der Helm, E. (2009). Overnight therapy? The role of sleep in emotional brain processing. *Psychological Bulletin*, *135*(5), 731–748.

Yoo, S. S., Gujar, N., Hu, P., Jolesz, F. A., & Walker, M. P. (2007). The human emotional brain without sleep—A prefrontal amygdala disconnect. *Current Biology*, *17*(20), R877–R878.

Zhong, X., Hilton, H. J., Gates, G. J., Jelic, S., Stern, Y., Bartels, M. N., et al. (2005). Increased sympathetic and decreased parasympathetic cardiovascular modulation in normal humans with acute sleep deprivation. *Journal of Applied Physiology*, *98*(6), 2024–2032.

Zohar, D., Tzischinsky, O., Epstein, R., & Lavie, P. (2005). The effects of sleep loss on medical residents' emotional reactions to work events: A cognitive-energy model. *Sleep*, *28*(1), 47–54.

Preface

Sleep and Affect provides a one-stop source for state-of-the-art theory, methods, and clinical science regarding the interrelations between sleep and affect. Supported by a strong empirical backbone focused on contemporary research outlining links between sleep and affect, this book offers a guide to (1) state-of-the art research methods for studying sleep and affect and (2) contemporary clinical research supporting clinical decision making. As such, this book informs both research design and the treatment of patients suffering from problems related to sleep and affect. In order to provide clear guidance for researchers, the book also outlines future directions for the continued advancement of our understanding of sleep and affect.

A plethora of work has shed light on the relationships between sleep and specific psychological disorders (e.g., anxiety and mood disorders). Although this is a critical area of inquiry, fewer publications have more broadly summarized the connections between sleep and affect outside of the context of disorder. Understanding the interaction between sleep and affect from neurobiological, cognitive, and behavioral approaches can reveal how this interplay influences human development, as well as how these factors play a role in the etiology and maintenance of psychological disorders, thereby informing both prevention and clinical intervention.

Sleep and Affect addresses this critical gap by reviewing the state-of-the-science on the interrelations between sleep and affect. The first major section of the book provides an overview of neurobiological, behavioral, and functional definitions of key terminology related to sleep and affect. This section is followed by a detailed discussion of the methods used to study sleep and affect. The third section of the book outlines theory and research pertaining to sleep within the specific domains of positive and negative affect. The final section of the book focuses on sleep and affect in specific populations, including children, adolescents, older adults, and people suffering from pain. The book concludes with a synthesis and discussion of future directions for research.

This first edition book aims to balance the breadth and depth of available information to provide a resource that is accessible for upper-level seminars within academic settings, while serving as a practical guide for established scholars studying and treating conditions related to the interplay between sleep and affect.

Kimberly A. Babson
Matthew T. Feldner

Acknowledgements

The editors would like to thank all the authors who contributed to this book and acknowledge the following individuals for their assistance in the reviewing and development of this volume: Lisa Baldini, Marcel Bonn-Miller, Melynda Casement, Emily Mischel, and Steven Woodward.

Contributors

Noemi Agagianian Department of Psychology, Haverford College, Haverford, Pennsylvania, USA

Sanford H. Auerbach Boston University School of Medicine, Sleep Disorders Center, Boston, Massachusetts, USA

Kimberly A. Babson National Center for PTSD, VA Palo Alto Health Care System, Menlo Park, and Department of Psychiatry and Behavioral Sciences, Stanford School of Medicine, Stanford, California, USA

Delwyn J. Bartlett Centre for Integrated Research and Understanding of Sleep (CIRUS), Woolcock Institute of Medical Research, Sydney Medical School, University of Sydney, Sydney, New South Wales, Australia

Emily D. Bastarache Department of Human Development, Cornell University, Ithaca, New York, USA

Willoughby Britton Department of Psychiatry and Human Behavior, Brown University Medical School, Providence, Rhode Island, USA

Sarah J. Bujarski Department of Psychological Science, University of Arkansas, Fayetteville, Arkansas, USA

Matthew P. Buman School of Nutrition and Health Promotion, Arizona State University, Phoenix, Arizona, USA

Linda E. Carlson Department of Oncology, University of Calgary, Calgary, Alberta, Canada

Julie Carrier Functional Neuroimaging Unit, University of Montreal Geriatric Institute; Center for Advanced Research in Sleep Medicine (CARSM), Hôpital du Sacré-Cœur de Montréal, and Department of Psychology, University of Montreal, Montreal, Quebec, Canada

Jamie Cassoff Attention Behavior and Sleep Lab, Douglas Mental Health University Institute, and Department of Psychology, McGill University, Montreal, Quebec, Canada

Joshua Cisler Brain Imaging Research Center, Psychiatric Research Institute, University of Arkansas for Medical Sciences, Little Rock, Arkansas, USA

Katherine C. Couturier Naval Submarine Medical Research Laboratory, Groton, Connecticut, USA

Christopher L. Drake Sleep Disorders and Research Center, Henry Ford Health System, Detroit, Michigan, USA

Courtney Dutton Department of Psychological Science, University of Arkansas, Fayetteville, Arkansas, USA

Jeff Dyche Department of Psychology, James Madison University, Harrisonburg, Virginia, USA

Christopher P. Fairholme   Department of Psychiatry and Behavioral Sciences, Stanford University, Stanford, California, USA

Matthew T. Feldner   Department of Psychological Science, University of Arkansas, Fayetteville, Arkansas, and Laureate Institute for Brain Research, Tulsa, Oklahoma, USA

Sheila N. Garland   Department of Family Medicine and Community Health, Perelman School of Medicine, University of Pennsylvania, Philadelphia, Pennsylvania, USA

Roberta E. Goldman   Department of Psychiatry and Human Behavior, Brown University Medical School, Providence, Rhode Island, USA

Reut Gruber   Department of Psychiatry, McGill University, and Attention Behavior and Sleep Lab, Douglas Mental Health University Institute, Montreal, Quebec, Canada

Alina A. Gusev   VA New England Healthcare System, Boston, Massachusetts, USA

Ian B. Hickie   Clinical Research Unit, Brain and Mind Research Institute, University of Sydney, Sydney, Australia

Pascal Hot   Université de Savoie, Laboratoire de Psychologie et Neurocognition (CNRS UMR-5105), Chambéry Cedex, France

Jeanine Kamphuis   Department of Forensic Psychiatry, Mental Health Services, Assen, The Netherlands

M. Kate Hall   Department of Psychology, James Madison University, Harrisonburg, Virginia, USA

Andreas Keil   Department of Psychology and Center for the Study of Emotion and Attention, University of Florida, Gainesville, Florida, USA

Simon D. Kyle   School of Psychological Sciences, University of Manchester, Manchester, UK

Marike Lancel   Department of Forensic Psychiatry, and Centre for Sleep and Psychiatry Assen, Mental Health Services, Assen, The Netherlands

Rachel Manber   Department of Psychiatry and Behavioral Sciences, Stanford University, Stanford, California, USA

Eleanor L. McGlinchey   Columbia University Medical Center, New York State Psychiatric Institute, New York, New York, USA

Patrick J. McNamara   Department of Neurology, Boston University School of Medicine; VA New England Healthcare System, Boston, Massachusetts, and Graduate School Dissertation Chair, Northcentral University, Prescott Valley, Arizona, USA

Kerri L. Melehan   Centre for Integrated Research and Understanding of Sleep (CIRUS), Woolcock Institute of Medical Research, Sydney Medical School, University of Sydney, Sydney, New South Wales, and Department of Respiratory and Sleep Medicine, Royal Prince Alfred Hospital, Sydney, Australia

Christopher B. Miller   Centre for Integrated Research and Understanding of Sleep (CIRUS), Woolcock Institute of Medical Research, Sydney Medical School, University of Sydney, Sydney, New South Wales, Australia, and Institute of Neuroscience & Psychology, University of Glasgow, UK

Jared Minkel Department of Psychiatry and Behavioral Sciences, Duke University Medical Center, Durham, North Carolina, USA

Emily Mischel Department of Psychological Science, University of Arkansas, Fayetteville, Arkansas, USA

Vladimir Miskovic Department of Psychology, State University of New York at Binghamton, Binghamton, New York, USA

Anthony D. Ong Department of Human Development, Cornell University, Ithaca, New York, USA

Jason C. Ong Department of Behavioral Sciences, Rush University Medical Center, Chicago, Illinois, USA

Soukaina Paquin Attention Behavior and Sleep Lab, Douglas Mental Health University Institute, Montreal, Quebec, Canada

Samantha Phillips School of Arts and Sciences, Duke University, Durham, North Carolina, USA

Vivek Pillai Sleep Disorders and Research Center, Henry Ford Health System, Detroit, Michigan, USA

Umberto Prunotto Dreamboard, Inc., Alba, Italy

Rébecca Robillard Clinical Research Unit, Brain and Mind Research Institute, University of Sydney, Sydney, Australia

Timothy Roehrs Sleep Disorders and Research Center, Henry Ford Hospital, and Department of Psychiatry and Behavioral Neuroscience, School of Medicine, Wayne State University, Detroit, Michigan, USA

Thomas Roth Sleep Disorders and Research Center, Henry Ford Hospital, and Department of Psychiatry and Behavioral Neuroscience, School of Medicine, Wayne State University, Detroit, Michigan, USA

Brian A. Sharpless Department of Psychology, Washington State University, Pullman, Washington, USA

J. Scott Steele Brain Imaging Research Center, Psychiatric Research Institute, University of Arkansas for Medical Sciences, Little Rock, Arkansas, USA

Andrew Steptoe Department of Epidemiology and Public Health, University College, London, UK

Merrill S. Wise Methodist Healthcare Sleep Disorders Center, Memphis, Tennessee, USA

Shawn D. Youngstedt School of Nutrition and Health Promotion; College of Nursing and Health Innovation, Arizona State University, and Phoenix VA Healthcare System, Phoenix, Arizona, USA

Isabella Zsoldos Université de Savoie, Laboratoire de Psychologie et Neurocognition (CNRS UMR-5105), Chambéry Cedex, France

PART 1

DEFINITIONS

CHAPTER 1

Neurophysiology of Sleep and Circadian Rhythms[‡]

Jeff Dyche*, Katherine C. Couturier[†], and M. Kate Hall*

*Department of Psychology, James Madison University, Harrisonburg, Virginia, USA
[†]Naval Submarine Medical Research Laboratory, Groton, Connecticut, USA

During the past half century, behavioral scientists have generally recognized three stages of consciousness: Non-Rapid Eye Movement (NREM) sleep, rapid eye movement (REM) sleep, and wakefulness (National Sleep Foundation, 2014). These three stages are choreographed in predictable processes across a 24-h period, which is the circadian rhythm of the sleep/wake cycle. This cycle depends on multifaceted switching machinery, all of which are modulated by neural mechanisms. The past 50 years have also brought huge advances in neuroscience methodology. Some obvious examples include expensive neuroimaging devices such as functional magnetic resonance imaging (fMRI) and positron emission tomography (PET) scans that have added knowledge on brain functioning and behavior that was heretofore impervious to human understanding (Chee et al., 2006; Drummond & Brown, 2001; Germain, Nofzinger, Kupfer, & Buysse, 2004; Maquet et al., 1996; Ruottinen et al., 2000). But even with additional progress in more traditional electrophysiological and neuroanatomical techniques, sleep remains one of the great mysteries of modern behavioral neuroscience. We spend about a third of our lives asleep yet the exact biological mechanism underlying sleep is not completely described in the literature. Still, with the 1990s being dubbed the "decade of the brain," neuroscience research flourished, and with that momentum continuing as we near the middle of the second decade of the twenty-first century, scientists

[‡]The views expressed in this article are those of the author and do not necessarily reflect the official policy or position of the Department of the Navy, Department of Defense, nor the U.S. Government.

have been making enormous progress in understanding the mechanisms that control sleep and wakefulness.

In this chapter, we focus on four main areas of the central nervous system: the brainstem, spinal cord, medulla oblongata, and pons. Specifically, nerve cells that form the brainstem, the oldest part of our brain, are located just on top of the spinal cord and extend to the near middle of our brain, looking somewhat like the stem of an ice cream cone with a large overflowing scoop. The spinal cord, the major communication passageway for the motor and sensory portions of our brain, progresses from the inferior portions in the lower back to superior regions near the neck, and then it begins to widen, swell if you will, as it eases into the enigmatic confines of our cranium. This "swelling" of nerve cells is the beginning of the brainstem, is called the medulla oblongata, and is located at the bottom (inferior region) of the brainstem. As you might guess, the medulla isn't where conscious thought occurs, but it is a very important center of behavior as it regulates functions such as breathing, swallowing, and vomiting—automatic behaviors that we don't have to (or sometimes don't want to) think about. Only 1.5 in. in length, this oblong material is conical in shape. The bottom smaller portion is basically continuous with our spinal cord while the top wider portion is connected to the bridge of the brainstem called the pons (literally "bridge" in Latin, as it connects the lower and upper regions of the brainstem). However, as primitive and as small as the medulla is, it also plays an important role in the functioning of sleep. Animal models have demonstrated that if we lesion the medial (middle) portion of the medulla, REM sleep is altered. When that lesion extends into the pons, REM sleep may be completely eliminated. If that part of a person's brain was damaged, what are the behavioral ramifications of such a sleep change? How is NREM sleep implicated? And are circadian rhythms modulated by the same or different mechanisms? What are some of the myths and misconceptions of fatigue? These questions and more will be discussed in this chapter.

If I didn't wake up, I'd still be sleeping. *Yogi Berra*

OVERVIEW

Normal sleep is divided into two basic groups. One is Rapid Eye Movement sleep or "REM," famous for dream mentation and muscle paralysis and the other is the less celebrated yet equally important "non-REM" sleep or "NREM." NREM sleep is further divided into increasingly deeper stages of sleep: stage N1, stage N2, and stage N3. The latter is also referred to as the deepest stage of sleep where the brain produces large, synchronous "delta" waves sometimes referred to as "slow wave sleep" (Siegel, 2002). While REM sleep is not formally sub-divided, research has demonstrated that REM sleep is comprised of phasic and tonic components (Carskadon & Dement, 2011). *Phasic* REM sleep is a sympathetically driven state characterized by the stereotypical REMs with some distal twitching of the face and limbs. The *tonic* portion of REM is more parasympathetically enervated and involves little or no eye movements or twitching of distal muscle groups. Most REM sleep that occurs early in the night is tonic and later episodes are more phasic. The individual role of each subdivision is not yet understood. Of course a paramount feature of REM, whether tonic or phasic, is paralysis of the more proximal skeletal muscles.

Sleep is a state that may not be described as being truly unconscious as even in the deepest levels of sleep there is some monitoring of the external environment. Most, for example, can awaken at the sound of a baby crying or of a loud shout of the sleeping person's first name, so the brain must be aware albeit at a reduced level. Indeed, sleep researchers typically define sleep in a multistaged fashion in that there is (1) a reduction in awareness of the environment, (2) lowered motility and muscular activity, (3) partial suppression of voluntary behavior, and, to distinguish sleep from coma or a true loss of consciousness, it is (4) reversible (Everson, 2009).

One of the most important points when discussing the physiology of sleep is that sleep is a state where the brain is very active. To fall asleep is not the central nervous system slowing across a period of time, perhaps in reaction to a long day's work of cogitating and muscle contractions that lead to a passive quiescence where consciousness is lost due to the brain slowing inexorably under the workload. No, it doesn't work that way. In fact, there are certain regions of the brain that work very rigorously to generate sleep (Lu, Greco, Shiromani, & Saper, 2000; McGinty & Sterman, 1968; Nauta, 1946; Sherin et al., 1996; Siegel, 2002; von Economo, 1930). The absence of these regions in laboratory animals leads to profound insomnia and, in certain cases, death (Lu et al., 2000; Nauta, 1946). Similarly, destroying other regions of the brain in animals leads to coma-like states (Lindsley, Schreiner, Knowles, & Magoun, 1950). We will outline those structures that not only generate the major sleep categories, NREM and REM, but also what brain structures are vital for wakefulness. You will notice an important relationship between the sleep generating structures, the waking generating circuitry and the wake reducing circuitry.

Sleeping is no mean art: for its sake one must stay awake all day. ***Friedrich Nietzsche***

WAKEFULNESS

In the early twentieth century, most scientists who studied the brain believed that wakefulness was maintained due to sensory input to the brain. If that input was removed, then loss of consciousness, or sleep, would occur. This is, of course, a passive view of how sleep is generated. In 1916, a Vienna psychiatrist named Baron Constantine von Economo had been evaluating patients with an interesting form of brain infection he termed "encephalitis lethargica" that profoundly impacted the normal sleep-wake cycle (von Economo, 1930). Unlike normal sleepers who sleep 7-8h each day, most of his patients slept excessively long (up to 20h a day, an amount normally seen only in bats, the longest sleeping animal known). Autopsies on such patients revealed that these hypersomnolent individuals had lesions in the most rostral region (the superior portions) of the reticular formation (RF; Figure 1.1). This was the first clue that the midbrain portion of the brainstem contained some wake-promoting circuits. Moreover, a few of his patients actually experienced an opposite phenomenon as they were profoundly insomniac and slept just a few hours a day (i.e., as much as a typical horse, who are very short sleepers). These patients also had lesions but away from the brainstem proper in the anterior portions of the hypothalamus and in the adjacent basal forebrain (BF), suggesting to von Economo that these regions were involved in the induction of sleep.

Of course nowadays, based on animal literature and a few pathological conditions in humans (e.g., encephalitis lethargica), it is known that the waking state depends on the active working neurons within the RF region of the brainstem (Moruzzi & Magoun, 1949; Siegel, 2002). The RF, in turn, communicates with the higher functions of the brain via an intricate system of neurons that project from the RF called the ascending reticular activating system (ARAS). The ARAS initially moves to the switchboard of the sensory system, the thalamus, and from there it makes its way through various synapses in all lobes of the cerebrum, without which consciousness is not possible without at least partial regeneration of pathways via a plasticity driven recovery of function. This interesting

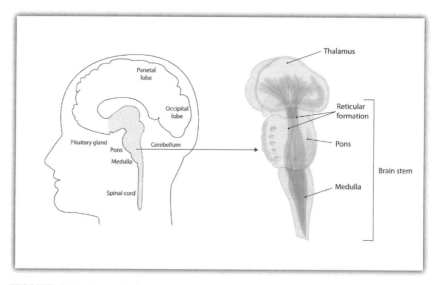

FIGURE 1.1 The reticular formation runs up the brainstem. This structure was damaged in von Economo's patients. wikicommons.

discovery of how the ARAS functioned was first demonstrated on cats by Moruzzi and Magoun in 1949 and is considered to be one of the most important concepts in the sleep-wake field (Siegel, 2002).

Incidentally, numerous psychological and physiological discoveries in the wake of Moruzzi and Magoun's discovery contributed to much more than just sleep science knowledge. Indeed, many psychologists with neurobiology backgrounds and neurobiologists with psychology backgrounds began collaborating with anatomists and physicians and the new interdisciplinary field of neuroscience was born. Therefore, sleep science might lay at least partial claim to being the progeny of one of the largest scientific disciplines in the world based on the total membership of the Society for Neuroscience (SfN, 2013).

To return to the anatomy and function of the ARAS—the sine qua non for arousal (and therefore consciousness)—it is further divided into two pathways in order to maintain normal alertness. The first dorsal (lower) branch consists of cholinergic neurons extending from the mesopontine tegmentum in the more caudal regions of the RF. This was actually anticipated by the early Moruzzi and Magoun paper but not confirmed until later (Siegel, 2002). There are two nuclei in the mesopontine tegmentum that generate large amounts of acetylcholine whose projections terminate in the thalamus (Jones and Beaudet, 1987). From the thalamus, the major sensory intersection of the brain, arousal information then projects widely to the cortex (Figure 1.2). This thalamic fan-like projection to the cortex is most likely mediated by the parabrachial nucleus (PB) and its glutamate productivity, the most common excitatory neurotransmitter in the brain as demonstrated by Lu, Sherman, Devor, and Saper (2006). The second pathway of the ARAS actually bypasses the thalamus, originating in the ventral portion of the brainstem and small regions of the lateral hypothalamus. This pathway includes imperative arousal centers such as the locus coeruleus (LC), a major producer of the excitatory neurotransmitter norepinephrine. Dopamine neurons are involved as well and begin their trajectory in the ventral periaqueductal gray matter (vPAG) within the midbrain. Dorsal and medial raphe nuclei (RN) are serotonin producers along this pathway while the mammillary bodies in the posterior hypothalamus produce histamine. In general, all these groups of monoamine-producing neurons fire more during wakefulness than during sleep. Of all of these groups in this second ARAS pathway, most research indicates that the vPAG is most actively involved with arousal (Lu, Jhou, & Saper, 2006) thereby implicating dopamine as a highly important modulator of consciousness.

Another region of arousal is often considered part of the second pathway described above, as all the monoamine-producing neurons project to this location, but it seems to have an augmenting effect on the entire ARAS so we will mention it separately. This region is the lateral hypothalamic area (LHA). It has been known at least since before WWII that the lateral

8 1. NEUROPHYSIOLOGY OF SLEEP AND CIRCADIAN RHYTHMS

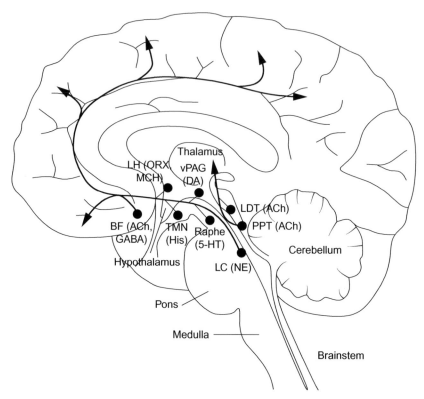

FIGURE 1.2 **This diagram of the ascending reticular activating pathway shows key neuronal projections that maintain alertness.** The cholinergic pathway (yellow) originates from cholinergic (ACh) neurons in the upper pons, the pedunculopontine (PPT), and laterodorsal tegmental nuclei (LDT). The second pathway (red) innervates the cerebral cortex and this pathway originates from neurons in the monoaminergic cell groups, including the tuberomammillary nucleus (TMN) which contains histamine (His), the vPAG which contains dopamine (DA), the dorsal and median RN which contain serotonin (5-HT), and the LC neurons which contain noradrenaline (NA). This pathway also receives modulations from neuropeptidergic neurons in the lateral hypothalamus (LHA) which contain hypocretin/orexin (ORX) or melanin-concentrating hormone (MCH), and from basal forebrain (BF) neurons that contain GABA or ACh. Saper, Scammell, and Lu (2005).

hypothalamus is meditated in wakefulness as discovered by Ranson (1939) who lesioned this important structure in monkeys and found them to be profoundly sleepy and/or in a virtual coma after recovery. More recent studies indicate the LHA contains a newly discovered peptide called hypocretin (also "orexin") that has been implicated in narcolepsy studies in which a loss of these cells produced strong sleep tendencies in numerous animal models (Chemelli et al., 1999; Mochizuki et al., 2004). It is through these two pathways that consciousness in humans and alertness in lower animals is maintained.

Many are curious as to where the switch from wakefulness to sleep occurs. Next, we will discuss the areas involved in NREM sleep and how the mutual inhibition of "sleep-promoting" and "wake-promoting" areas result in the transition between wake and sleep, commonly referred to as the "flip-flop" switch (Fuller, Gooley, & Saper, 2006; Saper et al., 2005).

> But I have promises to keep, and miles to go before I sleep, and miles to go before I sleep. **Robert Frost**

NREM SLEEP

It may not be surprising, after learning about the role of the ARAS and its importance in maintaining wakefulness, that somehow "turning off" this system would be important to the initiation and maintenance of sleep. Indeed, that is how it seems to work. Moreover, the arousal structures that are inhibited by sleep-promoting neurons also seem to disturb these same sleep processes to wake a sleeping brain (Saper et al., 2005). There may be several structures or systems that have the capability to contribute to this, but we will focus first on the initiation of NREM sleep.

Although sleep is a behavior that, ultimately, involves much of the brain, when it comes to putting us to sleep and quieting the ARAS, there is one brain structure that plays a role perhaps above them all—the ventrolateral preoptic area (VLPO) in the anterior hypothalamus (Figure 1.3). Damage to this small part of the brain in rats and cats causes profound insomnia, and sometimes death (McGinty & Sterman, 1968; Nauta, 1946). Sterman and Clemente (1962) discovered, in turn, that mild electrical currents to the preoptic area of the hypothalamus caused drowsiness, EEG Synchrony (typical of NREM sleep), and sometimes even put the animal completely to sleep. There is ample evidence that this part of the hypothalamus sends strong inhibitory signals via GABA and galanin projections to structures such as the LC, RN, and the mammillary bodies, all structures known to produce wakefulness. Therefore, the VLPO may induce sleep by suppressing the natural wake-promoting mechanism in the brain (McGinty & Szymusiak, 2000; Mohns, Karlsson, & Blumberg, 2006; Saper, Chou, & Scammell, 2001).

In addition to projecting powerful inhibitory messages to the ARAS, the neurons of the VLPO receive monoaminergic afferents from the same areas of the ARAS which it inhibits (Brown, Basheer, McKenna, Strecker, & McCarley, 2012; Chou et al., 2002). It is this process of mutual inhibition between the wake-promoting circuitry of the ARAS and the sleep-promoting areas of the VLPO that has led researchers to call it a "flip-flop" switch (Fuller et al., 2006; Lu et al., 2006; Nauta, 1946; Saper et al., 2005).

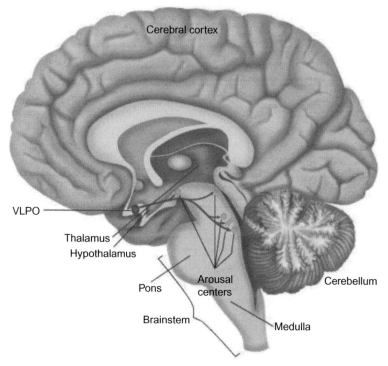

FIGURE 1.3 As a sleep-promoting structure, the VLPO projects inhibitory GABA to areas of the brain responsible for maintaining alertness. As such, NREM sleep ensues. Harvard Medical School Division of Sleep Medicine, 2008. http://healthysleep.med.harvard.edu.

There are several other studies that have confirmed the VLPOs role in sleep (Lu et al., 2000; Sherin et al., 1996). For example, it has been found that fos protein, which is a by-product of neural activity, has been found in the VLPO during sleep; while the subsequent lesioning of those specific neurons producing the fos protein suppressed sleep (Sherin, Elmquist, Torrealba, & Saper, 1998). Moreover, when animals are sleep deprived for almost a day and then allowed to sleep again, the VLPO neurons fire at an especially high rate, perhaps indicating a neural basis for the well-known "rebound" effect of sleep deprivation (Szymusiak et al., 1998).

Another structure that mediates NREM sleep is the BF (Berridge & Foote, 1996; Lee, Hassani, Alonso, & Jones, 2005; von Economo, 1930). This small structure includes nuclei in the preoptic nucleus, nucleus basalis, and portions of the diagonal band made famous by Paul Broca's disfluent patients (Kalat, 2013). This area has largely cholinergic projections to the hippocampus and the thalamus that can create bursts of EEG activity associated with waking. The BF is composed of numerous forebrain nuclei, extending caudolaterally from the medial septum to the ventral pallidum (VP), and projects reciprocally to the same wake-promoting structures as does the GABA-producing VLPO (Semba, 2000).

> The only time I have problems is when I sleep. *Tupac Shakur*

REM SLEEP

Of all the sleep stages, REM is probably the most well-known (or at least the most talked about by the general public). This is the stage most often associated with dreaming, and although researchers have found that dreaming can occur in all stages of sleep, dreams that occur during REM are found to be more vivid and easier to recall (Aserinsky & Kleitman, 1953).

REM, or "rapid eye movement," sleep was discovered by chance by a graduate student named Eugene Aserinsky, under the tutelage of Nathaniel Kleitman at the University of Chicago in the early 1950s (Dement & Vaughan, 1999; Siegel, 2002). At the time, Kleitman was interested in the relationship between slow eye movements and depth of sleep. It was during sessions of monitoring these eye movements in sleeping subjects that Aserinsky began to notice that occasionally some people who had been asleep for hours began to exhibit REMs, with EEG recordings that looked almost identical to the waking pattern of low voltage, fast waves (Dement & Vaughan, 1999). From here, the investigation into REM sleep function began.

The location of the REM generator was discovered by French scientist Michel Jouvet, who systemically lesioned sections of the cat brainstem, finding the area necessary for the neural control of sleep to be located in the pons, specifically the pontine tegmentum (Jouvet, 1962; Siegel, 2002). EEG recordings of the area displayed waves associated with the pontine component of a pattern of electrical potentials associated with REM sleep, called PGO waves. These waves occur in highest amplitude right before the onset of REM and in lower amplitude bursts during REM itself (Bizzi & Brooks, 1963; Jouvet, 1962). They are called PGO waves because of where recordings are found, first in the pons, then spreading to the lateral geniculate nucleus of the thalamus, and lastly to the occipital cortex—which perhaps indicates a relationship to the imagery we see in dreams (Ponto-Geniculo-Occipital) (Bizzi & Brooks, 1963). Further, PGO waves seem to occur concurrently with REMs that correspond with gaze direction associated with dream imagery (Steriade & McCarley, 2005). In cats with the isolated pontine lesion, only the pontine component of the PGO waves was seen, which gives further evidence that REM sleep originates in the pontine tegmentum.

In our discussion of NREM sleep, we indicated the role of the "flip-flop" switch in allowing for the transition between sleep and wake states. Research has demonstrated that this switch is also responsible for the transition to REM sleep (Lu et al., 2006). While in NREM, the mutually inhibitory regions involved in the switch include projections between the VLPO and wake-promoting monoaminergic regions and the neurons of the lateral hypothalamus; REM sleep involves the VLPO projections to the pontine tegmentum (Lu et al., 2006).

In the transition to REM, it appears that a system of GABA-ergic and glutamatergic "REM-on" and "REM-off" neuronal systems are what operate the circuitry in transitioning to REM sleep (Lu et al., 2006; Luppi et al., 2011). The VLPO sends inhibitory projections to the areas of the pontine tegmentum (as well as the ventrolateral periaqueductal gray area) not active during REM, thus "REM-off" neurons. These neurons then project to other "REM-on" areas of the brain, including glutamatergic neurons that send inhibitory messages to the spine, contributing to the muscle atonia associated with REM sleep (Clement, Sapin, Berod, Fort, & Luppi, 2011; Fuller & Lu, 2009; Lu et al., 2006). In fact, recent evidence suggests that damage to the glutamatergic regions projecting to the spine may have some involvement in REM behavior disorder, a sleep dysfunction that exhibits signs of REM sleep without muscle atonia, as well as the cataplexy associated with the most well-known REM disorder, narcolepsy (Clement et al., 2011; Luppi et al., 2011). As mentioned earlier, narcolepsy is also associated with a deficiency in hypocretin, an excitatory neuropeptide in the lateral hypothalamus associated with maintaining wakefulness and believed to play an important role in stabilizing the sleep/wake "flip-flop" switch (Chemelli et al., 1999; Saper et al., 2001).

Sleep is one of the many biological rhythms that ebb and flow across a 24-h period (Scheer & Shea, 2009; body temperature is another example). When a rhythm of a biological process cycles during such a time it is referred to as a circadian rhythm ("circa" is Latin for about a day). Even though the sleep-wake cycle is often used as an example of such a rhythm, it was not considered important to the early sleep researchers (Dement & Vaughan, 1999). This was due to fact that early sleep scientists often used cats as a model for sleep. Studying an animal that can sleep anytime anywhere had its advantages of course. But cats are one of the few animals that have poor circadian oscillation and can sleep at any time, as any cat owner can attest. However, as scientists turned more to studying other animals (including humans), it became evident that there are other processes and brain circuitry that can make sleep easier to obtain at certain times of the day, and, conversely, when wakefulness is easier to maintain.

There is a time for many words, and there is also a time for sleep. *Homer*

CIRCADIAN RHYTHMS

Although there are times during the day that we feel wide awake and others when we feel tired, most of us do not simply fall asleep randomly, without warning. Humans generally wake and sleep on a fairly predictable daily schedule. How does our body know when to sleep? While many contributing and complicating factors come into play, the underlying drives can be thought of in a simplified, two-process model first proposed in 1982 by Dr. Alexander Borbély, a Swiss sleep researcher (Borbély, 1982). This model suggests that our propensity to sleep can be thought of as the interaction between two different processes: the length of time awake (Process S; Figure 1.4) and our intrinsic circadian rhythm (Process C). Process S is the more straightforward of the two: an exponential increase in sleep propensity (as derived from slow wave activity) the longer one is awake. Conversely, that propensity declines in an exponential fashion with sleep. If this were the only factor influencing our drive for sleep, we would wake up alert and get progressively more and more tired as the day went on.

The other part of this model, Process C, reflects the complicated biology of our own internal clocks: Without external stimuli, the human biological

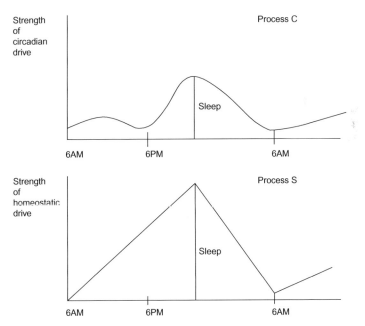

FIGURE 1.4 Graphs illustrating the two-process model of sleep propensity: Process S, determined by amount of time spent awake; and Process C, defined as our internal circadian rhythm.

clock runs with a periodicity slightly longer than 24h (Dijk & Lockley, 2002). The period of regular circadian oscillation in the absence of external cues is referred to as the "free-running" period or rhythm (Dunlap, Loros, & DeCoursey, 2004). We can experimentally observe this regulated daily rhythm in the suprachiasmatic nucleus (SCN), a small collection of hypothalamic cells just above the optic chiasm. This locus of "clock cells" coordinates the timing of peripheral molecular oscillators present in a wide range of tissues throughout the body (Relogio et al., 2011). An assortment of clock genes regulates individual variation between these tissues using positive and negative feedback loops to produce rhythmic activity, including gene transcription and phosphorylation, as well as post-transcriptional processing (Prasai, Pernicova, Grant, & Scott, 2011; Relogio et al., 2011; Reppert & Weaver, 2002). Under normal circumstances, the SCN maintains a regular 24-h cycle through the influence of external stimuli referred to as *zeitgebers* (German: "time-giver"). Zeitgebers may include eating schedules, activity levels, and social cues. In humans, dark/light shifts are far and away the most significant (Dunlap et al., 2004).

Light first affects our circadian rhythm when it bounces off our retina, the tissue at the back of our eyes. Unlike visual information, this process occurs independently of rods and cones (Thapan, Arendt, & Skene, 2001). Circadian light responses are proposed to be mediated by a third class of receptor called intrinsically photoreceptive retinal ganglion cells (Berson, Dunn, & Takao, 2002). These cells contain a photopigment called melanopsin which responds primarily to short-wavelength light of 446-483nm in the blue region of the light spectrum as indicated by melatonin suppression and induction of phase-shifting (Cajochen et al., 2005; Lockley, Brainard, & Czeisler, 2003; Thapan et al., 2001).

Entrained by photic input, the SCN prompts arousal, likely through direct projections to the posterior hypothalamic arousal systems (Abrahamson, Leak, & Moore, 2001). After light exposure, the SCN also sends signals directly to the pineal gland, causing the suppression of melatonin (Cajochen et al., 2005; Lockley et al., 2003). Circadian influence can be seen on a variety of systems within the human body. The most commonly studied in research scenarios are sleep-wake patterns, core body temperature, and melatonin and cortisol secretion.

Researchers can use these rhythmic markers to create a profile of each individual's circadian phase at any given point in time. Generally, the pattern shows a peak of melatonin secretion followed 1-2h later by sleep and the nadir of cortisol secretion, then the core body temperature reaches minimum 6-8h later accompanied by rising cortisol levels, a trough in serum melatonin and awakening around 2h later (Dunlap et al., 2004). Further data may be gathered on sleep stages, sleep onset, and sleep efficiency using polysomnography and actigraphy. These devices are useful not only in measuring sleep in its normal state but also during intentional restriction of sleep in laboratory settings. Measuring how sleep changes when restricted is just one outcome of brief or chronic sleep deprivation.

Normally, our circadian rhythm works to keep our internal biological processes in line with the world around us; however, there are several common situations that throw us out of sync. This disconnect between our internal circadian rhythm and our externally imposed schedule is referred to as "circadian misalignment" or "circadian dyssynchrony." Anyone who has traveled over multiple time zones has experienced the grogginess and difficulty of adapting to a sudden change in schedule: being wide awake in the middle of the night, struggling to wake in the morning. This stems from changing time zones too quickly for our finely tuned internal triggers for sleep and wakefulness to keep up (Sack, 2010). The American Academy of Sleep Medicine (AASM) refers to this misalignment as "time zone change (jet lag) syndrome" (American Academy of Sleep Medicine, 2005).

Circadian misalignment also rears up in night or rotating shift workers (Kolla & Auger, 2011). By nature of their work hours, these workers receive light and dark signals at odds with their desired schedule. Furthermore, on days off they often revert to a normal day-night schedule to keep up with social obligations. As a result, their internal circadian rhythms are often completely out of line with their sleep-wake scheduling, even after years of shift work (Arendt, 2010). These workers are at risk for "shift work sleep disorder" (American Academy of Sleep Medicine, 2005). This misalignment is not just of academic interest: trend analysis suggests that even when initial risk is assumed to be similar, there remains an increased risk of accidents during evening and night shifts (Folkard & Akerstedt, 2004).

> Men who are unhappy, like men who sleep badly, are always proud of the fact.
> **Bertrand Russell**

SLEEP DEPRIVATION

Many of us have felt "punch-drunk" after staying up all night; but little did we know that sleep deprivation really is like being drunk! Indeed, a NASA review of performance risks found significant evidence that performance deficits after 17h of wakefulness were comparable to performance at a blood alcohol level of 0.05% (Whitmire et al., 2009). Performance after restricted sleep of 4-5h per night for a week, or after 18h of wakefulness was comparable to performance with a blood alcohol level of 0.1% (Whitmire et al., 2009). Others have shown decrements in performance on a psychomotor test after 17h of sustained wakefulness to be similar to performance with a blood alcohol concentration of 0.05% (Dawson & Reid, 1997). More specifically, in a study examining driving performance, lane-keeping ability after one night without sleep was akin to a BAC of 0.07% (Fairclough & Graham, 1999).

Total sleep deprivation can affect one's evaluation of risk and subsequent decision-making (McKenna, Dickinson, Orff, & Drummond, 2007). Sleep deprivation may have a greater effect on monotonous tasks relative to complex and engaging tasks, although the latter are not immune (Lim & Dinges, 2008). Other studies have suggested that while critical or logical reasoning may be preserved after sleep deprivation, innovative, or flexible strategizing often deteriorates (Harrison & Horne, 2000b).

Sleep deprivation may result in a decrease in word fluency and a diminished ability to create novel or innovative language (Harrison & Horne, 1998). Other investigations confirm the decrements in multiple complex language tasks after sustained wakefulness (Pilcher et al., 2007). This could translate, in real-world situations, to a difficulty communicating in a crisis.

Memory can also be affected by sleep deprivation. People can have trouble recalling how recently something happened (Harrison & Horne, 2000a). Attention, working memory, and vigilance are impaired after sleep deprivation across multiple studies, whether they are basic skills tests or more elaborate simulations meant to approximate real-life situations (Ahola & Polo-Kantola, 2007). Sleep deprivation may also affect people's mood, causing them to react inappropriately. Harrison and Horne describe "irritability, impatience, childish humor, lack of regard for normal social conventions, and inappropriate interpersonal behaviors" in the setting of sleep deprivation, as well as "behavioral disinhibition" and occasional periods of "euphoria" (Harrison & Horne, 2000b). Such mood lability can be entertaining, but it is certainly not always desired, particularly in work or crisis situations.

All of these performance concerns can add up to serious consequences. After the 1989 massive oil spill caused by the grounding of the *Exxon Valdez*, the National Transportation Safety Board determined that fatigue and poor rest received by the crew was a direct cause of the accident (National Transportation Safety Board, 1990). After another tragic accident, the space shuttle *Challenger* explosion, the human factors analysis by the Presidential Commission that investigated the accident noted "irregular working hours and insufficient sleep" to be a potentially significant contributing factor (Rogers Commission, 1986). Clearly, we cannot take these risks lightly.

Beyond the short term, sleep deprivation may have far-reaching effects on health and well-being. Nowhere is this more evident than in studies of shift workers. Many long-term health consequences have been linked to shift work, including obesity, metabolic syndrome, diabetes, cardiovascular disease, stroke, heart attack, depression, and cancer (Brown et al., 2009; Esquirol et al., 2009; Garaulet, Ordovas, & Madrid, 2010; Pan,

Schernhammer, Sun, & Hu, 2011; Scheer, Hilton, Mantzoros, & Shea, 2009; Schernhammer et al., 2001; Tüchsen, Hannerz, & Burr, 2006). These long-term effects increase in prevalence not just in shift workers with clinically diagnosed sleep disorders, but in evening and night shift workers overall compared to day shift workers (Drake, Roehrs, Richardson, Walsh, & Roth, 2004). Experimental studies have suggested a misalignment of circadian and metabolic cycles as a possible explanation, beyond simple sleep disturbance. In particular, one study found suppressed leptin, elevated mean arterial blood pressure, and an abnormally high postprandial glucose response when sleep-wake cycles were maximally out of sync with circadian rhythm (Scheer et al., 2009).

Finally, in keeping with the theme of this text, sleep deprivation has been implicated in emotional responses across numerous experiments and labs. In short, sleep seems to clearly have an impact on emotional regulation (Walker & van Der Helm, 2009). Imaging studies in sleep-deprived subjects have implicated the amygdala as a probable modulator (Cahill & McGaugh, 1998). The amygdala has long been implicated in basic emotional responses but such responses may be attenuated by the prefrontal cortex (Kensinger & Schacter, 2006) which implies that the frontal lobes, if normally active, can modulate emotional expression. Predictably, in the face of sleep deprivation, there is a measureable decrease in the amount of prefrontal activation (hypofrontality) that may allow for a person to express emotions in ways a well-rested individual would not (Walker & van Der Helm, 2009). Additionally, there is evidence that sleep restriction activates the amygdala (Yoo, Gujar, Hue, Jolesz, & Walker, 2007) which may produce heightened emotional output across the emotional spectrum. To take just one example, if you have ever experienced "giddiness" after pulling an all nighter, then now you may know why.

In summary, there are many brain and endocrine structures that are normally activated to maintain homeostasis. Sleep is maintained by numerous brain structures and neurotransmitters, many of which are altered in sleep-deprived conditions. We highlighted the various stages of sleep and what the biological underpinnings are of each. We also overviewed how the circadian rhythms are an adjunct to the normal sleep cycle. But your biological rhythms and sleep processes may uncouple, resulting in the malaise of circadian dysrhythmia, which is most simply expressed in transmeridian travel (jet lag). Also, when a person is sleep deprived, there are many biological and behavioral variables that are compromised, sometimes to a great degree, which can sometimes go unnoticed by the deprived individual. Notably, one of the compromised behaviors might be the expression of emotion which is controlled by various structures of the limbic system (e.g., the amygdala) which work in concert with the higher order structures of the brain to maintain emotional regulation. Later chapters in this book will further elaborate on the emotional regulation of sleep.

References

Abrahamson, E. E., Leak, R. K., & Moore, R. Y. (2001). The suprachiasmatic nucleus projects to posterior hypothalamic arousal systems. *Neuroreport, 12*, 435–440.

Ahola, P., & Polo-Kantola, P. (2007). Sleep deprivation: Impact on cognitive performance. *Neuropsychiatric Disease and Treatment, 3*, 553–567.

American Academy of Sleep Medicine. (2005). *International classification of sleep disorders: Diagnostic and coding manual* (2nd ed.). Westchester, IL: American Academy of Sleep Medicine.

Arendt, J. (2010). Shift work: Coping with the biological clock. *Occupational Medicine, 60*, 10–20.

Aserinsky, E., & Kleitman, N. (1953). Regularly occurring periods of eye motility, and concomitant phenomena, during sleep. *Science, 118*, 273–274.

Berridge, C. W., & Foote, S. L. (1996). Enhancement of behavioral and electroencephalographic indices of waking following stimulation of noradrenergic beta-receptors within the medial septal region of the basal forebrain. *Journal of Neuroscience, 16*, 6999–7009.

Berson, D. M., Dunn, F. A., & Takao, M. (2002). Phototransduction by retinal ganglion cells that set the circadian clock. *Science, 295*, 1070–1073.

Bizzi, E., & Brooks, D. C. (1963). Pontine reticular formation: Relation to lateral geniculate nucleus during deep sleep. *Science, 141*(3577), 270–272.

Borbély, A. A. (1982). A two process model of sleep regulation. *Human Neurobiology, 1*, 195–204.

Brown, R. E., Basheer, R., McKenna, J. T., Strecker, R. E., & McCarley, R. W. (2012). Control of sleep and wakefulness. *Physiological Reviews, 92*, 1087–1187.

Brown, D., Feskanich, D., Sanchez, B., Rexrode, K., Sxhernhammer, E., & Lisabeth, L. (2009). Rotating night shift work and the risk of ischemic stroke. *American Journal of Epidemiology, 169*, 1370–1377.

Cahill, L., & McGaugh, J. L. (1998). Mechanisms of emotional arousal and lasting declarative memory. *Trends in Neurosciences, 21*, 294–299.

Cajochen, C., Munch, M., Kobialka, S., Krauchi, K., Steiner, R., Oelhafen, P., et al. (2005). High sensitivity of human melatonin, alertness, thermoregulation, and heart rate to short wavelength light. *The Journal of Clinical Endocrinology and Metabolism, 90*, 1311–1316.

Carskadon, M. A., & Dement, W. C. (2011). Monitoring and staging human sleep. In M. H. Kryger, T. Roth, & W. C. Dement (Eds.), *Principles and practice of sleep medicine* (5th ed., pp. 16–26). St. Louis: Elsevier Saunders.

Chee, M. W., Chuah, L. Y., Venkatraman, V., Chan, W. Y., Philip, P., & Dinges, D. F. (2006). Functional imaging of working memory following normal sleep and after 24 and 35 h of sleep deprivation: Correlations of fronto-parietal activation with performance. *NeuroImage, 31*, 419–428.

Chemelli, R. M., Willie, J. T., Sinton, C. M., Elmquist, J. K., Scammell, T., Lee, C., et al. (1999). Narcolepsy in *orexin* knockout mice: Molecular genetics of sleep regulation. *Cell, 98*, 437–451.

Chou, T. C., Bjorkurn, A. A., Gaus, S. E., Lu, J., Scammell, T. E., & Saper, C. B. (2002). Afferents to the ventrolateral preoptic nucleus. *The Journal of Neuroscience, 22*, 977–990.

Clement, O., Sapin, E., Berod, A., Fort, P., & Luppi, P. (2011). Evidence that neurons of the sublaterodorsal tegmental nucleus triggering paradoxical (REM) sleep are glutamatergic. *Sleep, 34*, 419–423.

Dawson, D., & Reid, K. (1997). Fatigue, alcohol and performance impairment. *Nature, 388*, 235.

Dement, W. C., & Vaughan, C. (1999). *The promise of sleep: A pioneer in sleep medicine explores the vital connection between health, happiness, and a good night's sleep*. New York, NY: Delacorte Press.

Dijk, D. J., & Lockley, S. W. (2002). Integration of human sleep-wake regulation and circadian rhythmicity. *Journal of Applied Physiology, 92*, 852–862.

Drake, C., Roehrs, T., Richardson, G., Walsh, J., & Roth, T. (2004). Shift work sleep disorder: Prevalence and consequences beyond that of symptomatic day workers. *Sleep, 27*, 1453–1462.

REFERENCES

Drummond, S. P., & Brown, G. G. (2001). The effects of total sleep deprivation on cerebral responses to cognitive performance. *Neuropsychopharmacology, 26,* S68–S73.

Dunlap, J. C., Loros, J. J., & DeCoursey, P. J. (2004). *Chronobiology: Biological timekeeping.* Sunderland: Sinauer Associates.

Esquirol, Y., Bongard, V., Mabile, L., Jonnier, B., Soulat, J., & Perret, B. (2009). Shift work and metabolic syndrome: Respective impacts of job strain, physical activity, and dietary rhythms. *Chronobiology International, 26,* 544–559.

Everson, C. A. (2009). Comparative research approaches to discovering the biomedical implications of sleep loss and sleep recovery. In C. J. Amlaner, & P. M. Fuller (Eds.), *Basics of sleep guide* (2nd ed., pp. 237–248). Westchester, IL: Sleep Research Society.

Fairclough, S., & Graham, R. (1999). Impairment of driving performance caused by sleep deprivation or alcohol: A comparative study. *Human Factors, 41,* 118–128.

Folkard, S., & Akerstedt, T. (2004). Trends in the risk of accidents and injuries and their implications for models of fatigue and performance [Review]. *Aviation, Space, and Environmental Medicine, 75,* A161–A167.

Fuller, P. M., Gooley, J. J., & Saper, C. B. (2006). Neurobiology of the sleep-wake cycle: Sleep architecture, circadian regulation, and regulatory feedback. *Journal of Biological Rhythms, 21*(6), 482–493.

Fuller, P. M., & Lu, J. (2009). Neurobiology of sleep. In C. J. Amlaner, & P. M. Fuller (Eds.), *Basics of sleep guide* (pp. 53–62). Westchester, IL: Sleep Research Society.

Garaulet, M., Ordovas, J., & Madrid, J. (2010). The chronobiology, etiology and pathophysiology of obesity. *International Journal of Obesity, 34,* 1667–1683.

Germain, A., Nofzinger, E. A., Kupfer, D. J., & Buysse, D. J. (2004). Neurobiology of non-REM sleep in depression: Further evidence for hypofrontality and thalamic dysregulation. *The American Journal of Psychiatry, 161,* 1856–1863.

Harrison, Y., & Horne, J. (1998). Sleep loss impairs short and novel language tasks having a prefrontal focus. *Journal of Sleep Research, 7,* 95–100.

Harrison, Y., & Horne, J. (2000a). Sleep loss and temporal memory. *Quarterly Journal of Experimental Psychology. A, Human Experimental Psychology, 53,* 271–279.

Harrison, Y., & Horne, J. (2000b). The impact of sleep deprivation on decision making: A review. *Journal of Experimental Psychology, 6,* 236–249.

Harvard Medical School Division of Sleep Medicine (2008). http://healthysleep.med.harvard.edu.

Jones, B. E., & Beaudet, A. (1987). Distribution of acetylcholine and catecholamine neurons in the cat brain stem studied by choline acetyltransferase and tyrosine hydroxylase immunohistochemistry. *Journal of Comparative Neurology, 261,* 15–32.

Jouvet, M. (1962). Recherches sur les structures nerveuses et le mecanismes responsables des differentes phases du sommeil physiologique. *Archives Italiennes de Biologie, 100,* 125–206.

Kalat, J. W. (2013). *Biological psychology.* Wadsworth: Cengage Learning.

Kensinger, E. A., & Schacter, D. L. (2006). Processing emotional pictures and words: Effects of valence and arousal. *Cognitive, Affective, & Behavioral Neuroscience, 6,* 110–126.

Kolla, B. P., & Auger, R. R. (2011). Jet lag and shift work sleep disorders: How to help reset the internal clock. *Cleveland Clinic Journal of Medicine, 78,* 675–684.

Lee, M. G., Hassani, O. K., Alonso, A., & Jones, B. E. (2005). Cholinergic basal forebrain neurons burst with theta during waking and paradoxical sleep. *Journal of Neuroscience, 25,* 4365–4369.

Lim, J., & Dinges, D. (2008). Sleep deprivation and vigilant attention. *Annals of the New York Academy of Sciences, 1129,* 305–322.

Lindsley, D. B., Schreiner, L. H., Knowles, W. B., & Magoun, H. W. (1950). Behavior and EEG changes following chronic brain stem lesions in the cat. *Electroencephalography and Clinical Neurophysiology, 2,* 483–498.

Lockley, S. W., Brainard, G. C., & Czeisler, C. A. (2003). High sensitivity of the human circadian melatonin rhythm to resetting by short wavelength light. *The Journal of Clinical Endocrinology and Metabolism, 88,* 4502–4505.

Lu, J., Greco, M. A., Shiromani, P., & Saper, C. B. (2000). Effect of lesions of the ventrolateral preoptic nucleus on NREM and REM sleep. *The Journal of Neuroscience, 20,* 3830–3842.

Lu, J., Jhou, T. C., & Saper, C. B. (2006). Identification of wake-active dopaminergic neurons in the ventral periaqueductal gray matter. *The Journal of Neuroscience, 26*(1), 193–202.

Lu, J., Sherman, D., Devor, M., & Saper, C. B. (2006). A putative flip-flop switch for control of REM sleep. *Nature, 441,* 589–594.

Luppi, P., Clement, O., Sapin, E., Gervasoni, D., Peyron, C., Leger, L., et al. (2011). The neuronal network responsible for paradoxical sleep and its dysfunctions causing narcolepsy and rapid eye movement (REM) behavior disorder. *Sleep Medicine Reviews, 15,* 153–163.

Maquet, P., Peters, J., Aerts, J., Delfiore, G., Degueldre, C., Luxen, A., et al. (1996). Functional neuroanatomy of human rapid-eye movement sleep and dreaming. *Nature, 383,* 163–166.

McGinty, D. J., & Sterman, M. B. (1968). Sleep suppression after basal forebrain lesions in the cat. *Science, 160,* 1253–1255.

McGinty, D., & Szymusiak, R. (2000). The sleep-wake switch: A neuronal alarm clock. *Nature Medicine, 6,* 510–511.

McKenna, B., Dickinson, D., Orff, H., & Drummond, S. (2007). The effects of one night of sleep deprivation on known-risk and ambiguous-risk decisions. *Journal of Sleep Research, 16,* 245–252.

Mochizuki, T., Crocker, A., McCormack, S., Yanagisawa, M., Sakurai, T., & Scammell, T. E. (2004). Behavioral state instability in orexin knock-out mice. *Journal of Neuroscience, 24,* 6291–6300.

Mohns, E. J., Karlsson, K.Æ., & Blumberg, M. S. (2006). The preoptic hypothalamus and basal forebrain play opposing roles in the descending modulation of sleep and wakefulness in infant rats. *European Journal of Neuroscience, 23,* 1301–1310. http://dx.doi.org/10.1111/j.1460-9568.2006.04652.x.

Moruzzi, G., & Magoun, H. W. (1949). Brain stem reticular formation and activation of the EEG. *Electroencephalography and Clinical Neurophysiology, 1,* 455–473.

National Sleep Foundation. (2014). *The sleep disorders.* Retrieved from, http://sleepdisorders.sleepfoundation.org/chapter-1-normal-sleep/an-introduction-to-normal-sleep/.

National Transportation Safety Board. (1990). *Grounding of the US Tankship Exxon Valdez on Bligh Reef, Prince William Sound near Valdez, Alaska, March 24, 1989.* Washington, DC: National Transportation Safety Board. Retrieved from, http://docs.lib.noaa.gov/noaa_documents/NOAA_related_docs/oil_spills/marine_accident_report_1990.pdf.

Nauta, W. J. H. (1946). Hypothalamic regulation of sleep in rats. An experimental study. *Journal of Neurophysiology, 9,* 285–314.

Pan, A., Schernhammer, E., Sun, Q., & Hu, F. (2011). Rotating night shift work and risk of type 2 diabetes: Two prospective cohort studies in women. *PLoS Medicine, 8,* e1001141. http://dx.doi.org/10.1371/journal.pmed.1001141.

Pilcher, J., McClelland, L., Moore, D., Haarmann, H., Baron, J., Wallston, T., et al. (2007). Language performance under sustained work and sleep deprivation conditions. *Aviation, Space, and Environmental Medicine, 78,* B25–B38.

Prasai, M. J., Pernicova, I., Grant, P. J., & Scott, E. M. (2011). An endocrinologist's guide to the clock. *The Journal of Clinical Endocrinology and Metabolism, 96,* 913–922.

Ranson, S. W. (1939). Somnolence caused by hypothalamic lesions in the monkey. *Archives of Neurology & Psychiatry, 41*(1), 1–23.

Relogio, A., Westermark, P. O., Wallach, T., Schellenberg, K., Kramer, A., & Herzel, H. (2011). Tuning the mammalian circadian clock: Robust synergy of two loops. *PLoS Computational Biology, 7,* e1002309. http://dx.doi.org/10.1371/journal.pcbi.1002309.

Reppert, S. M., & Weaver, D. R. (2002). Coordination of circadian timing in mammals. *Nature, 418,* 935–941.

Rogers Commission. (1986). *Report of the Presidential Commission on the Space Shuttle Challenger Accident.* Washington, DC: US Government Printing Office. Retrieved from, http://history.nasa.gov/rogersrep/genindex.htm.

Ruottinen, H. M., Partinen, M., Hublin, C., Bergman, J., Haaparanta, M., Solin, O., et al. (2000). An FDOPA PET study in patients with periodic limb movement disorder and restless leg syndrome. *Neurology, 54,* 502–504.

Sack, R. L. (2010). Jet lag. *New England Journal of Medicine, 362,* 440–447.

Saper, C. B., Chou, T. C., & Scammell, T. E. (2001). The sleep switch: Hypothalamic control of sleep and wakefulness. *Trends in Neurosciences, 24,* 726–731.

Saper, C. B., Scammell, T. E., & Lu, J. (2005). Review hypothalamic regulation of sleep and circadian rhythms. *Nature, 437,* 1257–1263.

Scheer, F., Hilton, M., Mantzoros, C., & Shea, S. (2009). Adverse metabolic and cardiovascular consequences of circadian misalignment. *Proceedings of the National Academy of Sciences of the United States of America, 106,* 4453–4458.

Scheer, F., & Shea, S. A. (2009). Fundamentals of the circadian system. In C. J. Amlaner, & P. M. Fuller (Eds.), *Basics of sleep guide* (2nd ed., pp. 199–210). Westchester, IL: Sleep Research Society.

Schernhammer, E., Laden, F., Speizer, F., Willett, W., Hunter, D., Kawachi, I., et al. (2001). Rotating night shifts and risk of breast cancer in women participating in the nurses' health study. *Journal of the National Cancer Institute, 93,* 1563–1568.

Semba, K. (2000). Multiple output pathways of the basal forebrain: Organization, chemical heterogeneity, and roles in vigilance. *Behavioural Brain Research, 115*(2), 117–141.

Sherin, J. E., Elmquist, J. K., Torrealba, F., & Saper, C. B. (1998). Innervation of histaminergic tuberomammillary neurons by GABAergic and galaninergic neurons in the ventrolateral preoptic nucleus of the rat. *The Journal of Neuroscience, 18,* 4705–4721.

Sherin, J. E., Shiromani, P. J., McCarley, R. W., & Saper, C. B. (1996). Activation of ventrolateral preoptic neurons during sleep. *Science, 271,* 216–219.

Siegel, J. (2002). *The neural control of sleep & waking.* New York, NY: Springer-Verlag.

Steriade, M. M., & McCarley, R. (2005). *Brain control of wakefulness and sleeping.* New York, NY: Springer.

Sterman, M. B., & Clemente, C. D. (1962). Forebrain inhibitory mechanisms, sleep pattern induced by basal forebrain stimulation in behaving cat. *Experimental Neurology, 6,* 103–117.

Szymusiak, R., Alam, N., Steininger, T. L., & McGinty, D. (1998). Sleep-waking discharge patterns of ventrolateral preoptic/anterior hypothalamic neurons in rats. *Brain Research, 803,* 178–188.

Thapan, K., Arendt, J., & Skene, D. J. (2001). An action spectrum for melatonin suppression: Evidence for a novel non-rod, non-cone photoreceptor system in humans. *The Journal of Physiology, 535,* 261–267.

Tüchsen, F., Hannerz, H., & Burr, H. (2006). A 12 year prospective study of circulatory disease among Danish shift workers. *Occupational and Environmental Medicine, 63,* 451–455.

von Economo, C. (1930). Sleep as a problem of localization. *The Journal of Nervous and Mental Disease, 71,* 249–259.

Walker, M. P., & van Der Helm, E. (2009). Overnight therapy? The role of sleep in emotional brain processing. *Psychological Bulletin, 135*(5), 731.

Whitmire, A., Leveton, L., Barger, L., Brainard, G., Dinges, D., Klerman, E., et al. (2009). Risk of performance errors due to sleep loss, circadian desynchronization, fatigue, and work overload. In John B. Charles, & Jancy C. McPhee (Eds.), *Human health and performance risks of space exploration missions: Evidence reviewed by the NASA Human Research Program, Lyndon B. Johnson Space Center* (p. 396). Houston, TX: National Aeronautics and Space Administration.

Yoo, S. S., Gujar, N., Hu, P., Jolesz, F. A., & Walker, M. P. (2007). The human emotional brain without sleep—A prefrontal amygdala disconnect. *Current Biology, 17,* R877–R878.

CHAPTER 2

Human Emotions: A Conceptual Overview

Andreas Keil and Vladimir Miskovic[†]

*Department of Psychology and Center for the Study of Emotion and Attention, University of Florida, Gainesville, Florida, USA
[†]Department of Psychology, State University of New York at Binghamton, Binghamton, New York, USA

FOUNDATIONS OF AFFECT AND EMOTION

Great works of art, music, and literature often touch us not so much by appealing to rational thought but through the "emotions" that they evoke. Even the oldest written testimonies of human culture contain abundant examples of authors discriminating between rational, calculated behavior and the primordial surge of feelings. Folk psychology and criminal justice endorse a similar distinction between the mind and the heart, and between premeditated action and the heat of passion. Although these constructs have considerable intuitive folk-psychological appeal, empirical scientists still face the tasks of defining, explaining, and predicting emotional processes in ways that are compatible with objective, reliable, and valid research. One major problem for an intuition-based, folk-psychology approach to emotions (the "ask them how they feel" method) is that the feeling aspect of an emotional response can be considered a philosophical *qualia* problem, in which the assessment of one individual's first-person experience by an external observer cannot be easily validated, unless other information is available or strong auxiliary assumptions are made (Mashour & LaRock, 2008). Later in this chapter, we discuss how this challenge is met using a variety of approaches available to researchers interested in emotional processes. In this section, we consider the fundamental principles that are widely used to organize the systematic study of emotions from a natural science perspective.

A Consensus Definition of "Emotions," as Seen Through the Lens of the Animal Model

A potential empirical basis defining an emotional response can be found in the defensive and appetitive behaviors shown by a variety of species in the animal kingdom, including invertebrate animals. Living organisms tend to possess an arsenal of mechanisms for successfully dealing with challenges imposed by the environment, ranging from successful reproduction to defense against predators. Many such mechanisms have evolved in different species. They are inherited as a "birth right," addressing different survival-related challenges, and they range in complexity with high malleability to environmental contexts (Mayr, 1974). For instance, although often regarded as simple organisms, bacteria are capable of flexibly using a spectrum of defensive processes to cope with different challenges presented by the host immune system (Hornef, Wick, Rhen, & Normark, 2002).

The immense number, flexibility, and diversity of physiological and behavioral mechanisms promoting survival are particularly apparent in mammals (e.g., Panksepp, 1998). Mammalian responses to environmental challenges and opportunities vary in scope, duration, sophistication, and level of organismic involvement. Yet, striking similarities exist within the class of mammalian behaviors that human observers tend to identify as "emotional": They are accompanied by pronounced changes in autonomic processes, including endocrine and reflex physiology; they help to orchestrate coordinated and purposeful behaviors; they often possess communicative aspects; they are phasic in nature and often show a defined time course. Often citing Darwin (1872) as one of the first to identify the systematic overlap between objective human emotional responses and responses observed in nonhuman animals, many prominent theorists have viewed human emotions as grounded in responses aimed at optimizing survival of the individual. As early as the nineteenth century, University of Chicago Philosopher John Dewey (1895) suggested that building on the conceptions of emotion promoted by Darwin and William James leads to the notion that emotions are action tendencies. He claimed that "Emotion is a mode of behavior, which is purposive, or has an intellectual content. [...] Certain movements, formerly useful, are reduced to tendencies to action or to attitudes, and when instinctively aroused into action, serve as means for realizing ends." Despite lively debates on what aspects of observable behavior versus inner mental processes should be emphasized in emotion theories (Averill et al., 1994), many current perspectives accept the core tenet that emotions can be viewed as purposeful processes in the context of survival, related to action.[1] In line with these views,

---

[1] Incidentally, the emphasis on the active component of emotion honors the word's lexical origin in the Latin root meaning "to move."

we broadly conceptualize emotions as adaptive action tendencies that occur in response to changes inside or outside the organism, specifically changes that challenge states and systems necessary for survival. This definition pragmatically focuses on phasic emotional responses with a well-defined time course related to a specific eliciting event. Examples of such states and systems are those underlying sexual reproduction, homeostasis, metabolism, and defense against bodily harm—processes typically discussed in the chapter on "motivation" in psychology and neuroscience textbooks.[2] Given this definition of emotions, concepts such as mood and temperament can be viewed as affective processes grounded in more generalized action dispositions. They differ from emotions in that they refer to an individual's tendency to display responses biased toward appetitive or aversive valence over extended periods of time and across multiple events. Across concepts of mood, temperament, and emotion, the semantic proximity between processes of motivation and affect is inherent in many historical and current models of emotional processes, and it is also apparent in the empirical research discussed next.

Emotion Science: Selected Milestones

Equipped with a broad consensus definition of emotional processes, we may now consider a few selected milestones in the history of affective science. Of course, a complete historical overview is beyond the scope of the present chapter, and we refer the reader to the many outstanding publications on the topic (e.g., Panksepp, 1998). Preceding the advent of psychology as a scientific discipline, systematic thinking on emotions was the domain of poetry and philosophy, both of which contributed greatly to the identification and specification of critical issues still relevant today: Can emotions be controlled? How are emotions related to an individual's mood and temperament? What is the relationship between bodily and mental events during emotional experience?

Views Based on Introspection and Observation

Surveying Homer, the Pentateuch, the Quran, or other ancient texts for specific notions on affective processes reveals a rich spectrum of implicit theories with respect to these questions, often emphasizing bodily and behavioral aspects of affect in addition to the subjective experience.

---

[2]More recently, some authors have emphasized the "primordial emotions," or those affective states primarily instigated by interoceptor-detected changes that are specific to the internal milieu of the organism (Denton, McKinley, Farrell, & Egan, 2009). The relation of these primordial emotions to biologic mechanisms of homeostasis may be more closely compared to emotional states triggered by perceptions of the external world (e.g., defensive reactions to predators or sexual lust).

The inclusion of behavior and physiology with self-reported emotional experience also occurs in systematic philosophical analyses of affective processes. For instance, in his letters on morals to his nephew Lucilius, the Roman Stoic Philosopher Seneca discusses the limitations of voluntary control of emotional expression and experience and the disjunction between mental and bodily processes. In the context of the blush, he takes a dim view of humans' ability to intentionally alter and falsely portray emotional responses: "We cannot forbid these feelings any more than we can summon them. Actors in the theater, who imitate the emotions, who portray fear and nervousness, who depict sorrow, imitate bashfulness by hanging their heads, lowering their voices, and keeping their eyes fixed and rooted upon the ground. They cannot, however, muster a blush; for the blush cannot be prevented or acquired [letter 11, paragraph 6]."

Although they took varying stances on the relationship between cognition and emotion, the founding figures of modern psychology have likewise embraced views of emotions that include observable behavior and physiological changes in addition to self-reported experience. In his famous account of emotions featured in most psychology textbooks (James, 1884), William James highlighted the physiological and behavioral responses to an emotionally engaging event as having primacy and temporal precedence over the subjective emotional experience. The thought experiment he proposed remains a worthwhile exercise for contemporary emotion researchers: "What kind of an emotion of fear would be left, if the feelings neither of quickened heart-beats nor of shallow breathing, neither of trembling lips nor of weakened limbs, neither of goose-flesh nor of visceral stirrings, were present, it is quite impossible to think. Can one fancy the state of rage and picture no ebullition of it in the chest, no flushing of the face, no dilatation of the nostrils, no clenching of the teeth, no impulse to vigorous action, but in their stead limp muscles, calm breathing, and a placid face? The present writer, for one, certainly cannot" (pp. 194–195).

The Emergence of Cognitive Approaches and Animal Models of Emotion

James's discussion of emotion, along with other early accounts, addresses two crucial questions occupying much of the current literature: Is there a gold standard for defining emotions? What kinds of data are relevant for describing and explaining emotional processes? Major portions of the twentieth-century discourse on emotion theory can be viewed as an ongoing debate on these issues, with the 1950s seeing an influx of physiological and early neuroscience-based accounts (Hebb, 1949; Weiskrantz, 1956) and the 1960s characterized by increasingly cognitive explanations of emotions (Arnold, 1968). Both movements have contributed to views held by many researchers today. These views tend to be linked to specific paradigms discussed in the upcoming section on current theories. Specifically, various *appraisal theories* of emotion have emphasized the role of higher-order cognitive processes, such as expectation and reasoning, for shaping emotions (Lazarus, 1991). Generally, these models propose that an initial event or stimulus leads to some form of cognitive processing (often referred to as an appraisal), which specifies the nature, scope, and intensity of the affective response (Smith & Ellsworth, 1985). For instance, a general form of physiological "arousal" might be cognitively interpreted as "sexual excitement" or "fearful anticipation," depending on the nature of the appraisal (Schachter & Singer, 1962). Although this particular notion has received little empirical support (Reisenzein, 1983), several lines of research have productively examined the time course and content of cognitive processes accompanying emotional responses (Scherer, 2009). Contemporary models of modification and regulation of emotional experience are often based on appraisal theory. Empirical research in this field has explored the role of different behavioral and cognitive strategies in changing the time course or intensity of emotions as quantified by a variety of indices.

With the increasing availability of (neuro)physiological and behavioral measurements, as well as new research paradigms emerging in the course of the twentieth century, emotion research underwent widespread diversification and specialization, producing a growing number of subdisciplines. One important area of research emerged from surgical lesion studies conducted on experimental animals (Cannon, 1929), leading to early comprehensive models of emotional processes inspired by the functional consequences of neuropathology (Papez, 1938). For example, Klüver-Bucy Syndrome was described in 1937, summarizing the behavioral consequences of brain lesions in monkeys (Klüver & Bucy, 1937). Following excision of the bilateral medial temporal lobes, the monkeys rapidly dropped down the social dominance hierarchy and displayed striking changes in behavior such as hypersexuality, lack of defensive reactivity, and increased oral exploration. Based on this success in localizing the brain structures mediating supposed socioemotional behavior, subsequent studies focused on exploring the significance of more specific medial temporal lobe structures (Jones & Mishkin, 1972). In these studies, the amygdaloid complex was eventually established as a critical hub structure in mammalian defensive responses (LeDoux, Thompson, Iadecola, Tucker, & Reis, 1983; Weiskrantz, 1956), leading to current paradigms in which behavioral neuroscientists have examined the functional circuitry of amygdaloid pathways involved in defensive responding down to a molecular level of analysis (Davis, Walker, Miles, & Grillon, 2010). In addition to systems and molecular neuroscience approaches using animal models, the neurophysiological tradition has also substantially impacted the field of human neuroimaging of affect,[3] a field that began its rapid growth in the 1980s and is now a large and active research area with many specializations, including human affective neuroscience and social neuroscience (Lindquist, Wager, Kober, Bliss-Moreau, & Barrett, 2012; Phan, Wager, Taylor, & Liberzon, 2004).

The Contributions of Psychophysiology and Human Neuroscience

Benefiting from emerging technical advances in data recording and digital signal processing, research in human psychophysiology (Lang, 1979), often conducted with psychiatric disorders in mind (Cuthbert et al., 2003), has capitalized on multidimensional arrays of data from bodily, behavioral, and physiological systems during an emotional challenge. This literature of the discipline has acknowledged that multiple measures, such as heart rate, skin conductance, endocrine changes, and reflex modulation (notably the eye blink startle response), are useful in describing the multifaceted affective response on the level of reflexive physiological systems (Graham & Clifton, 1966; Lang, Bradley, & Cuthbert, 1990). A major milestone in this field of study was the recognition that affective physiology tends to be nonmonolithic: Different indices of the same emotional episode, such as verbal self-report, observable behavior, and physiological (specifically autonomous) processes are seldom strongly correlated (Lang, 1993), suggesting that a single response variable may not be sufficient to adequately describe the emotional process under consideration. On the conceptual level, psychophysiological studies were the first to empirically examine the roles of perception and attention within the realm of emotional processes (Sokolov, 1963). Ranging from early studies based on animal behavior and analyses of heart rate change during novelty processing to multimethod, multimeasure research programs today, this research has characterized attentive orienting to a salient stimulus as one of the initial steps in a complex nexus of adaptive behaviors particularly relevant in humans (Bradley, 2009). Accordingly, a large portion of today's affective science relies on presenting external stimuli, manipulated in intensity and hedonic valence, to systematically study the unfolding of emotional responses, with levels of observation ranging from the molecular level (Martinez et al., 2013) to the level of social aggregates (Levenson & Ruef, 1992).

Even this incomplete overview shows that a multitude of disciplines are now involved in the study of emotions, focusing on different aspects of emotional processes in different contexts and using a wide range of

---

[3]Due largely to methodological limitations, researchers working with human participants have tended to perhaps overemphasize the contribution of cortical (relative to paleocortical and brainstem) contributions to emotion, leading to what some have called a "corticocentric myopia" in human research (Parvizi, 2009).

dependent variables. These different empirical approaches reflect different theoretical stances, which we address next.

CURRENT DEFINITIONS AND THEORETICAL APPROACHES

So far, we have illustrated that emotion is best viewed as a conceptually inclusive term that encompasses many facets of physiology, behavior, and experience. Commensurate with this rich database, a wide range of emotion theories exist, and many previous analyses have discussed ways of classifying them. One popular classification distinguishes theories that conceptualize emotions as grounded in a limited number of distinct states, such as fear, happiness, and sadness (the so-called basic emotions), from those that tend to conceptualize the spectrum of emotional processes in terms of a low-dimensional space consisting of more broadly defined axes such as pleasure/displeasure and intensity or emotional arousal (dimensional theories).

Basic emotions are held to be qualitatively distinct elementary building blocks (sometimes termed "natural kinds") of which the full spectrum of emotional processes is composed, and they are often thought to possess distinct anatomical and neurophysiological substrates (Panksepp, 1992). Researchers have intensely debated the existence and nature of basic emotions, discussing many crucial questions in emotion science (Levenson, 2011). For instance, does the presence of relatively distinct neural circuits for specific animal behaviors linked to fear, anger, sadness, and happiness represent sufficient evidence for the existence of basic emotions? Although there is no general agreement on the answer to this question, the research into the specific neural substrates of basic emotions has provided many relevant insights regarding the neuroanatomy, neurochemistry, and physiology of animal and human behavior. Because many larger-scale brain structures (e.g., the entirety of the amygdaloid complex) tend to display sensitivity to a wide range of stimuli and behaviors, the search for specificity has prompted basic-emotion researchers to characterize subcircuits and signaling pathways more tightly linked to one class of behaviors. This specification has, in turn, provided avenues for developing precise hypotheses on how the pharmacological or electrical modulation of microcircuits affects emotional states, generating valuable mechanistic knowledge (Panksepp, 2011). For example, the basolateral and central nuclei of the amygdala have emerged as key structures contributing to learned fears, and more primitive brain regions within the dorsal periaqueductal gray and the deep layers of the colliculi are important for the expression of unconditioned (innate) fears (Brandao, Anseloni, Pandossio, De Araujo, & Castilho, 1999).

Another question for basic-emotion theories is obvious because it concerns the number of basic emotions those theories include. Not surprisingly, this number varies as a function of the paradigm used and the measures of emotion examined, with some theories naming 6 (Levenson, 2011), 7 (Ekman, 1992), 8 (Plutchik, 2001), or 12 (Izard, see Blumberg & Izard, 1986) basic emotions. Reviews of the basic-emotion literature have also noted that, in addition to the number of basic emotions, the content of those emotions varies widely across the published literature. To rebut these criticisms, others have argued that the number and kind of emotions included in a given basic-emotion theory may be less important than the conceptual inferences and empirical research stimulated by the theory (Plutchik, 2001). In line with this notion, work focusing on facial expression of emotion has historically been grounded in basic-emotion theory.[4] Today, standardized picture sets used in studies of facial affect rely on basic emotions for naming and categorizing stimuli, an approach that has high face validity and intuitive appeal (e.g., Lundqvist, Flykt, & Öhman, 1998). In a similar vein, developmental work has successfully used basic-emotion approaches to describe how emotional language and self-report emerge in ontogeny, including interventions to foster healthy emotional development (Izard, Fine, Mostow, Trentacosta, & Campbell, 2002). Current discussions in the literature address the extent to which basic emotions are categories given by and found in nature (i.e., if they are natural kinds with ontological reality) or if they are better conceived of as social constructs (Barrett, 2012). This discussion leads to productive questions regarding emotional regulation and shaping through culture, as well as to the genetic control of the neural substrates, all of which are active research topics.

Dimensional perspectives on emotions have their historical roots in analyses of self-reported affect and introspections of emotional experience (Lindquist et al., 2012), as well as observations of affective experience (Wundt, 1917). Their core assumption is that the wealth of emotional processes is grounded in tendencies that can be reduced to a low-dimensional space. Often, this space includes a dimension of pleasure/displeasure, hedonic valence, or positive reinforcement value (i.e., a good-bad dimension), as well as a dimension of activity, intensity, or emotional arousal (Mehrabian, 1995; Watson, Clark, & Carey, 1988). The strengths of dimensional models of emotion relate to their description and prediction of quantitative and continuous aspects of emotion, as indexed by verbal report or by physiological measurements. For instance, the dimensions spanned by self-reports of hedonic valence and emotional arousal in the

---

[4]By contrast, Panksepp's approach to identifying discrete emotional systems has involved applying electrical stimulation to different diencephalic and brainstem structures of nonhuman animals (see Panksepp, 1998, for a comprehensive review).

bivariate model by Lang and colleagues (Lang, 1994) allow the quantification of appetitive and aversive tendencies in standardized situations that share key aspects with animal behavior and reflex physiology.[5] When plotting ratings of emotional arousal against ratings of hedonic valence given for a variety of situations, a boomerang-shaped pattern tends to emerge, with an appetitive and an aversive gradient representing the increased intensity of pleasant and unpleasant affects (Lang, Greenwald, Bradley, & Hamm, 1993). Interestingly, the unpleasant affect gradient is reliably steeper than the pleasant arm in this graph (Bradley, Codispoti, Sabatinelli, & Lang, 2001), mirroring the imbalance in the strength of aversive/avoidance behavior versus appetitive/approach behavior observed in rodents confronted with different levels of reward or punishment (Miller & Murray, 1952). A sizable literature is also devoted to the finding that calm stimuli reliably show a positivity offset (they are shifted toward pleasure on the hedonic valence dimension; Keil & Freund, 2009), again paralleling the finding that most animals show a bias toward appetitive behaviors at rest.

In contrast to basic-emotion theories, dimensional approaches tend to emphasize physiological substrates shared by different facets of emotional engagement (Cuthbert, Schupp, Bradley, Birbaumer, & Lang, 2000; Lane, Chua, & Dolan, 1999). Proponents of the dimensional approach hypothesize that being engaged by situations as diverse as social reward, sex, or a favorite meal shares fundamental aspects of appetitive processing, although the engagement differs in aspects related to the specific behavioral and physiological processes involved in each of these examples. By the same token, responses to threat, danger, pain, or social exclusion share core aspects in terms of their central and peripheral physiology. Notably, dimensional theories provide a framework for coactivation of appetitive and aversive tendencies in a given situation (jealousy is sometimes given as an example) and thus for complex affective states (Cacioppo & Gardner, 1999). Beyond providing two- or three-dimensional graphical affective spaces for emotional language, pictures, or sounds, many dimensional emotion theories make interesting predictions about the temporal dynamics and flexibility of emotional processes (Rolls, 1995; Sabatinelli, Lang, Bradley, Costa, & Keil, 2009). They often tend to emphasize parallel and multidirectional processes among sensory and motor processes, as well as semantic and memory systems, again giving rise to interesting network-oriented hypotheses with respect to measures such as brain connectivity (Keil et al., 2012).

---

[5]Other models conceptualize the affective space as consisting of orthogonal positive and negative affect systems, in which the level of activation (arousal) of each can vary independently (e.g., Cacioppo & Berntson, 1999).

Other distinctions between theories of emotion involve the putative mechanisms and processes (e.g., linear versus parallel) ascribed to emotions. Serial models of emotion tend to emphasize a linear temporal sequence between a stimulus, intermittent processes, and subsequent emotional reactivity (Gross, 1998; Smith & Ellsworth, 1985). Sometimes, an initial motive state is taken into account (Kreibig, Gendolla, & Scherer, 2012). We have already briefly touched upon appraisal theories, which emphasize the mediating role of cognitive constructs (the appraisals) in an emotional process, which is often described as a linear sequence of events (Moors, 2013). In a recent definition, appraisal is conceptualized as "a process that detects and assesses the significance of the environment for well-being" (Moors, Ellsworth, Scherer, & Frijda, 2013). This broad definition may be regarded as more inclusive than prominent past approaches, which strongly emphasized the "cognitive" aspects of appraisal (Arnold, 1968), highlighting interpretation and reasoning as sources of emotional processes. Accordingly, recent theoretical discussions (Moors, 2009) conclude that this broader conceptualization allows researchers to link findings from appraisal theory to other approaches. Such a related concept is the construct of "orienting," in which the organism responds to change with a characteristic pattern of behavioral, attentional, and physiological changes that foster information intake and action preparation (Bradley, 2009). Although possessing elements of temporal linearity, appraisal models or models based on orienting do not always propose fully linear sequences of separate elements (Scherer, 2009).

The aspect of nonlinearity and parallel processing is more obvious in parallel accounts of emotion, which often involve network models or so-called organismic theories (Damasio, 1998; Lewis & Liu, 2011). These accounts have their roots in cybernetic (Jirsa, Friedrich, Haken, & Kelso, 1994) and field theories of behavior (Lewin, 1951) and emphasize dynamic interactions between the organism and its environment over time. Hence, they often highlight an individual's active role in coping with challenge, while changing itself and the environment in the process. Many contemporary theories include elements of—or at least acknowledge—the parallel nature of emotional processing in highly complex physiological systems in which activation unfolds as a nexus of mutually interdependent dynamics rather than a stimulus-response chain (Pessoa & Adolphs, 2010). This trend is consistent with the realization that many behavioral and mental functions lack the linearity often suggested by textbook accounts of psychological processes. For example, vision was traditionally described as a hierarchical and sequential transmission of signals growing in complexity, starting with the retina and early sensory cortex. These structures were conceived of as passive data collection systems, not unlike photo cameras. Under conditions of natural viewing, however, an emerging consensus regards vision as an active sampling process, in which the ongoing behavioral and brain states of the observer strongly determine the lower-tier sensory processes underlying elementary visual perception (Schroeder, Wilson, Radman, Scharfman, & Lakatos, 2010). Recent advances in psychological and neuroscience research have provided models of how such adaptive interactions may be implemented in spatiotemporal neural dynamics (Engel, Fries, & Singer, 2001), and these same approaches have attracted the attention of researchers interested in emotion.

Finally, additional classifications of emotion theories distinguish between theories that treat emotions as natural kinds versus those that consider emotions to be sociocognitive constructs shaped by implicit and explicit cultural consensus. In a recent review, Gross and Barrett (2011) suggest a continuum of emotion theories that ranges from physiology-motivated basic-emotion theories to perspectives that emphasize cultural and social construction of self-report and behaviors related to emotion. Interestingly, these authors also include a discussion of how the conceptual views along this spectrum often require additional constructs such as the "regulation" of emotion. We briefly address this same issue later in this chapter. In concluding this brief overview, we remind the reader that the classifications presented are not exclusive, and their main goal is to provide some initial orientation regarding the existing approaches. Beyond organization of theories, the discussion of what theories share and do not share is also helpful in clarifying the underlying assumptions that often are not explicitly discussed in empirical papers, thus providing helpful information for researchers entering this complex and interesting field of study. The literature cited in this section may represent a starting point for asking fundamental questions in the context of one's own research.

THE DATA OF EMOTION: EMPIRICAL APPROACHES

Throughout this chapter, we express the view that human emotions are dynamic and complex phenomena best studied in a multivariate fashion, accepting the fact that there is no sole gold standard for emotional engagement. Self-reported affect, behavior, and physiology may be informative with respect to the emotional process under consideration, yet they may also be uncorrelated. In addition, we state that a consensus definition of emotions emphasizes that they consist of evolution-based action tendencies involving the entire organism. Thus, unsurprisingly, the data collected in typical studies of human emotion cover all aspects of biological, psychological, and behavioral reactivity. Here we illustrate this approach by discussing three selected paradigms widely used in human emotion research together with typical findings.

Challenge with Affective Stimuli

Since ancient times, confrontation with stimuli that are thought to be emotionally engaging has been used to evoke emotional processes in human beings. Galen of Pergamum famously diagnosed a case of "mad love" by repeatedly mentioning the name of a local theater actor to his patient while feeling her pulse. In past decades, systematic challenge with emotionally engaging stimuli has emerged as a dominant research strategy. Many sources of standardized affective stimuli are now available to researchers, some of which possess normative ratings or other normative information on their suitability to evoke certain responses in human participants. Sarah Bujarski and colleagues discuss these sources in detail in this volume. Some authors have discussed how the choice of the stimulus material affects the conceptual thrust of the research being undertaken, recommending that researchers be aware of the assumptions underlying a given set of stimuli (Dan-Glauser & Scherer, 2011). One assumption that may not be consistently emphasized in all research using standardized affective stimuli is that these stimuli often engage measurable response tendencies in observers/listeners, but the elicited responses may not represent the full set of behavior that is associated with the emotional process in many ecological situations. However, to the extent that daily life—at least in most of the Western world—is strongly characterized by audiovisual media consumption, a significant portion of contemporary humans' emotional behavior and experience occurs in the context of digital media presented on screens, which is in line with many laboratory research paradigms used in the field (Heim, Benasich, & Keil, 2013).

As an example, a large and growing body of research has capitalized on text, word, sounds, and pictures taken from the standardized sets developed at the University of Florida Center for Emotion and Attention. Prominently, the International Affective Picture System (IAPS; Lang, Bradley, & Cuthbert, 2005) has been extensively used in studies of visual emotion processing, manipulating picture content, timing, context, and many other variables, while collecting a spectrum of behavioral, physiological, and self-report measures. The IAPS is a collection of well over 1000 pictures and provides normative ratings regarding emotional arousal and hedonic valence, as well as dominance (which accounts for the least amount of variability in empirical studies) associated with each picture. Normative ratings are given for women and men separately and are based on large samples of American college students. Norms collected in other countries are also available (e.g., Ramirez et al., 1998). Affectively arousing IAPS pictures, both pleasant and unpleasant, have been shown to engage motivational reflex physiology indexing appetitive or aversive/defensive response tendencies interpreted as supporting a dimensional approach to emotional responses (Lang & Davis, 2006). The IAPS has also frequently been used in neurophysiological work (Keil et al., 2009). In a large body of studies, electrophysiological and hemodynamic imaging data have converged to demonstrate amplification of extended visual areas, deep structures (e.g., the amygdala), as well as frontal cortical areas, in response to pleasant and unpleasant pictures, as compared to neutral images (Lang & Bradley, 2010). These modulations have a strong positive linear relationship with self-reported emotional intensity or arousal, as well as with an array of autonomic indices that reflect the intensity of emotional engagement, such as the skin conductance response (Lang & Davis, 2006). By contrast, the phasic heart rate response, calculated as the change of R-wave rate over time, relative to stimulus onset, shows a more complex pattern: Viewing unpleasant pictures prompts strong heart rate deceleration (within 2–3 s of stimulus presentation), followed by moderate subsequent acceleration, and pleasant pictures tend to evoke less deceleration but greater subsequent acceleration (Bradley, 2009). Such a pattern is consistent with the heart's dual innervations by parasympathetic and sympathetic afferents and supports the perspective that emotions emerge as a complex interplay between systems maximizing the adaptive state of the organism vis-à-vis an engaging event.

Emotional Imagery

Reliving previously experienced emotional episodes and viewing text-driven cues conveying emotionally arousing content (for instance when reading a novel or play) have been shown to reliably produce self-reports of vivid imagery accompanied by physiological reactions (e.g., cardiac and facial electromyogram changes) consistent with affective engagement (Lang, 1979). Imagery of emotionally arousing material has been used extensively in the assessment and treatment of psychiatric disorders falling within the anxiety spectrum (McTeague et al., 2009, 2010). In this context, imagery can be regarded as an active memory representation, and emotional imagery specifically is characterized by containing aspects of motivational relevance for the organism.

Basic science studies of emotion have widely capitalized on the strength of emotional imagery. Imagery paradigms have been used to explore aspects of emotional memory, to examine responses to highly individualized situations, to avoid conflation of effects with sensory processing, and for many other applications. Even imagery prompted by simple word-cues engage facial, autonomic, and somatic reflexes, and these responses parallel reactions to actual pleasant or unpleasant events (Vrana, 1993). In affective neuroscience, a number of hemodynamic imaging studies have reported that mental imagery is associated with robust engagement of brain regions involved in processing input from the sensory modality matching the mental image (Kosslyn, 2005). Beyond sensory processing regions, imagery of affective content leads to engagement of motivational neural circuitry as well as motor regions involved in orchestrating behavioral responses to emotionally relevant external stimuli (Costa, Lang, Sabatinelli, Versace, & Bradley, 2010). In line with research using emotional pictures or conditioned stimuli, hemodynamic imaging studies of emotional imagery have shown activation of the insula and amygdala, as well as ventromedial prefrontal cortex deactivation during emotionally distressing imagery (Britton, Phan, Taylor, Fig, & Liberzon, 2005). During narrative imagery driven by short scripts, the hemodynamic signal was also reportedly enhanced in the supplementary motor area, lateral cerebellum, and left inferior frontal gyrus (Sabatinelli, Lang, Bradley, & Flaisch, 2006). In terms of specific emotional contents, Costa et al. (2010) observed that imagery of pleasant content selectively activated the nucleus accumbens (NAcc) and medial prefrontal cortex, and the amygdala was activated during imagery of emotionally engaging situations irrespective of hedonic valence. Conceptually, Lang (1979) has proposed that verbal event descriptions activate broad networks in the brain, coding not just sensory information, but also associated semantic and efferent response representations. Findings from recent electrophysiological studies with emotional imagery are consistent with this notion (Aftanas, Koshkarov, Mordvintsev, & Pokrovskaja, 1994), supporting the idea that emotional engagement through imagery—not unlike an overt emotional episode—involves widespread neural communication among networks representing the sensory, semantic, and motor processes that correspond to the content of the imagined scene.

Classical Conditioning

There are several routes of access to emotional systems. As mentioned, some stimuli appear to enjoy prepotent access to the neural circuitry that is relevant for orchestrating defensive or appetitive responses. For example, recurring predators are able to initiate fear responses in prey in the absence of previous experience; thus, cat fur and fox feces odors can instigate long-lasting freezing responses in laboratory rats that have never actually encountered such predators within their lifetimes (Rosen, 2004). However, many other cues or events can gain access to hard-wired emotional systems through learning. Classical (Pavlovian) conditioning, which involves the formation of associations between arbitrary stimuli and motivationally significant outcomes, is perhaps the most thoroughly studied laboratory method for determining experience-dependent modifications occurring within emotional systems.[6] In a typical conditioning

---

[6]More simple forms of learning (e.g., habituation and sensitization) also have unique molecular signatures that have been clarified in simple invertebrate organisms such as the sea slug *Aplysia californica*.

paradigm, participants encounter a neutral cue (e.g., a tone or light) that is then consistently paired in close temporal proximity with a motivationally engaging stimulus (e.g., electric shock or delivery of a reward).

The great success story of affective neuroscience within the past two decades has been a detailed mapping of fear (Maren & Quirk, 2004) and appetitive (Shuler & Bear, 2006) memory formation at the neuroanatomical, neurophysiological, neurochemical, and genetic levels of analysis. Recent advances in neuroscience technology (e.g., the availability of optogenetic probes) permits an unprecedented level of detail, even in terms of being able to activate specific cell assemblies representing affectively charged "engrams" (Liu et al., 2012). Similarly, functional magnet resonance imaging studies have documented conserved neural circuits that help to mediate aversive and appetitive conditioning in human participants. By recording electrical and magnetic aspects of neural processing in humans, we now have considerable evidence of the changes in sensory processing of affectively conditioned cues occurring at multiple spatiotemporal scales (Miskovic & Keil, 2012).

To summarize, we recommend that the primary focus of empirical emotion research be laid on observable behaviors, self-report, and physiological measures. In contrast, the phenomenology of affect (the actual feelings of the participants) is vastly complicated and emotion researchers have taken various stances. For example, Panksepp (1998) embraces a strong version of dual aspect monism that highlights a central role for raw affective experience, while others have more or less explicitly denied the capacity for raw affective experience in animals that do not have elaborate working memory abilities (LeDoux, 2012). The selection of paradigms and findings presented above aims to stimulate exploration of the literature cited along with critical questioning of the strengths and weaknesses of the respective paradigms given for addressing specific conceptual questions. Many more laboratory paradigms for inducing and observing emotional processes exist and they are expanded on in Chapter 5.

THE ROLE OF "REGULATION" IN EMOTION RESEARCH

One of the fastest growing areas in emotion research and theory may be the study of cognitive modification of emotional experience and behavior through regulatory processes. Often referred to as emotion regulation, this approach typically involves laboratory situations in which participants are invited to alter aspects of the intensity, direction, or duration of their emotional experience (Gross, 2013). Researchers interested in the voluntary control of emotional processes have also studied how facial expressions are modulated as participants attempt to disguise or manage emotional episodes—an approach that has been particularly fruitful in developmental studies (Saarni, 2000).

Because of the great potential for application in the domain of health and well-being, the idea of exerting flexible and voluntary control over one's feelings has considerable appeal. In fact, discussions of research on the interplay between sleep and emotion regulation can be found in several of the chapters in this volume (for example, see Chapter 19). Recent conceptual work in the area of emotion regulation has included the interesting question of agency and how it is related to definitions of emotion (Gross & Barrett, 2011). Here, the notion of a conscious self that is the agent of regulatory efforts acting on the emotional process implies a conceptual separation between a homuncular representation of the individual and the primordial urges that are being controlled (Cacioppo & Berntson, 2012). Although fully compatible with the folk-psychological distinction between the mind and the heart, several authors have argued that such a quasi-dualistic approach, separating the regulator's intentions from the regulated process, is at odds with conceptions of emotions as multilevel processes involving the entire organism. The quasi-dualistic approach particularly contrasts with models that see changes in cognitive systems as an integral part of the emotional process itself (Gross & Barrett, 2011).

As stated above, emotion regulation approaches have been successfully applied in the context of linear process models as well as appraisal theories (Hajcak & Nieuwenhuis, 2006). This is not surprising because these theories emphasize the role of a cognitive or perceptual process that acts as a mediator between an event and an eventual emotional response. Accordingly, one of the most dominant models of emotion regulation is based on a serial process model, proposing a sequence of situation, attention, appraisal, and response (Gross, 2013). The active role of the individual in modulating this sequence is then categorized as a function of when in the sequence the modulatory effort takes place. In contemporary empirical research, the concept of reappraisal has been particularly relevant as an "antecedent emotion regulation strategy" and refers to voluntarily altering the appraisal stage of the emotional process by changing assumptions, interpretations, or expectations about the eliciting stimulus (Gross, 1998). Potentially reflecting the great success of emotion regulation paradigms, recent approaches have included attempts to measure "unconscious regulation of affect." This approach may be seen as problematic in situations in which researchers do not manipulate the participants' efforts toward regulation, yet infer from measuring reactivity that the participants have spontaneously "downregulated" or "upregulated" their responses.

To the extent that many studies of emotion regulation have recorded multiple variables, including neurophysiological, peripheral, endocrine, behavioral, and self-report measures, this research has produced many interesting findings regarding the interaction between cognitive variables and other aspects of the emotional response. Importantly, recent work has replicated and extended the observation that the different domains and descriptors of emotional processes are not highly correlated (Mauss & Robinson, 2009). In addition, research manipulating the information given to participants about affective stimuli (e.g., a bloody movie is a documentary versus a well-made computer animation) has illustrated how the semantics of a stimulus rather than its mere appearance alone affect the emotional response (Foti & Hajcak, 2008). Such findings enable interpretations in the context of nonappraisal theories as well and provide stimulating avenues for research aiming to control the physical features of a stimulus, while manipulating its emotional content.

CONCLUSIONS AND ELEMENTS OF A COMPREHENSIVE APPROACH FOR THE STUDY OF EMOTION

We conclude this brief and selective conceptual review by restating the points we think are critical for a productive research program in the affective science of sleep. First, we suggest a pragmatic consensus definition of emotions grounded in animal models of motivation and focusing on observable and measurable phenomena. Conceptually, such a definition is in line with viewing emotions as evolutionary devices that assist humans in addressing challenges from the environment. Second, we recommend that the primary focus of empirical emotion research should be the collection of data describing observable behavior, including self-report data such as ratings of affect. Although self-reported affect is a useful and productive variable in this array, attempts to access the actual experience (i.e., the subjective feeling) of an individual and then to use it as a gold standard for emotion research have not been productive. In fact, the recognition that self-reported affect and observable emotional expressions may not be related is an important result of the conceptual and empirical discourse of the past decades. Among other consequences, this development has led to suggested changes in nomenclature, such as characterizing responses to threat as "defensive" rather than "fearful," highlighting the absence of a direct measure of experiential affect (feeling) in the animal (and ultimately human) model. The pragmatic approach presented throughout this chapter focuses on the three domains of emotion data accessible to empirical researchers: observable behavior, verbal report of affect, and physiological measurements. It is compatible with many theoretical flavors prevalent in emotion research today, and, in our view, it does not result in bias toward a specific emotion theory or family of theories. Although theoretically inclusive, the pragmatic approach is not compatible with a folk-psychological understanding of affect, however, because a multilevel, multidomain natural science perspective is at odds with the notion that, at the core of an emotional response, is a hidden but "true" unitary feeling state, the empirical key to which is yet to be discovered. Equipped with recent conceptual, empirical, and technical advances, multidomain analyses of affective processes are now possible at exquisite levels of detail. In combination with progress in the areas of study design and data analysis, these advances hold promise for developing mechanistic knowledge regarding the behavioral and physiological processes underlying human emotions in healthy individuals as well as in clinical populations.

References

Aftanas, L. I., Koshkarov, V. I., Mordvintsev, Y. N., & Pokrovskaja, V. L. (1994). Dimensional analysis of human EEG during experimental affective experience. *International Journal of Psychophysiology, 18*, 67–70. http://dx.doi.org/10.1016/0167-8760(84)90015-1.

Arnold, M. B. (1968). In defense of Arnold's theory of emotion. *Psychological Bulletin, 70*, 283–284.

Averill, J. R., Clore, G. L., LeDoux, J. E., Panksepp, J., Watson, D., Clark, L. A., et al. (1994). What influences the subjective experience of emotion? In R. J. D. Paul Ekman (Ed.), *The nature of emotion: Fundamental questions. Series in affective science* (pp. 377–407). New York, NY, US: Oxford University Press.

Barrett, L. F. (2012). Emotions are real. *Emotion, 12*, 413–429.

Blumberg, S. H., & Izard, C. E. (1986). Discriminating patterns of emotions in 10- and 11-year-old children's anxiety and depression. *Journal of Personality and Social Psychology, 51*, 852–857.

Bradley, M. M. (2009). Natural selective attention: Orienting and emotion. *Psychophysiology, 46*, 1–11.

Bradley, M. M., Codispoti, M., Sabatinelli, D., & Lang, P. J. (2001). Emotion and motivation II: Sex differences in picture processing. *Emotion, 1*, 300–319.

Brandao, M. L., Anseloni, V. Z., Pandossio, J. E., De Araujo, J. E., & Castilho, V. M. (1999). Neurochemical mechanisms of the defensive behavior in the dorsal midbrain. *Neuroscience and Biobehavioral Reviews, 23*, 863–875, S014976349900038X [pii].

Britton, J. C., Phan, K. L., Taylor, S. F., Fig, L. M., & Liberzon, I. (2005). Corticolimbic blood flow in posttraumatic stress disorder during script-driven imagery. *Biological Psychiatry, 57*, 832–840. http://dx.doi.org/10.1016/j.biopsych.2004.12.025, S0006-3223(04)01365-4 [pii].

Cacioppo, J. T., & Berntson, G. G. (1999). The affect system: Architecture and operating characteristics. *Current Directions in Psychological Science, 8*, 133–137. http://dx.doi.org/10.1111/1467-8721.00031.

Cacioppo, J. T., & Berntson, G. G. (2012). Is consciousness epiphenomenal? Social neuroscience and the case for interacting brains. *Euresis, 3*, 31–50.

Cacioppo, J. T., & Gardner, W. L. (1999). Emotion. *Annual Reviews in Psychology, 50*, 191–214.

Cannon, W. B. (1929). Bodily changes in pain, hunger, fear and rage. NY: Appleton.

Costa, V. D., Lang, P. J., Sabatinelli, D., Versace, F., & Bradley, M. M. (2010). Emotional imagery: Assessing pleasure and arousal in the brain's reward circuitry. *Human Brain Mapping, 31*, 1446–1457. http://dx.doi.org/10.1002/hbm.20948.

Cuthbert, B. N., Lang, P. J., Strauss, C., Drobes, D., Patrick, C. J., & Bradley, M. M. (2003). The psychophysiology of anxiety disorder: Fear memory imagery. *Psychophysiology, 40*, 407–422.

REFERENCES

Cuthbert, B. N., Schupp, H. T., Bradley, M. M., Birbaumer, N., & Lang, P. J. (2000). Brain potentials in affective picture processing: Covariation with autonomic arousal and affective report. *Biological Psychology, 52*, 95–111.

Damasio, A. R. (1998). Emotion in the perspective of an integrated nervous system. *Brain Research Brain Research Reviews, 26*, 83–86.

Dan-Glauser, E. S., & Scherer, K. R. (2011). The Geneva affective picture database (GAPED): A new 730-picture database focusing on valence and normative significance. *Behavioral Research Methods, 43*, 468–477. http://dx.doi.org/10.3758/s13428-011-0064-1.

Darwin, C. (1872). *The expressions of the emotions in man and animals.* Chicago, IL: University of Chicago Press.

Davis, M., Walker, D. L., Miles, L., & Grillon, C. (2010). Phasic vs sustained fear in rats and humans: Role of the extended amygdala in fear vs anxiety. *Neuropsychopharmacology, 35*, 105–135. http://dx.doi.org/10.1038/npp.2009.109, npp2009109 [pii].

Denton, D. A., McKinley, M. J., Farrell, M., & Egan, G. F. (2009). The role of primordial emotions in the evolutionary origin of consciousness. *Consciousness and Cognition, 18*, 500–514. http://dx.doi.org/10.1016/j.concog.2008.06.009, S1053-8100(08)00096-2 [pii].

Dewey, J. (1895). The theory of emotion. (2) The significance of emotions. *Psychological Review, 2*, 13–32.

Ekman, P. (1992). Are there basic emotions. *Psychological Review, 99*, 550–553. http://dx.doi.org/10.1037/0033-295x.99.3.550.

Engel, A. K., Fries, P., & Singer, W. (2001). Dynamic predictions: Oscillations and synchrony in top-down processing. *Nature Reviews Neuroscience, 2*, 704–716. http://dx.doi.org/10.1038/3509456535094565.

Foti, D., & Hajcak, G. (2008). Deconstructing reappraisal: Descriptions preceding arousing pictures modulate the subsequent neural response. *Journal of Cognitive Neuroscience, 20*, 977–988. http://dx.doi.org/10.1162/jocn.2008.20066.

Graham, F. K., & Clifton, R. K. (1966). Heart-rate change as a component of the orienting response. *Psychology Bulletin, 65*, 305–320.

Gross, J. J. (1998). Antecedent- and response-focused emotion regulation: Divergent consequences for experience, expression, and physiology. *Journal of Personality and Social Psychology, 74*, 224–237.

Gross, J. J. (2013). Emotion regulation: Taking stock and moving forward. *Emotion, 13*, 359–365. http://dx.doi.org/10.1037/a0032135, 2013-10205-001 [pii].

Gross, J. J., & Barrett, L. F. (2011). Emotion generation and emotion regulation: One or two depends on your point of view. *Emotion Review, 3*, 8–16. http://dx.doi.org/10.1177/1754073910380974.

Hajcak, G., & Nieuwenhuis, S. (2006). Reappraisal modulates the electrocortical response to unpleasant pictures. *Cognitive, Affective, and Behavioral Neuroscience, 6*, 291–297.

Hebb, D. (1949). *The organization of behavior: A neuropsychological theory.* New York: Wiley.

Heim, S., Benasich, A. A., & Keil, A. (2013). Distraction by emotion in early adolescence: Affective facilitation and interference during the attentional blink. *Frontiers in Developmental Psychology, 4*, 580. http://dx.doi.org/10.3389/fpsyg.2013.00580.

Hornef, M. W., Wick, M. J., Rhen, M., & Normark, S. (2002). Bacterial strategies for overcoming host innate and adaptive immune responses. *Nature Immunology, 3*, 1033–1040. http://dx.doi.org/10.1038/ni1102-1033, ni1102-1033 [pii].

Izard, C. E., Fine, S., Mostow, A., Trentacosta, C., & Campbell, J. (2002). Emotion processes in normal and abnormal development and preventive intervention. *Developmental Psychopathology, 14*, 761–787.

James, W. (1884). What is an emotion? *Mind, 9*, 188–205.

Jirsa, V. K., Friedrich, R., Haken, H., & Kelso, J. A. (1994). A theoretical model of phase transitions in the human brain. *Biological Cybernetics, 71*, 27–35.

Jones, B., & Mishkin, M. (1972). Limbic lesions and the problem of stimulus–reinforcement associations. *Experimental Neurology, 36,* 362–377.

Keil, A., Costa, V., Smith, J. C., Sabatinelli, D., McGinnis, E. M., Bradley, M. M., et al. (2012). Tagging cortical networks in emotion: A topographical analysis. *Human Brain Mapping, 33,* 2920–2931. http://dx.doi.org/10.1002/hbm.21413.

Keil, A., & Freund, A. M. (2009). Changes in the sensitivity to appetitive and aversive arousal across adulthood. *Psychology and Aging, 24,* 668–680.

Keil, A., Sabatinelli, D., Ding, M., Lang, P. J., Ihssen, N., & Heim, S. (2009). Re-entrant projections modulate visual cortex in affective perception: Directional evidence from granger causality analysis. *Human Brain Mapping, 30,* 532–540.

Klüver, H., & Bucy, P. C. (1937). Psychich blindness and other symptoms following bilateral temporal lobectomy in rhesus monkeys. *American Journal of Physiology, 119,* 352–353.

Kosslyn, S. M. (2005). Mental images and the brain. *Cognitive Neuropsychology, 22,* 333–347. http://dx.doi.org/10.1080/02643290442000130, 769488928 [pii].

Kreibig, S. D., Gendolla, G. H., & Scherer, K. R. (2012). Goal relevance and goal conduciveness appraisals lead to differential autonomic reactivity in emotional responding to performance feedback. *Biological Psychology, 91,* 365–375. http://dx.doi.org/10.1016/j.biopsycho.2012.08.007, S0301-0511(12)00172-X [pii].

Lane, R. D., Chua, P. M., & Dolan, R. J. (1999). Common effects of emotional valence, arousal and attention on neural activation during visual processing of pictures. *Neuropsychologia, 37,* 989–997.

Lang, P. J. (1979). A bioinformational theory of emotional imagery. *Psychophysiology, 16,* 495–512.

Lang, P. J. (1993). The three-systems approach to emotion. In N. Birbaumer, & A. Ohman (Eds.), *The structure of emotion: Psychophysiological, cognitive and clinical aspects.* Tubingen, Germany: Hogrefe & Huber.

Lang, P. J. (1994). The motivational organization of emotion: Affect-reflex connections. In S. H. M. Van Goozen, N. E. Van de Poll, & J. E. Sergeant (Eds.), *Emotions: Essays on emotion theory* (pp. 61–93). Hillsdale, NJ: Lawrence Erlbaum Associates.

Lang, P. J., & Bradley, M. M. (2010). Emotion and the motivational brain. *Biological Psychology, 84,* 437–450. http://dx.doi.org/10.1016/j.biopsycho.2009.10.007, S0301-0511(09)00225-7 [pii].

Lang, P. J., Bradley, M. M., & Cuthbert, B. N. (1990). Emotion, attention, and the startle reflex. *Psychological Review, 97,* 377–395.

Lang, P. J., Bradley, M. M., & Cuthbert, B. N. (2005). International affective picture system: Technical manual and affective ratings. Gainesville, FL: NIMH Center for the Study of Emotion and Attention.

Lang, P. J., & Davis, M. (2006). Emotion, motivation, and the brain: Reflex foundations in animal and human research. *Progress in Brain Research, 156,* 3–29.

Lang, P. J., Greenwald, M. K., Bradley, M. M., & Hamm, A. O. (1993). Looking at pictures: Affective, facial, visceral, and behavioral reactions. *Psychophysiology, 30,* 261–273.

Lazarus, R. S. (1991). Progress on a cognitive-motivational-relational theory of emotion. *American Psychologist, 46,* 819–834.

LeDoux, J. (2012). A neuroscientist's perspective on debates about the nature of emotion. *Emotion Review, 4,* 375–379.

LeDoux, J. E., Thompson, M. E., Iadecola, C., Tucker, L. W., & Reis, D. J. (1983). Local cerebral blood flow increases during auditory and emotional processing in the conscious rat. *Science, 221,* 576–577.

Levenson, R. W. (2011). Basic emotion questions. *Emotion Review, 3,* 379–386. http://dx.doi.org/10.1177/1754073911410743.

Levenson, R. W., & Ruef, A. M. (1992). Empathy: A physiological substrate. *Journal of Personality and Social Psychology, 63,* 234–246.

Lewin, K. (1951). Field theory in social science. Oxford, UK: Harpers.
Lewis, M. D., & Liu, Z. X. (2011). Three time scales of neural self-organization underlying basic and nonbasic emotions. *Emotion Review*, 3, 416–423.
Lindquist, K. A., Wager, T. D., Kober, H., Bliss-Moreau, E., & Barrett, L. F. (2012). The brain basis of emotion: A meta-analytic review. *Behavioral and Brain Sciences*, 35, 121–143. http://dx.doi.org/10.1017/S0140525X11000446, S0140525X11000446 [pii].
Liu, X., Ramirez, S., Pang, P. T., Puryear, C. B., Govindarajan, A., Deisseroth, K., et al. (2012). Optogenetic stimulation of a hippocampal engram activates fear memory recall. *Nature*, 484, 381–385. http://dx.doi.org/10.1038/nature11028, nature11028 [pii].
Lundqvist, D., Flykt, A., & Öhman, A. (1998). *Karolinska directed emotional faces*. Stockholm: Dept. of Neurosciences, Karolinska Hospital.
Maren, S., & Quirk, G. J. (2004). Neuronal signalling of fear memory. *Nature Reviews Neuroscience*, 5, 844–852. http://dx.doi.org/10.1038/nrn1535, nrn1535 [pii].
Martinez, R. C., Gupta, N., Lazaro-Munoz, G., Sears, R. M., Kim, S., Moscarello, J. M., et al. (2013). Active vs. reactive threat responding is associated with differential c-Fos expression in specific regions of amygdala and prefrontal cortex. *Learning and Memory*, 20, 446–452. http://dx.doi.org/10.1101/lm.031047.113, 20/8/446 [pii].
Mashour, G. A., & LaRock, E. (2008). Inverse zombies, anesthesia awareness, and the hard problem of unconsciousness. *Consciousness and Cognition*, 17, 1163–1168. http://dx.doi.org/10.1016/j.concog.2008.06.004, S1053-8100(08)00091-3 [pii].
Mauss, I. B., & Robinson, M. D. (2009). Measures of emotion: A review. *Cognition and Emotion*, 23, 209–237. http://dx.doi.org/10.1080/02699930802204677.
Mayr, E. (1974). Behavior programs and evolutionary strategies. *American Scientist*, 62, 650–659.
McTeague, L. M., Lang, P. J., Laplante, M. C., Cuthbert, B. N., Shumen, J. R., & Bradley, M. M. (2010). Aversive imagery in posttraumatic stress disorder: Trauma recurrence, comorbidity, and physiological reactivity. *Biological Psychiatry*, 67, 346–356. http://dx.doi.org/10.1016/j.biopsych.2009.08.023, S0006-3223(09)01025-7 [pii].
McTeague, L. M., Lang, P. J., Laplante, M. -C., Cuthbert, B. N., Strauss, C. C., & Bradley, M. M. (2009). Fearful imagery in social phobia: Generalization, comorbidity, and physiological reactivity. *Biological Psychiatry*, 65, 374–382.
Mehrabian, A. (1995). Framework for a comprehensive description and measurement of emotional states. *Genetic, Social, and General Psychology Monographs*, 121, 339–361.
Miller, N. E., & Murray, E. J. (1952). Displacement and conflict; Learnable drive as a basis for the steeper gradient of avoidance than of approach. *Journal of Experimental Psychology*, 43, 227–231.
Miskovic, V., & Keil, A. (2012). Acquired fears reflected in cortical sensory processing: A review of electrophysiological studies of human classical conditioning. *Psychophysiology*, 49, 1230–1241. http://dx.doi.org/10.1111/j.1469-8986.2012.01398.x.
Moors, A. (2009). Theories of emotion causation: A review. *Cognition & Emotion*, 23, 625–662.
Moors, A. (2013). On the causal role of appraisal in emotion. *Emotion Review*, 5, 132–140.
Moors, A., Ellsworth, P. C., Scherer, K. R., & Frijda, N. H. (2013). Appraisal theories of emotion: State of the art and future development. *Emotion Review*, 5, 119–124.
Panksepp, J. (1992). A critical role for "affective neuroscience" in resolving what is basic about basic emotions. *Psychological Review*, 99, 554–560.
Panksepp, J. (1998). *Affective neuroscience: The foundations of human and animal emotions*. New York, USA: Oxford University Press.
Panksepp, J. (2011). The basic emotional circuits of mammalian brains: Do animals have affective lives? *Neuroscience and Biobehavioral Reviews*, 35, 1791–1804. http://dx.doi.org/10.1016/j.neubiorev.2011.08.003, S0149-7634(11)00149-7 [pii].
Papez, J. W. (1938). A proposed mechanism of emotion. *Archives of Neurology and Psychiatry*, 38, 725–743.

Parvizi, J. (2009). Corticocentric myopia: Old bias in new cognitive sciences. *Trends in Cognitive Science, 13*, 354–359. http://dx.doi.org/10.1016/j.tics.2009.04.008, S1364-6613(09)00132-6 [pii].

Pessoa, L., & Adolphs, R. (2010). Emotion processing and the amygdala: From a 'low road' to 'many roads' of evaluating biological significance. *Nature Reviews Neuroscience, 11*, 773–783. http://dx.doi.org/10.1038/nrn2920, nrn2920 [pii].

Phan, K. L., Wager, T. D., Taylor, S. F., & Liberzon, I. (2004). Functional neuroimaging studies of human emotions. *CNS Spectrum, 9*, 258–266.

Plutchik, R. (2001). The nature of emotions—Human emotions have deep evolutionary roots, a fact that may explain their complexity and provide tools for clinical practice. *American Scientist, 89*, 344–350. http://dx.doi.org/10.1511/2001.28.739.

Ramirez, I., Hernandez, M. A., Sanchez, M., Fernandez, M. C., Vila, J., Pastor, M. C., et al. (1998). Spanish norms for the "International Affective Picture System". *Journal of Psychophysiology, 12*, 312–313.

Reisenzein, R. (1983). The Schachter theory of emotion: Two decades later. *Psychological Bulletin, 94*(2), 239–264.

Rolls, E. T. (1995). A theory of emotion and consciousness, and its application to understanding the neural basis of emotion. In M. S. Gazzaniga (Ed.), *The Cognitive Neurosciences* (pp. 1091–1106). Cambridge, MA, US: MIT Press.

Rosen, J. B. (2004). The neurobiology of conditioned and unconditioned fear: A neurobehavioral system analysis of the amygdala. *Behavioral and Cognitive Neuroscience Reviews, 3*, 23–41. http://dx.doi.org/10.1177/1534582304265945.

Saarni, C. (2000). Emotion regulation and coping. *International Journal of Psychology, 35*, 112.

Sabatinelli, D., Lang, P. J., Bradley, M. M., Costa, V. D., & Keil, A. (2009). The timing of emotional discrimination in human amygdala and ventral visual cortex. *Journal of Neuroscience, 29*, 14864–14868. http://dx.doi.org/10.1523/JNEUROSCI.3278-09.2009, 29/47/14864 [pii].

Sabatinelli, D., Lang, P. J., Bradley, M. M., & Flaisch, T. (2006). The neural basis of narrative imagery: Emotion and action. *Progress in Brain Research, 156*, 93–103. http://dx.doi.org/10.1016/S0079-6123(06)56005-4 S0079-6123(06)56005-4 [pii].

Schachter, S., & Singer, J. E. (1962). Cognitive, social, and physiological determinants of emotional state. *Psycholological Review, 69*, 379–399.

Scherer, K. R. (2009). Emotions are emergent processes: They require a dynamic computational architecture. *Philosophical Transactions of the Royal Society London, B: Biological Sciences, 364*, 3459–3474. http://dx.doi.org/10.1098/rstb.2009.0141, 364/1535/3459 [pii].

Schroeder, C. E., Wilson, D. A., Radman, T., Scharfman, H., & Lakatos, P. (2010). Dynamics of active sensing and perceptual selection. *Current Opinion in Neurobiology, 20*, 172–176. http://dx.doi.org/10.1016/j.conb.2010.02.010, S0959-4388(10)00032-2 [pii].

Shuler, M. G., & Bear, M. F. (2006). Reward timing in the primary visual cortex. *Science, 311*, 1606–1609.

Smith, C. A., & Ellsworth, P. C. (1985). Patterns of cognitive appraisal in emotion. *Journal of Personality and Social Psychology, 48*, 813–838.

Sokolov, E. N. (1963). *Perception and the conditioned reflex*. New York: Macmillan.

Vrana, S. R. (1993). The psychophysiology of disgust: Differentiating negative emotional contexts with facial EMG. *Psychophysiology, 30*, 279–286.

Watson, D., Clark, L. A., & Carey, G. (1988). Positive and negative affectivity and their relation to anxiety and depressive disorders. *Journal of Abnormal Psychology, 97*, 346–353.

Weiskrantz, L. (1956). Behavioral changes associated with ablation of the amygdaloid complex in monkeys. *Journal of Comparative and Physiological Psychology, 49*, 381–391.

Wundt, Q. (1917). *Grundriss der Psychologie (Outline of psychology)* (13th Ed.). Leipzig: Alfred Kroener Verlag.

CHAPTER

3

Sleep, Emotions, and Emotion Regulation: An Overview

Christopher P. Fairholme[1] *and Rachel Manber*[1]
[1]Department of Psychiatry and Behavioral Sciences, Stanford University, Stanford, California, USA

SLEEP, EMOTIONS, AND EMOTION REGULATION: AN OVERVIEW

Deficient sleep is ubiquitous and has been associated with a broad range of psychopathology (Harvey, Murray, Chandler, & Soehner, 2011). Difficulties with emotions and emotion regulation have similarly been identified as a potential transdiagnostic process across various forms of psychopathology (Kring & Sloan, 2010). Recently, this has led to an increased interest in the relationship between sleep and emotional experiences. The aim of this chapter is to provide a framework for understanding how sleep and emotions relate to one another. Because the relationship between sleep and emotions appears to be moderated by the presence of psychopathology (see Chapter 11), we focus on studies looking at individuals without a current psychiatric disorder. Understanding the relationship between sleep and emotions in the absence of psychopathology could provide insight into how such processes might be altered following the development of psychopathology.

In evaluating the relationships between sleep and emotional functioning, we must further elucidate and parse our definition of sleep deficiency along the following three dimensions: acute-chronic, intrinsic-extrinsic, and subjective-objective. For instance, insomnia is characterized by chronic intrinsic and subjective sleep deficiency, whereas sleep deprivation is characterized by acute extrinsic and objective manipulation of sleep opportunity, usually in the laboratory. The effect of sleep deficiency on emotional functioning and its role in the development and/or maintenance of psychopathology might vary depending upon each of these dimensions.

For instance, the effect of acute sleep deficiency on emotional functioning might be a proximal factor contributing to the etiology of emotional disorders. In contrast, chronic sleep deficiency might be a maintaining factor for emotional disorders largely independent of their etiology (i.e., regardless of when the sleep deficiency emerged relative to the onset of the psychopathology). There is no reason to assume that these effects are dependent on one another. For a subset of individuals, sleep deficiency might play an etiological role, but not a maintaining role. For another subset of individuals, sleep deficiency might not be involved in the etiology of their psychopathology, but it might have become a strong maintaining factor. Although the extant literature is still relatively new, we hope that increasing conceptual and definitional clarity in the literature may help further elucidate the precise nature of such complex, dynamic relationships between sleep and emotional experiences and psychopathology.

EMOTIONS AND EMOTION REGULATION

Emotions are functional and adaptive phenomena that help organize and motivate behavior in the service of an organism's goals. Emotions can interrupt ongoing cognitive or behavioral processes, direct attention to stimuli directly relevant to goal pursuit, and trigger action tendencies supporting the individual's goals (Frijda, 2007; Gross, 2014; Levenson, 1999). We consider emotions to be emergent phenomena, arising from a dynamic interaction of input across multiple modes of experience, including cognition, behavior, and physiological sensations (Barrett, 2013; Barrett, Mesquita, Ochsner, & Gross, 2007).

Conceptually, we consider emotion regulation as the attempted modification of any aspect of an emotional experience that can occur at various levels of conscious awareness and utilize varying degrees of effortful control. Broadly speaking, the act of emotion regulation may be aimed at amplifying, attenuating, maintaining, or even preventing emotional responding and can occur in the context of motivated behavior intended to promote or prevent desired or undesired future states, respectively. Gross (1998, 2014) provides a useful conceptual framework for understanding the dynamic process of emotional experience and its regulation. His model takes the emotion-generative process as its base, highlighting the dynamic nature of the process and delineating how emotion regulation efforts can be initiated at any point during the process of emotion generation. The emotion-generative process, which he terms the modal model of emotion, begins with some psychologically relevant situation that can be external (e.g., encountering a bear in the woods) or internal (e.g., a fleeting thought about a past or potential future failure). Attention is then brought to bear on specific aspects of the situation, and appraisals of the situation are generated based on their presumed relevance and impact upon an individual's currently salient goals. The emotional response generated by this unfolding process then helps to coordinate responses across multiple systems (i.e., behavioral, cognitive, physiological) in service of an individual's current motivationally salient goals. From this perspective, emotion regulation can occur at any point during the emotion generative process. For instance, emotion regulation can occur prior to encountering a given situation (e.g., turning down an invitation to a party to avoid social interaction), by altering the situation in some way to reduce its emotional impact (e.g., avoiding making eye contact when someone else initiates a conversation), by focusing or redirecting attention to alter a situation's emotional impact (e.g., focusing on what one will say next in order to avoid saying something "stupid"), by generating alternative appraisals of the situation to alter its emotional impact (e.g., "I don't care what Sally thinks about me anyway"), or altering one's response to the situation itself (e.g., resisting the urge to end the conversation and escape the situation). Any such attempts to modify the emotional response will alter the psychological situation as the dynamic emotion-generative process continues to unfold over time. Although this model has not yet been directly evaluated in the sleep and emotion literature, we believe that it provides a useful heuristic for evaluating the extant research, and we return to the modal model later in the chapter.

SLEEP AND ITS IMPACT ON AFFECT AND AFFECT REGULATION

A number of studies have utilized sleep restriction (reduced amount of total sleep) or sleep deprivation (no sleep) paradigms to evaluate the impact of sleep duration on affective functioning in healthy samples. As such, these studies evaluate the effect of acute, objective sleep deficiency attributable to external sources. The extent to which findings from these studies might generalize to other types of sleep deficiency or individuals with psychopathology remains unclear. However, a number of studies suggest that acute, objective, external sleep deficiency increases negative emotions and emotional reactivity among healthy participants.

Using a within-subject design, Larson, Durocher, Yang, DellaValla, and Carter (2012) found that one night of sleep deprivation led to increased subjective reports of state anxiety and increased cardiovascular reactivity to a mild stressor compared to a night of normal sleep, although increases in subjective anxiety were not correlated with increases in cardiac reactivity. Increases in stress reactivity following sleep curtailment may depend on the intensity of the stressor. Minkel et al. (2012) randomized healthy adults to receive either 48 hours of sleep deprivation or 9 hours of sleep opportunity while being continuously monitored in the sleep laboratory. The authors then examined each participant's subjective affective response to high- and low-stress experimental tasks using the Profile of Mood States. Compared to those provided with ample sleep opportunity, participants who were sleep deprived showed large increases in anger and anxiety ($d = 0.91$, $d = 0.81$, both p's < 0.01) and moderate increases stress and depression ($d = 0.61$, $d = 0.56$, $p < 0.05$, $p = 0.07$) when exposed to a low-stress but not to a high-stress experimental task. This finding suggests that insufficient sleep might lower the threshold for emotional reactivity by impairing the emotional appraisal process, potentially sensitizing individuals to the effects of events that may have otherwise been discounted as minor.

Further support for this hypothesis comes from the literature on posttraumatic stress disorder (PTSD). Bryant, Creamer, O'Donnell, Silove, and McFarlane (2010) assessed individuals during a hospital admission immediately following a traumatic event and found that self-reported sleep deficiency occurring during the 2 weeks prior to the traumatic event was associated with a threefold increase in new incident cases of psychiatric disorders after the trauma. Thus, it is possible that the relationship between poor sleep and emotional reactivity could follow a curvilinear trend, such that sleep deficiency increases emotional reactivity to low- and very high-intensity stressors, and reactivity to more moderate stressors remains comparable.

A seminal study by Zohar, Tzischinsky, Epstein, and Lavie (2005) examined the effects of externally imposed sleep loss on emotional reactivity to goal-enhancing and goal-disruptive events among medical residents. Reduced total sleep duration was associated with elevated negative affect in response to goal-disruptive events, but not in the absence of such events. That is, basal negative affect did not vary depending upon sleep the previous night. However, the opposite pattern was observed for positive affect. Reduced total sleep duration was associated with elevated basal positive affect (i.e., in the absence of a goal-enhancing event), but the affective benefit of goal-enhancing events was attenuated following nights with less sleep. An effect of reduced sleep duration (acute, objective, extrinsic sleep deficiency) on negative affect was only observed when a stressor with motivational significance was encountered, whereas reduced sleep duration directly impacted positive affect. One intriguing implication of this finding is that reduced sleep duration exerts its influence on affect, at least partially, through motivational mechanisms, potentially by increasing sensitivity to punishment or negative consequences. Given the intimate link between motivation and emotions (Frijda, 2007), one intriguing implication of this finding is that poor sleep might exert an indirect effect on psychopathology via its impact on motivation. It is possible that poor sleep impairs motivation and ultimately functioning by increasing the motivational salience of goal-disruptive events and reducing the motivational salience of goal-enhancing events. Unfortunately, relatively less literature has examined the motivational consequences of sleep disruption.

The three studies described so far suggest that insufficient sleep might exert relatively larger effects on emotion-regulatory and self-regulatory processes than on state or trait levels of affect. There is some additional evidence supporting an effect of sleep deficiency on emotion regulation. Yoo, Gujar, Hu, Jolesz, and Walker (2007) used functional magnetic resonance imaging (fMRI) to examine emotional responses to negative visual stimuli under conditions of sleep deprivation (one full night) compared to normal sleep. Consistent with previous findings that insufficient sleep leads to increased emotional reactivity, they found increased amygdala activation following exposure to negatively valenced pictures among individuals in the sleep deprivation group compared to a normal sleep control condition. They also found evidence for reduced downregulation of negative emotional reactions. Compared to the normal sleep controls, individuals in the sleep-deprived group evidenced reduced functional connectivity between the medial prefrontal cortex and the amygdala. This suggests the intriguing possibility that not only does insufficient sleep lead to increased emotional reactivity, thus increasing the chances that relatively minor stressors will lead to maladaptive emotional responses, but it is also associated with failure to downregulate the negative emotional experience, thus possibly exposing the individual to prolonged amygdala activation.

Baum et al. (2014) utilized an experimental design to examine the effect of sleep on affect and affect regulation among healthy adolescents (aged 14 to 17 years). They manipulated sleep across 2 weeks, following a baseline week, using a counterbalanced, crossover design. Participants were randomized to a sleep restriction (6.5 hours in bed a night for five consecutive nights) or a sleep extension (10 hours in bed a night for five consecutive nights) condition and then crossed over to the other condition the following week. At the end of each week, adolescents and their parents rated the adolescent's affect and affect regulation difficulties over the previous week. Adolescent's subjective ratings of anger and anxiety were significantly higher, and ratings of positive affect were significantly lower following sleep restriction, compared to sleep extension but ratings of depressed mood did not differ between the two conditions. Importantly, both adolescent and parent ratings of emotion regulation difficulties were significantly higher following sleep restriction compared to sleep extension. The study by Baum and colleagues is among the first to directly evaluate the effects of both restricting and extending sleep on emotion regulation. Moreover, the findings suggest that chronic partial sleep deprivation at a level naturally experienced on school and work days is detrimental to emotional experiences that could be part of the pathway linking stress to the development of psychopathology. More broadly, findings such as this suggest multiple pathways for sleep deficiency to impact emotions and possibly psychopathology.

Insufficient sleep might also impact emotions and psychopathology via synergistic effects with other vulnerability factors. A recent imaging study found evidence that insufficient sleep might amplify the impact of a trait vulnerability factor (i.e., trait anxiety) on emotional responding (Goldstein et al., 2013). The study examined anticipatory responding to negative and neutral stimuli under conditions of certainty or uncertainty following a night of sleep deprivation or normal sleep. Participants were provided with a cue indicating whether the stimuli would be negative or neutral, or the cue presented was ambiguous so that it did not indicate the valence of the stimuli that was to follow. Consistent with the findings from the above-described study conducted by Yoo et al. (2007), sleep deprivation increased emotional reactivity (as indexed by the magnitude of bilateral amygdala response) independent of cue type. Furthermore, following sleep deprivation, the right anterior insula, which under normal circumstances evidences increased activation to uncertainty or ambiguous cues, showed elevated responding under the two certainty conditions (negative and neutral cues) but not following ambiguous cues. In support of the idea that sleep deprivation synergistically interacts with other vulnerability factors, evidence of increased activation in the right anterior insula in response to uncertainty or ambiguous cues was the strongest among individuals high in trait anxiety.

Collectively, these findings suggest multiple pathways by which insufficient sleep might impact affective functioning. First, acute, objective, external sleep disruption can directly increase negative affect and reduce positive affect. Second, acute, objective, external sleep disruption impairs emotion regulation, potentially reducing resources for coping with increased emotional reactivity. Third, acute, objective, external sleep disruption might reduce the individual's ability to contextualize an emotional response, resulting in further amplification of emotional reactivity independent of current cue or context. It is noteworthy that a disconnect between the magnitude of emotional response and the current context represents a key clinical feature of a wide range of psychopathology, including anxiety disorders, depressive disorders, and insomnia.

AFFECT AND ITS IMPACT ON SLEEP

Compared to the extensive literature evaluating the effect of sleep on affective functioning, relatively fewer studies have evaluated the effect of affective functioning on sleep parameters. In general, research examining the relationship between sleep and emotion has conceptualized emotion as varying along a valence and an arousal dimension (Watson & Tellegen, 1985). The available literature does strongly indicate that arousal has an adverse effect on a range of sleep parameters, although operational definitions and uses of the term arousal vary substantially in this body of literature. As discussed earlier, emotions are multifaceted constructions, comprised of thoughts, physical sensations, and action tendencies. Similarly, arousal can refer to increased physiological and/or cognitive activity, both of which we consider to be aspects of emotional arousal.

Stress can be thought of as generating a range of emotions with negative valence and high arousal. Etiological theories have long acknowledged the potent role of stress in the pathogenesis of insomnia disorder (Spielman, Caruso, & Glovinsky, 1987), and the relationship between stress and poor sleep is well documented in the literature. Stress leads to increases in sleep latency and number and duration of awakenings, decrease in sleep efficiency, and alteration in sleep architecture, such as decreased percent time in rapid eye movement (REM) sleep and slow-wave sleep (SWS; Kim & Dimsdale, 2007). This literature suggests that sleep fragmentation (increased awakenings) is the most common consequence of stress (broadly defined to include both acute and chronic stress), whereas sleep latency is particularly responsive to acute stressors.

Additional support comes from studies that have used caffeine as an analog for examining the effects of physiological arousal on sleep. Caffeine offers a nice paradigm for examining the effects of physiological arousal, separate from valence. Using a within-subjects design, Bonnet and Arand (1992) examined the effects of thrice daily caffeine administrations on subjective (self-report) and objective (EEG) sleep parameters. For both self-report and EEG assessments, they found that acute caffeine ingestion decreased total sleep time, SWS (EEG only), sleep quality (self-report only), and sleep efficiency, while it increased sleep latency and nocturnal awakenings, supporting the notion that arousal adversely impacts sleep. Physiological arousal did not show any impact on latency to REM sleep or percentage of total sleep spent in REM, suggesting differential effects of arousal and valence on sleep architecture. However, valence is typically not directly assessed or accounted for in studies utilizing caffeine as a proxy for physiological arousal, leaving open the question of whether valence plays a role in sleep disruption above and beyond arousal.

Unfortunately, relatively fewer studies have examined the effects of valence on sleep parameters. A notable exception is a recent study that used polysomnography to examine the effects of a negative-emotion induction on sleep parameters (Vandekerckhove et al., 2011). Healthy controls were randomized to receive either neutral or failure feedback on a task presented as a test of general intelligence. The negative-emotion induction significantly reduced sleep efficiency, total sleep time, and percentage

REM sleep, while significantly increasing number of awakenings, total time awake, number of awakenings from REM, and latency to SWS.

Tang and Harvey (2004) conducted a pair of experiments to further explore the relationship of anxious arousal on sleep assessed via subjective reports and actigraphy. This study is noteworthy because they attempted to separately manipulate anxiety and arousal in order to examine how valence and arousal might uniquely impact sleep. In the first experiment, good sleepers were randomized to an anxious arousal group (told they would have to give a speech in front of an audience upon waking from a nap), a nonanxious cognitive arousal group (told they would have to write an essay upon waking from a nap), or a no manipulation group (just asked to take a nap). Importantly, manipulation checks supported the dissociation of anxiety and arousal across the three groups. Participants were then provided with a 60-minute nap opportunity, and their sleep was assessed with actigraphy and retrospectively by subjective report following the nap opportunity. Importantly, consistent with the hypothesized manipulations, the anxious arousal group was the only group to evidence a significant increase in anxiety, while cognitive arousal did not differ between the two arousal groups prior to the sleep opportunity. The anxious arousal group reported longer time to fall asleep and shorter total nap times than the no manipulation group (although interpretation of the total nap time finding is limited because total nap time was artificially restricted by the 60-minute sleep opportunity provided, and it is a direct function of time to fall asleep). There were no significant differences between the nonanxious cognitive arousal and control groups on sleep latency or total nap time, suggesting that valence exerts an effect on sleep above and beyond arousal, at least for cognitive arousal. In a second experiment, Tang and Harvey randomized good sleepers to an anxious arousal group (the same speech threat paradigm), a nonanxious physiological arousal group (caffeine pill), or a placebo control group. Both the anxious and physiological arousal groups reported taking longer to fall asleep and having shorter total nap times than the placebo control group; however, only the anxious arousal group reported poorer sleep quality than the control group.

Taken together, these findings suggest that valence and arousal might both contribute to poor sleep. Valence might be more likely to affect appraisals of the previous night's sleep and lead to subjective complaints of poor sleep, whereas arousal might be more likely to impair sleep latency and duration. Furthermore, valence and arousal appear to differentially affect sleep architecture, with valence exerting a relatively larger influence on REM sleep (Vandekerckhove et al., 2011) and arousal exerting its influence by reducing SWS (Bonnet & Arand, 1992). However, no study has directly examined the potential differential effects of valence and arousal on sleep architecture using polysomnography. A study utilizing polysomnography to extend the findings from Tang and Harvey (2004) would provide direct examination of this hypothesis.

AFFECT REGULATION AND ITS IMPACT ON SLEEP

Affective influences on sleep are not restricted to the individual's current affective state. Consistent with the modal model of emotions discussed earlier, emotions are generated and unfold over time. As they unfold, they are susceptible to implicit and explicit attempts to modify their expression and subjective experience (Gross, 2014). Such attempts to modify the experience and expression of emotions, which we consider to be emotion regulation, are also likely to affect sleep parameters. Certain emotion regulation strategies have been found to have more adverse consequences than others. For instance, attempts to suppress emotional responses or avoid the subjective experience of particular emotions altogether have been shown to have a range of maladaptive sleep consequences, whereas strategies promoting nonjudgmental acceptance of current emotional states seem to promote more positive outcomes. These are reviewed below.

A number of studies have evaluated the consequences of overreliance on maladaptive strategies such as worry and rumination to manage emotions. Thomsen and colleagues examined the relationship between the self-reported habitual tendency to engage in rumination following stressful events and self-reported sleep quality in a sample of undergraduates (Thomsen, Mehsen, Christensen, & Zachariae, 2003). Rumination was significantly correlated with self-reported sleep quality, and this relationship remained significant after adjusting for current negative mood, suggesting that emotion regulation strategies might impact sleep above and beyond current mood.

Another study extended this finding by inducing rumination following a stressful event (Guastella & Moulds, 2007). Following a midterm exam, undergraduates scoring high and low on a trait measure of rumination were randomly assigned to either a rumination or distraction condition. Prior to going to bed, individuals in the rumination condition were instructed to reflect on the test they took earlier that day and think about how they felt during the exam, how they performed, and the potential consequences of their performance, whereas individuals in the distraction condition were provided with neutral statements to think about. The next morning, all participants completed questionnaires assessing sleep quality. In the rumination condition, high-trait ruminators reported significantly worse sleep quality than low-trait ruminators. There was no significant difference in subjective sleep quality ratings for high- and low-trait ruminators in the distraction condition, suggesting that individual differences in how individuals respond to and attempt to manage negative emotions might contribute directly to poor sleep.

A recent study by Vandekerckhove et al. (2012) used polysomnography to examine the effect of two different emotion regulation strategies on various sleep parameters among healthy controls. Following an adaptation and baseline night in the lab, participants completed a series of cognitive tasks that were presented as a test of general intelligence, following which they received feedback that they had performed poorly (well below average) on the tasks. After being provided with an appropriate rationale, and just prior to bed, participants were randomized to write about the task and their performance on the task either using an "experiential" strategy (where they were asked to focus on emotions brought up by the task and their performance) or an "analytic" strategy (where they were asked to focus objectively and analyze their performance and its potential causes and consequences). Participants in the experiential strategy group took longer to fall asleep, but experienced significantly fewer awakenings, longer total sleep time, and higher sleep efficiency compared to the analytic strategy group. This suggests that processing an emotion prior to going to sleep might make it more difficult to fall asleep, but might help downregulate emotions enough so that they are less likely to disrupt sleep during the night, leading to increased sleep duration, reduced sleep fragmentation, and improved sleep quality.

There are a number of ways in which affective functioning can impact sleep. Cognitive and physiological arousal can lead to poor sleep by increasing sleep latency and sleep fragmentation, as well as decreasing sleep quality and total sleep time. Negatively valenced emotions have also been found to increase sleep fragmentation and reduce sleep quality, sleep efficiency, and total sleep time. Negative valence and high arousal appear to have differential effects on sleep architecture, with arousal reducing SWS and negative valence adversely affecting REM sleep. In addition to direct effects of valence and arousal on sleep, attempts to modify the expression or subjective experience of emotions also appear to affect sleep. It is possible that strategies aimed at reducing or "getting rid of" emotional responses appear to be associated with reduced sleep quality, more frequent awakenings, shorter total sleep time, and lower sleep efficiency compared to strategies promoting acknowledgement and an acceptance of an individual's current emotional state.

SLEEP AND AFFECT: BIDIRECTIONAL RELATIONS

The studies reviewed so far have primarily investigated unidirectional relationships, with sleep predicting affect or vice-a-versa. Comparatively fewer studies have directly examined the hypothesized bidirectional or reciprocal relationship between sleep and affect. Longitudinal evaluation of patterns of sleep and affective experience within individuals across time will help to shed light on the dynamic relationship between these two constructs. A number of the studies that have been conducted have relied upon mood ratings taken immediately upon waking and prior to bedtime; however, such an assessment schedule is likely to offer inflated estimates of the relationship between mood and sleep. Experience sampling methodology (ESM) is an alternative technology that might offer a more reliable estimate of the sleep-mood relationship than once-a-day mood measurements. ESM takes random samples of affective experience throughout the day. Aggregations of these repeated measurements will offer a more reliable and potentially more representative picture of daytime mood.

A recent study by de Wild-Hartmann et al. (2013) used ESM and a sleep diary to examine the relationship between nighttime sleep and daytime mood among 553 woman without current psychopathology. This study is different from previous studies described because the authors examined the covariation between naturally occurring variation in night-to-night sleep (as opposed to strong manipulations reducing sleep opportunity) and variations in day-to-day affect. Multiple indicators of poor subjective sleep (lower sleep quality, shorter total sleep time, longer sleep latency, increased sleep fragmentation) were prospectively associated with reduced levels of next day positive affect. In contrast, only lower sleep quality and longer sleep latency were prospectively associated with higher next-day negative affect. Given the potential buffering effects that positive affect can have upon negative affect (Wichers et al., 2007), poor sleep's deleterious effect on positive affect might represent another way in which poor sleep might indirectly influence negative mood and ultimately increase risk for psychopathology. The authors also evaluated prospective relationships of daily positive and negative affect on sleep parameters but found only weak associations. The only significant finding was that lower daily positive affect was prospectively associated with reduced sleep quality. Daily negative affect did not significantly predict any of the sleep parameters in this sample of healthy controls.

The de Wild-Hartmann et al. (2013) study found strong support for the hypothesis that poor sleep predicts reduced levels of positive affect and more modest support for the hypothesis that poor sleep predicts increased negative affect. It is noteworthy that, when the authors accounted for daily relationships between sleep and positive affect, negative affect did not predict any of the sleep parameters examined. The only significant prospective relationship between affect and sleep was the finding that reduced positive affect predicted poorer sleep quality. It is possible that the relationship between affect and sleep is strongest for emotions with high negative valence and high arousal, such as following stressful events. It is also possible that, in the absence of significant stressors, affect is much more likely to be impacted by than to impact sleep. This suggests that, in addition to direct effects, affect and sleep might have a synergistic effect on functioning following a stressor.

DISCUSSION

Sleep and emotions are related in a complex and dynamic way. Increasing evidence suggests that sleep impacts affective functioning in multiple ways. Insufficient and poor sleep appear to have a robust inverse relationship with positive affect. However, the relationship between sleep and negative affect is less clear, with a number of studies failing to find an association. Insufficient sleep (either acute or chronic partial sleep deprivation) has been shown to have a strong direct effect on negative affect, but this effect is not observed when examining daily variations in sleep quality or duration. However, multiple indirect pathways seem to lead from deficient sleep to negative emotion. Insufficient and/or poor sleep might indirectly increase negative affect by reducing the potential buffering effect of positive affect, altering the motivational salience of goal-relevant events, increasing sensitivity to punishment or barriers to goal attainment, and increasing emotional reactivity more broadly. Insufficient sleep also appears to impair emotion regulation and the individual's ability to interpret stressors in the context of current salient goals. Thus, a likely consequence of insufficient sleep is a reduced ability to cope with ongoing stressors and an increased likelihood of reacting negatively to relatively neutral or mildly stressful events, both of which might increase vulnerability to developing psychopathology.

An inherent difficulty in synthesizing the literature on the impact of poor or insufficient sleep on affect is the lack of coherence among definitions of affect. Indeed, there is an ongoing debate in the emotion-science literature about how to define affect. The modal model of emotions (Gross, 2014) might be particularly helpful for framing the complex relationship between sleep and affect because it highlights the process by which emotions are generated and unfold over time. According to this model, the emotion-generative process begins with the situation, broadly construed to capture the internal experience or construction of the situation, which could be based on external or internal cues. Attentional resources are then allocated to particular aspects of the situation and appraisals are assigned to help interpret and make meaning of the current situation. The appraisal results in an emotional response that could be experienced and/or expressed across multiple modes (cognition, behavior/action tendencies, and physiological sensations). The emotional response, which may or may not be subjectively and/or objectively observable, then feeds back

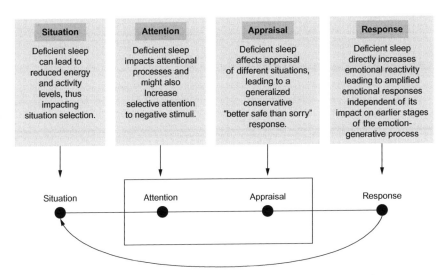

FIGURE 3.1 Hypothesized effects of deficient sleep on affect within the modal model of emotion.

into and modifies the current situation, and the process continues (Gross, 2014). This process is graphically depicted in the bottom half of Figure 3.1.

Sleep can impact emotional experiences at different stages of this modal model of the emotion-generative process (see Figure 3.1). Insufficient and poor sleep can lead to reduced energy and activity levels, which can influence the types of activities individuals engage in and the kinds of situations they encounter. For instance, insomnia patients often respond to the feeling of low energy by canceling obligations or avoiding activities that are usually enjoyable in favor of "resting" and conserving energy. Activity reduction (particularly reduced social contact) and failure to fulfill roles or obligations can each contribute to negative emotions. Insufficient sleep also affects attentional processes (Lim & Dinges, 2010), including working memory, as well as simple and complex attention. Such attentional impairments might contribute to preferential processing of negative or "threat-relevant" information, although this later point has not yet been directly evaluated (Espie, Broomfield, MacMahon, Macphee, & Taylor, 2006). In that way, insufficient and poor sleep could also affect the appraisal process. Insufficient sleep might also lead individuals to adopt more conservative, generalized interpretations (which would typically be more catastrophic or negative in tone than when well rested) in order to prioritize conserving energy and maintaining safety. As reviewed above, insufficient sleep is known to amplify emotional reactivity (particularly to neutral and mildly negative events), directly increasing the intensity of emotional responses, independent of sleep's impact on earlier stages of the emotion-generative process. The modal model of emotions can be useful in the study of how sleep quantity and quality impacts emotional functioning. It will be important to understand the effect of sleep at different points of the emotion-generative process because these effects might be different. Increased understanding of how sleep might affect the temporal dynamics of emotional responding might help shed further light on the complex, dynamic relationship between sleep and affect.

Affective functioning also influences various sleep parameters, although relatively less literature has examined the influence of affective states on sleep. Two widely agreed upon dimensions of emotion, valence and arousal, appear to have similar effects on global sleep parameters (increased sleep latency and fragmentation, decreased sleep quality and total sleep time) but potentially unique effects on sleep architecture, with arousal relating more strongly with SWS and valence with REM sleep. This is consistent with recent conceptual models emphasizing the importance of REM sleep for emotional processing (cf., Goldstein & Walker, 2014). Emotion regulation efforts also appear to influence sleep parameters, with the habitual tendency to rely on particular maladaptive strategies producing alterations to global sleep parameters. In particular, strategies that promote a hyperfocus on or overelaboration of emotional states (Guastella & Moulds, 2007) or detached processing of emotionally salient events (Vandekerckhove et al., 2012) appear to contribute to reduced quality sleep, increased fragmentation, and shorter total sleep time.

A significant challenge in understanding how affective functioning influences sleep is disentangling the unique contributions of arousal and valence. Studies that have examined the impact of affect on sleep have largely not examined the potentially unique effects of valence and arousal (with Tang and Harvey (2004) a notable exception). Some preliminary evidence suggests that valence and arousal might differentially impact sleep architecture; however, additional studies utilizing polysomnography are necessary to examine this hypothesis. One way to begin to disentangle the unique impact of arousal and valence would be to examine manipulations of both high-arousal positive (e.g., excitement) and high-arousal negative (e.g., anxiety) emotions on sleep parameters. Such data would provide direct evidence for the hypothesis that arousal and valence have differential effects on sleep. Such data might also help clarify some of the discrepancies in the literature examining the impact of stress on sleep (Kim & Dimsdale, 2007).

A growing literature suggests that emotion regulation might also have an impact on sleep parameters. Individual differences in the habitual use of particular strategies (such as rumination and worry) appear to be associated with sleep deficiency. Importantly, this relationship appears to be present above and beyond the influence of state affect. This suggests that interventions directly targeting maladaptive emotion regulation strategies might show improvements in sleep parameters. Indeed, some evidence supports this hypothesis. McGowan and Behar (2013) randomized participants with high levels of trait worry to 2 weeks of either a stimulus control (consisting of a daily 30-minute worry period occurring at an a priori time and place) or focused worry intervention (consisting of instructions to not avoid naturally occurring worry, because doing so might paradoxically increase worry and anxiety). Stimulus control resulted in significant reductions in worry, anxiety, general negative affect, and symptoms of insomnia.

Evidence suggests that, although processing negative emotions prior to sleep initially increases sleep latency, such efforts to downregulate negative emotions might be effective in reducing sleep fragmentation and improving sleep duration and sleep quality. If replicated, this finding could support utilizing interventions targeting emotion processing such as expressive emotional writing (Smyth, 1998) or even stimulus control for worry (McGowan & Behar, 2013) prior to a traditional "buffer zone" recommendation in sleep hygiene or stimulus control for insomnia instructions.

The empirical literature consistently supports a strong relationship between sleep deficiency and reduced positive affect and increased negative affect and difficulties with emotion regulation. Increasing evidence supports an impact of emotions and emotion regulation on sleep parameters and sleep architecture as well. Such data might have important clinical implications in treating individuals with deficient sleep or disorders characterized by emotion-processing deficits. Key areas for future research include studies examining (a) how chronic, intrinsic sleep deficiency (as is commonly seen among individuals with insomnia) impacts emotional functioning, (b) how arousal and valence might uniquely impact sleep architecture, and (c) the particular conditions under which emotion regulation influences sleep, independent of the impact of state affect.

References

Barrett, L. F. (2013). Psychological construction: The Darwinian approach to the science of emotion. *Emotion Review, 5*(4), 379–389. http://dx.doi.org/10.1177/1754073913489753.

Barrett, L. F., Mesquita, B., Ochsner, K. N., & Gross, J. J. (2007). The experience of emotion. *Annual Review of Psychology, 58*, 373–403. http://dx.doi.org/10.1146/annurev.psych.58.110405.085709.

Baum, K. T., Desai, A., Field, J., Miller, L. E., Rausch, J., & Beebe, D. W. (2014). Sleep restriction worsens mood and emotion regulation in adolescents. *Journal of Child Psychology and Psychiatry, 55*(2), 180–190. http://dx.doi.org/10.1111/Jcpp.12125.

Bonnet, M. H., & Arand, D. L. (1992). Caffeine use as a model of acute and chronic insomnia. *Sleep, 15*(6), 526–536.

Bryant, R. A., Creamer, M., O'Donnell, M., Silove, D., & McFarlane, A. C. (2010). Sleep disturbance immediately prior to trauma predicts subsequent psychiatric disorder. *Sleep, 33*(1), 69–74.

de Wild-Hartmann, J. A., Wichers, M., van Bemmel, A. L., Derom, C., Thiery, E., Jacobs, N., et al. (2013). Day-to-day associations between subjective sleep and affect in regard to future depression in a female population-based sample. *British Journal of Psychiatry, 202,* 407–412. http://dx.doi.org/10.1192/bjp.bp.112.123794.

Espie, C. A., Broomfield, N. M., MacMahon, K. M. A., Macphee, L. M., & Taylor, L. M. (2006). The attention-intention-effort pathway in the development of psychophysiologic insomnia: A theoretical review. *Sleep Medicine Reviews, 10*(4), 215–245. http://dx.doi.org/10.1016/j.smrv.2006.03.002.

Frijda, N. H. (2007). *The laws of emotion.* Mahwah: Lawrence Erlbaum Associates.

Goldstein, A. N., Greer, S. M., Saletin, J. M., Harvey, A. G., Nitschke, J. B., & Walker, M. P. (2013). Tired and apprehensive: Anxiety amplifies the impact of sleep loss on aversive brain anticipation. *Journal of Neuroscience, 33*(26), 10607–10615. http://dx.doi.org/10.1523/Jneurosci.5578-12.2013.

Goldstein, A. N., & Walker, M. P. (2014). The role of sleep in emotional brain function. *Annual Review of Clinical Psychology, 10,* 679–708. http://dx.doi.org/10.1146/annurev-clinpsy-032813-153716.

Gross, J. J. (1998). The emerging field of emotion regulation: An integrative review. *Review of General Psychology, 2*(3), 271–299. http://dx.doi.org/10.1037/1089-2680.2.3.271.

Gross, J. J. (2014). Emotion regulation: Conceptual and empirical foundations. In J. J. Gross (Ed.), *Handbook of emotion regulation* (2nd edn, pp. 3–20) New York, NY: Guilford Press.

Guastella, A. J., & Moulds, M. L. (2007). The impact of rumination on sleep quality following a stressful life event. *Personality and Individual Differences, 42*(6), 1151–1162. http://dx.doi.org/10.1016/j.paid.2006.04.028.

Harvey, A. G., Murray, G., Chandler, R. A., & Soehner, A. (2011). Sleep disturbance as transdiagnostic: Consideration of neurobiological mechanisms. *Clinical Psychology Review, 31*(2), 225–235. http://dx.doi.org/10.1016/j.cpr.2010.04.003.

Kim, E. J., & Dimsdale, J. E. (2007). The effect of psychosocial stress on sleep: A review of polysomnographic evidence. *Behavioral Sleep Medicine, 5*(4), 256–278. http://dx.doi.org/10.1080/15402000701557383.

Kring, Ann M., & Sloan, Denise M. (2010). *Emotion regulation and psychopathology: A transdiagnostic approach to etiology and treatment.* US: Guilford Press, 2010.

Larson, R. A., Durocher, J. J., Yang, H., DellaValla, J. P., & Carter, J. R. (2012). Influence of 24-hour sleep deprivation on anxiety and cardiovascular reactivity in humans. *Faseb Journal, 26,* 684.15.

Levenson, R. W. (1999). The intrapersonal functions of emotion. *Cognition and Emotion, 13*(5), 481–504. http://dx.doi.org/10.1080/026999399379159.

Lim, J., & Dinges, D. F. (2010). A meta-analysis of the impact of short-term sleep deprivation on cognitive variables. *Psycholical Bulletin, 136*(3), 375–389. http://dx.doi.org/10.1037/a0018883.

McGowan, S. K., & Behar, E. (2013). A preliminary investigation of stimulus control training for worry: Effects on anxiety and insomnia. *Behavior Modification, 37*(1), 90–112. http://dx.doi.org/10.1177/0145445512455661.

Minkel, J. D., Banks, S., Htaik, O., Moreta, M. C., Jones, C. W., McGlinchey, E. L., et al. (2012). Sleep deprivation and stressors: Evidence for elevated negative affect in response to mild stressors when sleep deprived. *Emotion, 12*(5), 1015–1020. http://dx.doi.org/10.1037/a0026871.

Smyth, J. M. (1998). Written emotional expression: Effect sizes, outcome types and moderating variables. *Journal of Consulting and Clinical Psychology, 66*(1), 174–184. http://dx.doi.org/10.1037//0022-006x.66.1.174.

Spielman, A. J., Caruso, L. S., & Glovinsky, P. B. (1987). A behavioral perspective on insomnia treatment. *Psychiatric Clinics of North America, 10*(4), 541–553.

Tang, N. K., & Harvey, A. G. (2004). Effects of cognitive arousal and physiological arousal on sleep perception. *Sleep, 27*(1), 69–78.

Thomsen, D. K., Mehsen, M. Y., Christensen, S., & Zachariae, R. (2003). Rumination – relationship with negative mood and sleep quality. *Personality and Individual Differences*, *34*(7), 1293–1301. http://dx.doi.org/10.1016/S0191-8869(02)00120-4, Pii S0191-8869(02)00120-4.

Vandekerckhove, M., Kestemont, J., Weiss, R., Schotte, C., Exadaktylos, V., Haex, B., et al. (2012). Experiential versus analytical emotion regulation and sleep: Breaking the link between negative events and sleep disturbance. *Emotion*, *12*(6), 1415–1421. http://dx.doi.org/10.1037/A0028501.

Vandekerckhove, M., Weiss, R., Schotte, C., Exadaktylos, V., Haex, B., Verbraecken, J., et al. (2011). The role of presleep negative emotion in sleep physiology. *Psychophysiology*, *48*(12), 1738–1744. http://dx.doi.org/10.1111/j.1469-8986.2011.01281.x.

Watson, D., & Tellegen, A. (1985). Toward a consensual structure of mood. *Psychological Bulletin*, *98*(2), 219–235.

Wichers, M. C., Myin-Germeys, I., Jacobs, N., Peeters, F., Kenis, G., Derom, C., et al. (2007). Evidence that moment-to-moment variation in positive emotions buffer genetic risk for depression: A momentary assessment twin study. *Acta Psychiatrica Scandinavica*, *115*(6), 451–457. http://dx.doi.org/10.1111/j.1600-0447.2006.00924.x.

Yoo, S. S., Gujar, N., Hu, P., Jolesz, F. A., & Walker, M. P. (2007). The human emotional brain without sleep – a prefrontal amygdala disconnect. *Current Biology*, *17*(20), R877–R878. http://dx.doi.org/10.1016/j.cub.2007.08.007.

Zohar, D., Tzischinsky, O., Epstein, R., & Lavie, P. (2005). The effects of sleep loss on medical residents' emotional reactions to work events: A cognitive-energy model. *Sleep*, *28*(1), 47–54.

PART 2

METHODS

CHAPTER 4

Methodology for the Assessment of Sleep

Christopher B. Miller,[*,†] *Simon D. Kyle,*[‡]
Kerri L. Melehan,[*,§] *and Delwyn J. Bartlett,*[*]

[*]Centre for Integrated Research and Understanding of Sleep (CIRUS), Woolcock Institute of Medical Research, Sydney Medical School, University of Sydney, Sydney, New South Wales, Australia
[†]Institute of Neuroscience & Psychology, University of Glasgow, UK
[‡]School of Psychological Sciences, University of Manchester, Manchester, UK
[§]Department of Respiratory and Sleep Medicine, Royal Prince Alfred Hospital, Sydney, Australia

Abbreviations

AASM	American Academy of Sleep Medicine
CPAP	continuous positive airway pressure
DISS	daytime insomnia symptom scale
DLMO	dim light melatonin onset
EEG	electroencephalography
EMG	electromyogram
EOG	electrooculogram
ESS	Epworth sleepiness scale
fMRI	functional magnetic resonance imaging
FOSQ	functional outcomes of sleep questionnaire
MRI	magnetic resonance imaging
MRS	magnetic resonance spectroscopy
MSLT	multiple sleep latency test
MWT	maintenance of wakefulness test
N1	NREM sleep stage 1
N2	NREM sleep stage 2
N3	NREM sleep stage 3 or slow-wave sleep
NREM	nonrapid eye movement
OSA	obstructive sleep apnea

PET	positron emission tomography
PSA	power spectral analysis
PSG	polysomnography
PSQI	Pittsburgh sleep quality index
REM	rapid eye movement
SCN	suprachiasmatic nucleus
SE	sleep efficiency
SOL	sleep onset latency
SWS	slow-wave sleep

Sleep is a complex behavior that has been described as "a reversible state of perceptual disengagement and unresponsiveness from the environment" (Carskadon & Dement, 2011, p. 16). In addition to perceptual disengagement, normal sleep also consists of closed eyes, postural recumbency, and relative stillness (Carskadon & Dement, 2011; Hirshkowitz, 2004). Although the primary function of sleep is currently unknown, many theories exist. Sleep is thought to be necessary for the repair of bodily wear and tear, memory encoding, and learning processes (Colrain, 2011). However, measuring sleep is difficult. The aim and focus of this chapter is to assess the most frequently used subjective and objective sleep measurements (see Table 4.1 for an overview). This chapter begins with a brief introduction to the regulation of sleep and then goes on to describe the measurement of sleep and daytime functioning.

Two main processes interact to regulate sleep and wakefulness. The first is the homeostatic drive for sleep, which is commonly referred to as sleep propensity or sleep need/debt (Borbély, 1982). Sleep homeostasis depends on the amount of time spent awake and can be quantified physiologically by the main objective measure of sleep, electroencephalography (EEG) (Achermann, Dijk, Brunner, & Borbély, 1993). Cognitive performance and specifically alertness are known to be sensitive to accumulating sleep pressure during the day (Van Dongen & Dinges, 2005). Homeostatic sleep pressure can only be reset through sleep. Sleep is also governed by the circadian rhythm. Internal biological rhythms have evolved to comply with the Earth's solar period of roughly 24 h (Hirshkowitz, 2004). Such rhythms are endogenously produced and therefore can operate without external time cues (Czeisler et al., 1999). The body uses photic and nonphotic cues to synchronize the internal clock to the light-dark environment (Czeisler et al., 1999). The suprachiasmatic nucleus (SCN) is known to control internal circadian timing. The SCN is the master clock and is located in the hypothalamus, directly above the optic chiasm (Dibner, Schibler, & Albrecht, 2010). Exposure to light synchronizes the SCN with the external

TABLE 4.1 Overview of the Methods used for the Assessment of Sleep

Method	Description	Pros	Cons
Polysomnography (PSG)	Objective measure of sleep and disorders of sleep	1. Gold standard measure of sleep and the disorders of sleep	1. Expensive 2. Difficult to access 3. Discomfort 4. First-night effect 5. Requires analysis and interpretation
Multiple sleep latency test (MSLT)	Objective measure of daytime sleepiness	1. Gold standard measure of severity of daytime sleepiness 2. Used in the diagnosis of narcolepsy	1. Requires PSG prior 2. In lab assessment with skilled staff member 3. Difficult to access 4. Ceiling effect 5. Lack of ecological validity 6. Requires analysis and interpretation
Maintenance of wakefulness test (MWT)	Objective evaluation of wakefulness and alertness under soporific settings	1. No ceiling effect and does not require PSG 2. Improved ecological validity over MSLT	1. In lab assessment with skilled staff member 2. Difficult to access 3. Requires analysis and interpretation
Actigraphy	Objective measure of movement used to infer sleep-wake behavior	1. Correlates with PSG 2. Can be used long-term 3. Ecologically valid 4. Target for treatment	1. Limited by battery length 2. Requires analysis and interpretation
Clinical interview	Objective measure of movement used to infer sleep-wake behavior	1. Accessible 2. Clinically valid 3. Useful prior to treatment or research	1. Lack of standardization 2. Limited without further measures of sleep

(*Continued*)

TABLE 4.1 Overview of the Methods used for the Assessment of Sleep—Cont'd

Method	Description	Pros	Cons
Sleep diary	Subjective measure of sleep-wake behavior	1. Accessible 2. Target for treatment 3. Easy to use and inexpensive	1. Poor correlation with PSG 2. Must remember to fill in 3. Subjective measure of sleep 4. Requires analysis and interpretation
Questionnaires	Subjective measures of sleep, wake, health, and cognition	1. Accessible 2. Target for treatment 3. Easy to use and inexpensive	1. Open to bias 2. Requires analysis and interpretation
Brain imaging	Objective measure of the brain and its functioning	1. Noninvasive method for investigating the brain 2. Can be used during sleep 3. Can measure brain activation and metabolism	1. Expensive 2. Difficult to access 3. Poor temporal resolution 4. Lack of validity 5. Requires analysis and interpretation
Cortisol	Hormone used to infer stress	1. Objective measure 2. Closely linked to the sleep cycle	1. Requires analysis 2. Difficult to access 3. Test requires further validation
Melatonin	Hormone used to infer circadian alignment	1. Objective measure 2. May be used to indicate change with treatment 3. Closely linked to the sleep cycle	1. Requires analysis 2. Difficult to access 3. Test requires further validation

light-dark cycle and serves as the primary circadian time giver for mammals (Czeisler et al., 1999; Stephan & Nunez, 1977).

PART 1: OBJECTIVE DIAGNOSTIC MEASURES OF SLEEP

Electroencephalography and polysomnography. Sleep and wake states are measured by EEG whereby electrodes on the scalp record electrical brain activity (Rechtschaffen & Kales, 1968). Electrical brain activity is the gold standard objective measurement of sleep (Kushida et al., 2005), and electrodes are placed according to standardized international criteria called the 10-20 placement system (Jasper, 1958), which aims to ensure reproducibility of EEG studies. In the 10-20 system, each electrode site is mapped with letters and numbers. The letters F, T, C, P, and O refer to scalp locations and stand for frontal, temporal, central, parietal, and occipital areas, respectively. Even numbers denote the right hemisphere and odd numbers denote the left. "Z" means 0 and stands for the midline of the head (Oostenveld & Praamstra, 2001).

Clinicians use a diagnostic sleep study to diagnose sleep disorders, including sleep-related breathing disorders, parasomnias, sleep-related seizure disorders, and periodic limb movement disorders (Kushida et al., 2005). The diagnostic sleep study is extensive and provides a significant amount of useful information, but it is also expensive, difficult to access, and often uncomfortable for the patient (Ancoli-Israel et al., 2003). The recording of brain activity by EEG is only one aspect of the overall diagnostic sleep study. The study can gather other information about the body during sleep, using polysomnography (PSG), which simply means "many sleep recordings" in addition to EEG, and the following measures can also determine sleep: eye movements measured via the electrooculogram (EOG) and muscle tone measured via the electromyogram (EMG) on the chin. Any sleep disturbance due to sleep disorders such as obstructive sleep apnea (OSA), periodic limb movement disorder, and parasomnias can be detected as part of the overall PSG assessment. Such disorders can be distinguished using a number of measures, including respiratory effort and airflow, snoring, body position, heart rate, oxygen saturation levels, and limb and jaw movements over the course of the night.

Sleep behavior can be scored according to standardized PSG criteria (Iber, Ancoli-Israel, Chesson, & Quan, 2007), through an evaluation of electrical wave-forms produced by the brain, eye movements, and muscle activity. Primarily, a sleep scientist evaluates brain electrical wave-form patterns for amplitude and frequency, as well as bursts of brain activity. Busts of brain activity include K-complexes (a single large amplitude waveform <2 Hz in frequency with a brief positive electrical peak followed by a slower negative component; Loomis, Harvey, & Hobart, 1938) and sleep spindles (a short burst of oscillatory electrical activity of sigma frequency waves at 12-16 Hz that may follow from a K-complex; Loomis et al., 1938), and they can be used to categorize sleep stages (see Figure 4.1). A trained and certified sleep polysomnographic technician normally divides sleep into distinct sleep stages that are then summarized and reported on by a specialized sleep physician. A hypnogram is used to give an overview of the staging of sleep for the entire night (see Figure 4.2).

In adults, sleep consists of two main phases: non-rapid eye movement (NREM) sleep and rapid eye movement (REM) sleep. In healthy sleepers, the brain transitions through these stages of sleep in approximately 90-min cycles. On average, an individual experiences four to five sleep cycles during the night. NREM sleep comprises three distinct sleep stages. At sleep onset, an initial short period (5-10 min) of stage 1 sleep (N1) occurs, characterized by theta waves (4-7 Hz) and an absence of alpha waves (8-12 Hz) on the EEG recording, followed by a longer period (~20 min) of stage 2 sleep (N2) characterized by a mixed frequency background featuring sleep spindles and K-complexes. Stage 3 sleep (N3), otherwise known as slow-wave sleep (SWS), is characterized by delta frequency wave formations (0.5-4.5 Hz) of at least 75 µV amplitude, and it lasts for approximately 30-40 min (Iber et al., 2007).

REM sleep is distinct and contains mixed frequency theta and beta EEG activity, with characteristic sawtooth-shaped waves, rapid eye movements, and muscle atonia. REM sleep usually occurs for the first time at

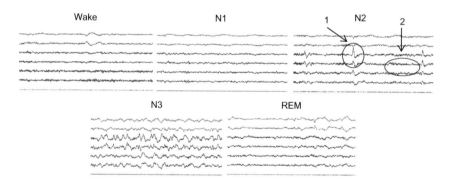

FIGURE 4.1 **Stages of sleep and scoring information.** Examples of periods of wakefulness and sleep objectively defined through electroencephalography (EEG) used on one healthy 22-year-old male as part of an overnight sleep study (polysomnography). Each example presents 30 s of data from the following seven electrodes (top to bottom): left eye, right eye (EOG: electrooculogram); C3-A2, C4-A1, O2-A1, O1-A1; and a single channel of chin EMG (electromyogram). The top left image displays the period of wakefulness. The top middle image displays stage 1 sleep (N1), and the top right image displays stage 2 sleep (N2). The bottom left image displays stage 3 sleep (N3 or SWS). Rapid eye movement (REM) sleep is displayed in the bottom right image. 1, a K-complex and 2, a sleep spindle.

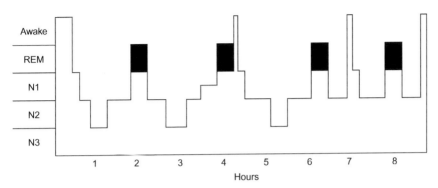

FIGURE 4.2 Hypnogram of scored human sleep staging. A normal hypnogram used to provide an overview of the progression of sleep stages during the night for an individual's sleep. The hypnogram is a continuous recording of electroencephalogram (EEG), electrooculogram (EOG), and electromyogram (EMG) measures for scoring sleep. The *y-axis* represents sleep stages (evaluated in 30-s epochs), including REM, N1, N2, and N3 or SWS The *x-axis* represents time during the night in hours.

the end of each sleep cycle, and the first REM period is generally short, lasting 5-10 min or less. REM sleep episodes then become increasingly longer throughout the sleep period (Aserinsky & Kleitman, 1953; Dement & Kleitman, 1957). The sleep cycles tend to alter their alignment as the sleep period progresses, with SWS dominant in the first third of the night and REM sleep more dominant toward the end of the sleep period (Carskadon & Dement, 2011). The proportion relative to total sleep time (TST) is as follows for the following sleep stages: N1 is approximately 2-5% of TST, N2 is 45-55%, N3 (SWS) is 20%, and REM sleep is 20-25% (Williams, Agnew, & Webb, 1964).

Overall, SWS is associated with homeostatic sleep pressure or the amount of time spent awake, and it can be quantified physiologically through EEG-derived slow-wave activity (SWA; spectral power in the 0.75-4.5 Hz bandwidth) (Achermann et al., 1993). Importantly, the majority of SWS is obtained during the first third of the night, and SWS is normally preserved by the brain under sleep-restricted schedules (Spiegel, Leproult, & Van Cauter, 1999). In addition, the onset of "deep" SWS occurs quickly in healthy individuals (normally after about 20 min of lighter N1 and N2 sleep), indicating the importance of this stage of sleep. Studies disrupting SWS have shown impairment in next day performance and endocrine secretion in healthy individuals (Spiegel et al., 1999). One of the primary functions of SWS may be to restore and repair both the brain and the body. On the other hand, REM sleep is more adaptable around SWS needs. REM sleep (named after the discovery of rapid eye movements during this sleep phase; Aserinsky & Kleitman, 1953) appears to be important for the consolidation of memory traces through increases in synaptic plasticity in the cortex of the brain (Diekelmann & Born, 2010). REM sleep is related to dream content, with people who awake during REM being more likely to report experiencing a dream relative to those who wake during other stages of sleep (Foulkes, 1962). Crucially, muscle paralysis or atonia takes place during REM sleep and prevents dream enactment. REM behavior disorder is characterized by a lack of muscle atonia and leads to patients acting out their dreams (Schenck, Hurwitz, & Mahowald, 1993). REM behavior disorder may also be a marker for the subsequent development of a neurodegenerative disorder such as Parkinson's disease (Schenck, Bundlie, & Mahowald, 1996).

Power spectral analysis of sleep. The standardized practice and scoring of sleep by PSG does not reflect all of the cortical processes that take place during the sleep period. EEG-defined sleep can also be analyzed using power spectral analysis (PSA), which reveals the microstructure of sleep. PSA is a mathematical quantification process for detecting periodicities in time series data (Perlis et al., 2001). PSA uses the fast Fourier transformation to group EEG rhythms into the following frequency bands: delta (0.5-4 Hz), theta (4-7 Hz), alpha (7-14 Hz), beta (15-30 Hz), and gamma (30-100+Hz) (Borbély, Baumann, Brandeis, Strauch, & Lehmann, 1981). The first Fourier analysis of EEG recordings was carried out in 1932 by Dietsch at the suggestion of Berger, the innovator of EEG recordings in humans (Achermann, 2009). PSA can be used to analyze the complex EEG signal, which is normally scored by hand and as an average of a 30-s period. The scorer may miss important aspects of this EEG signal if scoring over an 8-h recording. Quantitative methods such as PSA are essential for investigating the EEG signal in more detail. For example, slow-wave activation is a sensitive measure of sleep homeostasis or sleep pressure (Achermann, 2009), and NREM sleep is distinguished by both delta activity (0.5-4.5 Hz) and spindle frequency activity (12-14 Hz). REM sleep does not exhibit delta and spindle frequency activity. Further, alpha activity (~10 Hz) is present during the waking period before sleep onset (Achermann, 2009). Such quantification of EEG into frequency bands can allow PSA to give an overview of the structure of sleep, even when not scored visually by a technician. Alterations in the microstructure of sleep may not be detected via conventional visual scoring of sleep stages but are detected by PSA.

Multiple sleep latency test. The multiple sleep latency test (MSLT) was developed in the 1970s to measure the effects of sleep deprivation on sleepiness levels in young healthy individuals (Carskadon & Dement, 1982). Correlations between subjective measures of sleepiness and the MSLT suggested usage as an objective measure of daytime sleepiness, and these correlations led to the development of standardized MSLT testing protocols for both clinical and research purposes (Arand et al., 2005). The American Academy of Sleep Medicine (AASM) considerers the MSLT the *de facto* gold standard objective measure of excessive daytime sleepiness (Littner et al., 2005), and the test is primarily used to diagnose narcolepsy. Daytime sleepiness may be caused by an increased sleep propensity accumulated through a lack of adequate sleep or sleep disorders. In the MSLT, participants are given four or five daytime nap opportunities (tests), each lasting 20 min every 2 h, following an overnight PSG evaluation. Muscle activity, eye movements, heart rate, and EEG are monitored during the testing procedure. In a dark room, patients must lie in bed and go to sleep. If the patient remains awake for the full 20 min, then the test ends. If sleep occurs during the test, the clinician wakes the patient up after 15 min of sleep (Littner et al., 2005). The two main outcomes are the average time to fall asleep, which ranges from 0 to 20 min, and the presence or absence of REM sleep. Patients with narcolepsy reportedly have a mean of 3.1 min, while healthy controls have a mean of 11.6 min (Littner et al., 2005). The time to fall asleep is sensitive to a number of factors, including sleep deprivation, use of a stimulant or sedative medications, and the presence of sleep disorders such as narcolepsy and OSA (Kushida et al., 2005; Sunwoo et al., 2012). Although the MSLT is most often used to indicate narcolepsy compared with reported sleep propensity from fatigue and tiredness, it can assess REM-onset latency in those with moderate-to-severe OSA or excessive daytime sleepiness (Littner et al., 2005), in addition to profiling treatment effects (Thorpy, Westbrook, Ferber, & Fredrickson, 1992). Critics of the MSLT point out that the test exhibits a floor effect, meaning that the MSLT may not be sensitive enough to detect clinically significant changes due to treatment in severe populations (even when sleep latencies are doubled), as well as a ceiling effect imposed by the 20-min nap limit (limiting the usefulness in more alert individuals), and they assert that the sleep laboratory setting is not representative of the work place where sleep tendencies may be different (Arand et al., 2005).

Maintenance of wakefulness test. The maintenance of wakefulness test (MWT) is used to examine wakefulness and alertness in soporific settings (Arand et al., 2005) where high levels of vigilance and alertness are required for "fitness for duty" (Sullivan & Kushida, 2008). The MWT was developed to profile treatment changes in patients with severe levels of sleepiness and their ability to remain awake. By design, the MWT overcomes the three previously mentioned limitations with the MSLT concerning floor or ceiling effects and the lack of ecological validity (Arand et al., 2005). The ability to maintain wakefulness is the primary outcome calculated indirectly by measuring sleep latency (Mitler, Gujavarty, & Browman, 1982). Unlike the MSLT, the MWT does not necessarily require an overnight sleep study prior to the test. In a dimly lit room, the participants' EEG, EOG, and chin EMG are monitored, and he or she is asked to stay awake while seated in bed, with back and head supported by a bedrest for 40 min over four trials performed every 2 h. Physical activity and vocalizations are not allowed during the test (Littner et al., 2005). The reported outcome is the average length of time the patient is able to remain awake, with the first occurrence of 15s of sleep within a 30-s window being deemed the onset of sleep. The study is terminated after 40min or after three 30-s windows classified as stage N1 sleep or any 30-s window of any other stage of sleep. Studies have demonstrated mean decreased sleep latencies in those with excessive daytime sleepiness and increases in sleep latencies after treatment for OSA (Sullivan & Kushida, 2008).

Actigraphy. Sleep can also be profiled through the use of actigraphy, which allows patterns of light, sleep, and wake behavior to be assessed over days or weeks. Actigraphy is cost effective and more convenient than a full PSG (Ancoli-Israel et al., 2003), and it can be repeated across many nights to build an ecologically valid assessment of sleep without the first-night effect of PSG (Ancoli-Israel et al., 2003). The first-night effect is common with a PSG evaluation and is known to impair sleep in healthy good sleepers (Agnew, Webb, & Williams, 1966). Actigraphs are typically watch-like devices worn on the nondominant hand, and they use an accelerometer to record movement over a given threshold (see Figure 4.3). The wearer can also use an event marker to denote time in bed and awakenings during the night. Collected data is downloaded to a computer for the observation of rest and activity patterns across both night and day. Through analysis software, validated algorithms are applied for

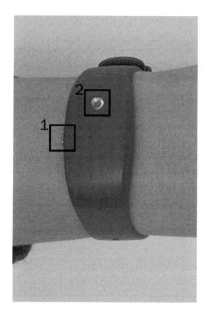

FIGURE 4.3 Picture of an actigraph. An actigraph worn on the wrist of the nondominant hand. 1, the event marker button; 2, the light sensor.

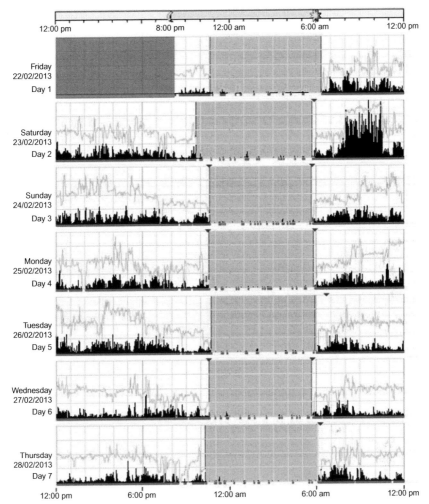

FIGURE 4.4 **Actogram of human rest-wake activity from an individual.** A scored actogram of human rest-wake activity and light levels from a 42-year-old female patient with insomnia undergoing bed restriction therapy. The x-axis represents time across the 24-h period (12:00 pm until 12:00 pm). The y-axis represents the day of the week (Friday until Thursday). Black lines represent activity levels. Red lines underneath indicate wakefulness. Lighter yellow lines above the black activity levels represent light levels. Light blue in the middle indicates a rest period. The dark blue triangles represent event markers for time in bed by the patient.

movement thresholds (Ancoli-Israel et al., 2003). Clinicians can then use these data to estimate wake and sleep parameters (see Figure 4.4).

The very first analog actigraphs were developed and used in the 1970s (for an overview, see Sadeh & Acebo, 2002). The devices initially underwent reliability testing for the assessment of sleep by Kripke, Mullaney,

Messin, and Wyborney (1978). The digital actigraphs currently in use have sufficient memory storage for long-term operation. The sampling rate (epoch) is normally standardized to once per minute, although the user can change the epoch through computer software (Sadeh & Acebo, 2002). A longer sampling rate can increase the battery length, but longer intervals are not generally used for research purposes because intervals of greater than 1 min may compromise data validity. Previously, actigraphy has demonstrated high agreement rates with PSG data for TST and sleep efficiency variables in healthy subjects (Kushida et al., 2001). However, these variables have been found to have lower agreement rates in patients with OSA and insomnia. Therefore, practice guidelines for the use of actigraphy suggest it should be used in combination with a sleep diary (Ancoli-Israel et al., 2003).

Actigraphy can provide an inexpensive way to follow up with patients experiencing varied sleep disorders. It is also very useful for patients with insomnia symptomology and treatment (see Figure 4.3) and in the diagnosis of circadian rhythm disorders when the patient's internal circadian timing does not significantly differ from the external environment (Morgenthaler et al., 2007). In advanced or delayed sleep phase disorders, the internal circadian clock of the patient may be ticking either too fast (advanced) or too slow (delayed). The strength of actigraphy is that it provides a noninvasive and long-term measure of activity and sleep. However, unlike PSG, actigraphy does not provide an assessment of sleep architecture and cannot provide a specific diagnosis of a sleep disorder, such as OSA.

PART 2: SUBJECTIVE DIAGNOSTIC MEASURES OF SLEEP

Clinical interview. A clinician can initially acquire an overview of the patient's sleep through a common unstructured clinical interview (Buysse, Ancoli-lsrael, Edinger, Lichstein, & Morin, 2006). Such clinical interviews are extremely useful for providing a baseline screening assessment and patient history prior to the start of treatment or research study. When taking a patient's history, a clinician normally asks questions to probe the following areas: family history, physical health history, previous medication and alcohol use, history of mental illness, and the timing and onset of sleep (Schramm et al., 1993). The clinical history can also include observations of the patient's sleep-wake behavior during the night reported by the partner of the patient, family, caregivers, or travelling companions. For example, reports of parasomnias, including the acting out of dreams and nightmares, can be described by the patient or others and documented in sleep logs or diaries (Blagrove, Farmer, & Williams, 2004).

A lack of common standardization reduces the reliability of these clinical assessments (Buysse et al., 2006). As a result, specifically designed structured interviews for sleep disorders have been used to probe symptoms of insomnia, idiopathic hypersomnia, sleep-wake schedule disorders, sleep-induced respiratory impairment, narcolepsy, restless leg syndrome, and periodic movement disorders, among other conditions (Ohayon, Guilleminault, Zulley, Palombini, & Raab, 1999; Schramm et al., 1993). One structured interview was initially developed and tested for reliability and validity according to DSM-III-R and DSM-IV criteria (Schramm et al., 1993). The interview can be used to evaluate sleep-wake disorders, and the average time for administration is approximately 20-30 min. The interview consists of the following: (1) A brief semistructured overview of physical and mental health and questions regarding OSA and narcolepsy; (2) A specific and structured inquiry of sleep disorder symptoms; and (3) A summary score sheet filled in at the end of the interview. The interviewer may omit irrelevant sections, depending on the response of the patient (Schramm et al., 1993), and physicians or health care professionals with no prior knowledge of sleep disorders can successfully administer it.

Sleep diary measures. Sleep can also be profiled through daily self-report measures of sleep. Sleep diaries are widely used in sleep science, and they are fundamental to understanding the subjective complaints of patients. Self-monitoring of sleep through a sleep diary enables nightly perceived metrics to be identified, and it normally includes the following estimated measures: sleep onset latency (SOL), wake-time after initial sleep onset (WASO), TST, total time spent in bed, sleep efficiency (SE: percentage of time spent asleep relative to the amount of time spent in bed), and a numerical estimation of overall sleep quality. Normally, patients are asked to complete the diary before they commence their day and refer back to the previous night's sleep with approximations to the nearest 5 min. Some patients may complete daily entries in the morning or in both the morning and evening. The morning assessment is both sufficient and preferred (Carney et al., 2012). Patients (especially those with insomnia) are advised to avoid filling in the sleep diary during the night, because this may further disrupt their sleep. Likewise, the use of a clock during the night to quantify periods of wakefulness should also be avoided. Thus, estimations of time are necessarily generalized to the nearest 5 min. When compared to gold standard PSG and actigraphy, sleep diaries tend to be less accurate because healthy participants tend to overestimate their TST, whereas patients with insomnia underestimate their sleep time (Lichstein et al., 2006; Maes et al., 2014).

Nevertheless, the sleep diary is the gold standard subjective assessment of sleep and an extremely important assessment of insomnia (Carney et al., 2012). The subjective complaint is considered primary, and clinicians typically quantify the progress of cognitive behavioral therapy for insomnia using primary sleep diary outcome metrics (SOL, TST, WASO, and %SE). However, a lack of sleep diary standardization has hampered research methodologies in measuring sleep (Carney et al., 2012), resulting in inconsistent results between studies and a difficulty in translating lab findings into clinical practice. To address this problem, the AASM established an expert consensus of sleep medicine experts in order to standardize a self-report sleep diary. The proposed consensus sleep diary is currently a live document available for use, but it still requires validation, testing, and refinement (Carney et al., 2012).

Questionnaire assessments. Sleep can be profiled subjectively through self-report questionnaire measures. Many questionnaires allow patients to report on sleep, quality of life, health-related quality of life, and also daytime functioning (Shahid, Wilkinson, & Marcu, 2012). This section focuses on a number of the most widely used and applicable questionnaires for the assessment of sleep and its disorders (Buysse et al., 2006).

Sleep. The Pittsburgh sleep quality index (PSQI; Buysse, Reynolds, Monk, Berman, & Kupfer, 1989) is one of the most widely used self-report measures for the assessment of sleep quality, having been cited well over 5000 times in the literature. This is a useful and general retrospective assessment of sleep quality and sleep disturbance over a 1-month period. Patients score 19 individual items that generate 7 component scores, including subjective sleep quality, sleep latency, sleep duration, sleep efficiency, sleep disturbance, use of sleep medication, and daytime dysfunction (Buysse et al., 1989). The first 4 items are open questions, while items 5-19 are rated on a 4-point Likert scale. Seven components are derived from the individual item scores, and a total score of overall sleep quality, ranging from 0 to 21, is obtained by adding the 7 component scores. A global PSQI score is used to differentiate between those with good and poor sleep (>5=poor sleeper). Clinicians can also use it with many patient and research populations to distinguish between good and poor sleepers. However, a disadvantage of the PSQI is the complexity of scoring the questionnaire.

Sleepiness. The Epworth sleepiness scale (ESS) profiles subjective estimations of excessive daytime sleepiness (Johns, 1992). The overall total score (out of 24) comprises a single factor, the propensity to fall asleep over the previous 2 weeks. The scale asks people to rate the likelihood that they would doze or fall asleep in eight different settings (sitting and reading, for example). Patients are asked to rate on a 0-3 scale (0 = "would never doze," 3 = "high chance") their chance of dozing within those eight situations. Scores above 16 are regarded as evidence of extreme sleepiness, as indicated in patients with idiopathic hypersomnolence or narcolepsy (Johns, 1992). In patients with moderate-to-severe OSA, the ESS is sensitive to change with continuous positive airway pressure (CPAP) treatment

(Antic et al., 2011). The ESS can also be used to profile difficulties for people who have restricted sleep opportunities or those who undertake shift work (Garbarino et al., 2002; Van Dongen, Baynard, Maislin, & Dinges, 2004). Generally, patients with insomnia score low on this questionnaire, but, clinically, if a score >10 is identified, the clinician should look for another sleep disorder or depression. Compared to objective tests of daytime sleepiness, the ESS shows weak correlation with other measures of sleep propensity with the MWT in those with narcolepsy ($r = -0.29$; Sangal, Mitler, & Sangal, 1999) and the MSLT in those with OSA ($r = -0.03$, $r = -0.15$; Chervin & Aldrich, 1999; Olson, Cole, & Ambrogetti, 1998).

The functional outcomes of sleep questionnaire (FOSQ; Weaver et al., 1997) can be used to profile functional status due to sleep loss or excessive daytime sleepiness and, in particular, OSA. The questionnaire probes the extent to which sleepiness or sleep disruption impairs five aspects of daily activities: general productivity, social outcomes, activity levels, vigilance, and sexual relationships. Respondents rate the difficulty of these activities on a 4-point or 6-point scale (no difficulty to extreme difficulty). The FOSQ has good reliability and internal consistency, and it can be used to map improvements due to CPAP treatment for OSA (Montserrat et al., 2001).

Fatigue. The Flinders fatigue scale (Gradisar et al., 2007) is a brief assessment for measuring daytime fatigue. The questionnaire uses seven items to evaluate fatigue over the previous 2 weeks. Six of the seven items are presented in Likert format, and responses range from 0 (not at all) to 4 (extremely). Using a multiple-item checklist to indicate more than one response, the fifth item measures the time of day when fatigue is experienced, and the sum of this response is reported. One further item focuses on whether the respondent attributes the fatigue specifically to sleep. All items are summed and a total overall fatigue score is calculated. Total scores range from 0 to 31, and higher scores indicate greater fatigue. The scale can also discriminate between good and poor sleepers, with poor sleepers scoring significantly higher (Gradisar et al., 2007).

Insomnia. The daytime insomnia symptom scale (DISS; Levitt et al., 2004) has been used previously to probe differences in sleep, rhythms, and affect in groups of patients with insomnia, as compared to healthy good-sleeping controls (Buysse et al., 2007; Levitt et al., 2004). The DISS can map time point changes during the acute phase (first 3 weeks) of a successful behavioral intervention for insomnia (Miller, Kyle, Marshall, & Espie, 2013). Specifically, the DISS uses 20 visual analogue scales (ranging from 0 to 10 on a 10-cm line) and asks the participant to mark where he or she agrees or disagrees with each of the statements at the specified moment in time. For example, question one asks, *"How alert do you feel"* with *"Very little"* at the extreme left of the 10cm line and *"Very much"* at the far right. These items form four factors (alert cognition, sleepiness/fatigue, positive mood, and negative mood). Previously, participants have completed the DISS at four assessment time points per day (at awakening, noon, early evening, and bedtime) for at least 1 week (Buysse et al., 2007; Levitt et al., 2004; Miller et al., 2013).

The insomnia severity index (Morin, 1993) is a short seven-item questionnaire that attempts to define the severity of both the nighttime and daytime impacts of insomnia over the past month (Morin, 1993). It has previously been validated as a useful clinical tool for diagnosis and measuring treatment outcome (Bastien, Vallières, & Morin, 2001). The items of the questionnaire evaluate the following: duration of sleep onset, sleep maintenance, early morning awakening, sleep dissatisfaction, daytime functioning, noticeability of sleep problems by others, and the distress caused by the sleep problem. Each item uses a 5-point Likert scale to capture a rating ($0 = no\ problem$, $4 = very\ severe\ problem$). Each item score is summed and yields a total score from 0 (no insomnia) to 28 (severe clinical insomnia). The total score is interpreted as follows: no insomnia (0-7), subthreshold insomnia (8-14), moderate clinical insomnia (15-21), and severe clinical insomnia (22-28). The questionnaire has also been used in population-based studies to quantify and evaluate insomnia symptomology in community samples. The instrument has shown high internal consistency for both clinical and community-based samples (Cronbach $\alpha \geq 0.90$).

PART 3: RESEARCH-FOCUSED MEASURES OF SLEEP

Neuroimaging of sleep. Magnetic resonance imaging (MRI) can be used to investigate *in vivo* anatomical changes and differences in the brain. Further brain-imaging techniques can also be used to evaluate cerebral blood flow with both functional magnetic resonance imaging (fMRI) and positron emission tomography (PET). In addition, PET and magnetic resonance spectroscopy (MRS) can help evaluate *in vivo* brain metabolism. Such techniques allow insights into human sleep, building on our understanding of sleep physiology and the brain. Within sleep disorders, brain imaging has been used to document disease progression and brain differences between healthy controls and those with insomnia, OSA, restless leg syndrome, periodic limb movement disorder, and parasomnias. However, the use of brain imaging in sleep research and sleep pathology is still in its infancy. In this section, we introduce the role of neuroimaging of the brain during periods of sleep and wakefulness.

Magnetic resonance imaging. MRI is used to build an anatomical image of the brain, and it is the basis for any subsequent functional imagining study. Specifically, MRI can be used to map structural brain-related changes (grey and white matter volumes) through voxel-based morphometry. Images of the brain are obtained using voxels, which are three-dimensional blocks that represent an area of the brain containing over a million brain cells

(Webb & Henkelman, 2008). Previously, researchers documented differences in brain structures using MRI; they found reduced grey matter volumes in the brains of those with sleep disorders (OSA, insomnia, REM behavior Disorder) as compared to healthy matched controls (Altena, Vrenken, Van Der Werf, vandenHeuvel, & VanSomeren, 2010; Hanyu et al., 2012; Joo et al., 2010). Sleep loss may therefore be responsible for such reductions in brain matter. Diagnostically, MRI during sleep can also be used to localize upper airway collapse in patients with OSA (Kim et al., 2014).

Functional magnetic resonance imaging. fMRI is a neuroimaging procedure that detects brain activation through changes in blood flow (Buonocore & Hecht, 1995). Primarily, fMRI uses the blood-oxygen-level-dependent (BOLD) contrast. This technique enables the identification of neural activity by detecting magnetic changes between oxygen-rich and oxygen-poor blood. The use of BOLD assumes that blood flow is coupled with the energy use of the brain region receiving the blood. For example, when an area of the cortex is activated, the blood flow to that area increases (Buonocore & Hecht, 1995). fMRI can thus be used to profile changes in the BOLD signals from the brain during both rest and sleep, although sleep in the fMRI environment is difficult due to noise (Lovblad et al., 1999). Based on the sleeping brain activity recorded by the fMRI, networks that undergo changes are identified (Picchioni, Duyn, & Horovitz, 2013), including the thalamocortical networks active during REM sleep (Wehrle et al., 2007), regions differentially activated in early and late N1 sleep (Picchioni et al., 2008), and areas of fluctuating activity during rest and sleep (Fukunaga et al., 2006). fMRI has also been used to profile the impact of sleep disorders, such as insomnia, during wakefulness (Spiegelhalder, Regen, Baglioni, Riemann, & Winkelman, 2013). Previous studies found reduced activation in the brains of patients with insomnia compared with controls during cognitive testing while awake in the scanner (Altena et al., 2008; Drummond et al., 2013; Stoffers et al., 2014), supporting reports of daytime impairments in individuals with insomnia. fMRI has good spatial resolution, but poor temporal resolution compared to EEG (Hall, Robson, Morris, & Brookes, 2013). This is because electrical signals detected by EEG are faster than the changes in blood flow detected by fMRI.

Positron-emission tomography. PET is an imaging technique that identifies chemical function in an organ. A radioactive tracer is used (oxygen-15; $H2\ ^{15}O$) and accumulates in areas with higher levels of chemical activity, which are then displayed as brighter areas on a scan. During sleep, cerebral blood flow of the brain can serve as a neuronal marker of synaptic activity for the PET (Maquet, 2000). Levels of glucose metabolism may also be estimated within the brain with PET, but glucose estimation requires a different compound [^{18}F]fluorodeoxyglucose, which has a longer half-life (110 min) (Maquet, 2000).

Clinicians have predominantly used PET to study aspects of SWS and REM sleep, specifically in healthy individuals (Braun et al., 1997). PET can compare differences in cortical processing during both sleep and wake, with localized decreases found in the sleeping brain during NREM sleep (Dang-Vu, 2012). For example, one study by Nofzinger et al. (2004) investigated the differences in patients with insomnia and healthy controls, and the authors found a smaller decrease in global glucose consumption in the people with insomnia during both sleep and wake when compared to the healthy controls, suggesting a higher arousal level during both of these states for those with insomnia.

Magnetic resonance spectroscopy. Brain metabolism can be measured through *in vivo* proton MRS, which is useful for understanding subtle changes due to disease and/or response to treatment. A number of metabolites are examined through this technique, including N-acetylaspartate (2.05 relative-unit parts per million (ppm)), creatine (3.02 ppm), choline (3.22 ppm), glutamate (2.35 ppm), glucose (3.43 ppm), *myo*-inositol (3.56 ppm), and lactate (1.33 ppm), among other measures (Rae, 2014). The sleeping brain and the consequences of sleep disorders have also been studied using MRS. Moderate oxygen desaturation can affect brain bioenergetic status during sleep (Rae et al., 2009). Brain neurochemistry also tends to be altered in those with OSA compared to healthy individuals (Bartlett et al., 2004; Kamba et al., 2001), with reduced creatine levels in the left hippocampal area correlating with increased OSA severity and deteriorating cognitive performance (Bartlett et al., 2004).

Combined fMRI/PET-EEG studies. Recent research has moved toward the concurrent use of EEG and fMRI because this combination allows high anatomical spatial resolution and high temporal resolution of neuronal activity. However, data collection can be troublesome due to artifacts from the magnet that can interfere with the EEG trace. Resulting statistical methods for data interpretation are also difficult. When combined, fMRI-EEG enables greater sensitivity and precision when defining brain activity during sleep (Picchioni et al., 2013). By using EEG to define the sleep stage at the time of the fMRI analysis, correlations between the sleep stage and functional activity of the brain are possible (Spoormaker et al., 2010). The PSA of EEG sleep frequency bands and the sleep microarchitecture (K-complexes or sleep spindles) may then be correlated with fMRI activation levels (Schabus et al., 2007), such as examining potential fMRI activation between two different types of sleep spindles (slow vs. fast; Schabus et al., 2007).

Hormones involved in sleep. The endocrine system uses hormones to signal changes related to sleep-wake timing and sleep stages. Many hormones can be measured as part of sleep research, and this section highlights two such hormones: cortisol and melatonin. Although many other hormones can be measured in sleep research, we have focused on the two most extensively studied in association with sleep and its disorders.

Cortisol. Sleep affects the alerting/stress hormone cortisol, and cortisol levels follow a strong circadian rhythm. Cortisol is the output from the hypothalamic-pituitary-adrenal axis, which is an endocrine system malleable to both bodily and external influence (Elder, Wetherell, Barclay, & Ellis, 2014). Cortisol secretion can be measured in saliva, urine, blood, or hair. Specifically, it has a 24-h circadian rhythm, with secretion increasing after awakening and peaking in the first 30 min of wakefulness. Cortisol then gradually declines over the course of the day and reaches its lowest point at around midnight. Levels remain low and begin to rise again in time for awakening (Follenius, Brandenberger, Bandesapt, Libert, & Ehrhart, 1992). During sleep, diminished cortisol secretion has been associated with SWS, suggesting cortisol may be involved in regulating sleep (Follenius et al., 1992). Sleep researchers have extensively studied the first cortisol sample on awakening and the resulting morning peak, referred to as the cortisol awakening response, and these levels are often disrupted in sleep and psychiatric disorders. For example, plasma and urinary cortisol concentrations are increased in patients with insomnia compared to healthy good-sleeping controls (Rodenbeck, Huether, Ruther, & Hajak, 2002; Vgontzas et al., 2001, 1998).

Melatonin. Melatonin is produced by the pineal gland and is controlled by the master circadian clock, the SCN (Dawson & Encel, 1993). Secretion of melatonin begins during the late evening (2 h before habitual bedtime), in time for sleep with the peak occurring during the middle of the night (Dijk & Cajochen, 1997). The onset of melatonin increase prior to sleep can be measured in saliva through the use of a dim light melatonin onset (DLMO) testing procedure, which is a marker of the patient's individual circadian timing (Zisapel, 2007). DLMO is a valuable aid in the diagnosis of a circadian rhythm disorder, and it is crucial for determining the time of melatonin administration during treatment (Keijzer, Smits, Duffy, & Curfs, 2014).

The onset of melatonin increase is highly coupled with subjective levels of sleepiness and is known to "prepare the body for sleep" (Cajochen, Krauchi, & Wirz-Justice, 2003). The onset of melatonin secretion is also known to be disrupted in sleep phase disorders. This produces a mismatch between the internal circadian rhythm of the body clock and the external environment by either having an early (advanced) or late (delayed) sleep period. The resulting effects can cause a degree of "social jet lag" and perhaps insomnia symptomology, specifically manifested by difficulty getting to sleep or waking too early (Cajochen et al., 2003). These disorders may be managed with melatonin supplements (although the timing, dose, and release rate must be carefully considered beforehand). The incorrect use of melatonin has the potential to exacerbate the sleep complaint. Melatonin supplements have been touted as a way to increase TST, but further evidence regarding effectiveness is required (Wade et al.,

2007). Behavioral interventions involving the correct timing and exposure to daylight (the primary resetting mechanism of the internal body clock) may be equally or more effective than melatonin (Rosenthal et al., 1990).

SUMMARY

Sleep and its impact on daytime functioning can be measured in a variety of ways through subjective self-report measures, including sleep dairies, questionnaire assessments, and interviews. Although PSG is the gold standard, other more complex objective measures are being used and can range from the relatively simple and inexpensive actigraphy to brain-imaging methods. When considering which measure of sleep to implement, the clinician should consider inherent data collection difficulties and cost. Defining the research goals prior to the implementation of a study enables the researcher to use the most effective methods to answer the primary question, while opening pathways for further research in associated areas. In this chapter, we summarize the well-used measures of sleep assessment to highlight and increase knowledge of these measures and make them more available for future research into both sleep and affect.

Acknowledgments

This work was supported by the National Health and Medical Research Council (NHMRC, Australia) Centre for Integrated Research and Understanding of Sleep (CIRUS), Grant No. 571421.

References

Achermann, P. (2009). EEG analysis applied to sleep. *Epileptologie, 26*, 28–33.
Achermann, P., Dijk, D.-J., Brunner, D. P., & Borbély, A. A. (1993). A model of human sleep homeostasis based on EEG slow-wave activity: Quantitative comparison of data and simulations. *Brain Research Bulletin, 31*(1–2), 97–113. http://dx.doi.org/10.1016/0361-9230(93)90016-5.
Agnew, H. W., Webb, W. B., & Williams, R. L. (1966). The first night effect: An EEG study of sleep. *Psychophysiology, 2*(3), 263–266. http://dx.doi.org/10.1111/j.1469-8986.1966.tb02650.x.
Altena, E., Van Der Werf, Y. D., Sanz-Arigita, E. J., Voorn, T. A., Rombouts, S.A.R.B., Kuijer, J. P. A., et al. (2008). Prefrontal hypoactivation and recovery in insomnia. *Sleep, 31*(9), 1271–1276.
Altena, E., Vrenken, H., Van Der Werf, Y. D., vandenHeuvel, O. A., & VanSomeren, E. J. W. (2010). Reduced orbitofrontal and parietal gray matter in chronic insomnia: A voxel-based morphometric study. *Biological Psychiatry, 67*(2), 182–185. http://dx.doi.org/10.1016/j.biopsych.2009.08.003.
Ancoli-Israel, S., Cole, R., Alessi, C., Chambers, M., Moorcroft, W., & Pollak, C. P. (2003). The role of actigraphy in the study of sleep and circadian rhythms. *Sleep, 26*(3), 342–392.

REFERENCES

Antic, N. A., Catcheside, P., Buchan, C., Hensley, M., Naughton, M. T., Rowland, S., et al. (2011). The effect of CPAP in normalizing daytime sleepiness, quality of life, and neurocognitive function in patients with moderate to severe OSA. *Sleep, 34*(1), 111–119.

Arand, D., Bonnet, M., Hurwitz, T., Mitler, M., Rosa, R., & Sangal, R. B. (2005). The clinical use of the MSLT and MWT. *Sleep, 28*(1), 123–144.

Aserinsky, E., & Kleitman, N. (1953). Regularly occurring periods of eye motility, and concomitant phenomena, during sleep. *Science, 118*(3062), 273–274.

Bartlett, D. J., Rae, C., Thompson, C. H., Byth, K., Joffe, D. A., Enright, T., et al. (2004). Hippocampal area metabolites relate to severity and cognitive function in obstructive sleep apnea. *Sleep Medicine, 5*(6), 593–596. http://dx.doi.org/10.1016/j.sleep.2004.08.004.

Bastien, C. H., Vallières, A., & Morin, C. M. (2001). Validation of the insomnia severity index as an outcome measure for insomnia research. *Sleep Medicine, 2*(4), 297–307. http://dx.doi.org/10.1016/S1389-9457(00)00065-4.

Blagrove, M., Farmer, L., & Williams, E. (2004). The relationship of nightmare frequency and nightmare distress to well-being. *Journal of Sleep Research, 13*(2), 129–136. http://dx.doi.org/10.1111/j.1365-2869.2004.00394.x.

Borbély, A. A. (1982). A two process model of sleep regulation. *Human Neurobiology, 1*(3), 195–204.

Borbély, A. A., Baumann, F., Brandeis, D., Strauch, I., & Lehmann, D. (1981). Sleep deprivation: Effect on sleep stages and EEG power density in man. *Electroencephalography and Clinical Neurophysiology, 51*(5), 483–493. http://dx.doi.org/10.1016/0013-4694(81)90225-X.

Braun, A. R., Balkin, T. J., Wesensten, N. J., Carson, R. E., Varga, M., Baldwin, P., et al. (1997). Regional cerebral blood flow throughout the sleep-wake cycle—An (H2O)-O-15 PET study. *Brain, 120*, 1173–1197. http://dx.doi.org/10.1093/brain/120.7.1173.

Buonocore, M. H., & Hecht, S. T. (1995). Functional magnetic-resonance-imaging depict the brain in action. *Nature Medicine, 1*(4), 379–381.

Buysse, D. J., Ancoli-Israel, S., Edinger, J. D., Lichstein, K. L., & Morin, C. M. (2006). Recommendations for a standard research assessment of insomnia. *Sleep: Journal of Sleep and Sleep Disorders Research, 29*(9), 1155–1173.

Buysse, D. J., Reynolds, C. F., Monk, T. H., Berman, S. R., & Kupfer, D. J. (1989). The Pittsburgh sleep quality index—A new instrument for psychiatric practice and research. *Psychiatry Research, 28*(2), 193–213. http://dx.doi.org/10.1016/0165-1781(89)90047-4.

Buysse, D. J., Thompson, W., Scott, J., Franzen, P. L., Germain, A., Hall, M., et al. (2007). Daytime symptoms in primary insomnia: A prospective analysis using ecological momentary assessment. *Sleep Medicine, 8*(3), 198–208. http://dx.doi.org/10.1016/j.sleep.2006.10.006.

Cajochen, C., Krauchi, K., & Wirz-Justice, A. (2003). Role of melatonin in the regulation of human circadian rhythms and sleep. *Journal of Neuroendocrinology, 15*(4), 432–437. http://dx.doi.org/10.1046/j.1365-2826.2003.00989.x.

Carney, C. E., Buysse, D. J., Ancoli-Israel, S., Edinger, J. D., Krystal, A. D., Lichstein, K. L., et al. (2012). The consensus sleep diary: Standardizing prospective sleep self-monitoring. *Sleep, 35*(2), 287–302. http://dx.doi.org/10.5665/sleep.1642.

Carskadon, M. A., & Dement, W. C. (1982). The multiple sleep latency test—What does it measure. *Sleep, 5*, S67–S72.

Carskadon, M. A., & Dement, W. C. (2011). Chapter 2—Normal human sleep: An overview. In M. H. Kryger, T. Roth, & W. C. Dement (Eds.), *Principles and practice of sleep medicine* (5th ed., pp. 16–26). Philadelphia, USA: Elsevier Saunders.

Chervin, R. D., & Aldrich, M. S. (1999). The Epworth sleepiness scale may not reflect objective measures of sleepiness or sleep apnea. *Neurology, 52*(1), 125. http://dx.doi.org/10.1212/wnl.52.1.125.

Colrain, I. M. (2011). Sleep and the brain. *Neuropsychology Review, 21*(1), 1–4. http://dx.doi.org/10.1007/s11065-011-9156-z.

Czeisler, C. A., Duffy, J. F., Shanahan, T. L., Brown, E. N., Mitchell, J. F., Rimmer, D. W., et al. (1999). Stability, precision, and near-24-hour period of the human circadian pacemaker. *Science, 284*(5423), 2177–2181. http://dx.doi.org/10.1126/science.284.5423.2177.

Dang-Vu, T. T. (2012). Neuronal oscillations in sleep: Insights from functional neuroimaging. *Neuromolecular Medicine, 14*(3), 154–167. http://dx.doi.org/10.1007/s12017-012-8166-1.

Dawson, D., & Encel, N. (1993). Melatonin and sleep in humans. *Journal of Pineal Research, 15*(1), 1–12. http://dx.doi.org/10.1111/j.1600-079X.1993.tb00503.x.

Dement, W., & Kleitman, N. (1957). Cyclic variations in EEG during sleep and their relation to eye movements, body motility, and dreaming. *Electroencephalography and Clinical Neurophysiology, 9*(4), 673–690. http://dx.doi.org/10.1016/0013-4694(57)90088-3.

Dibner, C., Schibler, U., & Albrecht, U. (2010). The mammalian circadian timing system: Organization and coordination of central and peripheral clocks. *Annual Review of Physiology, 72*, 517–549. http://dx.doi.org/10.1146/annurev-physiol-021909-135821.

Diekelmann, S., & Born, J. (2010). The memory function of sleep. *Nature Reviews Neuroscience, 11*(2), 114–126. http://dx.doi.org/10.1038/nrn2762.

Dijk, D. J., & Cajochen, C. (1997). Melatonin and the circadian regulation of sleep initiation, consolidation, structure, and the sleep EEG. *Journal of Biological Rhythms, 12*(6), 627–635. http://dx.doi.org/10.1177/074873049701200618.

Drummond, S. P. A., Walker, M., Almklov, E., Campos, M., Anderson, D. E., & Straus, L. D. (2013). Neural correlates of working memory performance in primary insomnia. *Sleep, 36*(9), 1307–1316. http://dx.doi.org/10.5665/Sleep.2952.

Elder, G. J., Wetherell, M. A., Barclay, N. L., & Ellis, J. G. (2014). The cortisol awakening response—Applications and implications for sleep medicine. *Sleep Medicine Reviews, 18*(3), 215–224.

Follenius, M., Brandenberger, G., Bandesapt, J. J., Libert, J. P., & Ehrhart, J. (1992). Nocturnal cortisol release in relation to sleep structure. *Sleep, 15*(1), 21–27.

Foulkes, W. D. (1962). Dream reports from different stages of sleep. *The Journal of Abnormal and Social Psychology, 65*(1), 14. http://dx.doi.org/10.1037/h0040431.

Fukunaga, M., Horovitz, S. G., van Gelderen, P., de Zwart, J. A., Jansma, J. M., Ikonomidou, V. N., et al. (2006). Large-amplitude, spatially correlated fluctuations in BOLD fMRI signals during extended rest and early sleep stages. *Magnetic Resonance Imaging, 24*(8), 979–992. http://dx.doi.org/10.1016/j.mri.2006.04.018.

Garbarino, S., De Carli, F., Nobili, L., Mascialino, B., Squarcia, S., Peneo, M. A., et al. (2002). Sleepiness and sleep disorders in shift workers: A study on a group of Italian police officers. *Sleep, 25*(6), 648–653.

Gradisar, M., Lack, L., Richards, H., Harris, J., Gallasch, J., Boundy, M., et al. (2007). The flinders fatigue scale: Preliminary psychometric properties and clinical sensitivity of a new scale for measuring daytime fatigue associated with insomnia. *Journal of Clinical Sleep Medicine: JCSM: Official Publication of the American Academy of Sleep Medicine, 3*(7), 722–728.

Hall, Emma L., Robson, Siân E., Morris, Peter G., Brookes, Matthew J. (2013). The relationship between MEG and fMRI. *NeuroImage*. http://dx.doi.org/10.1016/j.neuroimage.2013.11.005.

Hanyu, H., Inoue, Y., Sakurai, H., Kanetaka, H., Nakamura, M., Miyamoto, T., et al. (2012). Voxel-based magnetic resonance imaging study of structural brain changes in patients with idiopathic REM sleep behavior disorder. *Parkinsonism & Related Disorders, 18*(2), 136–139. http://dx.doi.org/10.1016/j.parkreldis.2011.08.023.

Hirshkowitz, M. (2004). Normal human sleep: An overview. *Medical Clinics of North America, 88*(3), 551–565. http://dx.doi.org/10.1016/j.mcna.2004.01.001.

Iber, C., Ancoli-Israel, S., Chesson, A., & Quan, S. F. (2007). *The AASM manual for the scoring of sleep and associated events: Rules, terminology, and technical specification* (1st ed.). Darien, IL: American Academy of Sleep Medicine.

Jasper, H. H. (1958). The ten twenty electrode system of the international federation. *Electroencephalography and Clinical Neurophysiology, 10*, 371–375.

Johns, M. W. (1992). Reliability and factor-analysis of the Epworth sleepiness scale. *Sleep, 15*(4), 376–381.

Joo, E. Y., Tae, W. S., Lee, M. J., Kang, J. W., Park, H. S., Lee, J. Y., et al. (2010). Reduced brain gray matter concentration in patients with obstructive sleep apnea syndrome. *Sleep, 33*(2), 235–241.

Kamba, M., Inoue, Y., Higami, S., Suto, Y., Ogawa, T., & Chen, W. (2001). Cerebral metabolic impairment in patients with obstructive sleep apnoea: An independent association of obstructive sleep apnoea with white matter change. *Journal of Neurology, Neurosurgery, and Psychiatry, 71*(3), 334–339. http://dx.doi.org/10.1136/jnnp.71.3.334.

Keijzer, H., Smits, M. G., Duffy, J. F., & Curfs, L. M. G. (2014). Why the dim light melatonin onset (DLMO) should be measured before treatment of patients with circadian rhythm sleep disorders. *Sleep Medicine Reviews, 18*(4), 333–339.

Kim, Y.-C., Lebel, R. M., Wu, Z., Ward, S. L., Davidson, K., Michael, C. K., et al. (2014). Real-time 3D magnetic resonance imaging of the pharyngeal airway in sleep apnea. *Magnetic Resonance in Medicine, 71*(4), 1501–1510.

Kripke, D. F., Mullaney, D. J., Messin, S., & Wyborney, V. G. (1978). Wrist actigraphic measures of sleep and rhythms. *Electroencephalography and Clinical Neurophysiology, 44*(5), 674–676. http://dx.doi.org/10.1016/0013-4694(78)90133-5.

Kushida, C. A., Chang, A., Gadkary, C., Guilleminault, C., Carrillo, O., & Dement, W. C. (2001). Comparison of actigraphic, polysomnographic, and subjective assessment of sleep parameters in sleep-disordered patients. *Sleep Medicine, 2*(5), 389–396. http://dx.doi.org/10.1016/S1389-9457(00)00098-8.

Kushida, C. A., Littner, M. R., Morgenthaler, T., Alessi, C. A., Bailey, D., Coleman, J., Jr., et al. (2005). Practice parameters for the indications for polysomnography and related procedures: An update for 2005. *Sleep, 28*(4), 499–521.

Levitt, H., Wood, A., Moul, D. E., Hall, M., Germain, A., Kupfer, D. J., et al. (2004). A pilot study of subjective daytime alertness and mood in primary insomnia participants using ecological momentary assessment. *Behavioral Sleep Medicine, 2*(2), 113–131. http://dx.doi.org/10.1207/s15402010bsm0202_3.

Lichstein, K. L., Stone, K. C., Donaldson, J., Nau, S. D., Soeffing, J. P., Murray, D., et al. (2006). Actigraphy validation with insomnia. *Sleep, 29*(2), 232–239.

Littner, M. R., Kushida, C., Wise, M., Davila, D. G., Morgenthaler, T., Lee-Chiong, T., et al. (2005). Practice parameters for clinical use of the multiple sleep latency test and the maintenance of wakefulness test—An American Academy of Sleep Medicine Report—Standards of practice committee of the American Academy of Sleep Medicine. *Sleep, 28*(1), 113–121.

Loomis, A. L., Harvey, E. N., & Hobart, G. A. (1938). Distribution of disturbance-patterns in the human electroencephalogram with special reference to sleep. *Journal of Neurophysiology, 1*, 413–430.

Lovblad, K. O., Thomas, R., Jakob, P. M., Scammell, T., Bassetti, C., Griswold, M., et al. (1999). Silent functional magnetic resonance imaging demonstrates focal activation in rapid eye movement sleep. *Neurology, 53*(9), 2193–2195.

Maes, J., Verbraecken, J., Willemen, M., De Volder, I., van Gastel, A., Michiels, N., et al. (2014). Sleep misperception, EEG characteristics and autonomic nervous system activity in primary insomnia: A retrospective study on polysomnographic data. *International Journal of Psychophysiology, 91*(3), 163–171. http://dx.doi.org/10.1016/j.ijpsycho.2013.10.012.

Maquet, P. (2000). Functional neuroimaging of normal human sleep by positron emission tomography. *Journal of Sleep Research, 9*(3), 207–231. http://dx.doi.org/10.1046/j.1365-2869.2000.00214.x.

Miller, C. B., Kyle, S. D., Marshall, N. S., & Espie, C. A. (2013). Ecological momentary assessment of daytime symptoms during sleep restriction therapy for insomnia. *Journal of Sleep Research, 22*(3), 266–272. http://dx.doi.org/10.1111/Jsr.12024.

Mitler, M. M., Gujavarty, K. S., & Browman, C. P. (1982). Maintenance of wakefulness test—A polysomnographic technique for evaluating treatment efficacy in patients with excessive somnolence. *Electroencephalography and Clinical Neurophysiology, 53*(6), 658–661. http://dx.doi.org/10.1016/0013-4694(82)90142-0.

Montserrat, J. M., Ferrer, M., Hernandez, L., Farre, R., Vilagut, G., Navajas, D., et al. (2001). Effectiveness of CPAP treatment in daytime function in sleep apnea syndrome—A randomized controlled study with an optimized placebo. *American Journal of Respiratory and Critical Care Medicine, 164*(4), 608–613.

Morgenthaler, T., Alessi, C., Friedman, L., Owens, J., Kapur, V., Boehlecke, B., et al. (2007). Practice parameters for the use of actigraphy in the assessment of sleep and sleep disorders: An update for 2007. *Sleep: Journal of Sleep and Sleep Disorders Research, 30*(4), 519–529.

Morin, C. M. (1993). *Insomnia: Psychological assessment and management*. New York, NY: Guilford Press.

Nofzinger, E. A., Buysse, D. J., Germain, A., Price, J. C., Miewald, J. M., & Kupfer, D. J. (2004). Functional neuroimaging evidence for hyperarousal in insomnia. *American Journal of Psychiatry, 161*(11), 2126–2129. http://dx.doi.org/10.1176/appi.ajp.161.11.2126.

Ohayon, M. M., Guilleminault, C., Zulley, J., Palombini, L., & Raab, H. (1999). Validation of the sleep-EVAL system against clinical assessments of sleep disorders and polysomnographic data. *Sleep, 22*(7), 925–930.

Olson, L., Cole, M., & Ambrogetti, A. (1998). Correlations among Epworth sleepiness scale scores, multiple sleep latency tests and psychological symptoms. *Journal of Sleep Research, 7*(4), 248–253. http://dx.doi.org/10.1046/j.1365-2869.1998.00123.x.

Oostenveld, R., & Praamstra, P. (2001). The five percent electrode system for high-resolution EEG and ERP measurements. *Clinical Neurophysiology, 112*(4), 713–719. http://dx.doi.org/10.1016/S1388-2457(00)00527-7.

Perlis, M. L., Kehr, E. L., Smith, M. T., Andrews, P. J., Orff, H., & Giles, D. E. (2001). Temporal and stagewise distribution of high frequency EEG activity in patients with primary and secondary insomnia and in good sleeper controls. *Journal of Sleep Research, 10*(2), 93–104. http://dx.doi.org/10.1046/j.1365-2869.2001.00247.x.

Picchioni, D., Duyn, J. H., & Horovitz, S. G. (2013). Sleep and the functional connectome. *NeuroImage, 80*, 387–396. http://dx.doi.org/10.1016/j.neuroimage.2013.05.067.

Picchioni, D., Fukunaga, M., Carr, W. S., Braun, A. R., Balkin, T. J., Duyn, J. H., et al. (2008). fMRI differences between early and late stage-1 sleep. *Neuroscience Letters, 441*(1), 81–85. http://dx.doi.org/10.1016/j.neulet.2008.06.010.

Rae, C. D. (2014). A guide to the metabolic pathways and function of metabolites observed in human brain H-1 magnetic resonance spectra. *Neurochemical Research, 39*(1), 1–36. http://dx.doi.org/10.1007/s11064-013-1199-5.

Rae, C., Bartlett, D. J., Yang, Q., Walton, D., Denotti, A., Sachinwalla, T., et al. (2009). Dynamic changes in brain bioenergetics during obstructive sleep apnea. *Journal of Cerebral Blood Flow and Metabolism, 29*(8), 1421–1428. http://dx.doi.org/10.1038/jcbfm.2009.57.

Rechtschaffen, A., & Kales, A. (1968). *A manual of standardized terminology, techniques and scoring system for sleep stages of human subjects*. Bethesda, USA: National Institute of Neurological Diseases and Blindness, Neurological Information Network. http://dx.doi.org/10.1234/12345678.

Rodenbeck, A., Huether, G., Ruther, E., & Hajak, G. (2002). Interactions between evening and nocturnal cortisol secretion and sleep parameters in patients with severe chronic primary insomnia. *Neuroscience Letters, 324*(2), 159–163. http://dx.doi.org/10.1016/S0304-3940(02)00192-1.

Rosenthal, N. E., Josephvanderpool, J. R., Levendosky, A. A., Johnston, S. H., Allen, R., Kelly, K. A., et al. (1990). Phase-shifting effects of bright morning light as treatment for delayed sleep phase syndrome. *Sleep, 13*(4), 354–361.

Sadeh, A., & Acebo, C. (2002). The role of actigraphy in sleep medicine. *Sleep Medicine Reviews, 6*(2), 113–124. http://dx.doi.org/10.1053/smrv.2001.0182.

REFERENCES

Sangal, R. B., Mitler, M. M., & Sangal, J. M. (1999). Subjective sleepiness ratings (Epworth sleepiness scale) do not reflect the same parameter of sleepiness as objective sleepiness (maintenance of wakefulness test) in patients with narcolepsy. *Clinical Neurophysiology, 110*(12), 2131–2135. http://dx.doi.org/10.1016/S1388-2457(99)00167-4.

Schabus, M., Dang-Vu, T. T., Albouy, G., Balteau, E., Boly, M., Carrier, J., et al. (2007). Hemodynamic cerebral correlates of sleep spindles during human non-rapid eye movement sleep. *Proceedings of the National Academy of Sciences of the United States of America, 104*(32), 13164–13169. http://dx.doi.org/10.1073/pnas.0703084104.

Schenck, C. H., Bundlie, S. R., & Mahowald, M. W. (1996). Delayed emergence of a parkinsonian disorder in 38% of 29 older men initially diagnosed with idiopathic rapid eye movement sleep behaviour disorder. *Neurology, 46*(2), 388–393. http://dx.doi.org/10.1212/WNL.46.2.388.

Schenck, C. H., Hurwitz, T. D., & Mahowald, M. W. (1993). REM sleep behaviour disorder: An update on a series of 96 patients and a review of the world literature. *Journal of Sleep Research, 2*(4), 224–231. http://dx.doi.org/10.1111/j.1365-2869.1993.tb00093.x.

Schramm, E., Hohagen, F., Grasshoff, U., Riemann, D., Hajak, G., Weess, H. G., et al. (1993). Test-retest reliability and validity of the structured interview for sleep disorders according to DSM-III—R. *The American Journal of Psychiatry, 150*(6), 867–872.

Shahid, A., Wilkinson, K., & Marcu, S. (2012). *STOP, THAT and one hundred other sleep scales.* New York: Springer.

Spiegel, K., Leproult, R., & Van Cauter, E. (1999). Impact of sleep debt on metabolic and endocrine function. *The Lancet, 354*(9188), 1435–1439. http://dx.doi.org/10.1016/S0140-6736(99)01376-8.

Spiegelhalder, K., Regen, W., Baglioni, C., Riemann, D., & Winkelman, J. W. (2013). Neuroimaging studies in insomnia. *Current Psychiatry Reports, 15*(11), 405.

Spoormaker, V. I., Schroter, M. S., Gleiser, P. M., Andrade, K. C., Dresler, M., Wehrle, R., et al. (2010). Development of a large-scale functional brain network during human non-rapid eye movement sleep. *Journal of Neuroscience, 30*(34), 11379–11387. http://dx.doi.org/10.1523/Jneurosci.2015-10.2010.

Stephan, F. K., & Nunez, A. A. (1977). Elimination of circadian rhythms in drinking, activity, sleep, and temperature by isolation of the suprachiasmatic nuclei. *Behavioral Biology, 20*(1), 1–16. http://dx.doi.org/10.1016/S0091-6773(77)90397-2.

Stoffers, D., Altena, E., van der Werf, Y. D., Sanz-Arigita, E. J., Voorn, T. A., Astill, R. G., et al. (2014). The caudate: A key node in the neuronal network imbalance of insomnia? *Brain, 137*(Pt. 2), 610–620. http://dx.doi.org/10.1093/brain/awt329.

Sullivan, S. S., & Kushida, C. A. (2008). MUltiple sleep latency test and maintenance of wakefulness test. *CHEST Journal, 134*(4), 854–861. http://dx.doi.org/10.1378/chest.08-0822.

Sunwoo, B. Y., Jackson, N., Maislin, G., Gurubhagavatula, I., George, C. F., & Pack, A. I. (2012). Reliability of a single objective measure in assessing sleepiness. *Sleep, 35*(1), 149–158. http://dx.doi.org/10.5665/sleep.1606.

Thorpy, M. J., Westbrook, P., Ferber, R., & Fredrickson, P. (1992). The clinical use of the multiple sleep latency test. *Sleep: Journal of Sleep Research & Sleep Medicine, 15*(3), 268–276.

Van Dongen, H. P. A., Baynard, M. D., Maislin, G., & Dinges, D. F. (2004). Systematic interindividual differences in neurobehavioral impairment from sleep loss: Evidence of trait-like differential vulnerability. *Sleep, 27*(3), 423–433.

Van Dongen, H. P. A., & Dinges, D. F. (2005). Sleep, circadian rhythms, and psychomotor vigilance. *Clinics in Sports Medicine, 24*(2), 237–249. http://dx.doi.org/10.1016/j.csm.2004.12.007.

Vgontzas, A. N., Bixler, E. O., Lin, H. M., Prolo, P., Mastorakos, G., Vela-Bueno, A., et al. (2001). Chronic insomnia is associated with nyctohemeral activation of the hypothalamic-pituitary-adrenal axis: Clinical implications. *Journal of Clinical Endocrinology & Metabolism, 86*(8), 3787–3794. http://dx.doi.org/10.1210/Jc.86.8.3787.

Vgontzas, A. N., Tsigos, C., Bixler, E. O., Stratakis, C. A., Zachman, K., Kales, A., et al. (1998). Chronic insomnia and activity of the stress system: A preliminary study. *Journal of Psychosomatic Research, 45*(1), 21–31. http://dx.doi.org/10.1016/S0022-3999(97)00302-4.

Wade, A. G., Ford, I., Crawford, G., McMahon, A. D., Nir, T., Laudon, M., et al. (2007). Efficacy of prolonged release melatonin in insomnia patients aged 55-80 years: Quality of age sleep and next-day alertness outcomes. *Current Medical Research and Opinion, 23*(10), 2597–2605. http://dx.doi.org/10.1185/030079907x233098.

Weaver, T. E., Laizner, A. M., Evans, L. K., Maislin, G., Chugh, D. K., Lyon, K., et al. (1997). An instrument to measure functional status outcomes for disorders of excessive sleepiness. *Sleep, 20*(10), 835–843.

Webb, S. W., & Henkelman, R. M. (2008). Physics of medical imaging. *Physics Today, 43*(4), 77–78. http://dx.doi.org/10.1063/1.2810532.

Wehrle, R., Kaufmann, C., Wetter, T. C., Holsboer, F., Auer, D. P., Pollmacher, T., et al. (2007). Functional microstates within human REM sleep: First evidence from fMRI of a thalamocortical network specific for phasic REM periods. *European Journal of Neuroscience, 25*(3), 863–871. http://dx.doi.org/10.1111/j.1460-9568.2007.05314.x.

Williams, R. L., Agnew, H. W., Jr., & Webb, W. B. (1964). Sleep patterns in young adults: An EEG study. *Electroencephalography and Clinical Neurophysiology, 17*(4), 376–381. http://dx.doi.org/10.1016/0013-4694(64)90160-9.

Zisapel, N. (2007). Sleep and sleep disturbances: Biological basis and clinical implications. *Cellular and Molecular Life Sciences, 64*(10), 1174–1186. http://dx.doi.org/10.1007/s00018-007-6529-9.

CHAPTER 5

The Elicitation and Assessment of Emotional Responding

Sarah J. Bujarski[*], Emily Mischel[*], Courtney Dutton[*], J. Scott Steele[†], and Joshua Cisler[†]

[*]Department of Psychological Science, University of Arkansas, Fayetteville, Arkansas, USA
[†]Brain Imaging Research Center, Psychiatric Research Institute, University of Arkansas for Medical Sciences, Little Rock, Arkansas, USA

The distinctions between the constructs of affect, emotion, and mood are central to the elicitation and assessment of emotional experience. Contemporary research conceptualizes affect as a superordinate category for both positively and negatively valenced states. A useful analogy in distinguishing between emotion and mood refers to mood as an "emotional climate" during which emotions are "fluctuating changes in emotional 'weather'" (APA, 1994, p. 763). This example highlights that one major distinction between emotions and mood relates to their duration. Indeed, contemporary perspectives on emotion posit that emotions occur over a relatively brief time course, with variability of responding occurring across each stage of emotional experience. Mood is conceptualized as having both a slower onset and decay, however. Research on emotional responding has consistently demonstrated that the experience of emotion is marked by a set of behavioral (e.g., facial expressivity), experiential (e.g., interpretations), and physiological (e.g., heart rate acceleration or deceleration) response tendencies (e.g., Gross, 1998), all of which influence how one reacts to a perceived emotion. In contrast, the more diffuse nature of mood is more likely to influence cognitions than actions (Davidson, 1994).

PRIMARY PERSPECTIVES

The vast research literature that encompasses the broad construct of emotion can make defining this seemingly simple term somewhat complex. As an organizational device, it is helpful to describe two major perspectives that dominate contemporary research approaches in the study of emotion elicitation. The dimensional perspective theorizes that emotions are relatively brief responses resulting from engagement with meaningful stimuli in the environment, which are organized by the direction of motivated behavior. In this perspective, valence refers to the degree of pleasantness or positivity versus unpleasantness or negativity associated with an emotion, and arousal refers to the strength of an experienced emotion (Hamann, 2012). Furthermore, the direction of behavior (i.e., approach or avoid) is determined by the valence of the emotion (e.g., negatively valenced stimuli elicit avoidances), and the intensity of the arousal determines the strength of the approach or avoidance behavior (Cacioppo & Gardner, 1999). Thus, the focus on motivated behavior in response to arousal stimuli highlights this perspective's reliance on the response tendencies (Gross, 1998) associated with encountering environmental stimuli to explain emotion.

In contrast, the discrete emotions perspective theorizes that emotions consist of unique, basic emotional states (e.g., happiness) that result from appraisals of stimuli in the environment (e.g., Frijda, 1986). Thus, although this perspective recognizes that labels (e.g., good vs. bad) are used to organize one's appraisals into consequent emotions, this perspective primarily focuses on the subjective experience of a specific, labeled, emotional experience.

A multitude of well-established laboratory methods have been developed to elicit emotion. The following section highlights a number of techniques that have been developed from both the dimensional and discrete emotions perspectives. Notably, many of these techniques have been used to provide empirical evidence in support of both perspectives, however, which highlights both the conceptual overlap of the dimensional and discrete perspectives and the complexity inherent in defining emotional experiences.

EMOTION ELICITATION: LABORATORY METHODS

Visual Stimuli

International Affective Picture System

The International Affective Picture System (IAPS) (Lang, Bradley, & Cuthbert, 1999) is a set of emotion-eliciting photographs categorized based on valence (i.e., positive vs. negative) and arousal (low vs. high).

The use of the IAPS in laboratory studies typically involves selecting a set of photographs that have been validated to elicit a particular emotional response and asking participants to view the photographs either in print form or with the use of a computer. Ratings of each photograph's valence and arousal gathered from initial research studies have resulted in the development of a set of norms (means and standard deviations) across a variety of age groups and cultures. For instance, a recent validation study in a Chilean population supports the reliability of the IAPS among participants of different cultures (Silva, 2011). Gender differences have also been replicated, with women reporting higher arousal and lower valence compared to men. Other research has demonstrated differences in valence and arousal ratings between men and women; however, these differences tend to be minimal (e.g., Mikels et al., 2005). In support of the dimensional theory of emotions, research on the IAPS has demonstrated that valence influences startle reactivity or the involuntary flexion of muscles to an unexpected stimulus, such that positively valenced pictures can inhibit startle and arousing negative pictures can accentuate startle (Cuthbert, Bradley, & Lang, 1996). In addition, data collected using the IAPS have shown that different discrete emotions (e.g., anger versus fear) relate to different ratings of valence and arousal. Indeed, the IAPS elicitation method is demonstrably effective at eliciting emotion across behavioral (e.g., facial), self-report, and physiological response tendencies (Mikels et al., 2005). Together, these results suggest that the IAPS is a useful and flexible methodology that can be used to elicit emotion varying in type (i.e., by specific emotion) and degree (i.e., by arousal).

Word Stimuli Timuli: Affective Norms for English Words

The Affective Norms for English Words (ANEW) (Bradley & Lang, 1999a, 1999b) consists of a set of English nouns, verbs, and adjectives that are rated on the dimensions of valence and arousal. In support of its validity, research has shown that positive and negative words relate to faster reaction times and higher accuracy compared to neutral words (e.g., Ali & Cimino, 1997). In addition to the established dimensional norms, a recent study has also compiled a discrete categorical characterization of the words in the ANEW database (Stevenson, Mikels, & James, 2007). The categorization of each word by discrete emotion allows researchers to more carefully control the specific emotion they wish to elicit, and when added to the dimensional norms, this approach allows researchers to elicit a specific emotion along a varying dimension of arousal.

Word Stimuli: The "Emotional" Stroop Task

The emotional Stroop task is frequently used to assess the influence of attention on information processing. The emotional Stroop effect refers to findings that individuals are slower to name the color of ink a word is printed in when that word is negative compared to neutral (e.g., Algom, Chajut, & Lev, 2004). This suggests that negative words are more likely to attract attention and, therefore, delay the processing of other stimulus information (i.e., the color of the ink). In addition to attention effects, other researchers have posited that disengaging from emotionally relevant stimuli is more difficult, which also serves to delay processing (e.g., Estes & Adelman, 2008). Thus, both the valence (negative vs. neutral) and the arousal level of the printed word are likely to impact how quickly and accurately the color of the ink can be processed.

Olfactory Stimuli

Empirical research has demonstrated that odors can be employed in effective emotion-elicitation techniques. Research posits that, through conditioning, odors can elicit a positively or negatively valenced emotional response via associative learning processes. Indeed, odors are theorized to be particularly effective stimuli for the recall of emotional memories (Herz, 1996). For example, individuals with dental procedure fears report negative emotions when presented with the odor of eugenol, a local anesthetic used by dentists (Robin, Alaoui-Ismaili, Dittmar, & Vernet-Maury, 1998). Similar to other stimuli that utilize sensory information to elicit emotion, olfactory stimuli not only result in self-reported experience of emotion, but physiological changes as well. For instance, pleasant and novel odors have been shown to decrease heart rate (Wilson, 2009). Finally, olfactory stimuli have been demonstrated to be as effective as audiovisual stimuli at eliciting emotion-relevant behavioral responses, such as facial expressions (de Groot, Semin, & Smeets, 2014).

Researchers have examined a variety of different odors in relation to eliciting discrete emotions. In regard to positive emotions, "green odor" derived from oak leaves is effective at eliciting positively valenced emotions (Sano, Tsuda, Sugano, Aou, & Hatanaka, 2002). H_2S is an odorant that has been used to elicit negatively valenced mood states; in fact, one recent study demonstrated that this smell facilitated faster reaction times in the identification of disgusted facial expressions (Seubert et al., 2010). Finally, exposure to sweat collected from individuals exposed to a fearful video has been shown to bias individuals toward interpreting ambiguous facial expressions as more fearful (Zhou & Chen, 2009).

Auditory Stimuli

Script-Driven Imagery

Script-driven imagery is a well-established method for eliciting emotion in the laboratory and involves participants writing brief narratives (Orr, Pitman, Lasko, & Herz, 1993; Pitman, Orr, Forgue, de Jong, & Claiborn, 1987). Typically, participants provide written accounts of both a neutral situation (e.g., getting dressed for work) and an experimental situation (e.g., trauma account). Participants are asked to include in their script components that relate to relevant response tendencies (i.e., behaviors, thoughts, and bodily arousal). Often, the script presentation is preceded by a baseline period and followed by a rehearsal period wherein participants are asked to imagine each script image as vividly as possible. Common outcome variables used with script-driven imagery procedures include neural responses (Frewen et al., 2011), self-reported distress, heart rate, skin conductance, and electromyogram data (Bauer et al., 2013; Hopper, Frewen, Sack, Lanius, & vander Kolk, 2007; Olatunji, Wolitzky-Taylor, Babson, & Feldner, 2009; Orr et al., 1998). Utilizing the script-driven imagery procedures, researchers have been able to elicit emotional states and physiological responding to better understand the experience of emotion in the context of disorders such as posttraumatic stress disorder (PTSD; Beckham et al., 2007; Olatunji, Babson, Smith, Feldner, & Connolly, 2009; Pineles et al., 2013).

International Affective Digital Sounds

The International Affective Digital Sounds (IADS) (Bradley & Lang, 1999a, 1999b) is a standardized set of 111 auditory stimuli that, like the IAPS, have been normalized on the dimensions of valence and arousal. Subsequently, the IADS has demonstrated the ability to elicit physiological responding and neural responding (in the auditory cortex) that differs for neutral vs. positively or negatively valenced sounds (Plichta et al., 2011). In an effort to examine the IADS with categorical emotions, Stevenson and James (2008) asked participants to rate all 111 sounds included in the IADS on five discrete emotions: happiness, fear, anger, sadness, and disgust. Results indicated that 42 (37.8%) sounds included in the IADS were labeled as a discrete emotion.

The IADS has also been normed with Spanish-speaking individuals (Redondo, Fraga, Padrón, & Piñera, 2008). In general, there were very high correlations between the Spanish and United States ratings for valence ($r = 0.918$) and arousal ($r = 0.903$). When examining the specific dimensions listed above, the researchers found that Americans typically endorsed higher valence than Spaniards, but Spaniards reported more arousal in response to the sounds. Although researchers must still address cultural considerations related to the universality of emotion, this study demonstrates that the IADS is a standardized method for eliciting emotion for both American and Spanish individuals, thus allowing for cross-cultural comparisons.

Musical Scores

In addition to digital sound clips, musical scores have been used to elicit emotion in laboratory studies. Khalfa, Isabelle, Jean-Pierre, and Manon (2002) presented individuals with 7-s musical excerpts known to elicit the following four emotions: (1) fear, (2) happiness, (3) sadness, and (4) peacefulness. Researchers measured self-reported emotional intensity and clarity, as well as the physiological index of skin conductance response (SCR), which measures arousal based on the amount of sweat produced on the fingers. These four emotions appear to exist on a continuum such that peacefulness (pleasant, calming), sadness (unpleasant, calming), happiness (pleasant, stimulating), and fear (unpleasant, stimulating) fall at different ends of the continuum (Khalfa et al., 2002). Results also demonstrated that musical scores depicting fear and happiness were associated with a greater SCR than were scores associated with sadness and peacefulness.

Although the existing literature demonstrates a connection between music and emotion, Baumgartner, Esslen, and Jäncke (2005) stressed the importance of studying the combined effect of music and pictures for a more ecologically valid assessment of emotional responding. Specifically, participants' self-report ratings were more accurate and prominent when sounds and pictures were presented together, followed by pictures only, and then music only. Similarly, psychophysiological measures (e.g., heart rate, skin conductance) demonstrated increases in responding in the combined and music only condition relative to the picture only condition.

Biological Stimuli

Hyperventilation

Depending on the sample, various tasks have been used to elicit anxiety in a clinical setting (Schmidt & Trakowski, 2004), and these tasks can be utilized in laboratory studies to examine phenomenon such as anxiety. Two interoceptive exposure tasks that have been used in laboratory studies are hyperventilation and breathing CO_2-enriched air. Hyperventilation can naturally lead to physiological symptoms that are similar to those experienced during a panic attack (e.g., heart palpitations and dizziness; Barlow, 2004). The act of hyperventilation removes CO_2 from the body at a rate faster than the body can replace it. The decrease in blood flow to the brain is thought to explain the dizziness, lightheadedness, and other similar symptoms commonly reported in association with panic (Barlow, 2004). The advantage of the hyperventilation task in emotion-relevant research is that it may be used as a way to model and study panic-relevant emotions, such as fear and anxiety, in a controlled environment.

The voluntary hyperventilation task has been utilized in many different laboratory studies to elicit the panic-relevant symptoms listed above in an effort to better understand the negative emotions involved in anxiety disorders, such as panic-relevant symptoms (Bunaciu et al., 2014; Hawks, Blumenthal, Feldner, Leen-Feldner, & Jones, 2011; Kossowsky, Wilhem, &

Schneider, 2013). The voluntary hyperventilation task lasts up to 5 min, and the researcher instructs participants to breathe at a rate such that they complete 30 respiratory cycles in 1 min. Voluntary hyperventilation has been widely used to study to a variety of factors, including trauma exposure (Badour et al., 2012) and nicotine withdrawal among individuals with PTSD (Feldner, Vujanovic, Gibson, & Zvolensky, 2008). One study found that, among trauma-exposed adolescents, the level of reported fear during the trauma predicted the level of reported anxiety experienced during the voluntary hyperventilation task (Badour et al., 2012), supporting a link between fear experienced during a trauma and later responding to bodily arousal.

Carbon Dioxide-Enriched Biological Challenge

In contrast to the voluntary hyperventilation task that lowers the level of CO_2 in the body, inhaling increasing levels of CO_2-enriched air has also shown the ability to elicit panic-relevant negative emotions such as fear and anxiety (Barlow, 2004). This laboratory task involves participants breathing in CO_2-enriched air (e.g., 5-35% CO_2) through a mask (see Lejuez, Forsyth, & Eifert, 1998 for a detailed description of one apparatus used to administer CO_2-enriched air) for a predetermined amount of time that varies as a result of CO_2 concentration. Early studies demonstrated that breathing CO_2-enriched air elicited anxiety symptoms similar to those reported during panic attacks (Barlow, 2004). This phenomenon has led to very interesting research confirming that inhalation of CO_2-enriched air can result in the same effects as decreasing CO_2 levels in the body through hyperventilation (Barlow, 2004).

One construct that has become integral in the study of panic-relevant symptoms in the laboratory is anxiety sensitivity. This refers to the fear of sensations thought to be related to anxiety and the belief that these sensations may have negative consequences (Reiss, 1991). The utilization of controlled laboratory procedures, such as the administration of CO_2-enriched air, to individuals with both high and low anxiety sensitivity has allowed researchers to begin to tease apart the cognitive and physical aspects of panic-relevant negative emotions (Blechert, Wilhelm, Meuret, Wilhelm, & Roth, 2013). Perceived control also seems to play a role in the level of anxiety reported by individuals, such that perceived control (e.g., the belief that they can stop an experimental procedure) is associated with less anxiety (Sanderson, Rapee, & Barlow, 1989). Based on this, Olatunji and colleagues studied the role of perceived control in recovery from a panic-relevant laboratory challenge (i.e., breathing CO_2-enriched air). During the recovery period following the CO_2 administration, a greater sense of perceived control predicted a decrease in anxiety for individuals with high anxiety sensitivity (Olatunji, Wolitzky-Taylor, et al., 2009).

Pharmacological Agents

Multiple pharmacological agents have been used in laboratory studies to elicit and study anxiety (Barlow, 2004) and depression (Homan, Drevets, & Hasler, 2014). Similar to the above-discussed voluntary hyperventilation task and breathing CO_2-enriched air, the administration of pharmacological agents has been studied as a method for eliciting anxiety-relevant symptoms in the laboratory. Researchers have utilized substances such as epinephrine, sodium lactate, and even caffeine to study negative emotions related to panic-relevant symptoms in a controlled laboratory environment (Barlow, 2004). van Zijderveld, TenVoorde, Veltman, and van Doornen (1997) intravenously administered three increasing dose levels of epinephrine for 15 min at each level, while monitoring participants' ratings of distress and panic symptoms. In this study, 66.6% of participants diagnosed with panic disorder experienced a panic attack during the epinephrine administration. This rate was compared to the results of previous studies (van Zijderveld et al., 1992) in which individuals without panic disorder endorsed increased anxiety during the epinephrine administration, but did not experience a panic attack (van Zijderveld et al., 1997).

The results of studies utilizing other pharmacological agents to elicit panic-like symptoms and negative emotions are similar to the ones conducted by van Zijderveld et al. (1997, 1992), in that the administration of these substances elicits panic attacks in a large proportion of individuals with a history of panic disorder, while contributing to increased anxiety without panic attacks in control groups. This is true for isoproterenol (Frohlich, Tarazi, & Dustan, 1969), yohimbine (Charney, Henider, & Breier, 1984), sodium lactate (Cowley & Arana, 1990), flumazenil (Nutt, Glue, Lawson, & Wilson, 1990), and caffeine (Charney, Henider, & Jatlow, 1985).

Although slightly different than administrating a substance to elicit anxiety-relevant symptoms in the lab, other methods are commonly used to promote affective experiences. For example, researchers have used pharmacological agents that deplete catecholamine in order to study the response to emotional stimuli in those in remission from major depressive disorder (Homan et al., 2014). This is not thought of as a way to elicit emotions in the laboratory, but it can help researchers study the relations between emotion and disorders such as major depression.

Blended Stimuli

Stressful Film Paradigm

The use of film clips to elicit emotion is another widely utilized method for laboratory induction procedures; in fact, meta-analyses suggest that films are one of the most effective methods for eliciting emotion (e.g., Westermann, Spies, Stahl, & Hesse, 1996), possibly because films serve as dynamic visual and auditory stimulation. Furthermore, film clips can be selected on the basis of either valence/arousal (dimensional) or discrete emotion (categorical) or a combination of both to elicit varying intensities of each basic emotion. Growing research has standardized sets of film clips that induce both positive and negative emotions (see Hewig et al., 2005, for ratings of Hollywood film clips for each discrete emotion). One category of film-based emotion-elicitation research, the stressful film paradigm, has demonstrated that stressful films can even serve as analogues for traumatic events, because they are capable of eliciting trauma-relevant emotions. Moreover, the induced affect from stressful films has been shown to have lasting effects. For instance, Verwoerd, Wessel, de Jong, and Nieuwenhuis (2009) sought to investigate whether difficulties disengaging from visual trauma-related stimuli presented in film clips would predict the maintenance of intrusive memories. These intrusive memories, in the form of unwanted visual images, in turn relate to the experience of distress. Their results suggested that the participants who were most distracted by the film clip images experienced more intrusive memories recorded in diaries over a 1-week period.

Virtual Reality

The advent of virtual reality technology represents both a new avenue for the treatment of emotion-related psychopathology (e.g., phobias) and a versatile methodological technique that has contributed new knowledge to the area of emotion research. For instance, many previous studies examining avoidance among spider phobics demonstrated that these individuals quickly averted their attention *away* from photographs of spiders (e.g., Hermans, Vansteenwegen, & Eelen, 1999), suggesting that fear relates to avoidance behavior. However, more recent research has demonstrated that fear- and anxiety-based disorders also relate to increased vigilance for threat and that, in virtual reality environments, spider phobics spend *more* time looking at the spiders (Rinck, Kwakkenbos, Dotsch, Wigboldus, & Becker, 2010). Thus, the use of virtual reality can enable researchers to better approximate emotional responding in real-world situations. Multifaceted virtual reality stimuli allow researchers to expose individuals to visual, auditory, and olfactory cues simultaneously, making the use of this technology highly effective as a treatment approach. Indeed, several virtual reality therapy worlds have been created to augment exposure therapy for PTSD, including "Virtual Vietnam" (Rothbaum, Hodges, Ready, Graap, & Alarcon, 2001) and "World Trade Center World" (Difede & Hoffman, 2002).

Laboratory Methods: Developmental Considerations

Modifications are often made to emotion-elicitation procedures to utilize them in developmentally appropriate ways for infant, child, and adolescent research. Importantly, limiting the intensity of stimuli used with children is necessary from an ethical standpoint, and researchers should avoid presenting extremely violent, distressing, or inappropriate content to youth (Brenner, 2000). This is especially important because, compared to adults, youth are less capable of self-regulating their emotions, and therefore, they may show increased emotional reactivity for longer periods of time following the elicitation of negative affect (Campos, Frankel, & Camras, 2004). Furthermore, the prefrontal cortex, a brain region that plays a large role in emotional understanding, is not fully developed in young children and adolescents; thus, youth show decreased ability to identify and label emotions, as compared to adults (Campos et al., 2004; Lieberman et al., 2007). Because of this decreased ability to identify and express emotions, children likely need increased time to process emotional stimuli and to self-report emotional reactions.

The use of films and musical scores for emotion elicitation is frequently employed with youth. Musical scores, particularly those used to elicit happiness and sadness, are common and elicit emotional reactions comparable to adults (Kelvin, Goodyer, Teasdale, & Brechin, 1999; von Leupoldt et al., 2007). However, some research has demonstrated that youth may be more likely to report feelings of happiness, as well as fewer feelings of anger when listening to the same scores as adults (Robazza, Macaluso, & D'urso, 1994). In addition, young children rate neutral films as somewhat more positive than older children and adults; however, nonneutral film ratings are otherwise similar across developmental level (von Leupoldt et al., 2007).

Despite the fact that many similar emotion-elicitation paradigms are used across the developmental spectrum, the validation of affect-elicitation methodology among youth generally lags behind adult research. For example, separate norms for three age groups of children (i.e., ages 7-9, 10-12, and 13-14) were developed for a subset of the IAPS stimuli (Lang et al., 1999). Results suggested that older children showed similar patterns of emotional reactivity and arousal ratings compared to adults; however, younger children rated unpleasant pictures as less arousing and pleasant or neutral pictures as more arousing than did older children and adults (McManis, Bradley, Berg, Cuthbert, & Lang, 2001). Similarly, the ANEW database (Bradley & Lang, 1999a, 1999b) does not have developed norms for use with youth. However, recent work has adapted words from the ANEW database to create a small database of emotional words for children. The stimuli include 81 words originally used with 9- to 11-year-olds. These words are all within a fourth- to fifth-grade reading level, and standardized valence ratings for each stimulus are available (Vasa, Carlino, London, & Min, 2006). The database has begun to be used in experimental paradigms to compare children's reactions to positive, negative, neutral, and threatening stimuli (Rubinsten & Tannock, 2010).

The IADS (Bradley & Lang, 1999a, 1999b) also does not include norms for youth, although sounds from this database have occasionally been used with children. For instance, Erlich, Lipp, and Slaughter (2013) incorporated pleasant and fear-relevant IADS stimuli into a study of infant processing of emotional sounds and found that infants show similar attentional and physiological responding to fear-relevant sounds, when compared to adults. A strength of using auditory, nonlinguistic, emotional stimuli with youth is that these stimuli do not rely on language development and therefore can be used with research in very young children.

Due to developmental differences in language abilities, emotional Stroop tasks are sometimes modified to use pictures rather than words among youth (Eschenbeck, Kohlmann, Heim-Dreger, Koller, & Leser, 2004), although more traditional word Stroop tasks are also commonly used with older youth (Dalgleish et al., 2003). In fact, research by Kindt and Brosschot (1999) suggests that Stroop tasks using words, rather than pictures, more effectively elicit fear-relevant responding to threat stimuli in children 8-12 years of age.

Script-driven imagery tasks have not been previously used in youth. Recently, empirical research has begun to evaluate the effectiveness of a script-driven imagery procedure among trauma-exposed adolescents, however (Mischel & Leen-Feldner, 2013). Developmental considerations associated with adapting this procedure for use with adolescents have included increased experimenter assistance in script development. In order to ensure that trauma scripts for this procedure include bodily sensations, action tendencies, emotional responses, and rich sensory information, researchers assist adolescents in identifying and integrating those topics into the script. Additionally, adolescents often require assistance in putting trauma-related events, emotional responses, and physiological responses into a sequence, and they may report this information in a disorganized manner without researcher assistance.

Some methodological differences between adult and youth emotion-elicitation procedures exist due to ethical and safety concerns specific to children. For example, only one group of researchers has used carbon dioxide-enriched air (CO_2) breathing tasks with youth (Pine et al., 1998, 2000, 2005). Compared to adult studies, which use either prolonged 5% CO_2-enriched air breathing or short-term 35% CO_2-enriched air breathing (Cavallini, Perna, Caldirola, & Bellodi, 1999), research with children has only used a 5% CO_2-enriched air breathing task. An alternative to the CO_2 breathing procedure, voluntary hyperventilation, is more commonly used with youth to elicit fear related to panic-like symptoms. Voluntary hyperventilation procedures with youth typically involve a 2-3 min breathing task during which children are instructed to breathe at a rate of 30 breaths per minute (Leen-Feldner, Feldner, Berstein, McCormick, & Zvolensky, 2005; Leen-Feldner, Feldner, Tull, Roemer, & Zvolensky, 2006). Although adult procedures are similar, they are typically longer in duration and have repeated exercises, with participants often being instructed to engage in fast-paced breathing for a total of 8-15 min (Marshall et al., 2008).

Laboratory Methods: Additional Considerations

The introduction of increasingly sophisticated research technology, such as virtual reality, offers a number of advantages for emotion-focused research. Indeed, research suggests that naturalistic settings can ensure confidence in the external validity of data derived from investigations of emotion and its associated processes (Wilhelm & Grossman, 2010). Unfortunately, laboratory settings may not always be able to closely approximate the natural environment. Instead, they may balance a relative disadvantage for external validity, but increase internal validity due to more sophisticated assessment. The use of equipment that allows for the measurement of brain activity during emotion-elicitation tasks is a clear example of this balance between internal and external validity in the area of emotion.

The majority of the emotion-elicitation strategies discussed in this chapter can be conducted in the context of brain-scanning equipment, such as functional magnetic resonance imaging (fMRI), in order to measure emotion-related brain activity. An fMRI analysis can be used to evaluate brain activity in response to stimuli by measuring changes in blood flow, with increased blood flow indicating an increase in activity in that area of the brain. A study by Lanius et al. (2002) adapted a commonly used methodology for measuring psychophysiological response to emotionally relevant scripts delivered in a scanner context. Similarly, fMRI studies of emotion have used many of the same pictorial stimuli paradigms as those used in non-fMRI related research, such as the IAPS. Nonetheless, the fMRI scanner context and properties of the neurovasculature (i.e., nerves and blood vessels) response necessitate special considerations when designing such experiments.

Although the principle consideration of an experiment conducted in an fMRI scanner is the same as a non-fMRI experiment, such that the task evokes the emotion of interest, researchers must take additional measures to ensure that the data are acquired safely and amenable to analysis. An fMRI scanner uses very strong magnetic fields, some of which cannot be turned off without great expense, to obtain measurements of brain function. Therefore, proper precautions must be taken to ensure a participant is free of any magnetic material prior to placing him or her in the scanner. The fMRI scanner is also a loud and confined space, and subjects sensitive to sound or prone to claustrophobia are unable to participate, unless an open fMRI scanner is available. The loudness of the scanner context also makes the use of audio stimuli challenging, though not impossible. Additionally, head motion can create intractable noise in the signal so participants must be able to remain motionless for the duration of the scan, which would preclude traditional behavior approach/avoidance tests in a scanner context or other paradigms necessitating or measuring overt behavioral indices of emotion. Although the vast majority of individuals are able to tolerate the scanning environment, for the aforementioned reasons, the duration of the tasks is held to as short as possible, while still providing sufficient statistical power for analyses.

The use of fMRI also requires the inclusion of either control groups (who do not receive the emotion-eliciting stimuli) or the use of both control and experimental stimuli for the same participant. To this end, any process of interest must be isolated as much as possible using control-experimental contrasts. For example, if one were interested in the brain's response to disgusted faces, a valid design would show the participant disgusted faces, but also neutral faces to control for the fact that the participant is looking at a face. The neural response to the disgusted and the neutral faces would then be compared to obtain a better-controlled measure of how the brain responds to disgust.

TRAJECTORIES OF EMOTIONAL EXPERIENCE

Selection of Assessment Time Point

Although the discrete emotions perspective draws attention to the brief nature of emotion, a limited view of this theory may inadvertently suggest that an emotion occurs instantaneously and can be measured at only one time point. On the contrary, considering that the experience of emotion relates to a multitude of response tendencies across multiple systems (e.g., neural, psychological), the unfolding of emotion is a dynamic process with multiple opportunities for assessment. Recently, researchers have lamented that some areas of emotion research have focused too exclusively on peak emotional intensity, thereby artificially chunking emotions into static states (Raz et al., 2012). An additional consideration in determining both the when and the how of assessing emotion is the interindividual variance in affective style (Davidson, Jackson, & Kalin, 2000). Affective style refers to individual differences in the experience of emotion in relation to awareness (Swinkels & Giuliano, 1995), intensity (Larsen, Diener, & Cropanzano, 1987), and reactivity (Gohm & Clore, 2000). Each of these facets of an emotion pertains to the temporal course of an experienced emotion. For instance, research has suggested that the subjective experience of the intensity of an emotion is correlated with its duration (Sonnemans & Frijda, 1994).

Attentional focus plays a pivotal role in both the temporal nature of emotional experience and the modulation of emotional responding. Indeed, the degree of attention afforded to a stimulus might be involved in the generation of emotion, its maintenance, and its alteration (Gross, 2001). Thus, attentional deployment has implications for the rise time, peak, and recovery from an emotional experience. One study making this assertion found that sadness ratings decreased more slowly when participants were instructed to focus on their own emotions while watching a sadness-inducing film (Joormann & Siemer, 2004). Somewhat in contrast with the theory that attentional focus is required for emotional responding, other studies indicate that emotional responses can occur when stimuli are not the focus of attention. For instance, intense emotional stimuli can produce physiological responding even in the presence of a distracting task (Muller, Anderson, & Keil, 2008). Taken together, evidence supports an integrated theory such that stimuli can elicit an emotion in the absence of direct attentional focus, but that the duration of the emotional experience is related to the degree of attention on the emotion subsequent to the initial response (Freund & Keil, 2013).

Converging research in the areas of affective style and the temporal course of emotions is referred to as affective chronometry (Davidson, 1998) and focuses on individual differences in (1) enhancement rate, or how rapidly affect increases from a baseline time point; and (2) decay rate, or how rapidly affect decreases from its peak. Related research employs a broader perspective of time course and refers to emotion dynamics as the latency, rise time, magnitude, duration, and offset of responses across behavioral, experiential, and psychological domains (Gross, 2001). Incorporating individual differences in emotional experience, the more broad temporal course of emotion, and the domains in which an emotion is expressed (or not) offers researchers the potential to assess emotion across several times points, using a variety of assessment strategies.

Selection of Assessment Strategy

Physiological Assessment

Physiological assessment strategies have been utilized to measure arousal associated with an emotional experience expressed by changes in the autonomic nervous system (ANS). The ANS is the primary system evaluated by physiological measurements in the context of emotional responding. Activation of the ANS in response to an emotionally relevant stimulus relates to responses in both the electrodermal (i.e., sweat glands) and cardiovascular (i.e., blood circulatory system) areas. Two of the most commonly utilized physiological measurements of emotion are skin conductance and heart rate. Both skin conductance and heart rate can be readily assessed using nonintrusive instruments that often include the topical placement of electrodes to different areas of the body. Typically, a baseline period is established wherein the individual's physiological arousal is measured prior to stimulus presentation to allow a point from which to measure change. Indeed, physiological techniques are useful in assessing emotions because of their documented relations to arousal. A number of studies have demonstrated that skin conductance level increases in a linear and systematic fashion in relation to the rated arousal of emotional stimuli (e.g., Lang, Greenwald, Bradley, & Hamm, 1993). Further, heart rate acceleration has been associated with increased arousal in response to both positive and negative stimuli (Vrana & Lang, 1990). Interestingly, and particularly relevant to the time course of emotion, the impact of emotional arousal on electrodermal responses, such as skin conductance, tends to be positively correlated, whereas cardiovascular responses, such as heart rate, tend to habituate as the duration of the emotion increases (Brouwer, van Wouwe, Muhl, van Erp, & Toet, 2013). Skin conductance measures may therefore provide useful information regarding the enhancement rate of an emotion as arousal increases, whereas arousal effects on heart rate changes may rely on the valence of the elicited emotion. Finally, physiological responding may also be useful in determining the offset of an emotion as arousal-related responding begins to return to a previously established baseline.

Self-Report Assessment

Although the retrospective self-report of emotional experiences has been criticized for its reliance on accurate memory for distal emotional events, real-time or immediate self-report has been posited to provide important information related to the subjective experience of an emotion. For instance, ecological momentary assessment (EMA) involves the use of daily monitoring of target events (e.g., emotions) in order to minimize retrospective recall bias and allow multiple within-subject observations (Shiffman, Stone, & Hufford, 2008). The use of electronic devices in the context of EMA can even provide the advantage of digitized reminders to self-report. Generally, the self-report assessment of emotion involves requesting individuals to rate the intensity of their emotion at its peak moment. Importantly, it should be recognized that subjective intensity is a complex construct that has been empirically demonstrated to encompass five dimensions: (1) duration of the emotion and the delay of its onset and peak, (2) perceived bodily changes, (3) the recollection and re-experience of the emotion, (4) the strength of the action tendency and resulting behavior, and (5) belief changes and influence on long-term behavior (Sonnemans & Frijda, 1994). Whether individuals take into account each of these dimensions when rating their emotions is uncertain, and, in fact, individual differences account for significant variability in emotional intensity (Sonnemans & Frijda, 1995).

Self-reported ratings of peak emotion intensity often utilize rating scales for multiple emotions. Visual analog scales (VASs; Freyd, 1923) and self-assessment manikins (SAMs; Bradley & Lang, 1994) represent two common self-report ratings of emotion. VASs require individuals to rate their emotion by indicating a position along a continuous line between two end-points, often rating emotions on a scale from 0 to 100. Anchors for each end-point are often provided so that individuals can indicate the currently experienced emotion within the context of intensity (e.g., 0 = no sadness at all, 100 = the most sad you could ever imagine being). The SAM utilizes pictorial representations of the pleasure (i.e., valence), arousal, and dominance (i.e., degree of control in the situation) associated with an emotion experience. The ability to relatively quickly rate emotions using VAS and SAM ratings offers the advantage of measuring emotions in real-time as the emotion peaks, or shortly after stimulus presentation. An additional advantage inherent in self-report measures of peak emotion lies in their low cost and the ability to utilize this method within a variety of contexts (e.g., laboratory and therapy setting) and among a variety of populations (e.g., clinical and nonclinical).

fMRI Assessment

fMRI is also able to aid our understanding of the time course of affective processes by measuring the time course of neural regions implicated in emotion. Goldin, McRae, Ramel, and Gross (2008) presented participants with 15-s film clips depicting either neutral nature scenes (to obtain a baseline) or highly negative disgust-inducing clips, such as surgical procedures, vomiting, and animal slaughter. They then examined the time courses of activity in amygdalae and insulae, brain regions that respond to the negative stimuli (Schienle, Schafer, Stark, Walter, & Vaitl, 2005). To define these time courses, mean BOLD signals (the flow of oxygen to different areas of the brain) were assessed at each measurement point following presentation of the stimulus. This time course allows for examining chronological patterns of change in neural response to emotion. For the negative stimuli, compared to the neutral stimuli, both left and right amygdalae showed a greater initial response, which was maintained throughout the presentation of the stimuli and did not decay until the stimuli were removed. Activity in both insulae was initially only slightly higher for the negative stimuli but increased throughout the duration of the exposure. These findings may suggest that the emotional response to negatively valenced stimuli is maintained throughout exposure, as evidenced by the constant and heightened amygdala activity. The increasing insula activity suggests that individuals may actually become more aware (Craig, 2009) of this negative emotion the longer the stimulus is presented (at least up to 15 s). These findings highlight the viability of fMRI as a research tool for learning about the time course of affective processes.

Behavioral Assessment

Behavior is a primary component of emotion-relevant response tendencies; thus, behavioral measures of responding inform the assessment of emotion. The coding of facial expressions is one method utilized to behaviorally measure the experience of emotion. The facial action coding system (FACS; Ekman & Friesen, 1978) is a psychometrically sound behavioral measure of expressed emotion that measures movement of individual facial muscles, including movements underlying expressions of basic emotions (i.e., fear, disgust, sadness, happiness, surprise, and anger). The FACS therefore allows for the coding and differentiation of overlapping negative affective states. The movements of the facial muscles can be scored at the onset of the emotion, the apex (i.e., when the movement was held at the highest intensity), or the offset. Thus, the FACS can be a useful method for measuring an emotion throughout the entirety of its duration. Finally, eye-tracking methodology has been utilized as a behavioral indicator of attention toward emotion-relevant stimuli. Indeed, extensive research has demonstrated a clear attention bias for threat in visual detection tasks (LoBue, Matthews, Harvey, & Stark, 2014), suggesting that negatively valenced stimuli may be processed more quickly than nonthreatening, positively valenced stimuli.

Developmental Considerations in Assessment

Similar to adult research, developmental research often utilizes self-report measures of emotion. VASs (Freyd, 1923) and SAMs (Bradley & Lang, 1994) are commonly used with youth, and research suggests that children's ratings on these measures resemble adult patterns of responding (McManis et al., 2001). Indeed, research suggests that children are capable of self-reporting emotional states (Ialongo, Edelsohn, & Kellam, 2001). Furthermore, the number and range of words to describe emotion that youth understand and use increases with age (Ridgeway, Waters, & Kuczaj, 1985), and even young children appear to begin using and understanding emotion description words around age 2-3 years (Betherton & Beeghly, 1982). Young children may provide more extreme affective response reports than older children or adults, however (Chambers & Johnston, 2002). Thus, other-reports by teachers, parents, or siblings may be more important sources of information regarding very young children and infants (Achenbach, McConaughy, & Howell, 1987; Johnston & Murray, 2003).

The physiological and neurological assessment of child affective responses may differ somewhat from adults. In regard to physiological measurement, research suggests that, although youth display increased arousal in response to fear or stress in patterns similar to adults, they may produce different patterns of heart rate deceleration or variability (Silvetti, Drago, & Ragonses, 2001). Furthermore, difficulty with children sitting still for long periods of time can cause artifacts in data due to movement, or child participants may require shorter baseline and measurement periods. Neurological methodology may also differ in the context of youth emotional responding. For example, neural signals of regulation appear shortly after they do for adults and also last for shorter periods of time (Dennis & Hajcak, 2009). Additionally, in comparison to adults, youth display different areas and patterns of brain activation when viewing emotional stimuli (Batty & Taylor, 2006). For instance, many studies indicate that adults show increased amygdala activation for fearful faces compared to neutral faces, and children tend to demonstrate greater amygdala activation for neutral, compared to fearful, faces (Thomas et al., 2001). Finally, the ongoing development of the prefrontal cortex throughout adolescence corresponds to an increasing amount of activity in the frontal areas of the brain when presented with emotional stimuli (Yurgelun-Todd & Killgore, 2006). Therefore, the time period and location of focus for neuroimaging data analysis in youth may differ from those for adults.

ELICITATION METHOD AND ASSESSMENT STRATEGY CONSIDERATIONS

Matching Method to Emotion

Two important methodological considerations are crucial for developing a thorough methodology aimed at better understanding the phenomena of emotion: (1) the assessment method chosen to elicit the emotion and (2) the time at which to assess the emotion. In regards to the first consideration, both the discrete emotions perspective and the dimensional theory of emotion are highly applicable. Studies relying on the discrete emotions perspective suggest that the use of briefly presented stimuli may be particularly useful in eliciting a discrete emotion (e.g., fear). Further, consistent with the dimensional theory of emotion, briefly presented stimuli, such as pictures, have been shown to effectively elicit a specified valence and arousal. However, if a researcher aims to elicit a range of emotions (e.g., negative emotions) at varying levels of intensity, or a single emotion at a high intensity, then briefly presented stimuli may not be sufficient. In this case, script-driven imagery and stressful films are potentially useful for eliciting more than one discrete emotion (e.g., sadness and fear), with the added advantage of experimentally manipulating the degree to which one is elicited compared to another (Badour

& Feldner, 2013). Further, because longer-lasting stimuli may undergo more prolonged or extensive processing due to their longer duration, they may potentially extend the higher range of arousal intensity that can be achieved.

The second important consideration is associated with methodology and relates to the timing of assessment. Here, we should consider emotion dynamics because they relate to the time course of an emotional response. Although briefly presented stimuli may effectively elicit a peak emotion, the expected duration of the emotional response may be time-limited. Thus, the peak of the elicited discrete emotion may be readily assessed, but an evaluation of rise and decay may not be feasible. In reference to the dimensional aspects of emotion, briefly presented or relatively less intense stimuli may diminish the discernable rise and decay of the emotion. In fact, relatively less intense stimuli that are associated with a lower degree of arousal may create a ceiling effect in which an emotional response does not rise significantly above baseline. If a methodology requires the assessment of emotion across its duration, that brief, lower intensity stimuli will likely not be sufficient.

The Impact of Sleep on Both Emotion Elicitation and Assessment

Mounting research suggests that sleep-deprived individuals evidence compromised coping skills (e.g., Killgore et al., 2008). Changes in emotional reactivity resulting from sleep deprivation appear to be associated with the amount of deprivation. The duration of sleep deprivation and the stage of sleep that is interrupted provide important information that can be used to better understand why sleep and emotion are so highly associated. Rapid eye movement (REM) sleep has been theorized to play an important role in emotion, such that brain systems involved in emotion are selectively activated during this phase of sleep (Rosales-Lagarde et al., 2012). In fact, empirical research has suggested that the interruption of REM sleep, compared to the interruption of non-REM sleep, differentially impacts emotional reactivity (Rosales-Lagarde et al., 2012). In a paradigm utilizing threatening pictures from the IAPS in an fMRI scanner, this study suggested that those with REM-sleep deprivation evidenced greater emotional responding compared to those whose non-REM sleep had been interrupted (Rosales-Lagarde et al., 2012). These results may provide evidence for the hypothesis that REM-sleep deprivation interferes with emotion regulation, such that reactivity to a stimulus may not decrease even if it has been previously encountered and regulated.

Finally, in regard to assessment strategy, sleep deprivation has been demonstrated to impact self-report, as well as neurological, physiological, and behavioral responding. For instance, sleep-deprived college students have been shown to perceive and rate their performance and concentration on a cognitive task higher overall compared to non-sleep-deprived students (Pilcher & Walters, 1997). This suggests that sleep deprivation may result in self-report biases. Also, research has demonstrated increased neural reactivity to negative stimuli following sleep deprivation; for instance, increased activity is evident in the amygdala (Yoo, Hu, Gujar, Jolesz, & Walker, 2007), a brain structure associated with threat detection. Laboratory studies utilizing physiological measurements have also shown increased reactivity. One such study utilized pupillary responses as a physiological indicator of reactivity to negative pictures, demonstrating that a full night of sleep deprivation was associated with larger pupil diameter—an indicator of increased reactivity in response to threatening information (Franzen, Buysse, Dahl, Thompson, & Siegle, 2009). In contrast, research also has demonstrated blunted reactivity to emotional cues following sleep deprivation. For instance, a recent study demonstrated that facial reactions in responses to stimuli were slower among undergraduates restricted to a 4-h night of sleep (Schwarz et al., 2013).

CONCLUSIONS

Technological advances continue to advance our understanding of the dynamic nature of emotional experiences. Digital devices allow for fast, real-time assessment, and virtual reality lets us create entire worlds immersed in emotionally relevant stimuli. Although this is indeed an exciting time for researchers utilizing laboratory methods, determining which method to use may remain a difficult task. However, the variety of elicitation strategies should be viewed as complimentary, not competing, because each provides incremental value to the understanding of an emotional experience. Thus, when developing a research protocol, an understanding of the available approaches can inform which technique will best address the question of interest. This chapter provides an overview of the commonly used tasks to guide this decision. Finally, with emerging research suggesting the importance of emotion dynamics across the time course and the overlapping but distinct variance influencing an emotional experience by subjective, physiological, and behavioral response domains, the researcher must consider multimodal assessments whenever possible. Ideally, a greater degree of methodological consistency across research designs will contribute to a more parsimonious theory of emotional experiences in the future.

References

Achenbach, T. M., McConaughy, S. H., & Howell, C. T. (1987). Child/adolescent behavioral and emotional problems: Implications of cross-informant correlations for situational specificity. *Psychological Bulletin, 101,* 213–232.

Algom, D., Chajut, E., & Lev, S. (2004). A rational look at the emotional Stroop phenomena: A generic slowdown, not a Stroop effect. *Journal of Experimental Psychology: General, 133,* 323–338. http://dx.doi.org/10.1037/0096-3445.133.3.323.

Ali, N., & Cimino, C. (1997). Hemispheric lateralization of perception and memory for emotional verbal stimuli in normal individuals. *Neuropsychology, 11,* 114–125. http://dx.doi.org/10.1037/0894-4105.11.1.114.

American Psychiatric Association. (1994). *Diagnostic and statistical manual of mental disorders* (4th ed.). Washington, DC: American Psychiatric Association.

Badour, C., & Feldner, M. (2013). Trauma-related reactivity and regulation of emotion: Associations with posttraumatic stress symptoms. *Journal of Behavior Therapy and Experimental Psychiatry, 44,* 69–76. http://dx.doi.org/10.1016/j.jbtep.2012.07.007.

Badour, C., Feldner, M., Blumenthal, H., Bujarski, S., Leen-Feldner, E., & Babson, K. (2012). Specificity of peritraumatic fear in predicting anxious reactivity to a biological challenge among traumatic event-exposed adolescents. *Cognitive Therapy Research, 36,* 397–406. http://dx.doi.org/10.1007/s10608-011-9380-0.

Barlow, D. H. (2004). *Anxiety and its disorders: The nature and treatment of anxiety and panic* (2nd ed.). New York: The Guilford Press.

Batty, M., & Taylor, M. J. (2006). The development of emotional face processing during childhood. *Developmental Science, 9,* 207–220. http://dx.doi.org/10.1111/j.1467-7687.2006.00480.x.

Bauer, M. R., Ruef, A. M., Pineles, S. L., Japuntich, S. J., Macklin, M. L., Lasko, N. B., et al. (2013). Psychophysiological assessment of PTSD: A potential research domain criteria construct. *Psychological Assessment, 25,* 1037–1043. http://dx.doi.org/10.1037/a0033432.

Baumgartner, T., Esslen, M., & Jäncke, L. (2005). From emotion perception to emotion experience: Emotions evoked by pictures and classical music. *International Journal of Psychophysiology, 60,* 34–43. http://dx.doi.org/10.1016/j.ijpsycho.2005.04.007.

Beckham, J., Dennis, M., McClernon, J., Mozley, S., Collie, C., & Vrana, S. (2007). The effects of cigarette smoking on script-driven imagery in smokers with and without posttraumatic stress disorder. *Addictive Behaviors, 32,* 2900–2915. http://dx.doi.org/10.1016/j.addbeh.2007.04.026.

Betherton, I., & Beeghly, M. (1982). Talking about internal states: The acquisition of an explicit theory of mind. *Developmental Psychology, 19,* 906–921.

Blechert, J., Wilhelm, F., Meuret, A., Wilhelm, E., & Roth, W. (2013). Experiential, autonomic, and respiratory correlates of CO_2 reactivity in individuals with high and low anxiety sensitivity. *Psychiatry Research, 209,* 566–573. http://dx.doi.org/10.1016/j.psychres.2013.02.010.

Bradley, M., & Lang, P. J. (1994). Measuring emotion: The self-assessment manikin and the semantic differential. *Journal of Behavior Therapy and Experimental Psychiatry, 25,* 49–59.

Bradley, M., & Lang, P. J. (1999a). Affective Norms for English Words (ANEW): Stimuli, instruction manual and affective ratings. Technical report C-1. Gainsville, FL: The Center for Research in Psychophysiology, University of Florida.

Bradley, M., & Lang, P. J. (1999b). International Affective Digitized Sounds (IADS): Stimuli, instruction manual and affective ratings. Technical report B-2. Gainesville, FL: The Center for Research in Psychophysiology, University of Florida.

Brenner, E. (2000). Mood induction in children: Methodological issues and clinical implications. *Review of General Psychology, 4,* 264–283. http://dx.doi.org/10.1037/1089-2680.4.3.264.

Brouwer, A., van Wouwe, N., Muhl, C., van Erp, J., & Toet, A. (2013). Perceiving blocks of emotional pictures and sounds: Effects on physiological variables. *Frontiers in Human Neuroscience, 7,* 1–10. http://dx.doi.org/10.3389/fnhum.2013.00295.

Bunaciu, L., Leen-Feldner, E. W., Blumenthal, H., Knapp, A. A., Badour, C. L., & Feldner, M. T. (2014). An experimental test of the effects of parental modeling on panic-relevant escape and avoidance among adolescents. *Behavior Therapy, 45,* 517–529. http://dx.doi.org/10.1016/j.beth.2014.02.011.

Cacioppo, J. T., & Gardner, W. L. (1999). Emotion. *Annual Review of Psychology, 50,* 191–214. http://dx.doi.org/10.1146/annurev.psych.50.1.191.

Campos, J. J., Frankel, C. B., & Camras, L. (2004). On the nature of emotion regulation. *Child Development, 75,* 377–394. http://dx.doi.org/10.1111/j.1467-8624.2004.00681.x.

Cavallini, M. C., Perna, G., Caldirola, D., & Bellodi, L. (1999). A segregation study of panic disorder in families of panic patients responsive to the 35% CO_2 challenge. *Biological Psychiatry, 46,* 815–820. http://dx.doi.org/10.1016/S0006-3223(99)00004-9.

Chambers, C. T., & Johnston, C. (2002). Developmental differences in children's use of rating scales. *Journal of Pediatric Psychology, 27,* 27–36. http://dx.doi.org/10.1093/jpepsy/27.1.27.

Charney, D. S., Henider, G. R., & Breier, A. (1984). Noradrenergic function in panic anxiety. Effects of yohimbine in healthy subjects and patients with agoraphobia and panic disorder. *Archives of General Psychiatry, 41,* 751–763.

Charney, D. S., Henider, G. R., & Jatlow, P. (1985). Increased anxiogenic effects of caffeine in panic disorders. *Archives of General Psychiatry, 42,* 233–243.

Cowley, D., & Arana, G. (1990). The diagnostic utility of lactate sensitivity in panic disorder. *Archives of General Psychiatry, 47,* 277–284. http://dx.doi.org/10.1001/archpsyc.1990.01810150077012.

Craig, A. D. (2009). How do you feel—Now? The anterior insula and human awareness. *Nature Reviews. Neuroscience, 10,* 59–70. http://dx.doi.org/10.1038/nrn2555.

Cuthbert, B., Bradley, M., & Lang, P. (1996). Probing picture perception: Activation and emotion. *Psychophysiology, 33,* 103–111. http://dx.doi.org/10.1111/j.1469-8986.1996.tb02114.x.

Dalgleish, T., Taghavi, R., Neshat-Doost, H., Moradi, A., Canterbury, R., & Yule, W. (2003). Patterns of processing bias for emotional information across clinical disorders: A comparison of attention, memory, and prospective cognition in children and adolescents with depression, generalized anxiety, and posttraumatic stress disorder. *Journal of Clinical Child & Adolescent Psychology, 32,* 10–21. http://dx.doi.org/10.1207/15374420360533022.

Davidson, R. J. (1994). On emotion, mood, and related affective constructs. In P. Ekman, & R. J. Davidson (Eds.), *The nature of emotion: Fundamental questions* (pp. 51–55). New York: Oxford University Press.

Davidson, R. (1998). Affective style and affective disorders: Perspectives from affective neuroscience. *Cognition & Emotion, 12,* 307–330. http://dx.doi.org/10.1080/026999398379628.

Davidson, R., Jackson, D., & Kalin, N. (2000). Emotion, plasticity, context, and regulation: Perspectives from affective neuroscience. *Psychological Bulletin, 126,* 890–909. http://dx.doi.org/10.1037/0033-2909.126.6.890.

de Groot, J., Semin, G., & Smeets, M. (2014). I can see, hear, and smell your fear: Comparing olfactory and audiovisual media in fear communication. *Journal of Experimental Psychology. General, 143,* 825–834. http://dx.doi.org/10.1037/a0033731.

Dennis, T. A., & Hajcak, G. (2009). The late positive potential: A neurophysiological marker for emotion regulation in children. *Journal of Child Psychology and Psychiatry, 50,* 1373–1383. http://dx.doi.org/10.1111/j.1469-7610.2009.02168.x.

Difede, J., & Hoffman, H. (2002). Virtual reality exposure therapy for World Trade Center posttraumatic stress disorder: A case report. *Cyberpsychology & Behavior, 5,* 529–535. http://dx.doi.org/10.1089/109493102321018169.

Ekman, P., & Friesen, W. (1978). *The facial action coding system: A technique for the measurement of facial movement.* Palo Alto, CA: Consulting Psychologists Press.

Erlich, N., Lipp, O., & Slaughter, V. (2013). Of hissing snakes and angry voices: Human infants are differentially responsive to evolutionary fear-relevant sounds. *Developmental Science, 16*, 894–904.

Eschenbeck, H., Kohlmann, C., Heim-Dreger, U., Koller, D., & Leser, M. (2004). Processing bias and anxiety in primary school children: A modified emotional stroop colour-naming task using pictorial facial expressions. *Psychology Science, 46*, 451–465.

Estes, Z., & Adelman, J. (2008). Automatic vigilance for negative words is categorical and general. *Emotion, 8*, 453–457. http://dx.doi.org/10.1037/a0012887.

Feldner, M., Vujanovic, A., Gibson, L., & Zvolensky, M. (2008). Posttraumatic stress disorder and anxious and fearful reactivity to bodily arousal: A test of the mediating role of nicotine withdrawal severity among daily smokers in 12-hr nicotine deprivation. *Experimental and Clinical Psychopharmacology, 16*, 144–155. http://dx.doi.org/10.1037/1064-16.2.144.

Franzen, P., Buysse, D., Dahl, R., Thompson, W., & Siegle, G. (2009). Sleep deprivation alters pupillary reactivity to emotional stimuli in healthy young adults. *Biological Psychology, 80*, 300–305. http://dx.doi.org/10.1016/j.biopsycho.2008.10.010.

Freund, A., & Keil, A. (2013). Out of mind, out of heart: Attention affects duration of emotional experience. *Cognition & Emotion, 27*, 549–557.

Frewen, P., Dozois, D., Neufeld, R., Densmore, M., Stevens, T., & Lanius, R. (2011). Neuroimaging social emotional processing in women: fMRI study of script-driven imagery. *Social Cognitive and Affective Neuroscience, 6*, 375–392. http://dx.doi.org/10.1093/scan/nsq047.

Freyd, M. (1923). The graphic rating scale. *Journal of Educational Psychology, 14*, 83–102.

Frijda, N. (1986). *The emotions.* Cambridge: Cambridge University Press.

Frohlich, E. D., Tarazi, R. C., & Dustan, H. P. (1969). Hyperdynamic beta-adrenergic circulatory state: Increased beta-receptor responsiveness. *Archives of Internal Medicine, 123*, 1–7. http://dx.doi.org/10.1001/archinte,1969.00300110003001.

Gohm, C., & Clore, G. (2000). Individual differences in emotional experience: Mapping available scales to processes. *Personality and Social Psychology Bulletin, 26*, 679–697. http://dx.doi.org/10.1177/0146167200268004.

Goldin, P., McRae, K., Ramel, W., & Gross, J. (2008). The neural basis of emotion regulation: Reappraisal and suppression of negative emotion. *Biological Psychiatry, 63*, 577–586. http://dx.doi.org/10.1016/j.biopsych.2007.05.031.

Gross, J. (1998). Antecedent- and response-focused emotion regulation: Divergent consequences for experience, expression, and physiology. *Journal of Personality and Social Psychology, 74*, 224–237. http://dx.doi.org/10.1037/0022-3514.74.1.224.

Gross, J. (2001). Emotion regulation in adulthood: Timing is everything. *Current Directions in Psychological Science, 10*, 214–219.

Hamann, S. (2012). Mapping discrete and dimensional emotions onto the brain: Controversies and consensus. *Trends in Cognitive Sciences, 16*, 458–466. http://dx.doi.org/10.1016/j.tics.2012.07.006.

Hawks, E., Blumenthal, H., Feldner, M., Leen-Feldner, E., & Jones, R. (2011). An examination of the relation between traumatic event exposure and panic-relevant biological challenge responding among adolescents. *Behavior Therapy, 42*, 427–438. http://dx.doi.org/10.1016/j.beth.2010.11.002.

Hermans, D., Vansteenwegen, D., & Eelen, P. (1999). Eye movement registration as a continuous index of attention deployment: Data from a group of spider anxious students. *Cognition & Emotion, 13*, 419–434.

Herz, R. S. (1996). A comparison of olfactory, tactile and visual as associated memory cues. *Chemical Senses, 21*, 614–615.

Hewig, J., Hagemann, D., Seifert, J., Gollwitzer, M., Naumann, E., & Bartussek, D. (2005). A revised film set for the induction of basic emotions. *Cognition & Emotion, 19*, 1095–1109. http://dx.doi.org/10.1080/02699930541000084.

Homan, P., Drevets, W., & Hasler, G. (2014). The effects of chatecholamine depletion on the neural response to fearful faces in remitted depression. *International Journal of Neuropsychopharmacology, 14*, 1–10.

Hopper, J., Frewen, P., Sack, M., Lanius, R., & vander Kolk, B. (2007). The responses Script-Driven Imagery Scale (RSDI): Assessment of state posttraumatic symptoms for psychobiological and treatment research. *Journal of Psychopathology and Behavioral Assessment, 29*, 249–268. http://dx.doi.org/10.1007/s10862-007-9046-0.

Ialongo, N. S., Edelsohn, G., & Kellam, S. (2001). A further look at the prognostic power of young children's reports of depressed mood and feelings. *Child Development, 72*, 736–747. http://dx.doi.org/10.1111/1467-8624.00312.

Johnston, C., & Murray, C. (2003). Incremental validity in the psychological assessment of children and adolescents. *Psychological Assessment, 15*, 496–507. http://dx.doi.org/10.1037/1040-3590.15.4.496.

Joormann, J., & Siemer, M. (2004). Memory accessibility, mood regulation, and dysphoria: Difficulties in repairing sad mood with happy memories? *Journal of Abnormal Psychology, 113*, 179–188. http://dx.doi.org/10.1037/0021-843X.113.2.179.

Kelvin, R. G., Goodyer, I. M., Teasdale, J. D., & Brechin, D. (1999). Latent negative self-schema and high emotionality in well adolescents at risk for psychopathology. *Journal of Child Psychology and Psychiatry, 40*, 959–968. http://dx.doi.org/10.1111/1469-7610.00513.

Khalfa, S., Isabelle, P., Jean-Pierre, B., & Manon, R. (2002). Event-related skin conductance responses to musical emotions in humans. *Neuroscience Letters, 328*, 145–149. http://dx.doi.org/10.1016/S0304-3940(02)00462-7.

Killgore, W., Kahn-Greene, E., Lipizzi, E., Newman, R., Kamimori, G., & Balkin, T. (2008). Sleep deprivation reduces perceived emotional intelligence and constructive thinking skills. *Sleep Medicine, 9*, 517–526. http://dx.doi.org/10.1016/j.sleep.2007.07.003.

Kindt, M., & Brosschot, J. (1999). Cognitive bias in spider-phobic children: Comparison of a pictorial and a linguistic spider Stroop. *Journal of Psychopathology and Behavioral Assessment, 21*, 207–220. http://dx.doi.org/10.1023/A:1022873331067.

Kossowsky, J., Wilhem, F., & Schneider, S. (2013). Responses to voluntary hyperventilation in children with separation anxiety disorder: Implications for the link to panic disorder. *Journal of Anxiety Disorders, 27*, 627–634. http://dx.doi.org/10.1016/j.janxdis.2013.08.001.

Lang, P., Bradley, M., & Cuthbert, B. (1999). *International Affective Picture System (IAPS): Technical manual and affective ratings*. Gainesville: University of Florida, Center for Research in Psychophysiology.

Lang, P., Greenwald, M., Bradley, M., & Hamm, A. (1993). Looking at pictures: Affective, facial, visceral, and behavioral reactions. *Psychophysiology, 30*, 261–273. http://dx.doi.org/10.1111/j.1469-8986.1993.tb03352.x.

Lanius, R., Williamson, P., Boksman, K., Densmore, M., Gupta, M., Neufeld, R., et al. (2002). Brain activation during script-driven imagery induced dissociative responses in PTSD: A functional magnetic resonance imaging investigation. *Biological Psychiatry, 52*, 305–311. http://dx.doi.org/10.1016/S0006-3223(02)01367-7.

Larsen, R., Diener, E., & Cropanzano, R. (1987). Cognitive operations associated with individual differences in affect intensity. *Journal of Personality and Social Psychology, 53*, 767–774. http://dx.doi.org/10.1037/0022-3514.53.4.767.

Leen-Feldner, E. W., Feldner, M. T., Berstein, A., McCormick, J. T., & Zvolensky, M. J. (2005). Anxiety sensitivity and anxious responding to bodily sensations: A test among adolescents using a voluntary hyperventilation challenge. *Cognitive Therapy and Research, 29*, 593–609. http://dx.doi.org/10.1007/s10608-005-3510-5.

Leen-Feldner, E. W., Feldner, M. T., Tull, M. T., Roemer, L., & Zvolensky, M. J. (2006). An examination of worry in relation to anxious responding to voluntary hyperventilation among adolescents. *Behaviour Research and Therapy, 44*, 1803–1809. http://dx.doi.org/10.1016/j.brat.2005.12.014.

Lejuez, C. W., Forsyth, J. P., & Eifert, G. H. (1998). Devices and methods for administering carbon dioxide-enriched air in experimental and clinical settings. *Journal of Behavior Therapy and Experimental Psychiatry, 29*, 239–248. http://dx.doi.org/10.1016/S005-7916(98)00018-4.

Lieberman, M. D., Eisenberger, N. I., Crockett, M. J., Tom, S. M., Pfeifer, J. H., & Way, B. M. (2007). Putting feelings into words: Affect labeling disrupts amygdala activity in response to affective stimuli. *Psychological Science, 18*, 421–428. http://dx.doi.org/10.1111/j.1467-9280.2007.01916.x.

LoBue, V., Matthews, K., Harvey, T., & Stark, S. L. (2014). What accounts for the rapid detection of threat? Evidence for an advantage in perceptual and behavioral responding from eye movements. *Emotion, 14*, 816–823. http://dx.doi.org/10.1037/a0035869.

Marshall, E. C., Zvolensky, M. J., Vujanovic, A. A., Gregor, K., Gibson, L. E., & Leyro, T. M. (2008). Reactivity to the voluntary hyperventilation challenge predicts distress tolerance to bodily sensations among daily cigarette smokers. *Experimental and Clinical Psychopharmacology, 16*, 313–321. http://dx.doi.org/10.1037/a0012752.

McManis, M., Bradley, M., Berg, K., Cuthbert, B., & Lang, P. (2001). Emotional reactions in children: Verbal, physiological, and behavioral responses to affective pictures. *Psychophysiology, 38*, 222–231. http://dx.doi.org/10.1017/S0048577201991140.

Mikels, J., Fredrickson, B., Larkin, G., Lindberg, C., Maglio, S., & Reuter-Lorenz, P. (2005). Emotional category data on images from the International Affective Picture System. *Behavior Research Methods, 37*, 626–630. http://dx.doi.org/10.3758/BF03192732.

Mischel, E. M., & Leen-Feldner, E. W. (2013). *Validation of a script-driven imagery procedure among traumatic event exposed adolescents.* (in preparation).

Muller, M., Anderson, S., & Keil, A. (2008). Time course of competition for visual processing resources between emotional pictures and foreground task. *Cerebral Cortex, 18*, 1892–1899. http://dx.doi.org/10.1093/cercor/bhm215.

Nutt, D., Glue, P., Lawson, C., & Wilson, S. (1990). Flumazenil provocation of panic attacks: Evidence for altered benzodiazepine receptor sensitivity in panic disorder. *Archives of General Psychiatry, 47*, 917–925. http://dx.doi.org/10.1001/archpsyc.1990.01810220033004.

Olatunji, B., Babson, K., Smith, R., Feldner, M., & Connolly, K. (2009). Gender as a moderator of the relation between PTSD and disgust: A laboratory test employing individualized script-driven imagery. *Journal of Anxiety Disorders, 23*, 1091–1097. http://dx.doi.org/10.1016/j.janxdis.2009.07.012.

Olatunji, B., Wolitzky-Taylor, K., Babson, K., & Feldner, M. (2009). Anxiety sensitivity and CO_2 challenge anxiety during recovery: Differential correspondence of arousal and perceived control. *Journal of Anxiety Disorders, 23*, 420–428. http://dx.doi.org/10.1016/j.janxdis.2008.08.006.

Orr, S. P., Lasko, N., Metzger, L., Berry, N., Ahern, C., & Pitman, R. (1998). Psychophysiologic assessment of women with posttraumatic stress disorder resulting from childhood sexual abuse. *Journal of Consulting and Clinical Psychology, 66*, 906–913. http://dx.doi.org/10.1037/0022-006X.66.6.906.

Orr, S. P., Pitman, R. K., Lasko, N. B., & Herz, L. R. (1993). Psychophysiological assessment of posttraumatic stress disorder imagery in World War II and Korean combat veterans. *Journal of Abnormal Psychology, 102*, 152–159.

Pilcher, J., & Walters, A. (1997). How sleep deprivation affects psychological variables related to college students' cognitive performance. *Journal of American College Health, 46*, 121–126. http://dx.doi.org/10.1080/07448489709595597.

Pine, D. S., Coplan, J. D., Papp, L. A., Klein, R. G., Martinez, J. M., Kovalenko, P., et al. (1998). Ventilatory physiology of children and adolescents with anxiety disorders. *Archives of General Psychiatry, 55*, 123–129. http://dx.doi.org/10.1001/archpsyc.55.2.123.

Pine, D. S., Klein, R. G., Coplan, J. D., Papp, L. A., Hoven, C. W., Martinez, J., et al. (2000). Differential carbon dioxide sensitivity in childhood anxiety disorders and nonill comparison group. *Archives of General Psychiatry, 57*, 960–967. http://dx.doi.org/10.1001/archpsyc.57.10.960.

Pine, D. S., Klein, R. G., Roberson-Nay, R., Mannuzza, S., Moulton, J. L., Woldehawariat, G., et al. (2005). Response to 5% carbon dioxide in children and adolescents. *Archives of General Psychiatry, 62*, 73–80. http://dx.doi.org/10.1001/archpsyc.62.1.73.

Pineles, S., Suvak, M., Liverant, G., Gregor, K., Wisco, B., Pitman, R., et al. (2013). Psychophysiological reactivity, subjective distress, and their associations with PTSD diagnosis. *Journal of Abnormal Psychology, 122*, 635–644. http://dx.doi.org/10.1037/a0033942.

Pitman, R. K., Orr, S. P., Forgue, D. F., de Jong, J. B., & Claiborn, J. M. (1987). Psychophysiologic assessment of posttraumatic stress disorder imagery in Vietnam Combat veterans. *Archives of General Psychiatry, 44*, 970–975.

Plichta, M. M., Gerdes, A. B. M., Alpers, G. W., Harnish, W., Brill, S., Wieser, M. J., et al. (2011). Auditory cortex activation is modulated by emotion: A functional near-infrared spectroscopy (fNIRS) study. *NeuroImage, 55*, 1200–1207. http://dx.doi.org/10.1016/j.neuroimage.2011.01.011.

Raz, G., Winetraub, Y., Jacob, Y., Kinreich, S., Maron-Katz, A., Shaham, G., et al. (2012). Portraying emotions at their unfolding: A multilayered approach for probing dynamics of neural networks. *NeuroImage, 60*, 1448–1461. http://dx.doi.org/10.1016/j.neuroimage.2011.12.084.

Redondo, J., Fraga, I., Padrón, I., & Piñera, A. (2008). Affective ratings of sound stimuli. *Behavior Research Methods, 40*, 784–790. http://dx.doi.org/10.3758/BRM.40.3.784.

Reiss, S. (1991). Expectancy model of fear, anxiety, and panic. *Clinical Psychology Review, 11*, 141–153. http://dx.doi.org/10.1016/0272-7358(91)90092-9.

Ridgeway, D., Waters, E., & Kuczaj, S. A. (1985). Acquisition of emotion-descriptive language: Receptive and productive vocabulary norms for ages 18 months to 6 years. *Developmental Psychology, 21*, 901–908. http://dx.doi.org/10.1037//0012-1649.21.5.901.

Rinck, M., Kwakkenbos, L., Dotsch, R., Wigboldus, D., & Becker, E. (2010). Attentional and behavioural responses of spider fearfuls to virtual spiders. *Cognition & Emotion, 24*, 1199–1206. http://dx.doi.org/10.1080/02699930903135945.

Robazza, C., Macaluso, C., & D'urso, V. (1994). Emotional reactions to music by gender, age, and expertise. *Perceptual and Motor Skills, 79*, 939–944. http://dx.doi.org/10.2466/pms.1994.79.2.939.

Robin, O., Alaoui-Ismaili, O., Dittmar, A., & Vernet-Maury, E. (1998). Emotional responses evoked by dental odors: An evaluation from autonomic parameters. *Journal of Dental Research, 77*, 1638–1646. http://dx.doi.org/10.1177/00220345980770081201.

Rosales-Lagarde, A., Armony, J., del Rio-Portilla, Y., Trejo-Martinez, D., Conde, R., & Corsi-Cabrera, M. (2012). Enhanced emotional reactivity after selective REM sleep deprivation in humans: An fMRI study. *Frontiers in Behavioral Neuroscience, 6*, 1–13. http://dx.doi.org/10.3389/fnbeh.2012.00025.

Rothbaum, B., Hodges, L., Ready, D., Graap, K., & Alarcon, R. (2001). Virtual reality exposure therapy for Vietnam veterans with posttraumatic stress disorder. *Journal of Clinical Psychiatry, 62*, 617–622. http://dx.doi.org/10.4088/JCP.v62n0808.

Rubinsten, O., & Tannock, R. (2010). Mathematics anxiety in children with developmental dyscalculia. *Behavioral and Brain Functions, 6*, 1–13. http://dx.doi.org/10.1186/1744-9081-6-46.

Sanderson, W. C., Rapee, R. M., & Barlow, D. H. (1989). The influence of illusion of control on panic attacks induced by 5.5% carbon dioxide enriched air. *Archives of General Psychiatry, 46*, 157–162. http://dx.doi.org/10.1001/archpsyc.1989.01810020059010.

Sano, K., Tsuda, Y., Sugano, H., Aou, S., & Hatanaka, A. (2002). Concentration effects of green odor on event-related potential (P300) and pleasantness. *Chemical Senses, 27*, 225–230. http://dx.doi.org/10.1093/chemse/27.3.225.

Schienle, A., Schafer, A., Stark, R., Walter, B., & Vaitl, D. (2005). Gender differences in the processing of disgust- and fear-inducing pictures: An fMRI study. *NeuroReport, 16*, 277–280. http://dx.doi.org/10.1097/00001756-200502280-00015.

Schmidt, N., & Trakowski, J. (2004). Interoceptive assessment and exposure in panic disorder: A descriptive study. *Cognitive and Behavioral Practice, 11*, 81–92. http://dx.doi.org/10.1016/S1077-7229(04)80010-5.

Schwarz, J., Popp, R., Haas, J., Zulley, J., Geisler, P., Alpers, G., et al. (2013). Shortened night sleep impairs facial responsiveness to emotional stimuli. *Biological Psychology, 93*, 41–44. http://dx.doi.org/10.1016/j.biopsycho.2013.01.008.

Seubert, J., Kellermann, T., Loughead, J., Boers, F., Brensinger, C., Schneider, F., et al. (2010). Processing of disgusted faces is facilitated by odor primes: A functional MRI study. *NeuroImage, 53*, 746–756. http://dx.doi.org/10.1016/j.neuroimage.2010.07.012.

Shiffman, S., Stone, A. A., & Hufford, M. R. (2008). Ecological momentary assessment. *Annual Review of Clinical Psychology, 4*, 1–32. http://dx.doi.org/10.1146/annurev.clinpsy.3.022806.091415.

Silva, J. (2011). International Affective Picture System (IAPS) in Chile: A cross-cultural adaptation and validation study. *Terapia Psicológica, 29*, 251–258.

Silvetti, M. S., Drago, F., & Ragonses, P. (2001). Heart rate variability in healthy children and adolescents is partially related to age and gender. *International Journal of Cardiology, 81*, 169–174. http://dx.doi.org/10.1016/S0167-5273(01)00537-X.

Sonnemans, J., & Frijda, N. (1994). The structure of subjective emotional intensity. *Cognition & Emotion, 8*, 329–350. http://dx.doi.org/10.1080/02699939408408945.

Sonnemans, J., & Frijda, N. (1995). The determinants of subjective emotional intensity. *Cognition & Emotion, 9*, 483–506. http://dx.doi.org/10.1080/02699939508408977.

Stevenson, R., & James, T. (2008). Affective auditory stimuli: Characterization of the International Affective Digitized Sounds (IADS) by discrete emotional categories. *Behavior Research Methods, 40*, 315–321. http://dx.doi.org/10.3758/BRM.40.1.315.

Stevenson, R., Mikels, J., & James, T. (2007). Characterization of the Affective Norms for English Words by discrete emotional categories. *Behavior Research Methods, 39*, 1020–1024. http://dx.doi.org/10.3758/BF03192999.

Swinkels, A., & Giuliano, T. (1995). The measurement and conceptualization of mood awareness: Monitoring and labeling one's mood states. *Personality and Social Psychology Bulletin, 21*, 934–949. http://dx.doi.org/10.1177/0146167295219008.

Thomas, K. M., Drevets, W. C., Whalen, P. J., Eccard, C. H., Dahl, R. E., Ryan, N. D., et al. (2001). Amygdala response to facial expressions in children and adults. *Biological Psychiatry, 49*, 309–316. http://dx.doi.org/10.1016/S0006-3223(00)01066-0.

van Zijderveld, G. A., TenVoorde, B., Veltman, D., & van Doornen, L. (1997). Cardiovascular, respiratory and panic reactions to epinephrine in panic disorder patients. *Biological Psychiatry, 41*, 249–251. http://dx.doi.org/10.1016/S0006-3223(96)00421-0.

van Zijderveld, G. A., Van Doornen, L., Orlebeke, O., Snieder, H., Van Faassen, I., & Tilders, F. (1992). The psychophysiological effects of adrenaline infusions as a function of trait anxiety and aerobic fitness. *Anxiety Research, 4*, 257–274. http://dx.doi.org/10.1080/08917779208248795.

Vasa, R. A., Carlino, A. R., London, K., & Min, C. (2006). Valence ratings of emotional and non-emotional words in children. *Personality and Individual Differences, 41*, 1169–1180. http://dx.doi.org/10.1016/j.paid.2006.03.025.

Verwoerd, J., Wessel, I., de Jong, P., & Nieuwenhuis, M. (2009). Preferential processing of visual trauma-film reminders predicts subsequent intrusive memories. *Cognition & Emotion, 23*, 1537–1551. http://dx.doi.org/10.1080/02699930802457952.

von Leupoldt, A., Rhode, J., Beregova, A., Thordsen-Sorensen, I., Nieden, J. Z., & Dahme, B. (2007). Films for eliciting emotional states in children. *Behavior Research Methods, 39*, 606–609. http://dx.doi.org/10.3758/BF03193032.

Vrana, S., & Lang, P. (1990). Fear imagery and the startle-probe reflex. *Journal of Abnormal Psychology, 99*, 189–197. http://dx.doi.org/10.1037/0021-843X.99.2.189.

Westermann, R., Spies, K., Stahl, G., & Hesse, F. (1996). Relative effectiveness and validity of mood induction procedure: A meta-analysis. *European Journal of Social Psychology, 26,* 557–580. http://dx.doi.org/10.1002/(SICI)1099-0992(199607)26:4<557::AID-EJSP769>3.0.CO;2-4.

Wilhelm, F., & Grossman, P. (2010). Emotions beyond the laboratory: Theoretical fundaments, study design, and analytic strategies for advanced ambulatory assessment. *Biological Psychology, 84,* 552–569. http://dx.doi.org/10.1016/j.biopsycho.2010.01.017.

Wilson, D. A. (2009). Olfaction as a model system for the neurobiology of mammalian short-term habituation. *Neurobiology of Learning and Memory, 92,* 199–205. http://dx.doi.org/10.1016/j.nlm.2008.07.010.

Yoo, S., Hu, P., Gujar, N., Jolesz, F., & Walker, M. (2007). A deficit in the ability to form new human memories without sleep. *Nature Neuroscience, 10,* 385–392. http://dx.doi.org/10.1038/nn1851.

Yurgelun-Todd, D. A., & Killgore, W. D. S. (2006). Fear-related activity in the prefrontal cortex increases with age during adolescence: A preliminary fMRI study. *Neuroscience Letters, 406,* 194–199. http://dx.doi.org/10.1016/j.neulet.2006.07.046.

Zhou, W., & Chen, D. (2009). Fear-related chemosignals modulate recognition of fear in ambiguous facial expressions. *Psychological Science, 20,* 177–183. http://dx.doi.org/10.1111/j.1467-9280.2009.02263.x.

CHAPTER 6

Methodological Considerations When Integrating Experimental Manipulations of Sleep and Emotion

Jared Minkel* and Samantha Phillips[†]

*Department of Psychiatry and Behavioral Sciences, Duke University Medical Center, Durham, North Carolina, USA
[†]School of Arts and Sciences, Duke University, Durham, North Carolina, USA

Interest in the interplay between emotion and sleep deprivation has never been higher. Sleep has been shown to play an important role in nearly every facet of emotional functioning studied to date, including emotional memory, emotional expression, interpretation of emotional signals, and biological substrates of emotion, with important implications for health and well-being. Much of our knowledge about the sleep-emotion connection has come from laboratory-based experiments, in which sleep duration is reduced arbitrarily. In the real world, sleep is curtailed for a reason, either because of a competing goal, such as work demands or entertainment, or because of a medical or psychological factor such as stress or pain. Any observed associations with sleep loss might be related to one of these factors, in addition to (or even instead of) sleep itself. Only by experimentally manipulating sleep can we be reasonably sure that the observed changes in emotion are indeed related to less sleep.

This is an exciting time to conduct such studies because both affective science and sleep research are fairly advanced but have developed independently. This history has produced a massive deficit in what we know about the relationships between sleep and emotion relative to what we can learn using currently available techniques. In this chapter, we review some of the methods used in sleep-deprivation and emotion research, with an emphasis on the unique challenges of applying them to a sleep-deprived sample. The chapter is organized around methods of emotion elicitation, measurement of emotional function, and experimental design. We end with a summary of key principles that are relevant across multiple methods.

EMOTION ELICITATION

Early studies of affect change during sleep deprivation simply measured mood at various times throughout an ongoing experiment. For example, one of the most cited studies of sleep and mood was conducted by Dinges et al. (1997), who restricted sleep to an average of about 5h per night and measured mood three times per day. The authors made no attempt to experimentally manipulate mood, however. Although this approach provided a useful background for ways in which sleep deprivation alters baseline emotional processes, more recent studies have emphasized experimentally manipulating emotional states as well as sleep duration. This newer approach provides a stronger scientific framework for exploring the ways in which sleep loss influences different aspects of emotional functioning. Below is an overview of some of the methods of emotional elicitation developed in affective science and now used within sleep-deprivation studies. We discuss these methods with special emphasis on additional considerations for using them within a sleep-deprivation experiment.

Visual Stimuli

Emotion can be elicited by many sensory stimuli, such as the smell of freshly baked cookies, the sound of a baby crying, or the taste of spoiled milk. Experimentally, visual stimuli, namely photographs and video clips, tend to be used because they can be reproduced with fidelity and presented for specific time periods. Below is a review of how such stimuli have been used in emotion research and special considerations for using these stimuli with sleep-deprived participants.

Photographs

Throughout affective science, the most common method for eliciting emotion is the use of still images, generally from the International Affective Picture System (IAPS; Lang, Bradley, & Cuthbert, 2008). This approach has many advantages, including standardized, validated stimuli that can easily be shared with other laboratories to replicate results. There are some unique challenges to using still images in a sleep-deprived sample, however. Perhaps the most important is that attentional problems can potentially confound the study because sleep-deprived participants may actually see less of the stimuli due to their eyes being closed more often than the eyes of rested participants. The longer the still image is displayed, the more likely a participant is to experience a microsleep or attentional lapse during the task. Many experimental protocols include significant intertrial intervals of a blank screen that may also be particularly problematic for sleep-deprived participants. Researchers can partially address this problem by using shorter display times for photographs, but an additional validity check, such as time-locked eye-tracking, is still recommended to ensure participants are viewing the stimuli.

Some researchers have argued that lapses of attention demonstrated during sleep deprivation are due to boredom (Harrison & Horne, 2000), but this hypothesis has not been rigorously evaluated and appears to be inconsistent with findings from emotion studies. Participants have demonstrated difficulties maintaining attention even in response to the most intense positive and negative photographs available in the IAPS database. These images include erotic depictions of nude models and graphic images of bodily injury and human waste. Participants did not report boredom while viewing these images, but they were clearly impaired, demonstrating slow eyelid closures and even sudden head drops indicative of sleep intrusions. Therefore, researchers must not assume that the interesting nature of an emotional task will prevent participants from falling asleep while completing it.

Film Clips

Short film clips can also elicit emotions. Although these clips are a bit more laborious to obtain and administer, they are more engaging than still images and less likely to be associated with lapses of attention or microsleep. In our previous pilot work, the participants who demonstrated significant problems maintaining wakefulness to photographs demonstrated no difficulty while watching videos a few minutes later. Robert Levenson and James Gross have developed a set of film clips that elicit specific emotional states such as anger, amusement, and disgust (Gross & Levenson, 1995; Rottenberg, Ray, & Gross, 2007). A set of six short film clips for eliciting emotions is also available for download from James Gross's laboratory at Stanford (http://spl.stanford.edu/resources.html), but there is not currently a film clip database equivalent to the IAPS database, which includes over 900 images.

Several potential pitfalls are associated with using film clips to elicit emotions. First, the intensity of the emotion elicited varies widely. For example, clips for eliciting disgust produce very intense responses, but those intended to elicit sadness are often less provocative. In addition, some of the film clips are now out-of-date and may produce unintended responses. One clip for eliciting anger comes from the movie "Witness" from 1985. The scene depicts some local boys bullying a group of Amish people, but the clothing and hairstyles of the bullies can elicit amusement. It may therefore be more effective to use clips from more contemporary movies. Clips that are unfamiliar to the participants tend to be the most effective at eliciting emotion, and foreign and independent films are often very useful resources in this regard.

A potential problem in using film clips to elicit emotion during a sleep-deprivation experiment is that participants are often allowed to watch movies and television throughout a night of extended wakefulness. Upon testing the following day, the sleep-deprived group may have been recently exposed to much more television than the control participants who were not given such an opportunity. Depending on what the sleep-deprived participants watched, this additional viewing could result in amplified or dampened responses. In our previous studies, we have eliminated access to television, requiring participants to read, complete puzzles, or interact with staff members to pass the time (Minkel et al., 2012). Although we were initially concerned that participants would become agitated and possibly even withdraw early from the study, this was not a problem. Nevertheless, participants should know how they will spend the night of sleep deprivation before they enter the study.

Behavioral Tasks

Both film and video are fairly passive methods for eliciting emotion; they do not directly concern the goals of the participant and are generally not personally relevant. An alternative that is more engaging and more intense is to provide tasks that elicit emotion. Below is a review of the behavioral tasks that have been used to elicit emotion in sleep-deprivation experiments to date.

Directed Emotional Expressions

An alternative to measuring spontaneous facial displays, which a participant makes naturally while viewing a film clip or photograph, is to instruct participants to make emotional displays in response to stimuli. Schwarz et al. (2013) used this method in a sleep-deprivation experiment. They instructed participants to react to emotional stimuli (either positive or negative pictures) with facial expressions that either matched or did not match the valence of the picture. For example, in a trial with the expression matched to the stimulus, a pleasant photograph was shown, and the participant was instructed to smile. In a trial with the expression not matching the stimulus, an unpleasant photograph was shown, and the participant was instructed to smile. In the second study, sleep-deprived participants had slower facial reactions regardless of whether the expression matched or did not match the stimuli. One advantage of this approach is that it ensures that the participant makes an expression during each trial. When measuring natural emotional expressions, not all participants have measurable responses. Overall, however, this method is fairly similar to the passive viewing paradigms described above.

Anger Inductions

Tasks that induce anger are particularly useful because eliciting this emotion using visual stimulation alone can be quite difficult. In sleep-deprivation experiments, inductive methods provide the added benefit that such procedures tend to be more highly arousing than passive viewing paradigms and therefore less likely to be affected by the participant falling asleep midstimulation. One effective method is the use of tasks such as the Point Subtraction Aggression Paradigm (PSAP; Cherek, 1981), which allows participants to accumulate money that can then be stolen by other players. Participants press buttons to earn points to be later exchanged for money. The other player (really the computer) steals points from the participant at random intervals. Participants have three options: to keep earning points, to protect their points for a certain amount of time, or to steal points from the other player. Yet, they cannot keep these stolen points, so stealing points is considered an aggressive action. Stealing points also costs the player money, allowing one to quantify how valuable getting back at the other participant is to the stealing participant. The experimenter must create a believable cover story, however, and this is perhaps the greatest difficulty when using PSAP in a sleep-deprivation protocol. In a standard setting, the researcher can tell the participant that another participant arrived a few minutes earlier and is in another room. This is not particularly believable when a participant has been in a laboratory for over 24 h. Cote, McCormick, Geniole, Renn, and MacAulay (2013) found a very clever solution. They told their Canadian participants that they were competing online against participants in Australia who were just ending a midnight shift. At present, this is the only experimental manipulation of anger to be used in a sleep-deprivation experiment.

Stressors

Stress is often studied separately from emotion, but it is included within the affective sciences. It is an important construct in psychology and medicine, and initial studies suggest that sleep deprivation alters fundamental stress processes related to the hypothalamic-pituitary-adrenal (HPA) axis and cardiovascular reactivity (see the "Psychophysiological Measurement" section below). The term "stress" has been defined as a response to a threat to one's physical or psychological integrity (McEwen & Seeman, 2003). Eliciting stress therefore requires exposing participants to such a threat.

The most common stress task is the Trier Social Stress Task (TSST) developed by Kirschbaum, Pirke, and Hellhammer (1993). The TSST has been extensively validated and elicits moderate psychosocial stress, as evidenced by autonomic and HPA-axis stress responses (Granger, Kivlighan, El-Sheikh, Gordis, & Stroud, 2007). The TSST consists of an anticipation period during which participants prepare a speech and a 10-min test period during which participants deliver the speech and then perform mental arithmetic, all in front of a three-person panel. This task elicits stress responses by exposing the participant to a threat to his or her psychological integrity in the form of negative evaluation of one's intelligence. We used the TSST with sleep-deprived participants at the University of Pennsylvania (Minkel et al., 2014). This is one of the few tasks that may be more effective with a sleep-deprivation experiment than in traditional settings because participants are in the laboratory for such a long time. Most tests involving the TSST are completed soon after the participant completes a consent form, possibly elevating anticipatory anxiety. In our study, we completed this stress task on the second day of testing, after participants had been in the laboratory for at least 20 h. The participants had already completed numerous neurobehavioral performance batteries and were not reminded of the TSST until after we had collected two baseline saliva samples from them. Each participant's affective state was therefore not influenced by anticipatory anxiety during the baseline cortisol collection. The primary disadvantage of this task is that it is not conducive to repeated administrations because unpredictability and unfamiliarity are the task's key stress-producing features. If participants know what to expect, the TSST becomes less stressful. Thus, using such tasks requires that comparisons be done between subjects only, rather than within and between subjects, which would increase statistical power.

Stress can also be induced using performance tasks with time pressure and negative feedback about performance. We have used this approach and found it to have different strengths and weaknesses than the TSST (Minkel et al., 2012). Unlike the TSST, this approach can be administered on consecutive days with little change in effectiveness. It is also easy to create a plausible cover story so that participants are not aware that the primary purpose of the task is to induce stress. In our study, we told participants that the tasks they were performing were being considered by NASA to test readiness to perform space missions and that we needed their performance to be very good to ensure that it was relevant to astronauts. High-stress bouts included difficult tasks with negative feedback about performance, and low-stress bouts included easier tasks and positive feedback about performance. The disadvantages of this approach include a lack of previous validation and possibly less robust activation of cardiovascular and HPA-axis reactivity to the stressor. In general, a previously validated stress-induction technique is preferable, provided it works within the context of a sleep-deprivation experiment.

If cognitive tasks are used to induce stress, the researcher must choose tasks that are not sensitive to sleep loss. For example, attention-based tasks would be a poor choice because sleep-deprived participants would perform much worse, introducing an unintended confound. Alternatives include tasks that involve pure chance, such as guessing a correct number, or ones that are so difficult that performance is uniformly poor. For most methods of sleep deprivation, participants cannot be "blinded" to their condition. So, it may be useful to intentionally manipulate expectations to ensure that performance is not different due to expectancy effects. Sleep-deprived participants may believe a task is less fair or that their performance is not important because they are sleep deprived. Many stressors require participants to be invested in their performance and to believe that it reflects an important attribute, such as intelligence. If they can easily explain away poor performance, they may not experience stress.

MEASUREMENT OF EMOTIONAL RESPONSES

Affective processes in humans have many components, including subjective feelings, behavioral expressions, physiological changes, and brain activation patterns. There are rich resources for quantifying each of these aspects in different ways. When designing an experiment, one must consider both elicitation and measurement as carefully as possible. It is often tempting to try to measure as many aspects as possible, but this often results in a lot of data that are very difficult to understand. The researcher can benefit more from choosing a small number of features and measuring them clearly with a theoretical model or hypothesis in mind. Below is an overview of many of the methods of emotion measurement used in sleep-deprivation studies. Many additional methods are available from affective science, and new technological advances continue to provide exciting new tools for collecting emotion-related data. Nevertheless, many of the methods reviewed here have withstood the test of time and continue to provide valuable information on emotional functioning in a variety of contexts.

Subjective/Self-Report Measures

The core feature of an emotion is the participant's experience or subjective feeling associated with it. Thus, there is no substitute for asking participants to report their own emotional experiences. Measuring such subjective responses to emotion probes is usually essential for validating that the chosen stimuli work as intended in the given sample. This kind of validation is even important when using previously validated stimuli, such as images or video clips, because populations vary widely in their responses. The same stimuli will almost certainly elicit different responses in an undergraduate sample in suburban Florida than in a community sample from Philadelphia.

Visual Stimuli

The IAPS database was developed using the Self Assessment Manikin (SAM), a freely available questionnaire that uses cartoon images to help participants report their emotional experiences in terms of valence (how positive or negative the emotion was) and arousal (the intensity of the emotion). The SAM form is very useful and intuitive. It requires little explanation and participants generally have no difficulty completing it. For video clips, James Gross has developed a postfilm questionnaire that is clear and comprehensive (see Rottenberg et al., 2007). Using such questionnaires takes very little time and can provide valuable data for validity checks, even if subjective responses are not a primary interest in the study. It is important to test responses in a rested state to validate responses prior to making claims about responses in sleep-deprived participants. Comparisons to normative data may be tempting, but such comparisons ultimately do not allow the researcher to distinguish whether differences are due to sleep deprivation or to differences between samples.

Mood States

Researchers use numerous tools to allow participants to report their mood. One such tool is the Profile of Mood States (POMS; McNair, Lorr, & Droppleman, 1971), which is commonly employed in sleep-deprivation studies and provides scores on six subscales, including tension-anxiety, depression-dejection, anger-hostility, vigor-activity, fatigue-inertia, and confusion-bewilderment. The researcher combines these scores to provide an overall measure called total mood disturbance. The POMS has repeatedly been shown to be sensitive to sleep deprivation, causing many to conclude that sleep deprivation results in mood disturbance (because of the name of the summary scale). This can be misleading, however, given that increases in fatigue alone could cause a significant elevation in the total mood disturbance composite scale. Some of our early work demonstrated that partial sleep restriction (sleep restricted to 4h per night for five consecutive nights) was associated with significant changes only for fatigue, vigor, and confusion, and not for anger, anxiety, or depression (Minkel, 2010). This result suggests that partial sleep restriction does not necessarily produce clinically relevant mood disturbance as may be implied by a report of elevated total Mood Disturbance.

An alternative measure that may be preferable is the Positive and Negative Affect Scale (PANAS; Watson, Clark, & Tellegen, 1988), which includes an equal number of positive and negative items. The expanded form (PANAS-X) has 13 subscales that allow for a more comprehensive report of mood changes. Talbot, McGlinchey, Kaplan, Dahl, and Harvey (2010) reported significant changes in nearly all of the positive items of the PANAS as a result of sleep deprivation, suggesting that positive mood states may be particularly sensitive to sleep loss. Unless specifically comparing results to previous studies using the POMS, the PANAS or PANAS-X is probably more useful.

Behavioral Tasks

Self-report measures such as the PANAS or PANAS-X can also be used to report emotional states immediately following an anger induction or stressor task. The instructions can simply be modified as needed to indicate the most appropriate time frame.

Psychophysiological Measurement

Physiological responses to stimuli, such as skin conductance and heart rate, were reported even in very early studies of emotion. These biological correlates of emotional functioning likely provide evidence of the intensity of an emotion and are still used today because they are objective and relatively cheap to collect. Below, we review many of the physiological variables that have been reported in studies of sleep deprivation and emotion, discussing the use of these techniques with sleep-deprived participants.

Pupillometry

Changes in pupil diameter have emerged as a particularly interesting and novel method for assessing certain aspects of emotional arousal. Researchers have shown that the pupil dilates when participants view emotional stimuli (Bradley, Miccoli, Escrig, & Lang, 2008), and the degree of dilation can provide an index of emotional arousal. The measurement device is called a pupillometer, and it consists of a video camera with an infrared light source to track pupil size. This allows measurement of the magnitude of pupil diameter change and the time it takes for this change to occur. Pupils also constrict in response to light, so researchers must carefully control the luminance of the visual stimuli when using this method. Additionally, sleepiness can lead to pupillary oscillations, which could affect the data (Wilhelm, Wilhelm, Lüdtke, Streicher, & Adler, 1998). This approach has been used in sleep-deprivation studies by Franzen, Buysse, Dahl, Thompson, and Siegle (2009), who have reported significantly more pupil dilation in response to negative pictures in sleep-deprived participants.

One must be careful when interpreting these findings, however. It is tempting to conclude that sleep deprivation leads to increased emotional reactivity towards negative stimuli, as evidenced by increased pupil dilation. But the researcher must first show that other elements of emotion have the same coherence with pupillary changes after sleep deprivation as they do when the participant is rested. One cannot assume that negative photographs elicit more intense subjective responses, behavioral tendencies, or neural correlates of negative emotion based on these findings alone. Nevertheless, this study made an important contribution that will facilitate additional studies of emotional reactivity to positive and negative stimuli in the future.

Saliva Assays

Biochemical assays of saliva can provide objective measures of specific hormones and enzymes that are relevant to affective processes. The most common salivary assay is for cortisol, a hormone that has a well-characterized role in HPA-axis responses to stress. Cortisol concentrations have been found to peak about 20 min after exposure to an acute stressor and can be used as an objective correlate of the intensity of a biological stress response (Gunnar & Quevedo, 2007). Ideally, one would take saliva samples at regular intervals before, during, and after the onset of a stressor. This sampling schedule allows the researcher to estimate cortisol exposure over time with area-under-the-curve analyses (see Pruessner, Kirschbaum, Meinlschmid, & Hellhammer, 2003). We found that sleep deprivation elevates both resting salivary cortisol and cortisol reactivity to stress in healthy adults (see Figure 6.1; Minkel et al., 2014). This finding has important health implications because both major depression and obesity have been linked to cortisol exposure. Chronically elevated cortisol has also been associated with neuronal cell death, particularly in the hippocampus. Additional studies are needed to determine whether other aspects of the stress response, such as subjective distress and behavioral responses, are also elevated.

Alpha amylase is a somewhat newer, and more controversial, biomarker of stress reactivity. Unlike cortisol, alpha amylase is a digestive enzyme that is thought to provide a measure of autonomic arousal in response to stressors. It can be assayed from the same saliva samples used to provide cortisol, but the collection technique is slightly different (Granger et al., 2007). In our study of stress and sleep deprivation, we found no influence of sleep deprivation on alpha amylase concentrations at rest or in response to the stressor, but additional studies with larger samples might still produce positive results. Because it is relatively easy to test a given sample for both cortisol and alpha amylase, researchers should consider testing for both, as long as it is affordable.

Testosterone is a hormone that has been linked to aggression, and it can also be measured through saliva. Carré and McCormick (2008) found that salivary testosterone is elevated in those who are more prone to reactive aggression. Based on these findings, Cote et al. (2013) investigated testosterone and aggression in a sleep-deprivation experiment and found that both are reduced following sleep loss.

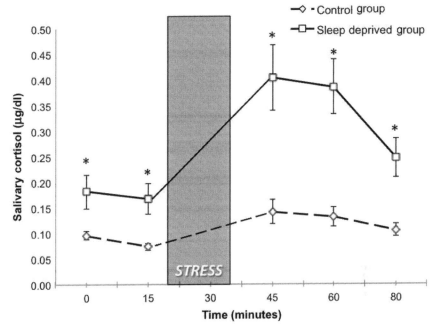

FIGURE 6.1 Salivary cortisol concentrations before and after the TSST. The gray box labeled "STRESSOR" indicates the timing of the TSST. Bars represent standard errors. Asterisks represent values that were significantly different ($p<0.05$) for sleep-deprived and control participants. Baseline cortisol was significantly elevated in the sleep-deprived group relative to the control group ($p<0.004$). Cortisol reactivity to the stressor was also significantly elevated for the sleep-deprived group, even after controlling for baseline differences (peak reactivity $p<0.001$). Figure reproduced with permission (Minkel et al., 2014).

Cardiovascular Measures

Cardiovascular measures have been used to detect physical stress responses under conditions of sleep deprivation. When Franzen et al. (2011) elicited stress in sleep-deprived participants, they used blood pressure and heart rate variability as the primary outcome variables. Such measures have excellent validity, are inexpensive to collect, and have important implications for health. Results from this study indicated that sleep deprivation elevated physiological stress in cardiovascular systems, which is consistent with the elevated salivary cortisol reported above. Together, these studies suggest that the human physiological response to psychosocial stress is greater when people have not slept.

Facial Displays

Facial displays of emotion are excellent dependent measures in affective science because, unlike many of the physiological measures reviewed above, they are directly linked to emotional function. Facial expressions are essential for communicating with others and have been extensively studied. In addition, they can be measured using electromyography (EMG), human scoring, or even automated computerized methods.

Electromyography

By using electrodes to record the electric activity of facial muscles, EMG can provide information on the timing and duration of positive and negative facial displays of emotion. Electrodes placed on the zygomatic major muscle, which is responsible for drawing the mouth into a smile, provide information on positive displays, and electrodes on the corrugator supercilli muscle, which is responsible for drawing the eyebrows and face into a frown, can provide information on negative displays (Dimberg, 1982). One study used this method to quantify reactions to the directed facial expressions task described above (see the "Behavioral Tasks" section), with and without sleep deprivation. In this experiment, the researchers used EMG to measure the time between the stimulus presentations (photographs of happy/angry faces or photographs of positive/negative scenes) and the facial responses of participants. Facial reactions to emotional stimuli were slower after partial sleep deprivation than they were for participants in the rested state (Schwarz et al., 2013). This suggests that facial expressions of emotion may be slower and less intense in people who are sleep deprived.

Human Scoring

In general, people are remarkably good at recognizing emotional expressions. With a little training, most research assistants can reliably score the valence and intensity of facial expressions of emotion (Kring & Sloan, 2007). Such scoring methods are validated first by ensuring reliability through interrater agreement (multiple judges arrive at the same result) and then by demonstrating that the scores have a strong relationship to the stimuli that elicited them (e.g., participants look happy in response to pleasant stimuli and sad in response to negative stimuli). Several validated scoring systems exist to ensure scoring is comprehensive and reproducible across laboratories. The most labor-intensive is called EMFACS and uses the Facial Action Coding System (FACS) developed by Ekman and Friesen (1978). This system involves looking at video frame-by-frame and scoring which parts of the face have moved. These are selected from a catalog of "action units" to describe what was happening at each time point. The action units are then used to quantify emotion-relevant facial displays (Friesen & Ekman, 1983).

A far less time-consuming alternative is the Facial Expression Coding System (FACES) method developed by Kring & Sloan, (2007). In this system, the experimenter scores video of facial expressions in real time and records the valence, intensity, and duration of each emotional expression. Experimenters also often rate the overall level of expressiveness. We used this system for scoring facial expressions in sleep-deprived participants and found it to be quite useful. The reliability for expressiveness was excellent for positive emotional displays, but negative displays were more subtle and difficult to score. It was also not easy to determine exactly when a facial expression ended, so the total number of facial displays was very difficult to score reliably. Nevertheless, this scoring system is relatively easy to use and provides high-quality data.

Automated Computer Scoring

Efforts are underway to develop automated computer algorithms for scoring facial expressions of emotion. David Dinges and Dimitris Metaxas are developing one such method for monitoring stress and emotion in the presence of sleep deprivation (Dinges et al., 2005). They have created multiple algorithms based on two-dimensional and three-dimensional tracking of facial expressions recorded during experimental manipulations. The system can currently measure fatigue based on slow eyelid closures, while quantifying emotional expressions (Jones et al., 2012). The system is not yet available publically, but other groups are working on similar algorithms, and progress appears to be quite rapid. Similar technology has begun to appear in commercial products largely for entertainment purposes. Once available, this approach will offer many important advantages over other scoring methods. In addition to requiring much less time on the part of the experimenter, the system will be able to report additional metrics, such as velocity of expression changes.

Neural Responses to Emotional Stimuli

A large number of studies report neural correlates of emotional stimuli, primarily using functional magnetic resonance imaging (fMRI). These studies are quite expensive (a typical 1-h fMRI scan costs about $500 to $600, depending on the institution), but interest in these findings is extraordinarily high. Among the reasons for the popularity of fMRI are that it is entirely noninvasive (no injections are needed), there is minimal risk to the participant from the magnetic field, and the method provides good temporal and spatial resolution of brain activity, even in subcortical structures. Positron emission tomography has some similar advantages, but it requires the injection of a radioactive contrast agent and has poorer spatiotemporal resolution. A handful of published studies have combined experimental sleep deprivation with fMRI measures of neural responses to emotion probes. We now briefly review these studies with a special emphasis on the unique difficulties involved in completing such scans on sleep-deprived participants.

Matthew Walker and colleagues at UC Berkeley have conducted most of the published fMRI studies that have integrated experimental manipulations of sleep duration with experimental manipulations of emotion. The first of these studies used positive and negative photographs from the IAPS database, and in both cases, brain regions associated with the generation of emotion were found to be more reactive following sleep deprivation. For negative stimuli, the amygdala demonstrated greater reactivity (Yoo, Gujar, Hu, Jolesz, & Walker, 2007). Activation in this region has been associated with fear in particular, suggesting that negative emotions should be more intense following sleep loss. Similarly, areas within the mesolimbic reward network, an area associated with positive emotion and reward, also showed stronger activation to positive photographs following sleep deprivation (Gujar, Yoo, Hu, & Walker, 2011). These findings clearly contradict some of the behavioral findings reviewed above, but no study has yet integrated findings on facial expressiveness with fMRI responses to photographs. Such studies will be important for making sense of the curious disconnect between brain and behavior that currently seems to exist.

The primary difficulty in using fMRI in any study of sleep deprivation is ensuring that participants are awake. Although one might assume that people cannot sleep during an MRI because it is cold, loud, and cramped, these factors are not enough to maintain wakefulness in someone who is sufficiently sleepy. Many scanners are now equipped with cameras for studying eye-tracking, and we have personally watched participants fall asleep, even when they were not experimentally sleep-deprived! Researchers must then have some method for catching those who fall asleep during the scan. A video camera of the participants' eyes is perhaps the best method because slow eyelid closures indicative of sleep onset are easy to observe in real time, even by untrained observers. If such video cameras are not available, the researcher can have the participant make some kind of behavioral response during the scan, such as pressing a button to indicate something about the stimulus. The researcher must then carefully review these behavioral data and remove trials during which participants failed to demonstrate that they were awake.

SPECIAL CONSIDERATIONS WHEN DESIGNING EXPERIMENTAL STUDIES OF SLEEP AND AFFECT

Environment

The most scientifically rigorous research studies of sleep deprivation and emotion involve manipulation of both sleep and affective states in a controlled, laboratory setting. Many of the most productive sleep laboratories are in hospital settings, where facilities are extremely expensive, requiring large grants. Less expensive alternatives include laboratories within nonmedical academic buildings or hotels. These settings are often perfectly appropriate for studies of sleep and emotion and can include low-risk biomedical procedures such as blood draws and saliva collection. A participant can still have a negative reaction to sleep deprivation, however. So, researchers should have a plan for dealing with an adverse medical or psychiatric event such as fainting, severe anxiety, or mania. The most important variables to control are light, noise, and temperature. When creating a sleep laboratory in a room with windows, the researcher can block out most natural light with standard materials such as black garbage bags, aluminum foil, and cardboard taped carefully over all glass. The environment should then be checked with a light meter at several times throughout the day and night to ensure minimal variability. Noise is not as easily controlled and can be a significant problem if construction, sirens, and other obtrusive sounds occur unexpectedly. These problems are best addressed by carefully planning where to create a new sleep laboratory, but they will likely contribute to some level of error in experiments. One final consideration is the control of food and water intake. In many sleep-deprivation experiments, participants can have a snack during the night to prevent feelings of hunger from interfering with performance. If possible, the experimenter should provide standard options for food and snacks, avoiding stimulating and sedating food and drinks. When studying emotion, one must be particularly careful to ensure that food and drinks have a minimal impact on affect and are as equal as possible in sleep-deprivation and control conditions.

Types of Sleep Manipulations

Within a controlled laboratory, several types of sleep manipulation exist, each with different advantages. The most common is acute total sleep deprivation during which participants are given no opportunity to sleep for a specified time. Generally, sleep is withheld for one full night, and testing occurs the following day. When using this method, the researcher must hold the time of day constant to control for circadian differences. This method is most popular because it is the most cost-effective way to get a significant result. In a laboratory environment, environmental variables such as light and temperature are held constant, and exercise, caffeine, and other common countermeasures to fatigue are restricted or entirely unavailable. A full night of sleep deprivation in such a setting is very intense and tends to produce large to extremely large effects in subjective feelings of fatigue and neurobehavioral performance on some tasks. Its primary limitation, however, is very low ecological validity. Although insufficient sleep is quite common in the general population, extended wakefulness without caffeine, bright light, or vigorous activity is extremely rare. Field studies (discussed below) have the highest ecological validity, but the accompanying loss of experimental control is costly. Chronic partial sleep restriction is perhaps the most viable alternative. When using this approach, researchers restrict sleep below normal levels, but sleep is not entirely eliminated. It is now well established that reduced sleep over multiple nights results in cumulative deficits in neurobehavioral performance and subjective mood states (Belenky et al., 2003; Van Dongen, Maislin, Mullington, & Dinges, 2003). This design significantly increases the cost of studies, because participants must stay in a laboratory for multiple nights. The increased relevance to real world sleep loss is a significant advantage. One study of sleep and emotion utilized such an approach and found significant reductions in facial expressiveness of emotion after one night of 4 h of sleep opportunity (Schwarz et al., 2013).

In addition to acute total sleep deprivation and chronic partial sleep restriction, one other method of manipulating sleep should be considered. Sleep fragmentation is in many ways the most elegant method, but it is also the most labor intensive. When utilizing this method, researchers briefly disrupt participants' sleep throughout the night using sounds or light (Roehrs, Merlotti, Petrucelli, Stepanski, & Roth, 1994). This method most closely approximates the sleep problems seen in clinical samples, including those in patients with medical or psychiatric problems. At present, we are not aware of any published studies that have reported experimental manipulations of stress or emotion in the context of a sleep fragmentation experiment, but these data would be extraordinarily valuable.

Research Designs

Perhaps the most important consideration in designing a study of sleep deprivation and emotion is determining a control group. In general, a within-subjects design is preferable because participants serve as their own controls, thus eliminating a significant source of variability. This is the design of choice in neuroimaging experiments for which controlling for individual differences in neuroanatomy and blood flow is particularly valuable. If choosing this approach, one must remember to counterbalance the order of sleep deprivation and wakefulness to ensure the observed results are not simply due to repetition. The primary disadvantage of this approach is that repeated administrations of stimuli are required. In studies of cognitive effects of sleep deprivation, one must consider learning effects. If not carefully controlled, an experiment can seem to indicate that sleep deprivation increases cognitive abilities, simply because people get better at a task with practice even if they also become more tired. In studies of emotion, one must consider the effect of repeated administrations on the affective response to the stimulus. Watching the exact same funny video clip 2 days in a row may reduce its potency. This problem can be avoided by using sets of stimuli that elicit similar responses. For video stimuli, researchers have identified "sister clips" that elicit nearly identical responses in participants. Counterbalancing the administration of these clips across participants can help ensure that group-level differences are related to sleep deprivation. Individual differences may be due to differences in sleep loss, however, or to differences in response to the stimulus that have nothing to do with sleep loss. If the primary goal of the study is to explain such individual differences, a larger number of stimuli can be helpful.

In some circumstances, the repeated administration of an affective probe is not feasible. The TSST, for example, relies on the unfamiliarity of the task to elicit a stress response. Completing it twice in 2 days would likely alter the experience of the second administration to a significant degree. In these cases, a within-subjects design is not a viable option, and responses between subjects must be compared instead. In these cases, the strongest design involves having sleep-deprived and control participants complete nearly identical procedures. Ideally, the control group would enter the laboratory at the same time of day and sleep in the laboratory. Although the unfamiliar environment may alter participant sleep, it can be quantified and compared to the sleep-deprived group. Many experiments allow the control participants to sleep at home, but this introduces an unfortunate confound because the sleep deprived group has also spent more time in the laboratory than the control group. If this design is used, additional analyses can be conducted to make a stronger argument that lack of sleep is specifically related to observed differences.

CONCLUSION

The methods reviewed here represent the current integration of affective science and sleep-deprivation research. Although the review does not provide an exhaustive list of everything that has been done, it comes surprisingly close. This is a very new area, and much more research is needed to more fully understand the ways in which sleep loss influences emotional functioning. The first challenge for the field is to include any validated emotion elicitation procedure within a sleep-deprivation protocol. The dominant method for measuring affect in sleep-deprivation experiments remains the periodic provision of a mood questionnaire. We have used this method and found it effective, but it can be supplemented with a rich variety of procedures, as reviewed above. Even a quick review of methods from affective science provides a wide range of ideas for projects that have not yet been done. As just one example, an extensive literature addresses the use of music and other sounds to elicit emotion. To the best of our knowledge, no experimental studies of sleep deprivation have elicited emotion using auditory stimuli. Similarly, experimental psychopathology is a flourishing field and has produced a number of innovative methods for eliciting clinically relevant emotion. Babson, Feldner, Trainor, and Smith (2009), for example, induced anxiety in a sleep-deprivation experiment by using a carbon-dioxide challenge originally developed for studying panic disorder. Future studies of sleep and emotion would do well to enrich protocols with some of these more active manipulations of affect that have already been validated by emotion researchers.

Similarly, sleep-deprivation research has just begun to use methods for quantifying various aspects of emotional responses. Here the challenge is to move beyond self-report to a more multimodal approach. Although few laboratories can afford fMRI, less-expensive methods, such as EEG and cardiovascular monitoring, can be quite valuable. Given that most of us now have a high-definition camera built into our phones, collecting video of behavioral data is easier than ever. Emotion-relevant behavior can also be quantified from behavioral tasks, such as the reactive aggression paradigm reviewed above. Most emotion researchers are eager to share the tasks they have developed and will provide them free of charge. Self-report measures should not be abandoned, however, because subjective feelings are an essential part of the emotional experience. Rather, these measures should be supplemented with additional objective measures of behavioral, physiological, and, when possible, neural measures of emotion as well.

Finally, we advise using the best methods of sleep manipulation that are economically feasible. Many studies of sleep deprivation and emotion allow participants to sleep at home or fail to effectively quantify sleepiness. When possible, counterbalanced within-subjects designs maximize power and reduce recruitment burden. If such an experiment design is not possible, participants in the control and experimental conditions should complete protocols that are as close to identical as possible. This will ensure that reported findings continue to be robust, replicable, and valid, contributing to our collective understanding of the affective consequences of insufficient sleep.

References

Babson, K. A., Feldner, M. T., Trainor, C. D., & Smith, R. C. (2009). An experimental investigation of the effects of acute sleep deprivation on panic-relevant biological challenge responding. *Behavior Therapy*, 40(3), 239–250. http://dx.doi.org/10.1016/j.beth.2008.06.001.

Belenky, G., Wesensten, N. J., Thorne, D. R., Thomas, M. L., Sing, H. C., Redmond, D. P., et al. (2003). Patterns of performance degradation and restoration during sleep restriction and subsequent recovery: A sleep dose-response study. *Journal of Sleep Research*, 12(1), 1–12. http://dx.doi.org/10.1046/j.1365-2869.2003.00337.x.

Bradley, M. M., Miccoli, L., Escrig, M. A., & Lang, P. J. (2008). The pupil as a measure of emotional arousal and autonomic activation. *Psychophysiology*, 45(4), 602–607. http://dx.doi.org/10.1111/j.1469-8986.2008.00654.x.

REFERENCES

Carré, J. M., & McCormick, C. M. (2008). Aggressive behavior and change in salivary testosterone concentrations predict willingness to engage in a competitive task. *Hormones and Behavior, 54*(3), 403–409. http://dx.doi.org/10.1016/j.yhbeh.2008.04.008.

Cherek, D. R. (1981). Effects of smoking different doses of nicotine on human aggressive behavior. *Psychopharmacology, 75*(4), 339–345.

Cote, K. A., McCormick, C. M., Geniole, S. N., Renn, R. P., & MacAulay, S. D. (2013). Sleep deprivation lowers reactive aggression and testosterone in men. *Biological Psychology, 92*(2), 249–256. http://dx.doi.org/10.1016/j.biopsycho.2012.09.011.

Dimberg, U. (1982). Facial reactions to facial expressions. *Psychophysiology, 19*(6), 643–647. http://dx.doi.org/10.1111/j.1469-8986.1982.tb02516.x.

Dinges, D. F., Pack, F., Williams, K., Gillen, K. A., Powell, J. W., Ott, G. E., et al. (1997). Cumulative sleepiness, mood disturbance and psychomotor vigilance performance decrements during a week of sleep restricted to 4-5 hours per night. *Sleep, 20,* 267–277.

Dinges, D. F., Rider, R. L., Dorrian, J., McGlinchey, E. L., Rogers, N. L., Cizman, Z., et al. (2005). Optical computer recognition of facial expressions associated with stress induced by performance demands. *Aviation, Space, and Environmental Medicine, 76*(Supplement 1), B172–B182.

Ekman, P., & Friesen, W. V. (1978). *Facial action coding system.* Palo Alto, CA: Consulting Psychologists Press.

Franzen, P. L., Buysse, D. J., Dahl, R. E., Thompson, W., & Siegle, G. J. (2009). Sleep deprivation alters pupillary reactivity to emotional stimuli in healthy young adults. *Biological Psychology, 80*(3), 300–305. http://dx.doi.org/10.1016/j.biopsycho.2008.10.010.

Franzen, P. L., Gianaros, P. J., Marsland, A. L., Hall, M. H., Siegle, G. J., Dahl, R. E., et al. (2011). Cardiovascular reactivity to acute psychological stress following sleep deprivation. *Psychosomatic Medicine, 73*(8), 679–682. http://dx.doi.org/10.1016/j.biopsycho.2008.10.010.

Friesen, W. V., & Ekman, P. (1983). *EMFACS-7: Emotional facial action coding system,* Unpublished manuscript, University of California at San Francisco, 2.

Granger, D. A., Kivlighan, K. T., El-Sheikh, M., Gordis, E. B., & Stroud, L. R. (2007). Salivary α-amylase in biobehavioral research. *Annals of the New York Academy of Sciences, 1098*(1), 122–144. http://dx.doi.org/10.1196/annals.1384.008.

Gross, J. J., & Levenson, R. W. (1995). Emotion elicitation using films. *Cognition & Emotion, 9*(1), 87–108. http://dx.doi.org/10.1080/02699939508408966.

Gujar, N., Yoo, S.-S., Hu, P., & Walker, M. P. (2011). Sleep deprivation amplifies reactivity of brain reward networks, biasing the appraisal of positive emotional experiences. *Journal of Neuroscience, 31*(12), 4466–4474. http://dx.doi.org/10.1523/jneurosci.3220-10.2011.

Gunnar, M., & Quevedo, K. (2007). The neurobiology of stress and development. *Annual Review of Psychology, 58,* 145–173. http://dx.doi.org/10.1146/annurev.psych.58.110405.085605.

Harrison, Y., & Horne, J. A. (2000). The impact of sleep deprivation on decision making: A review. *Journal of Experimental Psychology Applied, 6*(3), 236. http://dx.doi.org/10.1037/1076-898X.6.3.236.

Jones, C. W., Basner, M., Yu, X., Yang, F., Goel, N., Metaxas, D., et al. (2012). Unobtrusive tracking of slow eyelid closures as a measure of fatigue from sleep loss. *Sleep, 35,* A110.

Kirschbaum, C., Pirke, K. M., & Hellhammer, D. H. (1993). The 'Trier Social Stress Test'—A tool for investigating psychobiological stress responses in a laboratory setting. *Neuropsychobiology, 28*(1–2), 76–81. http://dx.doi.org/10.1159/000119004.

Kring, A. M., & Sloan, D. M. (2007). The facial expression coding system (FACES): Development, validation, and utility. *Psychological Assessment, 19*(2), 210–224. http://dx.doi.org/10.1037/1040-3590.19.2.210.

Lang, P. J., Bradley, M. M., & Cuthbert, B. N. (2008). *International affective picture system (IAPS): Affective ratings of pictures and instruction manual. Technical Report A-8.* Gainesville, FL: University of Florida.

McEwen, B. S., & Seeman, T. (2003). Stress and affect: Applicability of the concepts of allostasis and allostatic load. In R. Davidson, K. Scherer, & H. Godsmith (Eds.), *Handbook of affective sciences* (pp. 1117–1137). London: Oxford University Press.

McNair, D. M., Lorr, M., & Droppleman, L. F. (1971). *EITS manual for the profile of mood states*. San Diego, CA: Educational and Industrial Testing Service.

Minkel, J. (2010). *Affective consequences of sleep deprivation*. Publicly accessible Penn Dissertations, Paper 218.

Minkel, J. D., Banks, S., Htaik, O., Moreta, M. C., Jones, C. W., McGlinchey, E. L., et al. (2012). Sleep deprivation and stressors: Evidence for elevated negative affect in response to mild stressors when sleep deprived. *Emotion*, *12*(5), 1015. http://dx.doi.org/10.1037/a0026871.

Minkel, J., Moreta, M. C., Muto, J., Htaik, O., Jones, C. W., Basner, M., et al. (2014). Sleep deprivation potentiates hpa axis stress reactivity in healthy adults. *Health Psychology* http://dx.doi.org/10.1037/a0034219.

Pruessner, J. C., Kirschbaum, C., Meinlschmid, G., & Hellhammer, D. H. (2003). Two formulas for computation of the area under the curve represent measures of total hormone concentration versus time-dependent change. *Psychoneuroendocrinology*, *28*(7), 916–931. http://dx.doi.org/10.1016/S0306-4530(02)00108-7.

Roehrs, T., Merlotti, L., Petrucelli, N., Stepanski, E., & Roth, T. (1994). Experimental sleep fragmentation. *Sleep*, *17*(5), 438–443.

Rottenberg, J., Ray, R. D., & Gross, J. J. (2007). Emotion elicitation using films. In J. Coan & J. Allen (Eds.), *The handbook of emotion elicitation and assessment*. London: Oxford University Press.

Schwarz, J. F. A., Popp, R., Haas, J., Zulley, J., Geisler, P., Alpers, G. W., et al. (2013). Shortened night sleep impairs facial responsiveness to emotional stimuli. *Biological Psychology*, *93*(1), 41–44. http://dx.doi.org/10.1016/j.biopsycho.2013.01.008.

Talbot, L. S., McGlinchey, E. L., Kaplan, K. A., Dahl, R. E., & Harvey, A. G. (2010). Sleep deprivation in adolescents and adults: Changes in affect. *Emotion*, *10*(6), 831. http://dx.doi.org/10.1037/a0020138.

Van Dongen, H. P., Maislin, G., Mullington, J. M., & Dinges, D. F. (2003). The cumulative cost of additional wakefulness: Dose-response effects on neurobehavioral functions and sleep physiology from chronic sleep restriction and total sleep deprivation. *Sleep*, *26*(2), 117–129.

Watson, D., Clark, L. A., & Tellegen, A. (1988). Development and validation of brief measures of positive and negative affect: The PANAS scales. *Journal of Personality and Social Psychology*, *54*(6), 1063.

Wilhelm, B., Wilhelm, H., Lüdtke, H., Streicher, P., & Adler, M. (1998). Pupillographic assessment of sleepiness in sleep-deprived healthy subjects. *Sleep*, *21*(3), 258–265.

Yoo, S. S., Gujar, N., Hu, P., Jolesz, F. A., & Walker, M. P. (2007). The human emotional brain without sleep—A prefrontal amygdala disconnect. *Current Biology*, *17*(20), R877–R878. http://dx.doi.org/10.1016/j.cub.2007.08.007.

PART 3

EVIDENCE REGARDING SLEEP AND SPECIFIC TYPES OF AFFECT

SECTION 1

SLEEP AND NEGATIVE AFFECT

CHAPTER 7

The Interrelations Between Sleep and Fear/Anxiety: Implications for Behavioral Treatment

Kimberly A. Babson[*,†]

[*]National Center for PTSD, VA Palo Alto Health Care System, Menlo Park, California, USA
[†]Department of Psychiatry and Behavioral Sciences, Stanford School of Medicine, Stanford, California, USA

This chapter provides an overview of the interrelations between sleep and fear/anxiety and the implications of these interrelations for behavioral treatment. The chapter starts by introducing sleep disturbances commonly associated with fear/anxiety. After describing and differentiating between the key concepts of sleep, fear, and anxiety, the chapter reviews the research on this area. This research takes the form of (1) correlational studies, (2) studies examining the effect of poor sleep on fear/anxiety, and (3) studies examining the effect of fear/anxiety on sleep. The chapter considers experimental and clinical research within each of these broader categories. Next, the chapter examines the current theoretical models for the association between sleep and fear/anxiety, including psychosocial and neurobiological models. The chapter then discusses clinical implications, such as the importance of understanding the mechanisms underlying the relationships between sleep and fear/anxiety. The chapter concludes by considering directions for future research.

BACKGROUND AND DEFINITION OF KEY CONCEPTS

Lack of sleep has been recognized as one of the newest public health epidemics (CDC, 2014). The classification of sleep loss as a public health epidemic is not surprising given modern lifestyles, which are frequently characterized by increased work demands, shift work, elevations in daily stresses, and the ability to be constantly connected via Internet and mobile devices, all of which have been shown to reduce sleep quality.

Although the International Classification of Sleep Disorders (ICSD-2; American Academy of Sleep Medicine, 2005) has identified over 80 different disorders of sleep, insomnia remains the most common. Estimates suggest that more than 30% of Americans will experience symptoms of anxiety, with 10% of those individuals going on to develop an insomnia disorder (Roth, 2007). The specific classifications of insomnia vary across diagnostic manuals, but the essential features (or symptoms) of insomnia disorder are consistent. Based on the criteria presented in three major diagnostic manuals [Diagnostic and Statistical Manual for Mental Disorders-5 (DSM-5; American Psychiatric Association (APA), 2013); International Classification of Diseases-10 (ICSD; World Health Organization, 1992); American Academy of Sleep Medicines' Research Diagnostic Criteria (RDC; Edinger et al., 2004)], insomnia can be characterized as difficulty initiating or maintaining sleep (including early-morning awakening) or nonrestorative sleep that occurs for at least 1 month and results in significant functional impairment. Of those individuals with sleep problems, a significant proportion (40-50%) will also be diagnosed with a co-occurring psychological disorder (Breslau, Roth, Rosenthal, & Andreski, 1996; Buysse, Reynolds, Kupfer, & Thorpy, 1994; Ford & Kamerow, 1989). This research has resulted in increased recognition that sleep disturbance is a major risk factor for the development of mood and anxiety disorders.

Fear and anxiety are closely related emotional experiences that have developed as adaptive responses and defenses to perceived danger. Although fear and anxiety have some commonalities (i.e., strong negative feelings accompanied by bodily arousal), research has demonstrated that they have distinct physiological and conceptual forms. Fear can be conceptualized as a strong urge to defend oneself against present threatening stimuli, and behavioral fear response typically involves defending or removing oneself from the threatening stimulus (i.e., flight-or-fight response). In comparison, the DSM-5 defines anxiety as "the apprehensive anticipation of future danger or misfortune accompanied by a feeling of worry, distress, and/or somatic symptoms of tension. The focus of anticipated danger may be internal or external" (APA, 2013, p. 818). Therefore, fear is an intense emotional reaction to a present, current, or immediate threatening stimulus, and anxiety is an apprehension about potential future danger (Öhman, 2008). Fear and anxiety are emotional experiences that everyone encounters. However, a proportion of individuals develop disorders of fear/anxiety. Research has shown that disorders of fear/anxiety are the most frequently occurring form of psychopathology

(Kessler, Chiu, Demler, & Walters, 2005). Some even estimate that 29% of the general population will be diagnosed with a fear/anxiety disorder at some point in their lifetimes (Kessler et al., 2005).

Sleep disturbance and symptoms of fear/anxiety are robustly intertwined phenomena. In fact, approximately 74% of primary care patients with anxiety symptoms report clinically significant levels of sleep disturbance (Marcks, Weisber, Edele & Keller, 2010). Furthermore, 42% of individuals with sleep disturbances report elevated levels of anxiety (Mellinger, Balter, & Uhlenhuth, 1985). Although the association between sleep disturbance and both fear and anxiety is well documented, the nature of this relation is less well understood. The current research suggests that the relation between fear/anxiety and sleep is bidirectional in nature, such that fear/anxiety predicts sleep disturbances and sleep disturbances predict fear/anxiety (Babson & Feldner, 2010). The next section provides a detailed overview of these interrelations as indicated by correlational, experimental, and clinical research.

FEAR/ANXIETY IS ASSOCIATED WITH SLEEP DISTURBANCES

A large body of research shows that sleep loss is positively correlated with anxiety (Ford & Kamerow, 1989; Gregory et al., 2005; Gregory, Eley, O'Connor, & Plomin, 2004; Gregory & O'Connor, 2002), as well as tension, nervousness, and irritability (Oginska & Pokorski, 2006). Both human and animal studies have demonstrated a strong relation between both acute and chronic stress and disturbed sleep (see Van Reeth et al., 2000 for a review). For example, among a community-based sample of adults, individuals with insomnia had greater levels of anxiety compared to those without insomnia. In fact, those with insomnia were 17.35 times more likely to have clinical levels of anxiety compared to those without insomnia. In addition, self-reported anxiety scores increased in proportion to elevations in the frequency of insomnia (Taylor, Lichstein, Durrence, Reidel, & Bush, 2005). Similar relationships have been shown among college students, with students' perceived stress and anxiety accounting for the greatest amount of variance in explaining poor self-reported sleep quality (Lund, Reider, Whiting, & Prichard, 2010).

In addition to associations between stress and self-reported sleep problems, research has also demonstrated an association between fear/anxiety and objective indicators of disturbed sleep, such as polysomnography (PSG). For example, among a sample of older adults with insomnia, subclinical levels of both trait and state anxiety were associated with objectively measured waking after sleep onset, suggesting that, among older adults, low levels of elevated anxiety are associated with significant sleep fragmentation (Spira et al., 2008). This association extends outside of the elderly population. In fact, a meta-analysis of 177 studies conducted with patients with psychological disorders (including anxiety disorders) demonstrated that those with anxiety experienced sleep fragmentation, reduction in total sleep time and sleep efficiency, and increased latency to sleep onset. PSG results indicated normal slow-wave sleep, but the amount of time spent in NREM sleep was significantly reduced compared to healthy controls (Benca, Obermeyer, Thisted, & Gillin, 1992).

The relation between sleep and fear/anxiety has been shown to differ based on the method of sleep assessment and the type of anxiety or anxiety disorder investigated. These findings are drawn from a large body of research that is beyond the scope of the current chapter. Instead, this chapter provides a general overview of these differences. Although the self-reported assessment of sleep yields a consistent association between sleep and fear/anxiety, objective sleep assessment through PSG often provides divergent evidence. In fact, PSG findings are only consistent with self-reported sleep disturbances among individuals with generalized anxiety disorder (GAD). Among individuals with panic disorder (PD), PSG yields a reduction only in sleep efficiency, and research on individuals with posttraumatic stress disorder (PTSD) has resulted in mixed findings, with some studies demonstrating PSG differences and others yielding no differences (Pillai & Delahanty, 2012; Kobayashi, Huntley, Lavela, & Mellman, 2012). In comparison, in studies including individuals with a range of anxiety disorders, PSG results suggest lower slow-wave sleep and shortened REM latency (Papadimitriou & Linkowski, 2005). Sample characteristics (e.g., comorbid conditions, medication effects, small sample sizes) may account for these differences, but additional research is needed to better clarify the discrepant findings between self-reported and objective sleep assessments of individuals with anxiety disorders.

Summary

When considered in aggregate, correlational studies indicate that sleep and fear/anxiety are positively associated. Although this relationship is particularly robust among self-reported sleep disturbances, the relationship is less clear when objective indices of sleep are included. In such cases, in studies including individuals with a range of anxiety disorders or studies focused on individuals with GAD or subclinical levels of anxiety, PSG findings are consistent with self-report. However, for individuals with other anxiety disorders, such as PTSD, PSG findings are mixed. Future research is needed to better understand these discrepant findings.

SLEEP DISTURBANCE IS ASSOCIATED WITH ELEVATIONS IN FEAR/ANXIETY

Laboratory-based human models have demonstrated that acute sleep deprivation results in elevated fear/anxiety. In one of the first studies done in this area, Roy-Byrne, Uhde, and Post (1986) examined the effects of one night of sleep deprivation on responding to a panic-relevant biological challenge (hyperventilation) among 12 patients with PD, 10 patients with major depression, and 10 nonclinical volunteers. Following acute sleep deprivation, participants completed a laboratory-based assessment of reactivity to a 3-min voluntary hyperventilation procedure. The well-established hyperventilation procedure was utilized to index the effects of sleep deprivation on panic-relevant reactivity (see Rapee, 1995; Zvolensky & Eifert, 2001). A subgroup of patients with PD (58%) experienced a statistically significant increase in overall (self- and nurse-rated) anxiety levels compared to the control group. Moreover, 25% of the PD subgroup reported experiencing a panic attack following sleep deprivation. The effects of sleep deprivation on fear and anxiety have also been demonstrated among healthy individuals without psychological disorders. In this case, Sagaspe et al. (2006) experimentally induced a 36-h period of acute sleep deprivation among a sample of 12 young and healthy nonclinical participants. Results indicated that 36 h of acute sleep deprivation resulted in an increase in self-reported anxiety symptoms. In a separate study, Babson et al. (2009) tested the association between acute sleep deprivation and responding to panic-relevant biological challenge among nonclinical participants. One hundred and two participants were randomly assigned to either an experimental (24 h of acute sleep deprivation) or control (no sleep deprivation) group. On the days prior to and following the experimental (sleep) manipulation, participants completed a 5-min 10% carbon dioxide-enriched air, laboratory-based biological challenge. As predicted, sleep deprivation increased anxious and fearful responding to the challenge, indicating that acute sleep deprivation increases fearful responding to bodily arousal. In a follow-up study, this same team tested the effects of experimentally manipulated acute sleep deprivation on self-reported anxiety, depression-specific symptoms, and general distress within a healthy sample of adults. Results demonstrated that acute sleep deprivation increased anxiety, depression-specific symptoms, and general distress when compared to a normal-night-of-sleep control condition (Babson, Trainor, Feldner, & Blumenthal, 2010).

In addition to the experimental and quasiexperimental research outlined above, longitudinal evidence indicates that sleep deprivation is associated with an increased probability of developing an anxiety disorder (Ford & Kamerow, 1989; Neckelmann, Mykletun, & Dahl, 2007). Indeed, sleep problems in childhood and adolescence are related to elevated anxiety symptoms in adulthood (Gregory & O'Connor, 2002; Gregory, Eley, O'Connor, & Plomin, 2004; Gregory, Caspi, Eley, Moffitt, O'Connor, & Poulton, 2005; Ong, Wickramaratne, Tang, & Weissman, 2006). For instance, Gregory et al. (2005) examined the relationship between persistent sleep problems in childhood and the development of anxiety and/or depression in adulthood in a prospective longitudinal study. The participants were 1037 children who were assessed approximately every 2 years from ages 3 to 26. Persistent sleep problems were defined by parent reports of child sleep problems during age 5 and/or 7 years and also at age 9 years. Anxiety and depression were measured at ages 21 and 26 years via the Diagnostic Interview Schedule (Robins, Cottler, Bucholz, & Compton, 1995). Significantly more children with persistent sleep problems developed anxiety disorders by adulthood, compared to those without persistent sleep problems (46% and 33%, respectively). There was not a significant relationship between persistent childhood sleep problems and adulthood depression, however. Similar results were also observed among a population-based cohort study in which sleep and anxiety (amongst other variables) were assessed at baseline and 11 years later. Results demonstrated that individuals with chronic insomnia were at significantly greater risk for developing an anxiety disorder up to 11 years later (Neckelmann et al., 2007). This relation has been observed among specific disorders of anxiety as well. For example, research on exposure to traumatic events has demonstrated a temporal relationship between sleep problems and the development of PTSD following traumatic-event exposure. In this case, studies have suggested that insomnia 1-month after trauma exposure predicts the development of PTSD at post-trauma intervals of 6 weeks (Mellman, David, Kulick-Bell, Hebding, & Nolan, 1995), 6 months (Harvey & Bryant, 1998), and 1 year (Koren, Arnon, Lavie, & Klein, 2002).

Summary

Taken together, studies have demonstrated a strong temporal link between insomnia and the onset of fear/anxiety. Experimental and quasiexperimental studies of adults have shown that experimentally manipulated sleep deprivation results in elevated anxiety levels (and in some cases panic attacks) among individuals with anxiety disorders (Roy-Byrne et al., 1986). A similar pattern of findings has been observed among healthy nonclinical samples, for which the administration of a sleep deprivation protocol increased self-reported fear/anxiety (Babson et al., 2010; Sagaspe et al., 2006), as well as anxious and fearful responding to autonomic arousal (Babson et al., 2009). Results from longitudinal studies provide additional support for this temporal relationship, because research has shown that sleep problems in childhood and adolescence predict the onset of anxiety symptoms and anxiety disorders in adulthood (Gregory et al., 2002, 2005, 2004; Neckelmann et al., 2007).

FEAR/ANXIETY IS ASSOCIATED WITH SLEEP DISTURBANCE

Research on the relationship between fear/anxiety and sleep in humans has assumed epidemiological, experimental, and clinical designs. Epidemiological research has demonstrated an association between anxiety disorders and insomnia. In one such study, anxiety and insomnia appeared at the same time in 39% of cases, anxiety disorder preceded insomnia in 43% of cases, and insomnia preceded anxiety in 18% of cases (Ohayon & Roth, 2003). This work was consistent with a study conducted by Johnson, Roth, and Breslau (2006), which indicated that, among adolescents, anxiety disorders occurred prior to insomnia in 73% of cases. Experimental work also provides support for this temporal order. For example, experimental studies have demonstrated that experimentally induced stress (cognitive arousal) prior to bed resulted in delayed sleep onset (Gross & Borkovec, 1982; Haynes, Adams & Frazen, 1981). Finally, longitudinal research supports the idea that fear/anxiety results in disturbed sleep. Vahtera et al. (2007) conducted a 5-year prospective study among a population sample of 19,199 participants in order to examine the impact of significant stressful life events on the development of sleep problems among a healthy sample of adults. Results demonstrated that exposure to a highly stressful life event was followed by significant sleep disturbances in individuals who reported no sleep problems prior to the event. In addition, individuals with a vulnerability to anxiety (i.e., elevated baseline arousal and self-reported ratings of stress at baseline) had a significantly higher risk for the development of acute sleep problems after the stressful event. However, anxiety was shown to trigger acute sleep problems (up to 6-months postevent), but not chronic sleep problems (greater than 6-months postevent).

Summary

Experimental and clinical research has also provided support for the temporal order in which fear/anxiety causes insomnia. Experimental research among adults supports these findings, illustrating that induced stress (i.e., cognitive arousal) prior to bed results in delayed sleep onset (Gross and Borkovec, 1982; Haynes et al., 1981). Finally, longitudinal research indicates that individuals without sleep disturbance develop disturbed sleep subsequent to a significant stressful event (Vahtera et al., 2007). Altogether, the research reviewed here supports the notion that the relationship between fear/anxiety and sleep is bidirectional. In order to better understand the nature of this relationship, the theoretical models for the relationships between fear/anxiety and sleep need further examination.

THEORETICAL MODELS

Theoretical models for the relationships between insomnia and psychological symptoms vary according to specific orientation, but all models agree that this relationship is based on an increase in autonomic, cortical, cognitive, and emotional arousal (Espie, 2002; Harvey, 2002; Perlis, Giles, Mendelson, Bootzin, & Wyatt, 1997; Reimann et al., 2010). The manner in which these systems become activated accounts for the differences between models as described below.

Diathesis-Stress Model

The diathesis-stress model (Spielman, Caruso, & Glovinsky, 1987) was one of the first proposed models for explaining insomnia. This model suggests that insomnia is the result of the interactions between predisposing, precipitating, and perpetuating factors. Specifically, individuals with predisposing factors (e.g., genetic, physiological, psychological, environmental), when faced with precipitating factors (e.g., physiological or psychological stressors), tend to develop acute sleep disturbance. This acute disturbance is then maintained through perpetuating factors such as behavioral (increased time in bed, napping), cognitive (worry about not sleeping), and environmental (watch TV in bed) factors that maintain the sleep disturbance and the development of chronic insomnia. This model suggests that targeting precipitating or perpetuating factors, such as elevations in anxiety and cognitive and physiological arousal, would reduce symptoms of insomnia.

Stimulus Control Model

The stimulus control model (Bootzin, 1972) is a behavioral model of insomnia based on classical conditioning principles. This model posits that sleep is a conditioned response to a stimulus (i.e., the environment). Among individuals with insomnia, this conditioning becomes interrupted, and, instead of acting as a cue for sleep, the environment has now become a cue for increased arousal and wakefulness. Based upon this model, insomnia can be reduced by reestablishing the association between the environment and sleep. More specifically, for individuals with problematic levels of anxiety and insomnia, this model highlights the importance of eliminating anxiety (i.e., cognitive and physiological arousal) from the sleeping context in order to reduce the association between sleep and anxiety.

Cognitive Model

The cognitive model of insomnia (Harvey, 2003) focuses on how thoughts, beliefs, and feelings interfere with sleep, thus providing the context for the development of a maladaptive pattern. Specifically, Harvey suggests that insomnia develops from worry about sleep and the daytime consequences of sleep loss. This leads to an increase in cognitive and physiological arousal prior to sleep, which, in turn, results in increased wakefulness. Based upon this model, treatment interventions that target dysfunctional thoughts, beliefs, and worry result in decreased insomnia symptoms. The integration of this model is of critical importance, particularly for individuals with GAD and insomnia.

Neurocognitive Model

Perlis et al. (1997) has proposed a top-down approach to understanding the relationship between sleep and psychopathology. This model posits that insomnia is caused by an increase in cortical hyperarousal (observed through elevated sleep EEG readings). Elevated frequency EEG activity has been related to enhanced sensory processing and memory formation. This may account, in part, for sleep state misperception, in which individuals self-report being awake more than is demonstrated on PSG. For the individual, cortical hyperarousal is experienced as an increase in cognitive arousal. Such cognitive arousal may include intrusive thoughts prior to sleep and dysfunctional beliefs about sleep. Cognitive arousal then results in elevated levels of autonomic arousal, such as elevated heart rate, sweating, and increased blood pressure. A model proposed by Espie, Broomfield, MacMahon, and Taylor (2006) further suggests that the connection between elevated cognitive arousal and autonomic arousal results from dysfunctional affect regulation. Thus, among individuals who experience cognitive arousal (e.g., dysfunctional beliefs about sleep), only those who have a decrement in ability to regulate affective experiences go on to experience autonomic arousal and subsequent sleep disturbance.

In comparison to the top-down approach described above, Reimann et al. (2010) proposed a bottom-up model for explaining the relationship between insomnia and psychopathology. This model posits a diathesis-stress approach, such that a genetic dysfunction in sleep-wake regulation, when coupled with precipitating stressors (i.e., anxiety), results in concurrent sleep and emotional disturbances.

Neurobiological Model

Most recently, Buysse et al. (2011) have proposed a neurobiological model. This model incorporates the above-described behavioral and cognitive models, while incorporating neurological models described by Cano, Mochizuki, and Saper (2008). Specifically, this model suggests that insomnia is a disorder of sleep-wake regulation characterized by the simultaneous occurrence of both sleep- and wake-like neuronal activity within the limbic and parietal cortices, thalamus, and hypothalamic-brainstem, resulting in elevated levels of arousal within these regions during NREM sleep. This model suggests that sleep-state misperception is specifically due to the simultaneous activation of both sleep- and wake-like neuronal activity. Such a model implies that both behavioral and pharmacological interventions would be productive, because behavioral/cognitive treatments would address the physiological components, and medications would target neuronal activity.

Summary

Researchers have proposed several models to account for the onset and maintenance of insomnia and comorbid anxiety, but each model involves elevated cortical, cognitive, and autonomic arousal. The manner in which these domains interact and the impact of these interactions on vary by model; however, all models focus on ways to understand and decrease arousal. This specific focus on arousal within the underlying models of insomnia clearly shows how fear/anxiety can be intertwined so closely with insomnia.

PROPOSED MECHANISMS OF ACTION

In order to fully understand the relationship between fear/anxiety and sleep, researchers must investigate mechanisms that may be accounting for the relationship, at least in part. Neurobiological and cognitive-behavioral mechanisms have been proposed, and these mechanisms have received some empirical support.

Neurobiological

Neurobiological mechanisms may play a role in the relationship between sleep and fear/anxiety. The HPA-axis and the sympathoadrenomedullary system are integrated systems related to both stress and sleep. The primary physiological response to stress is the activation of the HPA-axis and the sympathoadrenomedullary system. These systems then activate the release of corticotropin-releasing hormone (CRH), which, in turn, triggers the release of adrenocorticotropin (ACTH) and glucocorticoids (e.g., cortisol), chemicals associated with fear and anxiety responding. The HPA-axis is also robustly associated with sleep structure. During the early phases of sleep, which are predominately comprised of slow-wave sleep, a persistent inhibition of the HPA-axis results in low levels of ACTH and cortisol. In comparison, during late-sleep, which is predominately comprised of REM sleep, significant elevations in HPA activity lead to increases in ACTH and cortisol (Balbo, Leproult, & Van Cauter, 2010). Although cortisol is primarily associated with stress, it also has a significant impact on sleep such that low cortisol levels are associated with sleep onset, and elevated cortisol levels are associated with wakefulness. In fact, the administration of HPA-axis mediators (i.e., cortisol, ACTH, CRH) has been shown to detrimentally impact sleep architecture, resulting in decreased slow-wave sleep and REM. Based on this model, stress-induced activation of the HPA axis, particularly during early phases of sleep, may significantly impact sleep (Van Reeth et al., 2000).

Cognitive-Behavioral

Researchers have suggested a number of cognitive-behavioral factors to account for the relation between fear/anxiety and sleep. Three of the most highly researched factors are presented herein. These include bedtime arousal, appraisal of the stressor, and coping.

Bedtime Arousal

Physiological and cognitive arousal at the time of bed has been associated not only with delayed sleep onset but also with poor overall sleep quality (Akerstedt, Kecklund, & Axelsson, 2007). Research has shown that, regardless of the frequency of daytime stressors, if an individual goes to sleep in a state of physiological or cognitive arousal, sleep disturbances will result. In comparison, if an individual experiences a stressful day but goes to sleep with low levels of physiological and cognitive arousal, then sleep disturbance in less likely to result (Akerstedt et al., 2007). An individual's ability to reduce physiological and cognitive arousal prior to bed on a stressful day is likely related to two additional factors, appraisal of the stressors and coping style, which might partially account for the relationship between sleep and fear/anxiety.

Appraisal of the Stressor

Research suggests that it is not the frequency of stressful events that impacts sleep, but rather the individual's appraisal of the perceived impact and intensity of the stressor. Individuals with a tendency to worry or ruminate (i.e., elevated levels of negative affect) are most at risk for having a negative appraisal of a stressor (major or minor), which, in turns, maintains cognitive and physiological arousal and results in sleep disturbance.

Coping Style

Similar to the appraisal of a stressor, an individual's coping style has been shown to partially account for the relationship between sleep and fear/anxiety. For example, individuals who use emotion-focused coping (i.e., focus on decreasing distress, not on solving the problem) tend to have greater rates of insomnia (Morin, Rodrigue, & Ivers, 2003). Therefore, interventions targeting insomnia should incorporate the teaching of strategies for effectively appraising stress and coping with stress and anxiety in an effective manner.

Summary

Understanding the factors that account for the relationship between sleep and fear/anxiety is critical for developing effective treatment and prevention models. This is a relatively young area of research, but researchers have already suggested a host of potential mechanisms, all of which require additional investigation. This chapter reviews the neurobiological, cognitive, and behavioral factors that likely interact to form the relationship between sleep and fear/anxiety. For example, an individual who perceives a daily stressor as having a large impact or intensity is likely to experience an increased stress response (activation of the HPA-axis, cortisol release). The consequences of the stress response (physiological and cognitive arousal) may increase worry and rumination, thus maintaining activation of the stress system. If an individual is likely to cope with this stress through emotion-focused coping, the physiological and cognitive arousal is likely to continue, resulting in elevated arousal prior to bed and, ultimately, poor sleep. The next-day consequences of poor sleep are then likely to sustain or increase anxiety responding to stressors, which may drive a vicious cycle and account for the bidirectional relationship between sleep and fear/anxiety.

IMPLICATIONS FOR INTERVENTION

Animal Models

Fear-conditioning models conducted in animals have provided a wealth of information used to create or adjust treatment models (Pace-Schott et al., 2009; Sanford, Silvestri, Ross, & Morrison, 2001; Wellman, Yang, Tang, & Sanford, 2008). Fear-conditioning experiments provide an analogue for the development of fear and anxiety symptoms, as well as the treatment of fear and anxiety. Specifically, during fear-conditioning experiments, rodents learn to fear a specific cue (e.g., tone) through a pairing of the cue with a fear-eliciting stimulus such as a foot shock. After several pairings, the rodent learns that the cue (e.g., tone) means a shock is coming, and it demonstrates a conditioned fear response to the cue, even when the shock no longer follows. The same principles appear to govern how individuals develop fear/anxiety symptoms. An analogue for treatment is also present within these studies in the form of trials focused on extinction learning. After presentation of the cue in a safe environment (i.e., without the presence of the shock), the rodent learns that the cue (tone) is no longer something to fear, and the stress response remits (i.e., extinction learning). Again, anxiety interventions are developed using this same principle. For this reason, fear-conditioning paradigms among animals provide an excellent analog for informing treatment.

This research has demonstrated that fear-conditioning experiments conducted in rodents result in disrupted sleep and changes in sleep architecture, as evidenced by a decrease in the number of REM episodes, REM duration, and REM percentage, and an increase in latency to REM (Sanford et al., 2001). Furthermore, results have shown that a fear-conditioned stimulus has the same impact on sleep architecture as the original unconditioned stimulus (Sanford, Yang, & Tang, 2003). Fear-conditioning experiments have also demonstrated that rats exhibiting extinction of contextual fear experience an increase in sleep compared to rats that continue to exhibit fear in response to the conditioned stimulus (Wellman et al., 2008). Sleep has also been shown to promote the generalization of fear extinction (Pace-Schott et al., 2009). Taken together, animal models suggest that a learned fear response (i.e., elevated cognitive and autonomic arousal) to a conditioned stimulus (e.g., objects, thoughts, feelings, physiological sensations) results in significant sleep disturbance, even when the original feared situation is no longer present. However, extinction of the conditioned fear responses can result in improved sleep, which can further fuel extinction learning. This bench research indicates that interventions to reduce fear/anxiety need to integrate anxiety and sleep interventions, because anxiety reduction improves sleep and sleep improves anxiety reduction.

Clinical Interventions

Early research on interventions for comorbid insomnia and anxiety assumed that, for those with an anxiety disorder, the presence of insomnia was merely a symptom of anxiety, and therefore, the successful treatment of the anxiety would resolve the sleep problems. For example, one study demonstrated that progressive muscle relaxation and anxiety management training resulted in a reduction in sleep onset latency and in an increase in overall sleep quality (Viens, De Konink, Mercier, St-Onge, & Lorrain, 2003). In a separate study, administering eszopliclone and escitalopram to individuals with GAD and insomnia resulted in improved sleep and daytime functioning, as well as lower symptoms of anxiety (Pollack et al., 2008). Finally, among individuals with PTSD, gold standard behavioral interventions for PTSD were shown to significantly reduce PTSD symptoms, but the impact on sleep symptoms was less pronounced. Although behavioral interventions for PTSD reduced sleep problems, the problems were not reduced to a subclinical level. Indeed, residual sleep disturbance is the most common symptom that does not fully remit with PTSD treatment (Galvoski, Monson, Bruce, & Resick, 2009). As this chapter highlights, recent research has indicated the bidirectional and co-occurring nature of sleep problems and fear/anxiety, however, suggesting that sleep disturbances should be independently targeted in order to result in optimal treatment for both the insomnia and the anxiety disorder (Uhde, Cortese, & Vedeniapin, 2009).

Cognitive behavioral therapy for insomnia (CBT-I) is currently the gold standard behavioral intervention for insomnia. Based on the models of insomnia outlined above, this intervention incorporates specific theoretical components in order to most effectively interrupt the insomnia pattern. Specifically, CBT-I is comprised of the following components: (1) sleep hygiene; (2) stimulus control, based on the stimulus control model; (3) sleep restriction, based on the neurobiological model and resetting of circadian rhythms; and (4) cognitive restructuring of maladaptive thoughts related to sleep, based on the cognitive model. A large body of research has demonstrated that CBT-I is an efficacious and effective treatment for insomnia (Morgenthaler et al., 2006; Morin, Culbert, & Schwartz, 1994). This evidence is comprised of work that has compared CBT-I to placebo (Edinger, Wohlgemuth, Radtke, Marsh, & Quillian, 2001; Manber et al., 2008; Rybarczyk et al., 2005) and delayed-treatment controls (Morin, Colecchi, Stone, Sood, & Brink, 1999). CBT-I has also been compared to pharmacotherapy interventions (specifically hypnotics), including temazepam (Morin et al., 1994), zolpidem (Jacobs, Pace-Schott, Stickgold, & Otto, 2004), and zopiclone (Siversten et al., 2006). These studies have combined to indicate that, when compared to pharmacotherapy, CBT-I produces comparable sleep-related improvements from pre- to post-intervention, and it also demonstrates greater long-term effects (Morin et al., 2009; Siversten et al., 2006). The significant evidence base for CBT-I has led to its recognition as a first-line treatment for insomnia by the National Institutes of Health (National Institutes of Health, 2005), as well as its widespread dissemination throughout the Veterans Health Administration (Manber et al., 2012).

In addition to the strong support for CBT-I for individuals with primary insomnia, recent work has demonstrated that CBT-I is an effective treatment for individuals with anxiety disorders such as PTSD (Germain,

Shear, Hall, & Buysse, 2007; Zayfert & DeViva, 2004). For example, in a pilot study of individuals with PTSD, Germain et al. (2007) demonstrated that a single 90-min sleep-focused intervention resulted in clinically significant improvements in sleep and a reduction in daytime PTSD symptom severity. Similarly, in studying a sample of veterans with PTSD, Ulmer, Edinger, and Calhoun (2011) compared CBT-I (with an additional nightmare component) to a usual care control. Their findings supported previous research indicating that CBT-I resulted in improved sleep and reduced daytime PTSD symptom severity, compared to those symptoms experienced by individuals in the usual care control condition. Although clinical trials of CBT-I for individuals with anxiety disorders are limited in number and, to date, have primarily focused on individuals with PTSD, these studies provide promising results for the efficacy of CBT-I in treating individuals with a wide range of anxiety disorders.

CONCLUSIONS AND FUTURE DIRECTIONS

Sleep loss has reached epidemic proportions, and experts now predict a continued increasing trend. Sleep is robustly associated with fear and anxiety among healthy individuals without clinical levels of anxiety and among individuals with anxiety disorders, including panic disorder, PTSD, GAD, and obsessive compulsive disorder. The nature of this relationship is bidirectional, and experimental and longitudinal research has provided significant empirical evidence supporting both temporal orders. Therefore, the relationship between fear/anxiety and sleep is likely cyclical, such that each maintains the other, resulting in a vicious cycle of fear, anxiety, and sleep disturbance. Although this area of research has benefited from increased investigation in recent years, it is still an area ripe for further research, as significant questions remain regarding the intertwined relationships between fear/anxiety and sleep. For example, future research might better illustrate the mechanisms that account for the bidirectional relationships between fear/anxiety and sleep. Such advances could help optimize treatments, while revealing the potential use of sleep interventions in preventing disorders of fear and anxiety.

References

Akerstedt, T., Kecklund, G., & Axelsson, J. (2007). Impared sleep after bedtime stress and worries. *Biological Psychology*, 76, 170–173. http://dx.doi.org/10.1016/j.biopsycho.2007.07.010.

American Academy of Sleep Medicine. (2005). *The international classification of sleep disorders (ICSD): Diagnostic and coding manual* (2nd ed.). Westchester, IL: American Academy of Sleep Medicine.

American Psychiatric Association. (2013). *Diagnostic and statistical manual of mental disorders* (5th ed.). Washington, DC: American Psychiatric Association.

Babson, K., & Feldner, M. (2010). Temporal relations between sleep problems and both traumatic event exposure and PTSD: A critical review of the empirical literature. *Journal of Anxiety Disorders, 24*, 1–15. http://dx.doi.org/10.1016/j.janxdis.2009.08.002.

Babson, K. A., Feldner, M. T., Trainor, C. D., & Smith, R. C. (2009). An experimental investigation of the effects of acute sleep deprivation on panic-relevant biological challenge responding. *Behavior Therapy, 40*, 239–250. http://dx.doi.org/10.1016/j.beth.2008.06.001.

Babson, K. A., Trainor, C. D., Feldner, M. T., & Blumenthal, H. (2010). A test of the effects of acute sleep deprivation on general and specific self-reported anxiety and depressive symptoms: An experimental extension. *Journal of Behavior Therapy and Experimental Psychiatry, 41*, 297–303. http://dx.doi.org/10.1016/j.jbtep.2010.02.008.

Balbo, M., Leproult, R., & Van Cauter, E. (2010). Impact of sleep and its disturbances on hypothalamo-pituitary-adrenal axis activity. *International Journal of Endocrinology, 2010*, 1–16. http://dx.doi.org/10.1155/2010/759234.

Benca, R. M., Obermeyer, W. H., Thisted, R. A., & Gillin, J. C. (1992). Sleep and psychiatric disorders: A meta-analysis. *Archives of General Psychiatry, 49*, 651–668. http://dx.doi.org/10.1001/archpsyc.1992.01820080059010.

Bootzin, R. (1972). Stimulus control treatment for insomnia. *Proceedings of the American Psychological Association, 7*, 395–396.

Breslau, N., Roth, T., Rosenthal, L., & Andreski, P. (1996). Sleep disturbance and psychiatric disorders: A longitudinal epidemiological study of young adults. *Biological Psychiatry, 39*, 411–418. http://dx.doi.org/10.1016/0006-3223(95)00188-3.

Buysse, D., Germain, A., Hall, M., Monk, T., & Nofzinger, E. (2011). A neurobiological model of insomnia. *Drug Discovery Today: Disease Models, 8*, 129–137. http://dx.doi.org/10.1016/j.ddmod.2011.07.002.

Buysse, D., Reynolds, C., Kupfer, D., & Thorpy, M. (1994). Clinical diagnoses in 216 insomnia patients using the international classification of sleep disorders (ICSD), DSM-IV and ICD-10 categories: A report from the APA/NIMH DSM-IV field trial. *Sleep: Journal of Sleep Research and Sleep Medicine, 17*, 630–637.

Cano, G., Mochizuki, T., & Saper, C. (2008). Neural circuitry of stress-induced insomnia in rats. *Journal of Neuroscience, 40*, 10167–10184. http://dx.doi.org/10.1523/JNEUROSCI.1809-08.2008.

Centers for Disease Control and Prevention. (2014). *Insufficient sleep is a public health epidemic.* Atlanta, Georgia: National Center for Chronic Disease and Prevention and Health Promotion, Division of Adult and Community Health. http://www.cdc.gov/features/dssleep/.

Edinger, J., Bonnet, M., Bootzin, R., Doghramji, K., Dorsey, C., Espie, C., et al. (2004). Derivation of research diagnostic criteria for insomnia: Report of an American Academy of Sleep Medicine work group. *Sleep, 27*, 1567–1596.

Edinger, J., Wohlgemuth, W., Radtke, R., Marsh, G., & Quillian, R. (2001). Cognitive behavioral therapy for treatment of chronic primary insomnia: A randomized controlled trial. *Journal of the American Medical Association, 285*, 1856–1864. http://dx.doi.org/10.1001/jama.285.14.1856.

Espie, C. (2002). Insomnia: Conceptual issues in the development, persistence, and treatment of sleep disorders in adults. *Annual Review of Psychology, 53*, 215–243.

Espie, C., Broomfield, N., MacMahon, K., & Taylor, L. (2006). The attention-intention-effort pathway in the development of psychophysiologic insomnia: A theoretical review. *Sleep Medicine Reviews, 10*, 215–245. http://dx.doi.org/10.1016/j.smrv.2006.03.002.

Ford, D., & Kamerow, D. (1989). Epidemiologic study of sleep disturbances and psychiatric disorders. *Journal of the American Medical Association, 262*, 1479–1484. http://dx.doi.org/10.1001/jama.1989.03430110069030.

Galvoski, T., Monson, C., Bruce, S., & Resick, P. (2009). Does cognitive-behavioral therapy for PTSD improve perceived health and sleep impairment? *Journal of Traumatic Stress, 22*, 197–204. http://dx.doi.org/10.1002/jts.20418.

Germain, A., Shear, K., Hall, M., & Buysse, D. (2007). Effects of a brief behavioral treatment for PTSD-related sleep disturbances: A pilot study. *Behaviour Research and Therapy, 45*, 627–632. http://dx.doi.org/10.1016/j.brat.2006.04.009.

Gregory, A. M., & O'Connor, T. G. (2002). Sleep problems in childhood: A longitudinal study of developmental change and association with behavioral problems. *Journal of the American Academy of Child and Adolescent Psychiatry, 41*(8), 964–971.

Gregory, A., Caspi, A., Eley, T., Moffitt, T., O'Connor, T., & Poulton, R. (2005). Prospective longitudinal associations between persistent sleep problems in childhood and anxiety and depression disorders in adulthood. *Journal of Abnormal Child Psychology, 33*, 157–163. http://dx.doi.org/10.1007/s10802-005-1824-0.

Gregory, A., Eley, T., O'Connor, T., & Plomin, R. (2004). Etiologies of the associations between childhood sleep and behavioral problems in a large twin sample. *Journal of the American Academy of Child and Adolescent Psychiatry, 43*, 748–757. http://dx.doi.org/10.1097/01.chi/0000122798.47863.a5.

Gross, R. T., & Borkovec, T. D. (1982). Effects of a cognitive intrusion manipulation on the sleep-onset latency of good sleepers. *Behavior Therapy, 13*, 112–116. http://dx.doi.org/10.1016/S0005-7894(82)80054-3.

Harvey, A. G. (2002). A cognitive model of insomnia. *Behaviour Research and Therapy, 40*, 869–893. http://dx.doi.org/10.1016/S0005-7967(01)00061-4.

Harvey, A. G. (2003). Attempted suppression of pre-sleep cognitive activity in insomnia. *Cognitive Therapy and Research, 27*, 593–602.

Harvey, A. G., & Bryant, R. A. (1998). The effect of attempted thought suppression in acute stress disorder. *Behaviour Research and Therapy, 36*, 583–590. http://dx.doi.org/10.1016/S0005-7967(98)00052-7.

Haynes, S., Adams, A., & Frazen, M. (1981). The effects of presleep stress on sleep-onset insomnia. *Journal of Abnormal Psychology, 90*, 601–606. http://dx.doi.org/10.1037/0021-843X.90.6.601.

Jacobs, G. D., Pace-Schott, E. F., Stickgold, R., & Otto, M. W. (2004). Cognitive behavior therapy and pharmacotherapy for insomnia: A randomized controlled trial and direct comparison. *Archives of Internal Medicine, 164*(17), 1888–1896. http://dx.doi.org/10.1001/archinte.164.17.1888.

Johnson, E., Roth, T., & Breslau, N. (2006). The association of insomnia with anxiety disorders and depression: Exploration of the direction of risk. *Journal of Psychiatric Research, 40*, 700–708. http://dx.doi.org/10.1016/j.jpsychires.2006.07.008.

Kessler, R., Chiu, W., Demler, O., & Walters, E. (2005). Prevalence, severity, and comorbidity of 12-month DSM-IV disorders in the national comorbidity survey replication. *Archives of General Psychiatry, 62*, 617–627. http://dx.doi.org/10.1001/archpsyc.62.6.617.

Kobayashi, I., Huntley, E., Lavela, J., & Mellman, T. (2012). Subjectively and objectively measured sleep with and without posttraumatic stress disorder and trauma exposure. *Sleep, 35*, 957–965. http://dx.doi.org/10.5665/sleep.1960.

Koren, D., Arnon, I., Lavie, P., & Klein, E. (2002). Sleep complaints as early predictors of posttraumatic stress disorder: A 1-year prospective study of injured survivors of motor vehicle accidents. *American Journal of Psychiatry, 159*, 855–857. http://dx.doi.org/10.1176/appi.ajp.159.5.855.

Lund, H., Reider, B., Whiting, A., & Prichard, R. (2010). Sleep patterns and predictors of disturbed sleep in a large population of college students. *Journal of Adolescent Health, 46*, 124–132. http://dx.doi.org/10.1016/j.jadohealth.2009.06.016.

Manber, R., Carney, C., Edinger, J., Epstein, D., Friedman, L., Haynes, P., et al. (2012). Dissemination of CBTI to the non-sleep specialist: Protocol development and training issues. *Journal of Clinical Sleep Medicine, 8*, 209–218. http://dx.doi.org/10.5664/jcsm.1786.

Manber, R., Edinger, J., Gress, J., San Pedro-Salcedo, M., Kuo, T., & Kalista, T. (2008). Cognitive behavioral therapy for insomnia enhances depression outcome in patients with comorbid major depressive disorder and insomnia. *Sleep, 31,* 489–495. http://dx.doi.org/10.5664/jcsm.1786.

Marcks, B. A., Weisber, R., Edele, M., & Keller, M. (2010). The relationship between sleep disturbance and the course of anxiety disorders in primary care patients. *Psychiatry Research, 178,* 487–492.

Mellinger, G., Balter, M., & Uhlenhuth, E. (1985). Insomnia and its treatment prevalence and correlates. *Archives of General Psychiatry, 42,* 225–232. http://dx.doi.org/10.1001/archpsyc.1985.01790260019002.

Mellman, T., David, D., Kulick-Bell, R., Hebding, J., & Nolan, B. (1995). Sleep disturbance and its relationship to psychiatric morbidity after hurricane Andrew. *American Journal of Psychiatry, 152,* 1659–1663.

Morgenthaler, T., Kraemer, M., Alessi, C., Friedman, L., Boehlecke, B., Brown, T., et al. (2006). Practice parameters for the psychological and behavioral treatment of insomnia: An update. An AASM Report. *Sleep, 29,* 1415–1419.

Morin, C., Colecchi, C., Stone, J., Sood, R., & Brink, D. (1999). Behavioral and pharmacological therapies for late-life insomnia: A randomized controlled trial. *Journal of the American Medical Association, 28,* 991–999. http://dx.doi.org/10.1001/jama.281.11.991.

Morin, C., Culbert, J., & Schwartz, S. (1994). Nonpharmacological interventions for insomnia: A meta-analysis of treatment efficacy. *American Journal of Psychiatry, 151,* 1172–1180.

Morin, C., Rodrigue, S., & Ivers, H. (2003). Role of stress, arousal, and coping skills in primary insomnia. *Psychosomatic Medicine, 65,* 259–267. http://dx.doi.org/10.1097/01.PSY.0000030391.09558.A3.

Morin, C., Vallieres, A., Guay, B., Ivers, H., Savard, J., Merette, C., et al. (2009). Cognitive-behavior therapy, singly and combined with medication, for persistent insomnia: Acute and maintenance therapeutic effects. *Journal of the American Medical Association, 301,* 2005–2015. http://dx.doi.org/10.1001/jama.2009.682.

National Institutes of Health. (2005). State of the science conference statement on manifestations and management of chronic insomnia in adults. *Sleep, 28,* 1049–1057.

Neckelmann, D., Mykletun, A., & Dahl, A. (2007). Chronic insomnia as a risk factor for developing anxiety and depression. *Sleep, 30,* 873–880.

Oginska, H., & Pokorski, J. (2006). Fatigue and mood correlates of sleep length in three age-social groups: School children, students, and employees. *Chronobiology International, 23,* 1317–1328. http://dx.doi.org/10.1080/07420520601089349.

Ohayon, M., & Roth, T. (2003). Place of chronic insomnia in the course of depressive and anxiety disorders. *Journal of Psychiatric Research, 37,* 9–15. http://dx.doi.org/10.1016/S0022-3956(02)00052-3.

Öhman, A. (2008). Fear and anxiety: Overlaps and dissociations. In M. Lewis & J. M. Haviland-Jones (Eds.), *Handbook of emotions*. (3rd ed.), (pp. 709–729). New York: Guilford.

Ong, S., Wickramaratne, P., Tang, M., & Weissman, M. (2006). Early childhood sleep and eating problems as predictors of adolesncent and adult mood and anxiety disorders. *Journal of Affective Disorders, 96,* 1–8. http://dx.doi.org/10.1016/j.jad.2006.05.025.

Pace-Schott, E., Milad, M., Orr, S., Rauch, S., Stickgold, R., & Pitman, R. (2009). Sleep promotes generalization of extinction of conditioned fear. *Sleep, 32,* 19–26.

Papadimitriou, G., & Linkowski, P. (2005). Sleep disturbance in anxiety disorders. *International Review of Psychiatry, 17,* 229–236. http://dx.doi.org/10.1080/09540260500104524.

Perlis, M., Giles, D., Mendelson, W., Bootzin, R., & Wyatt, J. (1997). Psychophysiological insomnia: The behavioural model of a neurocognitive perspective. *Journal of Sleep Research, 6,* 179–188. http://dx.doi.org/10.1046/j.1365-2869.1997.00045.x.

Pillai, V., & Delahanty, D. (2012). Sleep perception among individuals with posttraumatic stress disorder. *Sleep, 35,* 897–898. http://dx.doi.org/10.5665/sleep.1940.

Pollack, M., Kinrys, G., Krystal, A., McCall, V., Roth, T., Schaefer, K., et al. (2008). Eszopiclone coadministered with escitalopram in patients with insomnia and comorbid generalized anxiety disorder. *Archives of General Psychiatry, 65*, 551–562. http://dx.doi.org/10.1001/archpsyc.65.5.551.

Rapee, R. (1995). Psychological factors influencing the affective response to biological challenge procedures in panic disorder. *Journal of Anxiety Disorders, 9*, 59–74. http://dx.doi.org/10.1016/0887-6185(94)00028-9.

Reimann, D., Spiegelhalder, K., Feige, B., Voderholzer, U., Berger, M., Perlis, M., et al. (2010). The hyperarousal of insomnia: A review of the concept and its evidence. *Sleep Medicine Reviews, 14*, 19–31. http://dx.doi.org/10.1016/j.smrv.2009.04.002.

Robins, L. N., Cottler, L., Bucholz, K., & Compton, W. (1995). *The diagnostic interview schedule, version IV*. St. Louis, MO: Washington University.

Roth, T. (2007). Insomnia: Definition, prevalence, etiology, and consequences. *Journal of Clinical Sleep Medicine, 3*, S7–S10.

Roy-Byrne, P. P., Uhde, T. W., & Post, R. M. (1986). Effects of one night's sleep deprivation on mood and behavior in panic disorder. *Archives of General Psychiatry, 43*, 895–899. http://dx.doi.org/10.1001/archpsyc.1986.01800090085011.

Rybarczyk, B., Stepanski, E., Fogg, L., Lopez, M., Barry, P., & Davis, A. (2005). A placebo-controlled test of cognitive-behavioral therapy for comorbid insomnia in older adults. *Journal of Consulting and Clinical Psychology, 73*, 1164–1174. http://dx.doi.org/10.1037/0022-006X.73.6.116.

Sagaspe, P., Sanchez-Ortuno, M., Charles, A., Taillard, J., Valtat, C., Bioulac, B., et al. (2006). Effects of sleep deprivation on color-word, emotional, and specific stroop interference and on self-reported anxiety. *Brain and Cognition, 60*, 76–87. http://dx.doi.org/10.1016/j.bandc.2005.10.001.

Sanford, L. D., Silvestri, A. J., Ross, R. J., & Morrison, A. R. (2001). Influence of fear conditioning on elicited ponto-geniculo-occipital waves and rapid eye movement sleep. *Archives Italiennes de Biologie, 139*(3), 169–183.

Sanford, L., Yang, L., & Tang, X. (2003). Influence of contextual fear on sleep in mice: A strain comparison. *Sleep, 26*, 527–540.

Siversten, B., Omvik, S., Pallesen, S., Bjorvatn, B., Havik, O., Kvale, G., et al. (2006). Cognitive behavioral therapy vs zopiclone for treatment of chronic primary insomnia in older adults: A randomized controlled trial. *Journal of the American Medical Association, 295*, 2851–2858. http://dx.doi.org/10.1001/jama.295.24.2851.

Spielman, A., Caruso, L., & Glovinsky, P. (1987). A behavioral perspective on insomnia treatment. *Psychiatric Clinics of North America, 10*, 541–553.

Spira, A., Friedman, L., Aulakh, J., Lee, T., Sheikh, J., & Yesavage, J. (2008). Subclinical anxiety symptoms, sleep, and daytime dysfunction in older adults with primary insomnia. *Journal of Geriatric Psychiatry and Neurology, 21*, 56–60. http://dx.doi.org/10.1177/0891988707311043.

Taylor, D., Lichstein, K., Durrence, H., Reidel, B., & Bush, A. (2005). Epidemiology of insomnia, depression, and anxiety. *Sleep, 28*, 1457–1464.

Uhde, T., Cortese, B., & Vedeniapin, A. (2009). Anxiety and sleep problems: Emerging concepts and theoretical treatment implications. *Current Psychiatry Reports, 11*, 269–276.

Ulmer, C., Edinger, J., & Calhoun, P. (2011). A multi-component cognitive-behavioral intervention for sleep disturbance in veterans with PTSD: A pilot study. *Journal of Clinical Sleep Medicine, 7*, 57–68.

Vahtera, J., Kivimaki, M., Hublin, C., Korkeila, K., Suominen, S., Paunio, T., et al. (2007). Liability to anxiety and severe life events as predictors of new-onset sleep disturbances. *Sleep, 30*, 1537–1546.

Van Reeth, O., Weibel, L., Spiegel, K., Leproult, R., Dugovic, C., & Maccari, S. (2000). Interactions between stress and sleep: From basic research to clinical situations. *Sleep Medicine Reviews, 4*, 201–219. http://dx.doi.org/10.1053/smrv.1999.0097.

Viens, M., De Konink, J., Mercier, P., St-Onge, M., & Lorrain, D. (2003). Trait anxiety and sleep-onset insomnia: Evaluation of treatment using anxiety management training. *Journal of Psychosomatic Research, 54*, 31–37. http://dx.doi.org/10.1016/S0022-3999(02)00568-8.

Wellman, L. L., Yang, L., Tang, X., & Sanford, L. D. (2008). Contextual fear extinction ameliorates sleep disturbances found following fear conditioning in rats. *Sleep, 31*(7), 1035.

World Health Organization. (1992). *The ICD-10 classification of mental and behavioural disorders: Clinical descriptions and diagnostic guidelines* (Vol. 1). World Health Organization.

Zayfert, C., & DeViva, J. (2004). Residual insomnia following cognitive behavioral therapy for PTSD. *Journal of Traumatic Stress, 17*, 69–73. http://dx.doi.org/10.1023/B:JOTS.0000014679.31799.e7.

Zvolensky, M. J., & Eifert, G. H. (2001). A review of psychological factors/processes affecting anxious responding during voluntary hyperventilation and inhalations of carbon dioxide-enriched air. *Clinical Psychology Review, 21*, 375–400. http://dx.doi.org/10.1016/S0272-7358(99)00053-7.

CHAPTER 8

Nightmares and the Mood Regulatory Functions of Sleep

Patrick J. McNamara[*,†,‡], Umberto Prunotto[§], Sanford H. Auerbach[¶], and Alina A. Gusev[||]

[*]Department of Neurology, Boston University School of Medicine, Boston, Massachusetts, USA
[†]VA New England Healthcare System, Boston, Massachusetts, USA
[‡]Graduate School Dissertation Chair, Northcentral University, Prescott Valley, Arizona, USA
[§]Dreamboard, Inc., Alba, Italy
[¶]Boston University School of Medicine, Sleep Disorders Center, Boston, Massachusetts, USA
[||]VA New England Healthcare System, Boston, Massachusetts, USA

NIGHTMARES AND THE MOOD REGULATORY FUNCTIONS OF SLEEP

Nightmares are frightening dreams that most often occur during rapid eye movement (REM) sleep (Hartmann, 1984, 1998; Haynes & Mooney, 1975; Krakow, 2006). The REM-related nightmares typically appear during one of the late REM episodes that occur during the early morning period between 4 and 7 a.m. (Hartmann, 1984; Nielsen & Zadra, 2005; Spoormaker, Schredl, & van den Bout, 2006). The arousal associated with the nightmare may or may not be accompanied by signs of sympathetic surge (increased heart rate and blood pressure, as well as respiratory changes) and typically lasts between 5 and 30 min (Nielsen & Zadra, 2005). In addition, there is increased parasympathetic tone. Epidemiological studies (Belicki & Belicki, 1982; Bixler, Kales, Soldatos, Kales, & Healy, 1979; Haynes & Mooney, 1975; Levin, 1994; Ohayon, Morselli, & Guilleminault, 1997) indicate that 2-6% of the American population experiences nightmares at least once a week. That translates into approximately 18 million individuals in the US alone who experience frequent nightmares. To understand the nature of nightmares and the ways in which nightmares reveal the mood regulatory functions of sleep, we must first describe the characteristics of REM sleep.

REM PROPERTIES

The NREM-REM Cycle

REM sleep accounts for about 22% of total sleep time in humans. The other major form of human sleep is called non-REM and is composed of three key stages or phases of sleep, including N1, which is a transitional stage from wake into sleep; N2, which is a form of light sleep and which makes up the bulk of our sleep time; and N3 or slow-wave sleep (SWS), which appears to serve a variety of functions from physiologic repair to memory consolidation. Although the cortex is activated in REM, arousal thresholds in humans are variable in REM. A typical night of sleep will consist of NREM alternating with REM. As the night progresses, less time is devoted to SWS and more to REM. Thus, REM sleep tends to be more prominent toward the last part of the night (Carskadon & Rechtschaffen, 2000). REM sleep can be further divided into its tonic and phasic components.

The Tonic Component of REM

REM sleep is associated with the occurrence of some very odd physiologic events, including a desynchronized electroencephalogram (EEG), which is characteristic of the waking attentive state, and atonia of the antigravity muscles indicating that the sleeper is largely paralyzed during this phase of sleep. Other tonic features include hippocampal theta rhythm and predominant parasympathetic tone with pupillary constrictions and penile erections. Thermoregulation also decreases. The desynchronized EEG reflects widespread brain activation effects during REM.

The Phasic Component of REM Sleep

REM's phasic characteristics are equally mysterious and include bursts of REMs under the closed eyelids; myoclonic twitches of the facial and limb muscle groups; increased variability in heart rate, respiration, and blood pressure; and autonomic nervous system (ANS) discharges. The phasic aspects of REM, such as intermittent muscle twitching, ANS discharges, and REMs, occur in some mammals in association with bursts of pontine-geniculo-occipital (PGO) waves that activate the visual centers of the brain. As is the case with non-REM SWS deprivation, REM deprivation results in a rebound phenomenon wherein the individual spends more time in REM than usual until he or she catches up on some critical proportion of lost sleep. This REM rebound phenomenon may indicate that a certain amount of REM is required and must be made up if lost.

Brain Activation and REM

Brain activation patterns are significantly different for REM and NREM sleep. REM sleep demonstrates high activation levels in midline ventral paralimbic areas, including the medial prefrontal cortex, along with deactivation of the lateral prefrontal cortices (Maquet et al., 1996; Maquet & Franck, 1997; Nofzinger, Mintun, Wiseman, Kupfer, & Moore, 1997). Nofzinger et al. (1997, 2006) identify a midline anterior paralimbic area that includes both subcortical regions, such as the basal forebrain, hypothalamus, ventral striatum, and hippocampus, as well as paralimbic cortices, such as the insula, anterior cingulate, orbitofrontal cortex, and supplementary motor area. Reactivation of these paralimbic areas, as well as deactivation of dorsal prefrontal regions, with the transition into REM sleep has now been widely replicated (reviewed in Maquet et al., 2005; Nofzinger et al., 2006; Pace-Schott et al., 2009). In contrast, NREM and SWS are associated with deactivation of widespread areas of the cortex and subcortex, including those REM-activated anterior paralimbic areas (Dang-Vu et al., 2005, Kaufmann et al., 2006), a deactivation that correlates with the delta rhythms of SWS (Dang-Vu et al., 2005; Hofle et al., 1997). N2 NREM sleep involves higher activation levels in the cortex than SWS does, however (Kaufmann et al., 2006; Maquet, 1995), and we use N2 NREM sleep as a control or comparison state for our studies on REM sleep-associated emotional regulatory processes.

REM AND NIGHTMARES

It is difficult to see how a physiologic system such as REM could facilitate daytime mood regulation, but the evidence suggests that it does. We review some of that evidence below. On the other hand, it is easy to see how REM could facilitate nightmares. If the activation of the amygdala during REM becomes too intense, that activation could be associated with fear and stress. A simple hypothesis suggests that REM-related nightmares are due entirely to overactivation of the amygdala. But, if that simple explanation is to work, we need to know why the amygdala is not overactivated during every REM episode or why overactivation occurs only infrequently for most people and too frequently for individuals who suffer frequent nightmares. We need, in short, more information on nightmares themselves.

Phenomenology of Nightmares

For most persons with repeat nightmares, the dominant emotion is terror, overwhelming in its intensity. Frequent themes include the dreamer involved in aggressive encounters, being chased or attacked by unknown, threatening men or animals or supernatural beings, enclosure in unpleasant surroundings, dread due to imminent violent assault by an unknown stranger, and so on (Hartmann, 1984; Levin, 1994). Pursuit nightmares are fairly common and occur in 92% of women and 85% of men who report nightmares (Nielsen & Zadra, 2005). The prevalence of attack dreams may ranging from 67% to 90% (Harris, 1948; Nielsen & Zadra, 2005). Violence by the dreamer against others is uncommon even when the dreamer is being physically attacked, however.

Syndrome of Recurring Nightmares

Up to 50% of children between 3 and 6 years of age and 20% between 6 and 12 years experience frequent nightmares (American Psychiatric Association, 2000; Fisher, Pauley, & McGuire, 1989; MacFarlane, Allen, & Hoznik, 1954; Mindell & Barrett, 2002; Salzarulo & Chevalier, 1983; Simonds & Parraga, 1982; Vela-Bueno et al., 1985). Both the frequency of nightmares and the intensity of the distress they cause decline with age so that, by the time an individual reaches maturity, nightmares are only rarely experienced.

Frequent or repetitive nightmares are often associated with a recurrent theme, as well as increased frequency of dreaming. At least a subset of recurring nightmares has been linked to complex partial seizures (Penfield & Erickson, 1941; Solms, 1997). Some of these recurring nightmares have been considered to be manifestations of these seizures. On the other hand, recurring nightmares may be independent of any obvious epileptic activity. These latter recurrences tend to be less stereotyped and usually depict unpleasant conflicts and stresses that may vary over time (Cartwright, 1979; Zadra, 1996). Although some of these less well-stereotyped recurrences reflect significant trauma or psychopathology, many do not. Patients often report recurrent themes of being chased or threatened in some manner or being confronted with a natural disaster. Dreams with less recurrence and described according to recurrent themes or recurrent contents are not so clearly associated with psychopathology and may reflect adaptive functions (Domhoff, 1993).

SLEEP DISORDERS WITH NIGHTMARES

REM Behavior Disorder

First described in 1986, REM sleep behavior disorder (RBD) is considered to be a disorder of dream enactment. Although onset may occur at almost any age, it is most commonly encountered in middle-aged and older men, and it is associated with a loss of the muscle atonia that usually accompanies REM sleep. RBD can be seen in a variety of neurological conditions that have been associated with focal pontine lesions. It can also arise in patients with other sleep disorders, such as sleep apnea or narcolepsy. In addition, certain medications have been known to cause or, perhaps, provoke an underlying tendency to have RBD. Most commonly, RBD is now thought to be associated with a group of neurodegenerative disorders referred to as the synucleinopathies, which include Parkinson's disease and dementia with Lewy bodies. RBD may actually antecede the emergence of other symptoms by many years (Boeve et al., 2007; Olson, Boeve, & Silber, 2000).

Patients with RBD often present with a history of bad dreams or nightmares. Nightmares of RBD are vivid, action-filled, violent dreams that the dreamer acts out (due to disease-induced loss of REM-related atonia), sometimes resulting in injury to the dreamer or the sleeping partner (Schenck & Mahowald, 1996). Patients with idiopathic RBD typically complain of a history of vivid unpleasant dreams or nightmares and excessive movements in sleep. There appears to be enactment of violent dream content and a concomitant failure of REM-related muscle atonia. As with Jouvet, Vimont, Delorme, and Jouvet (1964; Jouvet, 1999) pontine-lesioned cats, which were thought to exhibit oneric behaviors normally under output inhibition, these patients appear to suffer from a similar disinhibition of selective brainstem motor pattern generators. The disinhibition, in turn, may be due to a pathologic process that affects pontine and some basal ganglia sites and other midline structures.

Like more typical nightmare content, nightmare content in RBD involves the patient under some sort of threat, directed either against himself or his wife or significant others. Most patients report that they repeatedly experienced this RBD-related nightmare of being attacked by animals, monsters, demons, or unfamiliar people. The dreamer attempts to fight back in self-defense. Fear rather than anger is the usual accompanying emotion reported. Despite these self-reported emotions in RBD dreams, content analyses of the dreams of RBD patients reveal a very high degree of aggression (Fantini & Ferini-Strambi, 2007). Compared to controls, patients with RBD report a very high frequency of aggression, expressed by various indicators, namely a higher percentage of *Dreams with at least one aggression* (66% vs. 15%), an increased aggression/friendliness interactions ratio (86% vs. 44%), and an increased aggressions/characters (A/C) ratio (0.81 vs. 0.12). According to Mahowald and Schenck (2000), the overall sleep architecture is typically intact, but most patients show increased SWS for their age. Schenck and Mahowald (1990) found that 28 of 65 patients they evaluated evidenced increased REM percent (>25% of total sleep time). In the Olson et al. study of 93 consecutive patients with RBD (Olson et al., 2000), 90 showed increased phasic activity in REM.

In summary, the nightmares exhibited by people with REM Behavior Disorder are much like the nightmares experienced by people without neurologic disease. They are under attack by anomalous agents such as dangerous strangers or animals, and they strive to defend themselves.

N2 NREM Nightmares

Some nightmares occur in association with NREM rather than REM sleep. Among these NREM nightmares are a small number that occur in N2 NREM sleep (Dement & Kleitman, 1957a, 1957b; Fisher, Byrne, Edwards, & Kahn, 1970; Vogel, Foulkes, & Trosman, 1966). They are very much like REM nightmares but appear to cause more distress than REM nightmares, perhaps because they are occurring in a very light sleep and so are more easily remembered.

Nightmares Associated with Narcolepsy

Narcolepsy is a disorder of sleep attacks in which an individual may suddenly experience an overwhelming sleepiness. A sleep attack can be triggered by intense emotion. In addition, the narcoleptic patient may experience terrifying sleep-onset (hypnagogic) or awakening (hypnopompic) hallucinations that are like short nightmares. The original use of the term nightmare seems to have denoted the combination of sleep paralysis and hypnagogic hallucinations (Mahowald & Ettinger, 1990). These hypnagogic hallucinations are usually associated with a degree of sleep paralysis and include seeing simple visual forms changing in size or more elaborate images of animals or people engaging in threatening actions. Other terrifying nightmare images may occur when the individual is attempting to wake up. Sleep paralysis is an inability to move during the transition into or out of sleep. When the patient is just waking up but cannot move, some terrifying images may appear to him or her. This sort of nightmare imagery occurs in more than 33% of the general population (Partinen, 1994).

MOOD DISTURBANCES, DEPRESSION, AND NIGHTMARES

Nightmares are most frequently associated with anxiety, depression, and sleep disturbances. Among individuals ($n = 718$) who seek medical help for a sleep disturbance of any kind, 26% report experiencing nightmares (Krakow, 2006). Moreover, among the patients who reported having nightmares, nightmares ranked among their major sleep problems, with 66% claiming that their bad dreams contributed to their disrupted sleep.

In addition, this subset of patients evidenced greater insomnia, sleep fragmentation, sleep-related daytime impairment, psychiatric history, and various medical conditions compared to the sleep patients who did not endorse having nightmares.

One frequently reported clinical correlate of recurrent nightmares is depression. Reduced REM sleep latency was identified as an objective indicator of depressive disorder and an inverse correlate of its severity 30 years ago (Kupfer & Foster, 1972). In subsequent years, reduced REM latency has proved to be one of the most robust and specific, if not exclusive, features of sleep in depressed patients (Benca, 2000). Other reported abnormalities in REM sleep include a prolonged duration of the first REM period, an increased density of eye movements, and an increased REM percentage of total sleep time. In short, there are strong physiological reasons to think of depression as a disorder of REM. Of course, the primary symptom in depression is mood dysfunction, thus linking REM once again to mood regulatory capacities. Other symptoms of depression include insomnia, pervasive anhedonia, negative or unreactive mood, appetite loss, and unpleasant dreams that sometimes develop into nightmares (Giles, Roffwarg, Schlesser, & Rush, 1986; McNamara, Auerbach, Johnson, Harris, & Doros, 2010). Interestingly, severely depressed people reportedly have poorer dream recall and more blunted dream affect (Armitage, Hudson, Trivedi, & Rush, 1995) than those who have less severe depression. Females, especially depressed females, report higher rates of unpleasant dreams and nightmares than other psychiatric patients or healthy subjects (Cartwright, 1991; McNamara, 2008). If depression is indeed a disorder involving too much REM, then, theoretically, deprivation of REM should improve mood. Prolonged selective REM deprivation by awakenings has improved mood (Vogel, Vogel, McAbee, & Thurmond, 1980). To our knowledge, selective REM deprivation has not been tried as a treatment for nightmares, however. The fact that most antidepressant drugs, across several different categories, exhibit robust suppression of REM sleep may support the therapeutic effects of REM deprivation, as well as the key role REM must play in mood regulation.

Recurrent nightmare sufferers with depression are thought to be at increased risk for suicidal ideation, although it is unclear whether the risk extends to actual attempts (Ağargün & Cartwright, 2003). Ağargün et al. (1998) examined the association between repetitive and frightening dreams and suicidal tendencies in patients with major depression. They reported that the patients with frequent nightmares, particularly the women, had higher mean suicide ideation scores and were more likely to be classified as suicidal. A prospective follow-up study in a sample drawn from the general population (Tanskanen et al., 2001) also reported that the frequency of nightmares is directly related to the risk of suicide.

Nightmares and PTSD

Recurrent nightmares occur in 40-71% of individuals with posttraumatic stress disorder or PTSD (Schreuder, Kleijn, & Rooijmans, 2000; Leskin, Woodward, Young, & Sheikh, 2002). When individuals with PTSD carry an additional neuropsychiatric diagnosis such as panic disorder, recurrent nightmares can occur in up to 96% (Leskin et al., 2002). Nightmares can, in fact, be considered a hallmark feature of PTSD. In fact, the importance of nightmares in PTSD can be highlighted by the fact that, in a recent practice guideline for the treatment of a nightmare disorder, the authors relied heavily on the PTSD literature. Unlike "typical" nightmares, the disturbing dreams in PTSD have content that is often linked to a specific memory of a traumatic incident in the person's life. The individual, in effect, relives the trauma with each nightmare. PTSD-related nightmares may occur in a REM episode throughout the sleep period instead of being confined to late REM episodes, as is the case with most nightmares (Nielsen & Zadra, 2005). Polysomnographic studies of PTSD patients have shown that they have poor sleep maintenance, increased eye movement density, decreased percentage of REM sleep, and increased tendency to have REM sleep at sleep onset (REM pressure). These data imply that people with PTSD may also experience increased amounts of SWS, although this has not yet been demonstrated.

Trauma-related nightmares are associated with frequent awakenings (Germaine & Nielsen, 2003; Woodward, Arsenault, Murray, & Bliwise, 2000) and are extremely distressing to the patient. One can only imagine how terrible it must feel to repeatedly undergo elements of a horrible traumatic experience such as combat or rape. In some sufferers, these nightmares have been found to recur for decades with little or no change (Coalson, 1995).

THEORIES OF THE NIGHTMARE

Freud on Nightmares

Freud's original dream theory postulated that the dream helped the sleeping person stay asleep because it allowed the person to hallucinate the satisfaction of desires and wishes while still asleep, and thus dreams protected sleep. Nightmares were hard to understand using the original Freudian theory of dream function, however. Nightmare images cannot be construed as wish fulfillments without engaging in a severe case of special pleading. Furthermore, given the fact that many people awaken from their nightmares, it seems hard to argue that nightmares protect sleep via hallucinated wish fulfillment.

Fisher on Nightmares

In one of the earliest theories of REM dreaming as mood regulation, Fisher and colleagues (1970) noted that REM nightmares occurring spontaneously in the laboratory were characterized by low or even absent ANS activation (e.g., heart rate, respiratory rate, etc.). Fisher and colleagues argued that the lowered ANS in some REM nightmares suggested that the normal function of REM was to desomatize or decouple ANS activation from emotional surges during REM. Nightmares result when the anxiety exceeds a certain threshold, and the REM desomatization mechanism breaks down, allowing autonomic activation to occur.

Hartmann on Nightmares

Hartmann, Russ, Oldfield, Sivan, and Cooper (1987) pointed out that nightmare patients exhibit "thin boundaries," so that they seem unable to effectively screen out intense stimuli. They also noted that these patients seemed unable to discriminate internal from external events as strongly as do individuals with thick boundaries. Nightmare patients therefore experienced more difficulty screening out irrelevant cognitive and emotional intrusions that can lead to nightmares during dreaming. Hartmann has proposed that nightmare images serve the function of contextualizing or framing overwhelming emotional experiences, which can then be subjected to processes of emotional integration. For example, a dream image of an oncoming tidal wave may turn overwhelming feelings of helplessness and fear into a vivid image that can be worked with in the dream.

Nielsen and Levin on Nightmares

Recently, Nielsen and Levin (2007) have proposed an affect and network model (AND) of nightmare production. It stipulates that nightmares result from dysfunction in a network of affective processes that, during normal dreaming, serves the adaptive function of *fear memory extinction*. As mentioned above, the idea is that dreams help to dampen down intense fear or anxiety. They do this by taking affect-laden dream memory elements and placing them in a dream context inimical to fear. These innocuous context images then facilitate the integration of fear memories into more abstract memory formats, thus reducing the affect and ANS arousal associated with the dream imagery. Given the normal adaptive function of fear extinction memory associated with normal dreaming, nightmares could possibly result in new learning that modulates previously learned fear-related conditioned responses. However, that modulation very likely amplifies fear-related responding rather than extinguishing it, given the dysfunction in the affect-processing system postulated by Levin and Nielsen. This latter possibility would predict that nightmares can get worse over time.

A Diathesis-Stress Model of Nightmares

We offer a simple heuristic model of amygdala overreactivity as the source of mood dysfunction and nightmares: (1) individuals most at risk for nightmares begin with a genetic endowment that predisposes them to an overreactive amygdala; (2) when stress-related events occur in an individual's life, the amygdala becomes very sensitive to stimulation. REM-related activations of the amygdala can now easily push it into an overactivated state, and a nightmare results. Whatever the sequence of neurophysiologic events that lead to a REM nightmare, REM-related activation of the amygdala is the key. Pharmacologic agents that inhibit stimulatory or noradrenergic receptors on the amygdala reduce the likelihood that a nightmare will arise during REM periods (Peskind, Bonner, Hoff, & Raskind, 2003). Amygdala reactivity during REM may also be linked to the ANS instabilities seen in REM. We are all familiar with the heart-pounding-terror-inducing physiologic arousal associated with nightmares. Given that these physiologic changes are likely related to the REM-related ANS instabilities mentioned above, a natural question to ask is whether these ANS "storms" contribute to nightmare distress. Interestingly, Fisher et al. (1970) found that nightmares were characterized by relatively low levels of ANS activation during REM sleep. In 60% (12 of 20) of the nightmares they studied with EEG and ANS-recording equipment, ANS activity was absent altogether. Fisher and colleagues suggested that normal REM dreaming possesses "a mechanism for tempering and modulating anxiety, for de-somatizing the physiological response to it ... [for] abolishing or diminishing the physiological concomitants' of fear and stress" (p. 770). Nightmares presumably result when that mechanism fails. Levin and Nielsen (2009) have recently put forward a similar hypothesis concerning REM nightmares. They suggest that dreams normally take fear memories and place them into a new narrative context so that the fear memory can be extinguished. Nightmares result when this normal function of dreaming breaks down.

THEORY AND TREATMENT STRATEGIES

Does any of this theorizing help clinicians treat the distress associated with REM nightmares? Two things might reduce nightmare frequency, intensity, and distress: a reduction in activation levels of the amygdala and a focus on working cognitively and imagistically with the content variables of nightmares that are linked with nightmare distress: "monstrous" or attacking figures and misfortunes. Cognitive-behavioral techniques, such as monitoring, relaxation, and exposure therapy, can accomplish the latter, and pharmacologic agents the former. A combination of pharmacologic and cognitive-behavioral treatment interventions is likely to be most effective at reducing distress from nightmares.

Several cognitive-behavioral techniques are effective in decreasing nightmare frequency. Monitoring nightmares, relaxation therapy, and exposure exercises decreased nightmare frequency and nightmare-induced fear in one study (Miller & Di Palato, 1983). Exposure treatment typically includes the following steps. The patient is given a self-help manual and instructed to write down the nightmare after awakening and to re-experience it in imagination. The re-experiencing under controlled conditions is thought to allow the patient learn that he has some control over the frightening imagery he experiences in nightmares. To supplement exposure exercises and to increase the patient's sense of mastery over the frightening nightmare content, the therapist can utilize cognitive restructuring techniques. Using these restructuring techniques, the patient takes the story line or the images (or both) of the nightmare and rewrites it, sometimes literally. Krakow et al. (2000) studied effects of imagery rehearsal therapy (IRT) by sexual assault survivors with PTSD and posttraumatic nightmares. These patients received two 3-h IRT sessions at 1-week intervals and a 1-h follow-up session 3 or 6 months later. In the first session, patients practiced pleasant imagery exercises and cognitive restructuring techniques for dealing with unpleasant images associated with nightmares and the original trauma. In the second session, participants wrote down a self-selected nightmare that was not too intense (preventing too much exposure) and then rewrote the story line of the nightmare in any way they wished. They had to keep rehearsing the new story line "mentally" at home. Results of the treatment showed that, relative to baseline assessments of the patients, IRT significantly reduced the number of nightmares per week and PTSD symptoms. Krakow and his group have shown that improvements after IRT are maintained for up to 30 months after treatment (Krakow, Kellner, & Pathak, 1995).

The only pharmacologic agent that has demonstrated any effectiveness in treating nightmares is Prazosin, an alpha-1 adrenergic antagonist. This drug is typically used to treat hypertension. In a placebo-controlled study (Raskind et al., 2006), 40 combat veterans with chronic PTSD and distressing trauma nightmares and sleep disturbance were randomized to evening Prazosin (13.3 ± 3 mg/day) or placebo for 8 weeks. Results showed that Prazosin was significantly superior to placebo for reducing trauma nightmares and improving sleep quality and global clinical status with large effect sizes. Prazosin shifted dream characteristics from those typical of trauma-related nightmares toward those typical of normal dreams. The beneficial effect was not due to the reduction of high autonomic reactivity because the authors found that blood pressure changes from baseline to the end of the study did not differ significantly between Prazosin and the placebo. This result strengthens the argument that Prazosin works at the level of the forebrain to reduce nightmare distress. As reviewed above, one central forebrain structure that is involved in nightmares is the amygdala.

Recent Findings/Current Research

REM sleep is known to modulate brain activity in limbic-prefrontal circuits (especially the amygdala) that normally mediate negative emotion. Replicated findings have shown decreased anterior-posterior EEG coherence at gamma frequencies during REM versus NREM sleep (Corsi-Cabrera et al., 2003; Corsi-Cabrera, Guevara, & del Río-Portilla, 2008; Pérez-Garci, del-Río-Portilla, Guevara, Arce, & Corsi-Cabrera, 2001), as well as in phasic versus tonic REM sleep (Corsi-Cabrera et al., 2008). Coherence decreases appear to reflect limbic and amygdalar activation during REM sleep that is particularly enhanced during phasic REM sleep. We predict that, as anterior-posterior coherence at gamma frequencies decreases, the overall and negative emotionality during REM sleep increases, as does the chance that a nightmare will emerge.

The REM sleep-related loss of anterior-posterior gamma coherence, as well as increased gamma power in anterior paralimbic areas during phasic REM, may reflect both diminished executive control and an enhancement of emotional intensity that are manifested in REM sleep dreams and nightmares. When operating normally, this same physiology may facilitate consolidation of emotional memories and thus mood regulation.

Aspects of REM and NREM sleep neurobiology have been implicated in facilitating the emotional memory consolidation crucial for all kinds of mood regulatory functions (Born & Wagner, 2004; Stickgold, Hobson, Fosse, & Fosse, 2001; Walker & Stickgold, 2006). REM sleep, in particular, appears to facilitate consolidation of emotional memories (Hu, Stylos-Allan, & Walker, 2006; Nishida, Pearsall, Buckner, & Walker, 2009; Wagner, Gais, & Born, 2001), especially when those memories involve negative emotions (Nishida et al., 2009; Wagner et al., 2001). Improved consolidation of negative memories has been correlated with prefrontal theta activity during REM in daytime naps (Nishida et al., 2009). REM sleep has also been associated with the modulation of negative emotional memory consolidation processes (Cartwright, Luten, Young, Mercer, & Bears, 1998; Germain, Buysse, & Nofzinger, 2008; Greenberg, Pillard, & Pearlman, 1972; Nielsen & Levin, 2007; Pace-Schott et al., 2009). For example, REM sleep appears to modulate consolidation and recall of negatively valenced memories (Nishida et al., 2009; Wagner et al., 2001) and thus exerts an ongoing and long-term impact on mood state. However, REM sleep also displays very fast-acting emotional regulatory responses as well. Evidence presented in this chapter suggests that REM sleep acts to dampen down the intensity of negative affect associated with fear. When this regulatory mechanism fails, nightmares emerge (Nielsen & Levin, 2007).

FUTURE DIRECTIONS

Up to now, most studies of dreams and nightmares have involved relatively small, homogenous samples (e.g. middle class college students). The new availability of relatively large datasets of dreams and nightmares on the web will dramatically enhance the numbers of dreams we have access to and therefore can analyze. Dreams in these large datasets come from all over the world and are comprised of diverse samples. For example, the site www.dreamboard.com contains over 200,000 posted dreams including tens of thousands of "nightmares" (unpleasant dreams marked by extreme fear or terror) from people all over the world. Many of these posted dreams constitute longitudinal dream series for which mood and dream content variables can be tracked over time. Preliminary analyses of these posted dreams and nightmares confirm sex differences in nightmare frequencies (women greater than men), but they also portend discovery of new associations of nightmare events with labile mood regulatory breakdowns that precede and signal the intense fear and terror associated with the nightmare event. Using these longitudinal dream series containing the occasional nightmare, we can begin to identify antecedent images and moods in dreams leading up to the nightmare, as well as dream content indices associated with damping down distress associated with frightening nightmare images.

References

Ağargün, M. Y., & Cartwright, R. (2003). REM sleep, dream variables and suicidality in depressed patients. *Psychiatry Research, 119*(1–2), 33–39. http://dx.doi.org/10.1016/S0165-1781(03)00111-2.

Ağargün, M. Y., Cilli, A. S., Kara, H., Tarhan, N., Kincir, F., & Oz, H. (1998). Repetitive frightening dreams and suicidal behavior in patients with major depression. *Comprehensive Psychiatry, 39*, 198–202, PMID: 9675503.

American Psychiatric Association. (2000). *Diagnostic and statistical manual of mental disorders* (4th ed.). Washington, D.C.: American Psychiatric Press (text rev.).

Armitage, R., Hudson, A., Trivedi, M., & Rush, A. J. (1995). Sex differences in the distribution of EEG frequencies during sleep: Unipolar depressed outpatients. *Journal of Affective Disorders, 34*(2), 121–129, PMID: 7665804.

Belicki, D., & Belicki, K. (1982). Nightmares in a university population. *Sleep Research, 11*, 116.

Benca, R. M. (2000). Mood disorders. In M. H. Kryger, T. Roth, & W. C. Dement (Eds.), *Principles and practice of sleep medicine* (3rd ed., pp. 1140–1148). Philadelphia, PA: W. B. Saunders, http://dx.doi.org/10.1002/1097-4679(198609)42:5<714::AID-JCLP2270420506>3.0.CO;2-K.

Bixler, E. O., Kales, A., Soldatos, C. R., Kales, J. D., & Healy, S. (1979). Prevalence of sleep disorders in the Los Angeles metropolitan area. *American Journal of Psychiatry, 136,* 1257–1262, PMID: 314756.

Boeve, B. F., Silber, M. H., Saper, C. B., Ferman, T. J., Dickson, D. W., Parisi, J. E., et al. (2007). Pathophysiology of REM sleep behaviour disorder and relevance to neurodegenerative disease. *Brain, 130*(11), 2770–2788. http://dx.doi.org/10.1093/brain/awm056.

Born, J., & Wagner, U. (2004). Memory consolidation during sleep: Role of cortisol feedback. *Annals of the New York Academy of Sciences, 1032*(1), 198–201. http://dx.doi.org/10.1196/annals.1314.020.

Carskadon, M. A., & Rechtschaffen, A. (2000). Monitoring and staging human sleep. In M. H. Kryger, T. Roth, & W. C. Dement (Eds.), *Principles and practice of sleep medicine* (3rd ed., pp. 1197–1216). Philadelphia, PA: W. B. Saunders.

Cartwright, R. (1979). The nature and function of repetitive dreams: A speculation. *Psychiatry, 42,* 131–137, PMID: 461586.

Cartwright, R. D. (1991). Dreams that work: The relation of dream incorporation to adaption to stressful events. *Dreaming, 1,* 3–9.

Cartwright, R., Luten, A., Young, M., Mercer, P., & Bears, M. (1998). Role of REM sleep and dream affect in overnight mood regulation: A study of normal volunteers. *Psychiatry Research, 81*(1), 1–8. http://dx.doi.org/10.1016/S0165-1781(98)00089-4.

Coalson, B. (1995). Nightmare help: Treatment of trauma survivors with PTSD. *Psychotherapy, 32,* 381–388, PMID: 7347751.

Corsi-Cabrera, M., Guevara, M. A., & del Río-Portilla, Y. (2008). Brain activity and temporal coupling related to eye movements during REM sleep: EEG and MEG results. *Brain Research, 1235,* 82–91. http://dx.doi.org/10.1016/j.brainres.2008.06.052.

Corsi-Cabrera, M., Miró, E., del-Río-Portilla, Y., Pérez-Garci, E., Villanueva, Y., & Guevara, M. A. (2003). Rapid eye movement sleep dreaming is characterized by uncoupled EEG activity between frontal and perceptual cortical regions. *Brain and Cognition, 51*(3), 337–345. http://dx.doi.org/10.1016/S0278-2626(03)00037-X.

Dang-Vu, T. T., Desseilles, M., Laureys, S., Degueldre, C., Perrin, F., Phillips, C., et al. (2005). Cerebral correlates of delta waves during non-REM sleep revisited. *NeuroImage, 28*(1), 14–21. http://dx.doi.org/10.1016/j.neuroimage.2005.05.028.

Dement, W., & Kleitman, N. (1957a). The relation of eye movements during sleep to dream activity: An objective method for the study of dreaming. *Journal of Experimental Psychology, 53*(5), 339–346. http://dx.doi.org/10.1037/h0048189.

Dement, W., & Kleitman, N. (1957b). Cyclic variations in EEG during sleep and their relation to eye movements, body motility and dreaming. *Electroencephalography & Clinical Neurophysiology, 9*(4), 673–690. http://dx.doi.org/10.1016/0013-4694(57)90088-3.

Domhoff, G. W. (1993). The repetition of dreams and dream elements: A possible clue to a function of dreams? In A. Moffitt, M. Kramer, & R. Hoffmann (Eds.), *The functions of dreaming* (pp. 293–320). Albany, NY: State University of New York Press.

Fantini, M. L., & Ferini-Strambi, L. (2007). REM-related dreams in REM behavior disorder. In D. Barrett & P. McNamara (Eds.), *Biological aspects: 1. The new science of dreaming* (pp. 185–200). Westport, CT: Praeger Publishers.

Fisher, C., Byrne, J., Edwards, A., & Kahn, E. (1970). A psychophysiological study of nightmares. *Journal of the American Psychoanalytic Association, 18,* 747–782. http://dx.doi.org/10.1177/000306517001800401.

Fisher, B. E., Pauley, C., & McGuire, K. (1989). Children's sleep behavior scale: Normative data on 870 children in grades 1 to 6. *Perceptual & Motor Skills, 68,* 227–236. http://dx.doi.org/10.2466/pms.1989.68.1.227.

Germain, A., Buysse, D. J., & Nofzinger, E. (2008). Sleep-specific mechanisms underlying posttraumatic stress disorder: Integrative review and neurobiological hypotheses. *Sleep Medicine Reviews, 12*(3), 185–195, PMID: 17997114.

Germaine, A., & Nielsen, T. A. (2003). Sleep pathophysiology in posttraumatic stress disorder and idiopathic nightmare sufferers. *Biological Psychiatry, 54,* 1092–1098. http://dx.doi.org/10.1016/S0006-3223(03)00071-4.

Giles, D. E., Roffwarg, H. P., Schlesser, M. A., & Rush, A. J. (1986). Which endogenous depressive symptoms relate to REM latency reduction? *Biological Psychiatry, 21*(5–6), 473–482. http://dx.doi.org/10.1016/0006-3223(86)90189-7.

Greenberg, R., Pillard, R., & Pearlman, C. (1972). The effect of dream (stage REM) deprivation on adaptation to stress. *Psychosomatic Medicine, 34*(3), 257–262, PMID: 4338295.

Harris, I. R. (1948). Observations concerning typical anxiety dreams. *Psychiatry, 11,* 301–309.

Hartmann, E. (1984). *The nightmare: The psychology and the biology of terrifying dreams.* New York: Basic Books, PMID: 18889234.

Hartmann, E. (1998). Nightmare after trauma as paradigm for all dreams: A new approach to the nature and function of dreaming. *Psychiatry, Interpersonal & Biological Processes, 61*(3), 223–228, PMID: 9823032.

Hartmann, E., Russ, D., Oldfield, M., Sivan, I., & Cooper, S. (1987). Who has nightmares? The personality of the lifelong nightmare sufferer. *Archives of General Psychiatry, 44,* 49–56, PMID: 3800584.

Haynes, S. N., & Mooney, D. K. (1975). Nightmares: Etiological, theoretical, and behavioral treatment considerations. *The Psychological Record, 25,* 225–236. http://dx.doi.org/10.1002/da.10151.

Hofle, N., Paus, T., Reutens, D., Fiset, P., Gotman, J., Evans, A. C., et al. (1997). Regional cerebral blood flow changes as a function of delta and spindle activity during slow wave sleep in humans. *Journal of Neuroscience, 17,* 4800–4808, PMID: 9169538.

Hu, P., Stylos-Allan, M., & Walker, M. P. (2006). Sleep facilitates consolidation of emotional declarative memory. *Psychological Science, 17*(10), 891–898. http://dx.doi.org/10.1111/j.1467-9280.2006.01799.x.

Jouvet, M. (1999). *The paradox of sleep: The story of dreaming.* Cambridge, MA: MIT Press.

Jouvet, D., Vimont, P., Delorme, F., & Jouvet, M. (1964). Study of selective deprivation of the paradoxical sleep phase in the cat. *Comptes Rendus des Seances de la Societe de Biologie et de Ses Filiales, 158,* 756–759.

Kaufmann, C., Wehrle, R., Wetter, T. C., Holsboer, F., Auer, D. P., Pollmächer, T., et al. (2006). Brain activation and hypothalamic functional connectivity during human non-rapid eye movement sleep: An EEG/fMRI study. *Brain, 129*(3), 655–667. http://dx.doi.org/10.1093/brain/awh686.

Krakow, B. (2006). Nightmare complaints in treatment-seeking patients in clinical sleep medicine settings: Diagnostic and treatment implications. *Sleep, 29*(10), 1313–1319, PMID: 17068985.

Krakow, B., Hollijield, M., Schrader, R., Koss, M., Tandberg, D., Lauriello, J., et al. (2000). Controlled study of imagery rehearsal for chronic nightmares in sexual assault survivors with PTSD: A preliminary report. *Journal of Truumatic Stress, 13,* 589–609. http://dx.doi.org/10.1023/A:1007854015481.

Krakow, B., Kellner, R., & Pathak, D. (1995). Imagery rehearsal treatment for chronic nightmares. *Behavioral Research Therapy, 33,* 837–843. http://dx.doi.org/10.1016/0005-7967(95)00009-M.

Kupfer, D. J., & Foster, F. G. (1972). Interval between onset of sleep and rapid-eye-movement sleep as an indicator of depression. *Lancet, 2*(7779), 684–686, PMID: 4115821.

Leskin, G. A., Woodward, S. H., Young, H. E., & Sheikh, J. I. (2002). Effects of comorbid diagnoses on sleep disturbance in PTSD. *Journal of Psychiatric Research, 36*(6), 449–452.

Levin, R. (1994). Sleep and dreaming characteristics of frequent nightmare subjects in a university population. *Dreaming, 4,* 127–137.

Levin, R., & Nielsen, T. (2009). Nightmares, bad dreams, and emotion dysregulation: A review and new neurocognitive model of dreaming. *Current Directions in Psychological Science, 18*(2), 84–88. http://dx.doi.org/10.1111/j.1467-8721.2009.01614.x.

MacFarlane, J. W., Allen, L., & Hoznik, M. P. (1954). A developmental study of the behavior problems of normal children between twenty-one and fourteen years. *Publications in Child Development, University of California, Berkeley, 2*, 1–222.

Mahowald, M. W., & Ettinger, M. G. (1990). Things that go bump in the night: The parasomnias revisited. *Journal of Clinical Neurophysiology, 7*(1), 119–143. http://dx.doi.org/10.1097/00004691-199001000-00009.

Mahowald, M. W., & Schenck, C. H. (2000). Diagnosis and management of parasomnias. *Clinical Cornerstone, 2*(5), 48–57. http://dx.doi.org/10.1016/S1098-3597(00)90040-1.

Maquet, P. (1995). Sleep function(s) and cerebral metabolism. *Behavioural Brain Research, 69*(1), 75–83. http://dx.doi.org/10.1016/0166-4328(95)00017-N.

Maquet, P., & Franck, G. (1997). REM sleep and amygdale. *Molecular Psychiatry, 2*(3), 195–196.

Maquet, P., Peters, J. M., Aerts, J., Delfiore, G., Degueldre, C., Luxen, A., et al. (1996). Functional neuroanatomy of human rapid-eye-movement sleep and dreaming. *Nature, 383*, 163–166. http://dx.doi.org/10.1038/sj.mp.4000239.

Maquet, P., Ruby, P., Maudoux, A., Albouy, G., Sterpenich, V., Dang-Vu, T., et al. (2005). Human cognition during REM sleep and the activity profile within frontal and parietal cortices: A reappraisal of functional neuroimaging data. *Progress in Brain Research, 150*, 219–595. http://dx.doi.org/10.1016/S0079-6123(05)50016-5.

McNamara, P. (2008). *Nightmares: The science and solution of those frightening visions during sleep*. Westport, CT: Praeger Press.

McNamara, P., Auerbach, S., Johnson, P., Harris, E., & Doros, G. (2010). Impact of REM sleep on distortions of self concept, mood and memory in depressed/anxious participants. *Journal of Affective Disorders, 122*(3), 198–207. http://dx.doi.org/10.1016/j.jad.2009.06.030.

Miller, W., & Di Palato, M. (1983). Treatment of nightmares via relaxation and desensitization: A controlled evaluation. *Journal of Consulting & Clinical Psychology, 51*, 870–877. http://dx.doi.org/10.1037/0022-006X.51.6.870.

Mindell, J. A., & Barrett, K. M. (2002). Nightmares and anxiety in elementary-aged children: Is there a relationship? *Child: Care Health & Development, 28*, 317–322. http://dx.doi.org/10.1046/j.1365-2214.2002.00274.x.

Nielsen, T. A., & Levin, R. (2007). Nightmares: A new neurocognitive model. *Sleep Medicine Reviews, 11*, 295–310. http://dx.doi.org/10.1016/j.smrv.2007.03.004.

Nielsen, T. A., & Zadra, A. L. (2005). Nightmares and other common dream disturbances. In M. Kryger, N. Roth, & W. C. Dement (Eds.), *Principles and practice of sleep medicine* (4th ed., pp. 926–935). Philadelphia, PA: W. B. Saunders.

Nishida, M., Pearsall, J., Buckner, R. L., & Walker, M. P. (2009). REM sleep, prefrontal theta, and the consolidation of human emotional memory. *Cerebral Cortex, 19*(5), 1158–1166. http://dx.doi.org/10.1093/cercor/bhn155.

Nofzinger, E. A., Mintun, M. A., Wiseman, M. B., Kupfer, D. J., & Moore, R. Y. (1997). Forebrain activation in REM sleep: An FDG PET study. *Brain Research, 770*, 192–201. http://dx.doi.org/10.1016/S0006-8993(97)00807-X.

Nofzinger, E. A., Nissen, C., Germain, A., Moul, D., Hall, M., Price, J. C., et al. (2006). Regional cerebral metabolic correlates of WASO during NREM sleep in insomnia. *Journal of Clinical Sleep Medicine, 2*(3), 316–322.

Ohayon, M. M., Morselli, P. L., & Guilleminault, C. (1997). Prevalence of nightmares and their relationship to psychopathology and daytime functioning in insomnia subjects. *Sleep, 20*, 340–348.

Olson, E. J., Boeve, B. F., & Silber, M. H. (2000). Rapid eye movement sleep behaviour disorder: Demographic, clinical and laboratory findings in 93 cases. *Brain, 123*, 331–339. http://dx.doi.org/10.1093/brain/123.2.331.

Pace-Schott, E. F., Milad, M. R., Orr, S. P., Rauch, S. L., Stickgold, R., & Pitman, R. K. (2009). Sleep promotes generalization of extinction of conditioned fear. *Sleep, 32*(1), 19.

REFERENCES

Penfield, W., & Erickson, T. C. (1941). *Epilepsy and cerebral localization. A study of the mechanism, treatment and prevention of epileptic seizures.* Baltimore, MD: C. C. Thomas, 10.1016/j.surneu.2004.05.043.

Pérez-Garci, E., del-Río-Portilla, Y., Guevara, M. A., Arce, C., & Corsi-Cabrera, M. (2001). Paradoxical sleep is characterized by uncoupled gamma activity between frontal and perceptual cortical regions. *Sleep, 1,* 24, PMID: 11204047.

Peskind, E. R., Bonner, L. T., Hoff, D. J., & Raskind, M. A. (2003). Prazosin reduces trauma-related nightmares in older men with chronic posttraumatic stress disorder. *Journal of Geriatric Psychiatry and Neurology, 16*(3), 165–171.

Raskind, M. A., Peskind, E., Hoff, D., Hart, K., Holmes, H., Warren, D., et al. (2006). Parallel group placebo controlled study of Prazosin for trauma nightmares and sleep disturbance in combat veterans with post-traumatic stress disorder. *Biological Psychiatry, 61*(8), 928–934.

Salzarulo, P., & Chevalier, A. (1983). Sleep problems in children and their relationship with early disturbances of the waking-sleeping rhythms. *Sleep, 6,* 47–51.

Schenck, C. H., & Mahowald, M. W. (1990). Polysomnographic, neurologic, psychiatric, and clinical outcome report on 70 consecutive cases with REM sleep disorder (RBD): Sustained clonazepam efficacy in 89.5% of 57 treated patients. *Cleveland Journal of Medicine, 57*(Suppl), S9–S23.

Schenck, C. H., & Mahowald, M. W. (1996). REM sleep parasomnias. *Neurology Clinics, 14,* 697–720.

Schreuder, B., Kleijn, W., & Rooijmans, H. (2000). Nocturnal re-experiencing more than forty years after war trauma. *Journal of Traumatic Stress, 13*(3), 453–463.

Simonds, J. F., & Parraga, H. (1982). Prevalence of sleep disorders and sleep behaviors in children and adolescents. *Journal of the American Academy of Child & Adolescent Psychiatry, 21,* 383–388.

Solms, M. (1997). *The neuropsychology of dreams.* Mahwah, NJ: Lawrence Erlbaum.

Spoormaker, V. I., Schredl, M., & van den Bout, J. (2006). Nightmares: From anxiety symptom to sleep disorder. *Sleep Medicine Reviews, 10*(1), 19–31.

Stickgold, R., Hobson, J. A., Fosse, R., & Fosse, M. (2001). Sleep, learning, and dreams: Off-line memory reprocessing. *Science, 294*(5544), 1052–1057.

Tanskanen, A., Toumilehto, J., Viinamaki, H., Vartiainen, E., Lehtonen, J., & Puska, P. (2001). Nightmares as predictors of suicide. *Sleep, 24,* 845–848.

Vela-Bueno, A., Bixler, E. O., Dobladez-Blanco, B., Rubio, M. E., Mattison, R. E., & Kales, A. (1985). Prevalence of night terrors and nightmares in elementary school children: A pilot study. *Research Communications in Psychology, Psychiatry, & Behavior, 10,* 177–188.

Vogel, G., Foulkes, D., & Trosman, H. (1966). Ego functions and dreaming during sleep onset. *Archives of General Psychiatry, 14*(3), 238–248.

Vogel, G. W., Vogel, F., McAbee, R. S., & Thurmond, A. J. (1980). Improvement of depression by REM sleep deprivation. New findings and a theory. *Archives of General Psychiatry, 37,* 247–253.

Wagner, U., Gais, S., & Born, J. (2001). Emotional memory formation is enhanced across sleep intervals with high amounts of rapid eye movement sleep. *Learning & Memory, 8*(2), 112–119.

Walker, M. P., & Stickgold, R. (2006). Sleep, memory, and plasticity. *Annual Review of Psychology, 57,* 139–166.

Woodward, S. H., Arsenault, N. J., Murray, C., & Bliwise, D. L. (2000). Laboratory sleep correlates of nightmare complaint in PTSD inpatients. *Biological Psychiatry, 48*(11), 1081–1087.

Zadra, A. (1996). Recurrent dreams: Their relation to life events. In D. Barrett (Ed.), *Trauma and dreams* (pp. 231–267). Cambridge, MA: Harvard University Press.

CHAPTER 9

Isolated Sleep Paralysis and Affect

Brian A. Sharpless

Department of Psychology, Washington State University, Pullman, Washington, USA

ISOLATED SLEEP PARALYSIS AND AFFECT

Among the various sleep-wake disorders, isolated sleep paralysis (ISP) is fairly unique. This uniqueness is due not only to its characteristic symptom constellations, but also to its purported role in mythology and folklore across time and place. Episodes of ISP are also riddled with affect (almost exclusively negatively valenced). For example, it is estimated that between 31.7% and 98.0% of individuals experience fear during ISP (e.g., Cheyne, Rueffer, & Newby-Clark, 1999; Mellman, Aigbogun, Graves, Lawson, & Alim, 2008; Spanos, McNulty, DuBreuil, & Pires, 1995). Even if not frightening, these episodes often leave sufferers convinced that something unusual has happened to them, or, alternately, they may awaken from these experiences in search of answers. Perhaps not surprisingly, humanity's many attempts to understand ISP have produced a panoply of causal theories ranging from the biological (e.g., discordance in Rapid eye movement (REM) sleep architecture; McNally & Clancy, 2005b) to the supernatural (e.g., demonic attacks, witchcraft, alien abductions as in Hufford, 1982). Some of these theories and appraisals have the potential to generate additional levels of affect that may manifest long after the actual ISP episode has ceased.

Before a more specific discussion of affect, this chapter presents a broad consideration of the phenomenon of ISP. Although empirical research on ISP is still in the beginning stages, a base of information is nonetheless accruing. However, as with many fields of research, intensive study of individual cases can still be very useful. Therefore, a clinical vignette may better ground the reader in a phenomenological understanding of ISP before specific research findings are discussed.

VIGNETTE

I've had it [ISP] for a while. But about three years ago it was happening like once a week because I was so stressed at the time, and I know I was underage, but I drank a little, and that stressed me out, so I think that might have caused it. I wasn't drinking at the times it happened, though. During a typical one [ISP episode], I felt something push on my chest, so I opened my eyes. I heard someone yelling or crying, and it sounded like it was coming from right beside my ear. I thought it was a ghost or something. It was yelling, "I'm gonna kill you!" and "I'm going to bring you to hell!" I couldn't move, and I was so scared… It looked like a woman with long black hair. I couldn't see her face, but she was sitting right in front of me and she was pushing on my chest. I really wanted to move, but I couldn't, and it felt like it took me 30 minutes to get out of it. After about 30 minutes, I tried focusing on moving my finger, and it worked, and I found I could move again. When it happens I can't go back to sleep because I am so scared and confused, and usually stay awake for at least an hour [parentheses mine].

This narrative was derived from a semistructured interview (Fearful Isolated Sleep Paralysis Interview (FISPI); Sharpless et al., 2010) conducted on a 22-year-old Japanese female. Some of the material was condensed from separate parts of the recording. This young woman experienced a total of 16 episodes over the course of her life and reported a great deal of distress and anxiety in conjunction with some interference in her overall functioning (i.e., for several days following her ISP episodes, she experienced anxiety and reported extreme difficulty sleeping). She displayed many of the paradigmatic features of ISP and, more specifically, *fearful ISP*. She also met diagnostic criteria for *recurrent fearful ISP*, a proposed diagnostic category that requires clinically significant distress and/or impairment as a result of episodes (see Table 9.1). When looking at her descriptions, one can clearly see the characteristic atonia and hallucinations taking place within the context of a clear sensorium. She was adamant that she was not "dreaming," but was conscious and awake. Unfortunately for her, these symptoms were also combined with high levels of fear, apprehension, and postepisode confusion. The ISP episodes interacted with her personal belief system as well, because she considered these ISP episodes to be actual nocturnal visitations from ghosts. This is all the more remarkable given that she had otherwise intact reality testing, was well-educated, and displayed no evidence of a formal thought disorder or recent substance use. She also did not have any medical conditions that may better explain her symptoms (e.g., narcolepsy, hypokalemia). Thus, her episodes would be classified as *isolated* sleep paralysis and not the less specific "sleep paralysis."

Interestingly, and in spite of the dramatic nature of these scary paroxysms, ISP is rarely assessed clinically or in research studies, and it is fairly absent from many general textbooks on medicine and psychopathology. This is unfortunate because not considering ISP as a diagnostic possibility

TABLE 9.1 Diagnostic Criteria for Isolated Sleep Paralysis, Fearful Isolated Sleep Paralysis Episodes, and Recurrent Fearful Isolated Sleep Paralysis

ICSD-2 RECURRENT ISOLATED SLEEP PARALYSIS CRITERIA[1]

A. The patient complains of an inability to move the trunk and all limbs at sleep onset or on waking from sleep.
B. Consciousness is preserved, and full recall is present.
C. Each episode lasts seconds to a few minutes.
D. Patient does not suffer from narcolepsy.

ICSD-3 RECURRENT ISOLATED SLEEP PARALYSIS CRITERIA[2]

A. A recurrent inability to move the trunk and all of the limbs at sleep onset or upon awakening from sleep.
B. Each episode lasts seconds to a few minutes.
C. The episodes cause clinically significant distress including bedtime anxiety or fear of sleep.
D. The disturbance is not better explained by another sleep disorder (especially narcolepsy), mental disorder, medical condition, medication, or substance use.

FEARFUL ISOLATED SLEEP PARALYSIS EPISODE[3]

A. A period of time at sleep onset or upon awakening during which voluntary movement is not possible, yet some degree of awareness is present.
B. The episode(s) of sleep paralysis is/are accompanied by significant fear, anxiety, or dread that may be associated with either the paralysis itself or the presence of hypnogogic (sleep onset) or hypnopompic (sleep offset) hallucinations.
C. The episode(s) of sleep paralysis is/are not better accounted for by the direct physiological effects of a substance (e.g., alcohol, drug of abuse, or medications).
D. Isolated sleep paralysis is not better accounted for by a general medical condition (e.g., narcolepsy, seizure disorder, hypokalemia) or other psychiatric diagnosis (e.g., sleep terror disorder).

RECURRENT FEARFUL ISOLATED SLEEP PARALYSIS[3]

A. At least two episodes of fearful Isolated Sleep Paralysis (as defined above) taking place in the past 6 months.
B. The episodes of sleep paralysis are accompanied by clinically significant distress and/or impairment.
C. The episodes of sleep paralysis are not better accounted for by the direct physiological effects of a substance (e.g., alcohol, drug of abuse, or medications).
D. The sleep paralysis episodes are not better accounted for by a general medical condition (e.g., narcolepsy, seizure disorder, hypokalemia) or other psychiatric diagnosis (e.g., sleep terror disorder).

[1] *American Academy of Sleep Medicine (2005).*
[2] *American Academy of Sleep Medicine (2014).*
[3] *Sharpless et al. (2010).*

can result in misdiagnosis. Should this young woman have presented for treatment complaining of visitations from ghosts that not only molested her in the night, but told her they were taking her to hell, the physician might have easily diagnosed her with more severe levels of pathology (e.g., a psychotic disorder). In fact, the risks of not accurately diagnosing

ISP have been noted several times in the literature (e.g., Douglass, Hays, Pazderka, & Russell, 1991; Gangdev, 2004; Shapiro & Spitz, 1976), but this phenomenon is not often in the forefront of clinicians' minds. This absence of regular assessment is all the more surprising because, as will be shown, sleep paralysis (SP) has been written about for over 2000 years.

DIAGNOSTIC CRITERIA

The first modern mental health professional to propose diagnostic criteria for ISP was Ernest Jones (Jones, 1949). His tripartite description of the *Nightmare* (what we would now term the *incubus* subcategory of ISP; Cheyne, Newby-Clark, & Rueffer, 1999; Cheyne, Rueffer, et al., 1999) consisted of agonizing angst, a sense of oppression and/or weight on the chest that interferes with breathing, and the conviction of a helpless paralysis. Thus, Jones considered the anxious or fearful appraisal of these episodes to be a core feature. However, Jones's use of the German term *angst* is much stronger than the rough English equivalent of anxiety or dread. The former is a concept that more accurately strikes at the core of one's being, but angst can be the experience of being confronted with the possibility of *nonbeing* (Kierkegaard, 1980; Sharpless, 2013). Jones was rightly adamant that the experience is not mere apprehension or worry.

Current diagnostic criteria are slightly different, but not all diagnostic systems even include ISP as a unique and codable entity. For instance, in the *Diagnostic Statistical Manual 5* (DSM-5; American Psychiatric Association, 2013), problematic cases of ISP are coded as either an "other specified sleep-wake disorder (code 307.49)" or, alternately, as an "unspecified sleep disorder (307.40)." In contrast, the *International Statistical Classification of Diseases and Related Health Problems 10* (ICD-10; World Health Organization, 2008) contains ISP as a separate codable entity (G47.53; as do both ICSD-2 (American Academy of Sleep Medicine, 2005) and ICSD-3 (American Academy of Sleep Medicine, 2014)). The diagnostic criteria for recurrent ISP, according to ICSD-2 and ICSD-3, can be found in Table 9.1. Notably, no affective response either during or after an episode was required for a diagnosis of ISP in ICSD-2, but this changed with ICSD-3. Furthermore, neither of the ICSDs specifies a minimal episode frequency or required temporal duration for recurrent ISP.

Sharpless et al. (2010) proposed a more specific definition of ISP that comes closer to Jones' earlier formulation and also specifies diagnostic thresholds. In *fearful ISP*, Sharpless made the occurrence of *clinically significant fear* during episodes a requirement. In order to meet criteria for the diagnostic entity of *recurrent fearful ISP*, an individual would have to meet episode frequency requirements and also experience clinically significant distress and/or impairment as a result of the ISP episodes (i.e., not just during episodes). Relevant criteria for fearful ISP episodes and recurrent fearful ISP can be found in Table 9.1.

PREVALENCE RATES

ISP is a surprisingly common phenomenon. Although the lack of a clear consensus on gold standard diagnostic instruments places the accuracy of certain published studies on somewhat questionable grounds (e.g., many studies do not rule out narcolepsy or other medical conditions, and thus they may be assessing SP and not ISP), ISP seems to occur much more frequently than many health professionals would surmise. A recent systematic prevalence study aggregated 35 empirical studies with a total N of 36,533 participants, and it found an overall lifetime SP rate of 20.8% (Sharpless & Barber, 2011). When this rate was dissected into more specific subgroups, 7.6% of the general population, 28.3% of undergraduate students, and 31.9% of psychiatric patients reported at least one episode. When all groups were analyzed by gender, slightly more women (18.8%) reported SP than men (15.7%). Lifetime rates of ISP are not yet known, but given the relatively low occurrence of narcolepsy (56.3 per 100,000 persons; Pelayo & Lopes, 2009), one can surmise that ISP rates are not drastically different from these rates of SP.

Racial minorities also appear to experience SP more than individuals of Caucasian ancestry, but the actual differences seem to be relatively smaller than initially believed (Sharpless & Barber, 2011). These data indicate that SP is far from a culture-bound disorder, and it has been ubiquitously found in various times and places.

To date, two studies have explored rates for fearful ISP. In a clinical sample of 39 patients with lifetime ISP (as determined by the FISPI; Sharpless et al., 2010), 69.2% also met lifetime fearful ISP episode criteria, and this proportion reduced to 43.6% for recurrent fearful ISP. In a sample of 156 students who were similarly diagnosed (Sharpless & Grom, in press) 75.64% met fearful ISP criteria, but only 15.38% suffered from recurrent fearful ISP. Therefore, the majority of ISP experients endorse clinically significant levels of fear, but a much smaller proportion experience significant distress and/or interference from their episodes.

ASSOCIATED COMORBIDITIES

In the traditional medical literature, SP has been most closely linked with narcolepsy (e.g., Wilson, 1928), and it has long been considered part of the "narcoleptic tetrad" (Pelayo & Lopes, 2009). DSM-5 removed SP from the narcolepsy criteria, however (American Psychiatric

Association, 2013). Apart from narcolepsy, SP and ISP have also been discussed in the context of certain other medical conditions such as hypertension (Bell, Hildreth, Jenkins, & Carter, 1988), hypokalemia, and Wilson's disease (Portala, Westermark, Ekselius, & Broman, 2002). SP rates also appear to increase through the course of pregnancy, whereas other parasomnias decrease (Hedman, Pohjasvaara, Tolonen, Salmivaara, & Myllyla, 2002).

Certain psychological conditions and symptoms have also been linked with ISP. Although early studies hypothesized that ISP was possibly a variant of panic disorder (another disorder evoking intense fear), it is more broadly associated with anxiety disorders or symptomatology, with higher levels of comorbidity in general (e.g., Otto et al., 2006; Sharpless et al., 2010; Szklo-Coxe, Young, Finn, & Mignot, 2007), and it may also be associated with depression (Szklo-Coxe et al., 2007). Whether this comorbidity indicates a vulnerability shared by each of these conditions or merely the fact that those with multiple diagnoses may have more disturbed sleep (making ISP more likely) is as yet unknown. Regardless, the construct of anxiety sensitivity (i.e., the tendency to experience fear- or anxiety-related symptoms due to the perceived possibility of negative social or health-related outcomes) appears to be one shared feature of ISP, panic disorder, and the other anxiety disorders (e.g., Ramsawh, Raffa, White, & Barlow, 2008; Sharpless et al., 2010).

Several studies have found ISP to be more prevalent in victims of trauma (e.g., Abrams, Mulligan, Carleton, & Asmundson, 2008; Mellman et al., 2008). Sharpless et al. (2010) found that the correlation between posttraumatic stress disorder (PTSD) and ISP only reached significance when fearful ISP criteria were used. Evidence indicates that those with ISP may be more prone to dissociation as well, and this could be additional evidence, albeit indirect, of the role of trauma in ISP (e.g., McNally & Clancy, 2005a; van der Kloet, Giesbrecht, Lynn, Merckelbach, & de Zutter, 2011). The nature of this causal relationship remains undetermined, however.

SLEEP PARALYSIS AND AFFECT

Cultural and Historical Contexts

Although SP was only formally named in 1928 (Wilson, 1928), one can find much earlier references to the phenomenon (e.g., Bond, 1753; Waller, 1816). The condition was rarely (if ever) discussed in positive or neutral affective terms. For instance, the ancient Greeks used the term *ephialtes* to denote hurricanes, nightmares, and things that can suddenly leap upon you (Haga, 1989). The Greeks sometimes associated *ephialtes* with the god

Pan, and they often used the phrase *pan-ephialtes* (Cheyne, Newby-Clark, et al., 1999; Cheyne, Rueffer, et al., 1999; Roscher, 2007). Pan himself was frequently portrayed as a frightening god whose screams could cause panic or even death to those who displeased him.

In ancient Rome, ISP was attributed to an incubus (male demon) or succubus (female demon), and physicians and laymen alike discussed their nocturnal physical and sexual assaults at length (e.g., Davies, 2003). Some of the more famous learned discussions of incubi/succubi can be found in the Malleus *Maleficarum* (Kramer & Sprenger, 1971), a fifteenth-century witch-hunting manual created at least partially in response to a papal bull.

Although scholars differ on the question of primacy, I argue that the earliest convincing description of ISP comes from Paulus Aegineta (Aegineta, 1844), a seventh-century Byzantine Greek physician. In his section "On Incubus, or Nightmare," he writes, "Persons suffering an attack experience incapability of motion, a torpid sensation in their sleep, a sense of suffocation, and oppression, as if from one pressing them down, with inability to cry out, or they utter inarticulate sounds. Some imagine often that they even hear the person who is going to press them down, that he offers lustful violence to them, but flies when they attempt to grasp him with their fingers" (p. 388). Aegineta's description contains many common components of ISP (e.g., atonia, breathing difficulties, pressure on the chest, sexual assault hallucinations), and it describes how the return of movement is often associated with episode cessation. His description is also quite similar to the clinical vignette presented above.

Some have argued that ISP has had a more direct impact on human culture (e.g., Jones, 1949). Although many of these claims are difficult to substantiate, if not outright impossible, sleep in general and dreams in particular seem to have played formative roles in societal and personal experiences. For instance, *oneiromancy*, or the use of dreams to prophesize future events, was a common practice throughout much of human history. Much like *augury* (interpreting events through the flights of birds), *oneiromancy* has probably had large impacts on world events. More specifically, though, Jones (1949) argues that experiences of Nightmare, or ISP, may be responsible for wide swaths of religious beliefs. He states, "For the true significance of the Nightmare to be properly appreciated, first by the learned professionals and then by the general public, would in my opinion entail consequences, both scientific and social, to which the term momentous might well be applied. What is at issue is nothing less than the very meaning of religion itself" (p. 8). Jones would claim that seeing images of deceased loved ones while having some degree of conscious awareness or seeing one's body from the outside (i.e., autoscopy) may have helped create the idea of a "soul" that continues on after one's mortal demise.

Regardless of the veracity of these larger claims, the experiences seen in ISP are indisputably consistent with any number of anomalous beliefs.

For instance, the idea of a nocturnally assaultive intruder is consistent with witches (Hufford, 1982), demons of various sorts (Davies, 2003), and malevolent creatures such as vampires and werewolves (Jones, 1949). The narrative descriptions of various encounters with ghosts, spirits, and shadow people also overlap with ISP symptomatology (e.g., Awadalla et al., 2004). Finally, many current descriptions of nocturnal alien abductions can be more parsimoniously explained as ISP episodes, when combined with certain other predisposing factors (e.g., memory distortion and fantasy-proneness; Clancy, 2007; Clancy, McNally, Schacter, Lenzenweger, & Pitman, 2002).

As shown in the section on prevalence, ISP is not bound to any one particular culture. Instead, it appears to be a fairly universal human phenomenon, but the cultural expressions of ISP experiences show differences that are potentially meaningful. This is especially the case in terms of the content of ISP hallucinations.

All of the ISP terms reviewed by this author connote some degree of fear and/or discomfort. As one example, the Japanese term for ISP is *kanashibari*, which means to be bound or fastened by metal (Arikawa, Templer, Brown, Cannon, & Thomas-Dodson, 1999). *Kanashibari* appears to emphasize the constrictive aspects of ISP atonia. Old Norse emphasizes the *oppressive* and possibly sexual aspects of ISP, with the *Mara* (a female supernatural being) suffocating her helpless victims at night (Davies, 2003). Along with the old English *Mare*, German *Mar*, and Irish *More*, the *Mara* are likely sources for our contemporary Nightmare. Interestingly, the original meaning of Nightmare appears to have been lost, as it was much more consistent with ISP than the current colloquial meaning (Sharpless, 2014). Similar to the *Mara*, ISP was experienced as the work of a malevolent witch in both Germany (*hexendrücken*; Cheyne, Newby-Clark, et al., 1999; Cheyne, Rueffer, et al., 1999) and Hungary (*boszorkany-nyomas*; Davies, 2003) who pressed her victims. And, as one last example, *guî yā chuáng*, or ghost oppression, can be found in Chinese sufferers (Awadalla et al., 2004). Although this is only a very small sampling from a great deal of indigenous terms, it hopefully conveys some of the different emphases that can be found according to locale.

Phenomenology and Specific Features of ISP and Their Relations to Fear

Atonia and related sensations. The core features of ISP deserve some mention because they can contribute to changes in affect, either alone or in combination. For instance, the paralysis of ISP is complete for all voluntary muscles except the eyes. Some report an inability to open their eyes at all during episodes, but the majority of sufferers are able to gaze about the room while their bodies are helpless and vulnerable. A general suppression of motor neurons occurs in ISP, with a well-understood biological basis primarily involving the medial medullary reticular formation (Marks, 2009).

Vocal chords (being voluntary muscles) are affected in ISP as well. Sufferers are neither able to call out for help nor verbally respond to the other constituent parts of ISP episodes. Over half of a clinical sample experienced this unexpected mutism as scary (Sharpless et al., 2010). However, even though articulate speech cannot be generated, a number of case reports describe inarticulate groans, especially near the ends of attacks (e.g., Macnish, 1834; Roscher, 2007).

Respiration can be affected as well, with corresponding sensations of smothering and/or pressure on the chest. Whether these feelings result from the paralysis or are perceptions based on the commonly high levels of anxiety sensitivity in ISP patients (e.g., Ramsawh et al., 2008) is not yet known, however, nor is the exact mechanism through which these respiration difficulties occur. Some evidence indicates that patients with ISP may have a higher body mass index (Sharpless et al., 2010), and given that most attacks occur during sleep in a supine position, the body weight pressure might lead to feelings of suffocation in ISP, just as in obstructive sleep apnea.

Though eye movements are not affected, some individuals report that they are unable to open their eyes during ISP. They instead strain and struggle, especially when experiencing the all-too-common auditory hallucinations. Even when individuals can look about the room, they can neither take action nor communicate their distress, but are left to merely watch the events transpiring around them, while attempting to anticipate what happens next. This conscious vulnerability can be especially terrifying when experiencing ISP hallucinations.

Hallucinations. The vast majority of people with ISP suffer from hallucinations. Estimates range from 75% to 88.5% (Cheyne, Newby-Clark, et al., 1999; Sharpless et al., 2010). Thus, only a minority of patients reports paralysis alone. Several relatively large studies have looked at the particular frequencies and patterns of hallucinations in student and clinical samples (e.g., Cheyne, Newby-Clark, et al., 1999; Cheyne, Ruetter, et al., 1999; Sharpless et al., 2010), and specific results can be found in Table 9.2. For ease of discussion, the many ISP hallucinations were grouped into tactile/kinesthetic sensations, auditory hallucinations, visual hallucinations, and the sensed presence.

Tactile/kinesthetic sensations and hallucinations constitute a wide variety of ISP experiences, some of which have already been discussed (e.g., suffocation, feelings of chest oppression). These may involve primarily tactile sensations such as feeling touched by an external agent, feeling the blankets being pulled off of you, or having strong sensations of heat or cold in one's extremities. Probably the most frightening kinesthetic

TABLE 9.2 Frequency of Isolated Sleep Paralysis Hallucinations and Experiences

Isolated Sleep Paralysis Symptom	Study	% of Sample
Auditory hallucinations	1, 2	11.37
	3	38.50
BODILY SENSATIONS AND KINESTHETIC/TACTILE HALLUCINATIONS		
Body pressure	1, 2	12.18
Pressure on chest/smothering	3	57.70
Try to speak or call out for help, but can't	3	61.50
Cold	3	23.10
Pain	3	19.20
Falling/flying/floating/spinning	3	34.60
Floating	1, 2	10.69
Feel like being touched	3	23.10
Feel like being strangled	3	19.20
Feel that body has moved/been moved	3	38.50
Leave or see body from the outside	3	30.80
Erotic/sexual feelings	3	19.20
The sensed presence	1, 2	15.00
	3	50.00
Visual hallucinations	1, 2	8.62
	3	34.60

Notes: 1 = Cheyne, Newby-Clark, et al. (1999), 2 = Cheyne, Rueffer, et al. (1999) (student sample only), and 3 = Sharpless et al. (2010). Both 1 and 2 are student samples, whereas 3 is a clinical sample of patients reporting panic attacks. Because not all of the categories from these three studies perfectly overlap, all of the experiences have been included separately.

hallucinations involve feeling physically or sexually assaulted. These experiences constitute some of the earliest depictions of ISP (Bond, 1753). The tactile/kinesthetic group also includes many illusory body movements such as falling, flying, spinning, or turning. Thus, for some individuals with ISP, their normal sense of proprioception is compromised. As shown in Table 9.2, the most frequent of these sensations in the clinical sample were an inability to call out for help and chest pressure, with the former eliciting the highest levels of distress.

Another unusual set of ISP somatic sensations/perceptions overlaps with dissociative phenomena. Specifically, out-of-body experiences (OBEs) in ISP are not uncommon. These may include perceiving one's own body from the outside (i.e., autoscopy) or feeling as if one's soul or self leaves the body in some way. Given the connections between ISP and trauma and/or PTSD (e.g., Abrams et al., 2008; Friedman & Paradis, 2002; Sharpless et al., 2010), these OBEs may not be surprising.

Auditory hallucinations can also occur, with clinical samples reporting them more than students (Table 9.2). These sounds run the gamut from environmental noises (e.g., scraping, water dripping) to inarticulate and articulate speech. Interestingly, people seem to experience the sounds in ISP as occurring outside of the body more often than in standard dreaming, and the fact that they occur with at least some degree of conscious awareness lends real world veracity to them. This veracity can lead to some disconcerting experiences. For instance, when the usually negative valence of ISP hallucinations is combined with the finding that the voices heard during hypnagogic (going to sleep) and hypnopompic (leading out of sleep) states are more commonly perceived as coming from known individuals (Jones, Fernyhough, & Laroi, 2010), it is easy to see how terrifying ISP can be. Several individuals have reported abusive speech and behavior from loved ones during these episodes.

The most vivid and terrifying experiences of ISP are probably visual hallucinations, however, especially those involving other people or entities (e.g., the woman with black hair in the vignette). Some can be fairly innocuous. For instance, inanimate objects, such as furniture, and various geometrical patterns may be hallucinated and, relatively speaking, elicit much less affect than animate objects. Before discussing these latter hallucinations, however, it is important to discuss the sensed presence, a common accompaniment to ISP.

Unlike inanimate objects, sensing and/or actually seeing another person/being during ISP is usually associated with a great deal of affect. As displayed in Table 9.2, this sensed presence is common among both clinical and nonclinical samples (Solomonova et al., 2008). These presences are rarely thought to be benevolent, and they are more frequently associated with ill-will and danger. Something bad feels immanent, and given that one is paralyzed and helpless during ISP, the sensed presence usually leads the sufferer to feel as if they are prey. Researchers have postulated that the *feeling* of the sensed presence is associated with some type of threat-activated vigilance system found in the activities of the limbic system and associated structures of the brain (Cheyne, 2001). The nature of the presence can vary widely, with the majority of individuals reporting a distinctly nonhuman presence (Sharpless, under review). This has been vividly captured in certain works of literature, such as Maupassant's *La Horla* (de Maupassant, 1945).

However, for many individuals with ISP, there is a transition from reliably *sensing* a presence to actually *seeing* one. The limited data indicate that a reliable temporal sequence to the phenomena may exist, such that sensing precedes seeing presences, with the process partially mediated by fear (Cheyne, Newby-Clark, et al., 1999; Cheyne, Rueffer, et al., 1999). Perhaps not surprisingly, these interactions with others evoke intense levels of negative affect (e.g., Girard & Cheyne, 2004; Parker & Blackmore, 2002; Sharpless et al., 2010).

As for who or what is seen, there is relatively little data beyond case studies and narratives in the historical and early medical records (discussed more below). The most common animals reportedly seen during ISP are cats (Davies, 2003), but bats (Nickell, 1995), chickens (Davies, 2003), and other creatures have been reported. As noted above, a panoply of ghosts, demons, and spirits can also be found in folklore and myth. In the one empirical study conducted on this topic, the author found that the majority of entities seen during ISP were nonhuman (Sharpless, under review). Recently deceased relatives were common among the *persons* seen by this sample of college students. In a sample of Cambodian refugees, Hinton, Pich, Chhean, and Pollack (2005) reported that some of these individuals saw Khmer Rouge uniforms on their hallucinations.

Importantly, factor analytic studies of individual ISP experiences have taken place, and the results are consistent across several studies, in addition to being quite interesting (Cheyne, 2005; Cheyne, Newby-Clark, et al., 1999; Cheyne, Rueffer, et al., 1999). Cheyne and colleagues identified a tripartite factor structure of intruder (e.g., sensed presence with visual, auditory, and tactile hallucinations), incubus (e.g., pressure on the chest, smothering sensations, pain, and death thoughts), and vestibular-motor (e.g., falling, floating, and OBEs) hallucinations. The first two categories are more strongly associated with fear, whereas the last can occasionally be associated with more pleasant affective states.

The Clear Sensorium in ISP and Appraisal of Episodes

Although the major components of REM sleep are clearly present during ISP (i.e., atonia and dreams/hallucinations), they occur in conjunction with wakeful consciousness. Many case reports indicate a standard wakefulness during episodes, but a recent study questions this assumption. Terzaghi, Ratti, Manni, and Manni (2012) analyzed spectral EEG data on a patient experiencing SP in the context of narcolepsy. They found that, during the SP episode, the patient was actually in a transitional state between sleep and wakefulness. This is consistent with Mahowald and Schenck's (2005) theory of dissociated states of mind (i.e., that wakefulness and the various sleep states are not necessarily exclusive). It is unclear if this finding holds for cases of ISP as well, but it is certainly possible.

Regardless of its relative level, *some* degree of consciousness is necessarily present during ISP episodes, allowing for a strong and direct influence on affect. Those experiencing ISP can perceive their environment, deliberate upon their experiences, and consider the possible consequences of their experiences (e.g., their mental or physical health) in a manner similar to when they are awake. Thus, they can make any number of appraisals on the various ISP experiences. More often than not, these appraisals tend toward the negative, permanent, and catastrophic.

Several specific negatively valenced appraisals have been found to occur during ISP episodes. For instance, individuals often report fear of imminent death (e.g., Arikawa et al., 1999). This could be due to the feelings of suffocation or fear of threatening hallucinations that are so often experienced during episodes. This appraisal is very similar to the catastrophic cognitions experienced during panic attacks (American Psychiatric Association, 2013). Several studies have also reported the catastrophic cognition that the temporary paralysis will in fact be permanent (e.g., Hinton et al., 2005; Koran & Raghavan, 1993; Ramsawh et al., 2008). These periepisode fears are very consistent with basic research findings on the high levels of anxiety sensitivity in ISP individuals (Ramsawh et al., 2008; Sharpless et al., 2010).

Negative appraisals and the affect that goes along with them do not necessarily stop with the cessation of the episode. Going back to the ancient writers, sufferers (and sometimes physicians) often feared that ISP was a forerunner to more serious conditions such as "the sacred disease" of epilepsy or apoplexy (bleeding from the internal organs and/or ischemia) (e.g., van Diemerbroeck, 1689). Contemporary sufferers are more likely to be apprehensive that they may be suffering from a serious and undiagnosed *mental* illness, as opposed to a physical one (Neal, Rich, & Smucker, 1994). Thus, reassurance and psychoeducation about ISP can be helpful (Sharpless & Barber, 2011). Some have even thought that ISP could lead to eventual death. Silimachus, a Roman follower of Hippocrates (cited in Aurelianus, 1950), believed that Nightmare (i.e., ISP) was the cause of many deaths in Rome in a manner similar to a plague or some sort of group hysteria. Documentation on this outbreak is far from ideal, but regardless, it speaks to the power of attributions for ISP.

Feelings of shame and embarrassment are also common in the wake of episodes (Neal et al., 1994; Otto et al., 2006), and these feelings may explain the relatively small percentage of people who actively seek out help in spite of the fact that episodes are troubling (Yeung, Xu, & Chang, 2005). This may especially be the case if there are erotic components to the hallucinations, and these are not uncommon (see Table 9.2).

As with other experiences that elicit strong affect (e.g., trauma, panic attacks), the locations and situations in which the episodes took place often become associated with fear and can lead to additional levels of apprehension and avoidance. Given that ISP typically occurs at night, the bedroom, bedroom furniture, and even darkness may become conditioned fearful stimuli (e.g., Alvaro, 2005). Bond (1753), the author of the first English-language treatise on Nightmare, was himself a sufferer. He reported that he often slept all night in a chair in order to avoid the characteristic situations in which he experienced these attacks.

Nonfearful Affect and Sleep Paralysis

Although the majority of ISP experiences appear to be laden with fear and anxiety, there are certain exceptions. For instance, people occasionally report feelings of bliss. Bliss is most commonly experienced during vestibular-motor hallucinations such as flying, floating, or spinning (Cheyne & Girard, 2007). Some individuals actively seek out these experiences through attempts at lucid dreaming and/or engaging in behaviors that make ISP more likely to occur (e.g., sleeping in a supine position). Interestingly, these vestibular-motor experiences appear to increase in frequency as ISP chronicity also increases. Cheyne (2005) hypothesized that individuals with more ISP experience may differentially focus on these less frightening hallucinations. Subsequent multiple regressions on these data revealed that only OBEs made independent contributions to bliss (Cheyne & Girard, 2007). Fear and bliss, not surprisingly, were also found to be negatively correlated. It has been speculated that certain ISP hallucinations may bear a relationship to spiritual beliefs (Hufford, 2005), and higher levels of spiritual beliefs were found in a sample of African-American individuals with ISP than in those without (Ramsawh et al., 2008). The causal nature of this relationship has not been established, however.

Feelings of sexual excitement also occur in ISP, and these feelings seem to be experienced as scary the majority of the time. Unfortunately, existing data do not allow for a more precise determination of the extent to which these erotic feelings are pleasant or scary (e.g., associated with a hallucinatory nocturnal sexual assault). Regardless, they are fairly common in general (see Table 9.1), with one study finding them present in 42.0% of an African-American sample (Paradis et al., 2009). So what about ISP could be erotic? There appear to be two main possibilities. First, these feelings could be reactions to sexual (and nonthreatening) hallucinations. Individuals have occasionally reported "intruders" or "incubi" that lack the malevolent characteristics usually associated with them and are instead perceived as sexually compelling. Second, these erotic sensations could be the result of some degree of direct stimulation or pressure on the sex organs (especially when sleeping in a prone position), which becomes subsequently incorporated into the overall ISP experience (e.g., see Yu, 2012). Nocturnal emissions have been reported during ISP as well (e.g., Roscher, 2007).

CLINICAL IMPAIRMENT AS A RESULT OF SLEEP PARALYSIS

A relatively unexplored area in the SP literature is the negative impact it has, if any, on its sufferers. Historically, a great many physicians and laymen have noted and cataloged the ill effects of sleep paralysis that go beyond the momentary periepisode fear and distress. The very fact that so many cultures, scientifically inclined or otherwise, attempted a number of cures for ISP is fairly compelling evidence that the condition was viewed as problematic (e.g., Aurelianus, 1950 van Diemerbroeck, 1689; Waller, 1816). These cures ranged from fairly innocuous (e.g., dietary restrictions) to quite invasive (e.g., regular venesection or drinking ammonia diluted with water).

Some of the impairment from ISP may be fairly obvious. Along with the above-mentioned avoidance behaviors and high levels of distress as a result of attacks, daytime sleepiness has also been reported (e.g., Jiménez-Genchi, Ávila-Rodríguez, Sánchez-Rojas, Terrez, & Nenclares-Portocarrero, 2009; Wing, Lee, & Chen, 1994). Worry about additional attacks occurs as well, with some individuals taking preventative measures in the hopes of staving off future ISP. Some are likely to have therapeutic impact (e.g., not sleeping on ones back), whereas others are more closely tied to superstitious beliefs (e.g., sleeping with foul-smelling objects or salt on your chest; Jones, 1949). Anomalous beliefs may be formed by ISP as well, or, alternately, they may be triggers for subsequent episodes. Those who ascribe culturally asynchronous supernatural or paranormal causes to ISP will likely find themselves at variance with the prevailing societal norms. This could lead to additional life difficulties, interpersonal problems, or increased isolation. Such beliefs may require psychoeducation about the known psychological and biological bases of ISP (e.g., Ohaeri, Adelekan, Odejide, & Ikuesan, 1992).

CONCLUSIONS

This chapter has surveyed the many historical, psychopathological, and experiential aspects of ISP. Historically, ISP has likely played at least a part in many beliefs most scientists would categorize as supernatural or paranormal. The existing empirical data dovetail well with these earlier beliefs. ISP was also seen to be distressing enough by many cultures in many time periods that it lead to a multitude of attempts to cure the condition and/or prevent future attacks. As for the phenomenon itself, all of ISP's constituent parts (atonia, hallucinations, clear sensorium) can contribute to any number of affective states, with the most pronounced being fear. Positive feelings can result as well, but these are relatively less common.

References

Abrams, M. P., Mulligan, A. D., Carleton, R. N., & Asmundson, G. J. G. (2008). Prevalence and correlates of sleep paralysis in adults reporting childhood sexual abuse. *Journal of Anxiety Disorders*, 22(8), 1535–1541. http://dx.doi.org.ezaccess.libraries.psu.edu/10.1016/j.janxdis.2008.03.007.

Aegineta, P. (1844). In F. Adams (Ed.), *The seven books of paulus aegineta*. Translated from the Greek with a commentary embracing a complete view of the knowledge possessed by the Greeks, Romans, and Arabians on all subjects connected with medicine and surgery (F. Adams, Trans.). London: Syndeham Society.

Alvaro, L. (2005). Hallucinations and pathological visual perceptions in Maupassant's fantastical short stories—A neurological approach. *Journal of the History of the Neurosciences*, 14, 100–115.

American Academy of Sleep Medicine. (2005). *International classification of sleep disorders: Diagnostic and coding manual* (2nd ed.). Darien, IL: American Academy of Sleep Medicine.

American Academy of Sleep Medicine. (2014). *International classification of sleep disorders: Diagnostic and coding manual* (3rd ed.). Darien, IL: American Academy of Sleep Medicine.

American Psychiatric Association. (2013). *Diagnostic and statistical manual of mental disorders: DSM-V* (5th ed.). Arlington, VA: American Psychiatric Association.

Arikawa, H., Templer, D. I., Brown, R., Cannon, W. G., & Thomas-Dodson, S. (1999). The structure and correlates of kanshibari. *The Journal of Psychology*, 133(4), 369–375.

Aurelianus, C. (1950). In I. E. Drabkin (Ed.), *On acute diseases and chronic diseases*. (I. E. Drabkin, Trans.). Chicago: University of Chicago Press.

Awadalla, A., Al-Fayez, G., Harville, M., Arikawa, H., Tomeo, M. E., Templer, D. I., et al. (2004). Comparative prevalence of isolated sleep paralysis in Kuwaiti, Sudanese, and American college students. *Psychological Reports*, 95(1), 317–322. http://dx.doi.org.ezaccess.libraries.psu.edu/10.2466/PR0.95.5.317-322.

Bell, C. C., Hildreth, C. J., Jenkins, E. J., & Carter, C. (1988). The relationship of isolated sleep paralysis and panic disorder to hypertension. *Journal of the National Medical Association*, 80(3), 289–294.

Bond, J. (1753). *An essay on the incubus, or night mare*. London: D. Wilson and T. Durham.

Cheyne, J. A. (2001). The ominous numinous: Sensed presence and 'other' hallucinations. *Journal of Consciousness Studies*, 8(5-7), 133–150.

Cheyne, J. A. (2005). Sleep paralysis episode frequency and number, types, and structure of associated hallucinations. *Journal of Sleep Research*, 14(3), 319–324. http://dx.doi.org.ezaccess.libraries.psu.edu/10.1111/j.1365-2869.2005.00477.x.

Cheyne, J. A., & Girard, T. A. (2007). Paranoid delusions and threatening hallucinations: A prospective study of sleep paralysis experiences. *Consciousness and Cognition*, 16(4), 959–974.

Cheyne, J. A., Newby-Clark, I. R., & Rueffer, S. D. (1999). Relations among hypnagogic and hypnopompic experiences associated with sleep paralysis. *Journal of Sleep Research*, 8, 313–317.

Cheyne, J. A., Rueffer, S. D., & Newby-Clark, I. R. (1999). Hypnagogic and hypnopompic hallucinations during sleep paralysis: Neurological and cultural construction of the nightmare. *Consciousness and Cognition*, 8(3), 319–337. http://dx.doi.org.ezaccess.libraries.psu.edu/10.1006/ccog.1999.0404.

Clancy, S. A. (2007). *Abducted: How people come to believe they were kidnapped by aliens*. Cambridge, MA: First Harvard University Press.

Clancy, S. A., McNally, R. J., Schacter, D. K., Lenzenweger, M. F., & Pitman, R. K. (2002). Memory distortion in people reporting abduction by aliens. *Journal of Abnormal Psychology*, 11(3), 455–461.

Davies, O. (2003). The nightmare experience, sleep paralysis and witchcraft accusations. *Folklore*, 114(2), 181.

de Maupassant, G. (1945). The Horla. In S. Commins (Ed.), *The best stories of Guy de Maupassant*. New York: Modern Library.

Douglass, A. B., Hays, P., Pazderka, F., & Russell, J. M. (1991). Florid refractory schizophrenias that turn out to be treatable variants of HLA-associated narcolepsy. *The Journal of Nervous and Mental Disease*, 179(1), 12–17.

REFERENCES

Friedman, S., & Paradis, C. (2002). Panic disorder in African Americans: Symptomatology and isolated sleep paralysis. *Culture, Medicine and Psychiatry, 26*(2), 179–198. http://dx.doi.org.ezaccess.libraries.psu.edu/10.1023/A:1016307515418.

Gangdev, P. (2004). Relevance of sleep paralysis and hypnic hallucinations to psychiatry. *Australasian Psychiatry, 12*(1), 77–80. http://dx.doi.org.ezaccess.libraries.psu.edu/10.1046/j.1039-8562.2003.02065.x.

Girard, T. A., & Cheyne, J. A. (2004). Individual differences in lateralisation of hallucinations associated with sleep paralysis. *Laterality, 9*(1), 93–111. http://dx.doi.org.ezaccess.libraries.psu.edu/10.1080/13576500244000210.

Haga, E. (1989). The nightmare—A riding ghost with sexual connotations. *Nordic Journal of Psychiatry, 43*(6), 515–520.

Hedman, C., Pohjasvaara, T., Tolonen, U., Salmivaara, A., & Myllyla, V. V. (2002). Parasomnias decline during pregnancy. *Acta Neurologica Scandinavica, 105*, 209–214.

Hinton, D. E., Pich, V., Chhean, D., & Pollack, M. H. (2005). The ghost pushes you down: Sleep paralysis-type panic attacks in a Khmer refugee population. *Transcultural Psychiatry, 42*(1), 46–77. http://dx.doi.org.ezaccess.libraries.psu.edu/10.1177/1363461505050710.

Hufford, D. J. (1982). *The terror that comes in the night: An experience-centred study of supernatural assault traditions*. Philadelphia, PA: University of Pennsylvania Press.

Hufford, D. J. (2005). Sleep paralysis as spiritual experience. *Transcultural Psychiatry, 42*(1), 11–45. http://dx.doi.org.ezaccess.libraries.psu.edu/10.1177/1363461505050709.

Jiménez-Genchi, A., Ávila-Rodríguez, V. M., Sánchez-Rojas, F., Terrez, B. E. V., & Nenclares-Portocarrero, A. (2009). Sleep paralysis in adolescents: The 'a dead body climbed on top of me' phenomenon in Mexico. *Psychiatry and Clinical Neurosciences, 63*(4), 546–549. http://dx.doi.org.ezaccess.libraries.psu.edu/10.1111/j.1440-1819.2009.01984.x.

Jones, E. (1949). *On the nightmare* (2nd Impression ed.). London, UK: Hogarth Press and the Institute of Psycho-analysis.

Jones, S. R., Fernyhough, C., & Laroi, F. (2010). A phenomenological survey of auditory verbal hallucinations in the hypnagogic and hypnopompic states. *Phenomenology and the Cognitive Sciences, 9*, 213–224.

Kierkegaard, S. A. (1980). In H. Hong & E. Hong (Eds.), *The concept of anxiety: A simple psychological orienting deliberation on the dogmatic issue of hereditary sin* (H. Hong & E. Hong, Trans.). Princeton, NJ: Princeton University Press.

Koran, L. M., & Raghavan, S. (1993). Fluoxetine for isolated sleep paralysis. *Psychosomatics, 34*(2), 184–187. http://dx.doi.org.ezaccess.libraries.psu.edu/10.1016/S0033-3182(93)71913-1.

Kramer, H., & Sprenger, J. (1971). In M. Summers (Ed.), *The malleus maleficarum* (N. Summers, Trans.). Mineola, NY: Dover Publications.

Macnish, R. (1834). *The philosophy of sleep* (First American ed.). New York: D. Appleton and Company.

Mahowald, M. W., & Schenck, C. H. (2005). Insights from studying human sleep disorders. *Nature, 437*, 1279–1285.

Marks, G. A. (2009). Neurobiology of sleep. In T. L. Lee-Chiong (Ed.), *Sleep medicine essentials* (pp. 5–10). Hoboken, NJ: John Wiley and Sons.

McNally, R. J., & Clancy, S. A. (2005a). Sleep paralysis in adults reporting repressed, recovered, or continuous memories of childhood sexual abuse. *Journal of Anxiety Disorders, 19*(5), 595–602. http://dx.doi.org.ezaccess.libraries.psu.edu/10.1016/j.janxdis.2004.05.003.

McNally, R. J., & Clancy, S. A. (2005b). Sleep paralysis, sexual abuse, and space alien abduction. *Transcultural Psychiatry, 42*(1), 113–122. http://dx.doi.org.ezaccess.libraries.psu.edu/10.1177/1363461505050715.

Mellman, T. A., Aigbogun, N., Graves, R. E., Lawson, W. B., & Alim, T. N. (2008). Sleep paralysis and trauma, psychiatric symptoms and disorders in an adult African American population attending primary medical care. *Depression and Anxiety, 25*(5), 435–440. http://dx.doi.org.ezaccess.libraries.psu.edu/10.1002/da.20311.

Neal, A. M., Rich, L. N., & Smucker, W. D. (1994). The presence of panic disorder among African American hypertensives: A pilot study. *The Journal of Black Psychology, 20*(1), 29–35.
Nickell, J. (1995). The skeptic-raping demon of Zanzibar. *Skeptical Briefs, 5*(4), 7.
Ohaeri, J. U., Adelekan, M. F., Odejide, A. O., & Ikuesan, B. A. (1992). The pattern of isolated sleep paralysis among Nigerian nursing students. *Journal of the National Medical Association, 84*(1), 67–70.
Otto, M. W., Simon, N. M., Powers, M., Hinton, D., Zalta, A. K., & Pollack, M. H. (2006). Rates of isolated sleep paralysis in outpatients with anxiety disorders. *Journal of Anxiety Disorders, 20*(5), 687–693. http://dx.doi.org.ezaccess.libraries.psu.edu/10.1016/j.janxdis.2005.07.002.
Paradis, C., Friedman, S., Hinton, D. E., McNally, R. J., Solomon, L. Z., & Lyons, K. A. (2009). The assessment of the phenomenology of sleep paralysis: The unusual sleep experiences questionnaire (USEQ). *CNS Neuroscience & Therapeutics, 15*(3), 220–226. http://dx.doi.org.ezaccess.libraries.psu.edu/10.1111/j.1755-5949.2009.00098.x.
Parker, J. D., & Blackmore, S. J. (2002). Comparing the content of sleep paralysis and dream reports. *Dreaming, 12*(1), 45–59. http://dx.doi.org.ezaccess.libraries.psu.edu/10.1023/A:1013894522583.
Pelayo, R., & Lopes, M. C. (2009). Narcolepsy. In T. L. Lee-Chiong (Ed.), *Sleep medicine essentials* (pp. 47–51). Hoboken, NJ: Wiley-Blackwell.
Portala, K., Westermark, K., Ekselius, L., & Broman, J. (2002). Sleep in patients with treated Wilson's disease: A questionnaire study. *Nordic Journal of Psychiatry, 56*(4), 291–297.
Ramsawh, H. J., Raffa, S. D., White, K. S., & Barlow, D. H. (2008). Risk factors for isolated sleep paralysis in an African American sample: A preliminary study. *Behavior Therapy, 39*(4), 386–397. http://dx.doi.org.ezaccess.libraries.psu.edu/10.1016/j.beth.2007.11.002.
Roscher, W. H. (2007). *Ephialtes: A pathological-mythological treatise on the nightmare in classical antiquity* [Ephialtes] (A. V. O'Brien Trans.) (Revised ed.). Putnam, CT: Spring Publishing (pp. 96–159).
Shapiro, B., & Spitz, H. (1976). Problems in the differential diagnosis of schizophrenia. *American Journal of Psychiatry, 133*(11), 1321–1323.
Sharpless, B. A. (2013). Kierkegaard's conception of psychology. *Journal of Theoretical and Philosophical Psychology, 33*(2), 90–106.
Sharpless, B.A. (2014). Changing conceptions of the Nightmare in medicine. *Hektoen International: A Journal of Medical Humanities. (Moments in History section)*. Retrieved from http://www.hekteninternational.org.
Sharpless, B. A. (under review). *Then sensed presences and hallucinations of others in isolated sleep paralysis.*
Sharpless, B. A., & Barber, J. P. (2011). Lifetime prevalence rates of sleep paralysis: A systematic review. *Sleep Medicine Reviews, 15*(5), 311. http://dx.doi.org/10.1016/j.smrv.2011.01.007.
Sharpless, B. A., & Grom, J. L. (in press). Isolated sleep paralysis: Fear, prevention, and disruption. *Behavioral Sleep Medicine.*
Sharpless, B. A., McCarthy, K. S., Chambless, D. L., Milrod, B. L., Khalsa, S. R., & Barber, J. P. (2010). Isolated sleep paralysis and fearful isolated sleep paralysis in outpatients with panic attacks. *Journal of Clinical Psychology, 66*(12), 1292–1306. http://dx.doi.org/10.1002/jclp.20724; 10.1002/jclp.20724.
Solomonova, E., Nielsen, T., Stenstrom, P., Simard, V., Frantova, E., & Donderi, D. (2008). Sensed presence as a correlate of sleep paralysis distress, social anxiety and waking state social imagery. *Consciousness and Cognition, 17*(1), 49–63. http://dx.doi.org.ezaccess.libraries.psu.edu/10.1016/j.concog.2007.04.007.
Spanos, N. P., McNulty, S. A., DuBreuil, S. C., & Pires, M. (1995). The frequency and correlates of sleep paralysis in a university sample. *Journal of Research in Personality, 29*(3), 285–305. http://dx.doi.org.ezaccess.libraries.psu.edu/10.1006/jrpe.1995.1017.

REFERENCES

Szklo-Coxe, M., Young, T., Finn, L., & Mignot, E. (2007). Depression: Relationships to sleep paralysis and other sleep disturbances in a community sample. *Journal of Sleep Research, 16*(3), 297–312. http://dx.doi.org.ezaccess.libraries.psu.edu/10.1111/j.1365-2869.2007.00600.x.

Terzaghi, M., Ratti, P. L., Manni, F., & Manni, R. (2012). Sleep paralysis in narcolepsy: More than just a motor dissociative phenomenon? *Neurological Sciences, 33*(1), 169–172. http://dx.doi.org.ezaccess.libraries.psu.edu/10.1007/s10072-011-0644-y.

van der Kloet, D., Giesbrecht, T., Lynn, S. J., Merckelbach, H., & de Zutter, A. (2011). Sleep normalization and decrease in dissociative experiences: Evaluation in an inpatient sample. *Journal of Abnormal Psychology, 121*(1), 140–150.

van Diemerbroeck, I. (1689). *The anatomy of human bodies, comprehending the most modern discoveries and curiosities in that art. To which is added a particular treatise of the small-pox and measles. Together with several practical observations and experienced cures.* (W. Salmom Trans.). London: W. Whitwood.

Waller, J. (1816). *A treatise on the incubus, or night-mare, disturbed sleep, terrific dreams, and nocturnal visions with the means of removing these distressing complaints.* London: E Cox and Son.

Wilson, S. A. K. (1928). The narcolepsies. *Brain, 51*, 63–109.

Wing, Y., Lee, S. T., & Chen, C. (1994). Sleep paralysis in Chinese: Ghost oppression phenomenon in Hong Kong. *Sleep, 17*(7), 609–613.

World Health Organization. (2008). *International statistical classification of diseases and related health problems* (10th Rev.). New York: World Health Organization.

Yeung, A., Xu, Y., & Chang, D. F. (2005). Prevalence and illness beliefs of sleep paralysis among Chinese psychiatric patients in China and the United States. *Transcultural Psychiatry, 42*(1), 135–143. http://dx.doi.org.ezaccess.libraries.psu.edu/10.1177/1363461505050725.

Yu, C. K. (2012). The effect of sleep position on dream experiences. *Dreaming, 22*(3), 212–221.

CHAPTER 10

Sleep and Repetitive Thought: The Role of Rumination and Worry in Sleep Disturbance

Vivek Pillai and Christopher L. Drake

Sleep Disorders and Research Center, Henry Ford Health System, Detroit, Michigan, USA

SLEEP AND REPETITIVE THOUGHT: THE ROLE OF RUMINATION AND WORRY IN SLEEP DISTURBANCE

Negative affect is widely recognized as a common precipitant of both subjective and objective sleep disturbance (Vandekerckhove & Cluydts, 2010). However, recent models of insomnia disorder attribute sleep disturbance more to the dysregulation of negative affect than to the mere emergence of negative affective states (Espie, 2002). A burgeoning and reliable body of research suggests that repetitive thought can preclude adaptive emotion regulation and prolong negative affect states (for a review, see Thomsen, 2006). Repetitive thought therefore represents a more critical ingredient in the etiology of sleep disturbance than previously thought.

Broadly defined, repetitive thought is the process of recurrently focusing attention on the self and on the environment (Segerstrom, Stanton, Alden, & Shortridge, 2003). As such, repetitive thought may be constructive or unconstructive, based on the function of the thought content and the nature of underlying affect states. In the context of psychopathology and hence this chapter, repetitive thought refers to the perseverative, intrusive activation of cognitive representations of stressful events or negatively valenced affect. Although the literature is teeming with numerous conceptualizations of unconstructive repetitive thought (for a review, see Watkins, 2008), two conceptually distinct forms have garnered the most research attention: worry and rumination.

WORRY: PHENOMENOLOGY AND ASSESSMENT

A cardinal feature of anxiety disorders such as generalized anxiety disorder, worry involves recurrent, intrusive thoughts or images about the potential negative outcomes signaled by a perceived threat (Borkovec, Robinson, Pruzinsky, & DePree, 1983). Commonly reported functions of worry include problem-solving, preparation for the worst, and determining ways to avoid feared stimuli (Borkovec & Roemer, 1995). However, nearly 70% of reported worries involve fears of unlikely or implausible events. Thus, any perceived benefits are typically overshadowed by the physiological and emotional costs of worrying (Borkovec, 1994). Studies indicate that worrying is associated with various indices of autonomic arousal such as increased skin conductance, increased skin conductance variability, and a high cortisol awakening response (CAR) (Schlotz, Hellhammer, Schulz, & Stone, 2004; Weise, Ong, Tesler, Kim, & Roth, 2013). Notably, daily levels of worry are associated with elevated heart rate and low heart rate variability (HRV) not only while waking, but also during the following sleep period (Brosschot, Gerin, & Thayer, 2006; Brosschot, Van Dijk, & Thayer, 2007). Similarly, worry is predictive of future anxiety and negative affect (Calmes & Roberts, 2007; Llera & Newman, 2010; McLaughlin, Borkovec, & Sibrava, 2007). Thus, worry is a perseverative thought mechanism by which cognitive manifestations of perceived threats, as well as the associated physiological and emotional arousal, are prolonged.

The Penn-State Worry Questionnaire (PSWQ; Meyer, Miller, Metzger, & Borkovec, 1990) is presently the most widely used instrument for measuring worry. This 16-item self-report questionnaire assesses the pervasiveness (e.g., "many situations worry me"), excessiveness (e.g., "I am always worrying about something"), and uncontrollability of worry ("I know I shouldn't worry, but I just can't help it"). The PSWQ has excellent internal consistency ($\alpha = 0.95$) and high test-retest reliability over an 8-10-week period ($r = 0.92$). Zero-order correlations between the PSWQ and sleep disturbance measures, such as the Pittsburgh Sleep Quality Index (PSQI: Buysse, Reynolds, Monk, & Berman, 1989), range from moderate ($r = 0.43$, $p < 0.01$) to high ($r = 0.67$, $p < 0.01$) in clinical samples (Swanson, Pickett, Flynn, & Armitage, 2011; Yook et al., 2008).

Another instrument, the Night-time Thoughts Questionnaire (NTQ), assesses the extent to which ("not at all" to "a lot") worrisome thoughts occur during the night (Watts, Coyle, & East, 1994). The NTQ is composed of six subscales: mental activity and rehearsal (e.g., "rehearsing important things I will do tomorrow"), thoughts about sleep (e.g., "wanting to sleep," "being tired tomorrow"), family and long-term concerns ("going over and over the same thing"), positive concerns and plans (e.g., "things I enjoy"), somatic preoccupations (e.g., "feeling too hot or too cold"), and work and recent concerns (e.g., "concerns about work"). In a recent study, participants categorized as "high worriers" based on the PSWQ scored significantly higher than controls on all factors of the NTQ, including the "thoughts about sleep" factor, implying an association between trait worry and sleep-related worry (Omvik, Pallesen, Bjorvatn, Thayer, & Nordhus, 2007).

Finally, based on clinical observation of insomnia patients, Tang and Harvey (2004a) recently developed the Anxiety and Preoccupation about Sleep Questionnaire (APSQ). The APSQ is a 10-item measure composed of two collinear factors: worries about the consequences of poor sleep (e.g., "I worry about how the amount of sleep I get is going to affect my health") and worries about the uncontrollability of sleep (e.g., "my failure to rectify my sleep problems troubles me a lot"). In two prior studies, the APSQ achieved high internal consistency ($\alpha = 0.92$-0.93), and it reliably distinguished between good and poor sleepers (Jansson-Fröjmark, Harvey, Norell-Clarke, & Linton, 2012; Tang & Harvey, 2004a).

WORRY AND SLEEP

Evidence suggesting that cognitive factors may play a more central role in the etiology of sleep disturbance than somatic arousal provided the initial impetus for the study of cognitive mechanisms in sleep disruption. Now a landmark in the field, an early study indicated that individuals with insomnia disorder were 10 times more likely to attribute their sleep disturbance to cognitive factors, including worrying, planning, or difficulty controlling thoughts, than to somatic complaints, such as sweating or shifting in bed (Lichstein & Rosenthal, 1980). As theoretical and empirical models of the worry were refined in the following decades, several studies recognized its deleterious impact on sleep (Harvey, 2002b; Watts et al., 1994). Today, nearly all insomnia models highlight worry as a precipitant of sleep disturbance (Espie, 2002; Harvey, 2005).

As Harvey (2005) observes, the bulk of prior research on sleep and worry comes from three groups of studies: exploratory studies, which examine the content of presleep cognition; correlational studies on the association between sleep parameters and worry measures; and experimental studies, which explore the impact of experimentally induced worry on sleep. Many studies on the so-called worry and sleep rely on instruments that measure diffuse cognitive activity that is not specific to worry; thus, these studies are beyond the scope of this chapter (for a review, see Harvey, 2005). Exploratory thought-sampling studies that attempt to distinguish worry from other forms of cognition point to generalized worrying, problem-solving, and preoccupation with sleep as the most pervasive themes in presleep cognitive activity (Fichten et al., 1998; Harvey, 2002a;

Kuisk, Bertelson, & Walsh, 1989; Nelson & Harvey, 2003; Watts et al., 1994). However, with one exception (Fichten et al., 1998), these studies focus on individuals with insomnia disorder. Therefore, it is presently unclear how generalizable these findings are to other patient populations, especially those for whom sleep disturbance is not the primary presenting problem.

Correlational studies, on the other hand, have shown that university students who endorse high levels of worry report significantly shorter habitual sleep durations (Kelly, 2002); that work-related worry is associated with poor sleep quality (Rodríguez-Muñoz, Notelaers, & Moreno-Jiménez, 2011); and that worrying about sleeplessness is related to self-reported sleep disturbances, including shorter total sleep time (TST), longer sleep onset latency (SOL), and prolonged wake time after sleep onset (WASO) among patients with long-term (>6 months) insomnia disorder (Jansson & Linton, 2006). Similarly, in a nationally representative sample of over 2000 participants, individuals with insomnia disorder reported significantly higher levels of worry than did a group of poor sleepers who reported at least one insomnia symptom (onset/maintenance/daytime impairment) but did not meet full diagnostic criteria (Jansson-Fröjmark et al., 2012). Both groups reported significantly higher worry than normal sleepers, however, implying a potential dose response relationship between worry and the severity or chronicity of sleep disturbance. Together these studies offer a coherent picture of the association between worry and sleep disturbance. Notably, with the exception of Jansson-Fröjmark et al. (2012) who assessed worry using the APSQ, all aforesaid investigators developed novel worry measures for their respective studies.

Because prior studies relied on trait measures of worry, they failed to capture any within-person variance or temporal effects in the association between worry and sleep. Specifically, common trait measures of worry, such as the PSWQ (Meyer et al., 1990), assess one's propensity for worry and thus offer little insight into whether worrying actually occurred during a particular sampling period. To the best of our knowledge, only one study has directly addressed the covariation between state worry and sleep. Using an ecological momentary assessment (EMA) design, Weise et al. (2013) showed that presleep worry (on a three-item scale: "worrying," "sweating," and "heart racing or pounding") was associated with self-reported and actigraphy-based WASO, lower actigraphy-based sleep efficiency (SE), and poor subjective sleep quality. Importantly, actigraphy-based WASO for participants who worried prior to sleep (~47 min) was nearly twice that for those who did not (~24 min). Though this study offers important and timely support for the relationship between the act of worrying and sleep disturbance, it suffers from a number of limitations, including the use of a nonstandardized worry measure that conflates cognitive and physiological correlates of worry. Furthermore, causality could not be inferred because levels of presleep worry were not experimentally manipulated.

Despite widespread acceptance of the role of worry in sleep disturbance, few experimental studies have investigated the impact of worry induction on sleep. Three of these studies (Hall, Buysse, Reynolds, Kupfer, & Baum, 1996; Gross & Borkovec, 1982; Tang & Harvey, 2004b) employed a speech threat paradigm in which participants were informed prior to going to sleep that they would have to deliver a speech upon waking. Presumably, the prospect of public speaking elicits worry that, in turn, disrupts sleep. Data suggest that such a worry induction task is associated with longer polysomnography (PSG) and self-reported SOLs and shorter PSG-based TST during an afternoon nap (Gross & Borkovec, 1982; Tang & Harvey, 2004b), as well as longer PSG-based SOLs and more frequent awakenings during the night (Hall, Buysse, Reynolds, Kupfer, & Baum, 1996). Notably, worry-induced SOL was in the clinically significant range (~54 min) in the Gross and Borkovec study. However, this study did not involve a manipulation check, rendering it impossible to ascertain whether participants engaged in worry in the presleep period (Gross & Borkovec, 1982). Similarly, Hall and colleagues assessed presleep "intrusive thoughts" using the Impact of Event Scale, an instrument more germane to rumination than worry (see the section on "Rumination: Phenomenology and Assessment"). Finally, Tang and Harvey (2004b) assessed task-induced cognition using the State Trait Anxiety Inventory-State Scale, an empirically validated measure of state anxiety (Hedberg, 1972), and not worry *per se* (see the section on "Rumination and Worry: Common and Distinguishing Features"). It is therefore unclear whether any of the above studies captured presleep worry or another form of cognitive activity.

Another classic study on "cognitive anxiety" and sleep followed a similar paradigm and deserves mention here (Lichstein & Fanning, 1990, p. 49). In a study involving insomnia patients and healthy controls, experimenters staged a PSG-equipment malfunction purported to potentially shock participants during an overnight sleep study. The insomnia group exhibited a significantly higher skin conductance response than did the control group, presumably because the former were more likely to worry about being accidentally shocked while going to bed. Furthermore, the percentage of PSG epochs scored as sleep was significantly lower among insomniacs than in controls. Once again, the protocol did not involve a manipulation check to determine whether participants actually engaged in presleep worry, however, and the impact of other forms of cognitive activity cannot be ruled out. A recent study on the role of repetitive thought forms in insomnia disorder aimed to address this particular limitation of the sleep and worry literature.

Carney, Harris, Moss, and Edinger (2010) assessed trait rumination and worry in a sample of 210 patients before monitoring their sleep via electronic sleep diaries for a period of 2 weeks. Patients with higher trait rumination scores reported significant sleep disturbance in parameters such as WASO and SE. Surprisingly, trait worry did not exert a main effect on sleep. As Carney and colleagues note, a troubling implication of these findings is that previous studies may have mischaracterized the repetitive thought processes observed in insomnia disorder. In other words, rumination and not worry may serve as the key etiological agent behind wake-promoting cognitive activity. Arguably, some of this confusion is attributable to the inherent similarities between rumination and worry, because both involve abstract, negatively valenced, perseverative thinking. A close look at the phenomenology of rumination is warranted to understand any conceptual and empirical differences between these two repetitive thought forms.

RUMINATION: PHENOMENOLOGY AND ASSESSMENT

Nolen-Hoeksema (1990) originally proposed the construct of rumination as an explanation for the gender disparity in the prevalence of depression and as a mechanism by which depressive episodes are exacerbated. Rumination refers to passively and repetitively focusing attention on the self or on negative affect. Empirical data support three mechanisms by which rumination prolongs negative affect: (1) intensifying negative affect states by calling attention to negative traits and memories, (2) inhibiting adaptive distraction from negative affect, and (3) interfering with problem-solving by usurping attentional resources required for coping (Wisco & Nolen-Hoeksema, 2008). As it was initially conceptualized, rumination involves recurrent thoughts about mood symptoms and not about specific life events or stressors (Nolen-Hoeksema, 1991). Indeed, in her initial description of this response style, Nolen-Hoeksema stressed that nonspecific rumination was more insidious than rumination anchored to a specific event. However, the concept of rumination has since been expanded beyond its roots in depression to capture recurrent, intrusive thoughts about past events and stressors across a variety of disorders, including posttraumatic stress disorder (PTSD) and insomnia disorder (Carney, Harris, Falco, & Edinger, 2013; Echiverri, Jaeger, Chen, Moore, & Zoellner, 2011; Watkins, 2008). This chapter adopts the latter, more transdiagnostic view of rumination.

Not unlike worry, rumination can be conceptualized as both a trait- and state-level construct. In an early study, 83% of a nonclinical sample of college students exhibited consistency in their "typical" cognitive response style (rumination vs. distraction) over a 30-day period (Nolen-Hoeksema, Morrow, & Fredrickson, 1993). Similarly, in a large sample of bereaved adults, scores on a trait measure of rumination did not vary significantly between a baseline assessment and a follow-up roughly 5 months later

(Nolen-Hoeksema, Parker, & Larson, 1994). Together, these studies suggest that a proportion of the population exhibits a trait disposition for rumination. However, data from daily or experience sampling studies suggest that the level of actual engagement in rumination varies significantly from day to day, as a function of life events or stressors, even among trait ruminators (Moberly & Watkins, 2008; Puterman, DeLongis, & Pomaki, 2010). Similar daily variations in levels of state rumination have also been reported in community (Wood, Saltzberg, Neale, Stone, & Rachmiel, 1990) and student samples (Lavallee & Campbell, 1995). A number of empirically validated measures of both state and trait ruminations are presently available in the literature.

The original standardized measure of rumination is called the Response Style Questionnaire, Rumination Scale (RRS: Nolen-Hoeksema & Morrow, 1991). The RRS consists of 22 items scored on a Likert-type scale that indicates how often ("almost never" to "almost always") one engages in ruminative thought. The RRS has shown high internal consistency ($\alpha = 0.88$) and acceptable test-retest reliability ($r = 0.62$) over a 26-week period (Bagby, Rector, Bacchiochi, & McBride, 2004). A recent factor analysis of the RRS by Treynor, Gonzalez, and Nolen-Hoeksema (2003) reveals three underlying components: intropunitive brooding (e.g., "what am I doing to deserve this"), reflective pondering about the causes of negative affect (e.g., "go someplace alone to think about your feelings"), and focusing on depressive symptoms (e.g., "think about how sad you feel"). Although the depressive symptom-focused subscale is highly collinear with depression, the brooding and reflective pondering subscales, known together as self-focused rumination, are less confounded and constitute the active components of the rumination construct (Armey et al., 2009; Bagby & Parker, 2001; Bagby et al., 2004).

For more stress-specific rumination, sleep studies (Thomsen, Mehlsen, Christensen, & Zachariae, 2003; Zoccola, Dickerson, & Lam, 2009) have turned to the Emotion Control Questionnaire-Rehearsal Scale (ECQ-R: Roger & Najarian, 1989). The ECQ-R is composed of 14 true-or-false items that assess whether or not one is generally likely to engage in perseverative, intrusive ideation about prior events and stressors (e.g., "I get worked up thinking about things that have upset me in the past"). Though conceptually congruent with rumination, we were unable to find any empirical data on the convergent validity of this instrument with respect to the RRS.

Two other scales of life event- or trauma-related rumination have emerged from the PTSD literature: the Revised Impact of Events Scale-Intrusion subscale (IES-I: Weiss, 2007) and, more recently, the Event Related Rumination Inventory (ERRI: Cann et al., 2011). The IES-I comprises eight items that measure the severity of recurrent, egodystonic ideation in response to stressors (e.g., "I thought about it when I didn't meant to"; "other things kept making me think about it"). Notably, this scale also includes sleep-specific items (e.g., "I had dreams about it"; "I had trouble falling asleep") that may inflate its correlation with sleep measures. The 20-item ERRI is similar to the IES-I in its focus on event-related rumination, with the exception that it also probes for the frequency of more deliberate forms of rumination aimed at adaptive coping or posttraumatic growth (e.g., "I thought about whether I have learned anything as a result of my experience"). Owing to their novelty, these scales have yet to receive adequate research use. However, given the well-established association between stress and sleep disturbance (Healey et al., 1981; LeBlanc et al., 2009), instruments such as the IES-I and the ERRI may help elucidate whether rumination mediates the association between stress exposure and sleep disruption.

A final rumination scale that deserves mention is the Daytime Insomnia Symptom Response Scale (DISRS). Developed and validated by Carney et al. (2013) for the assessment of rumination specific to daytime insomnia disorder symptoms, the DISRS is a 20-item scale that targets rumination about three domains of insomnia disorder complaints: cognitive and motivational problems (e.g., "think about how unmotivated you feel"; "think about how everything requires more effort than usual"), negative affect (e.g., "think about how irritable you feel"), and fatigue (e.g., "think about how tired you feel"). The DISRS exhibits high internal consistency ($\alpha = 0.93$-0.94) and good convergent validity. In a sample of patients with comorbid insomnia disorder and depression, the DISRS was significantly related to insomnia severity, even after controlling for depressive symptoms and depressive rumination on the RRS.

RUMINATION AND SLEEP

Although research on worry and sleep disturbance has grown steadily since the 1980s, rumination did not fall under the lens of sleep researchers until the past decade. Not surprisingly, we found only five studies on rumination and sleep disturbance. In 2003, Thomsen et al. (2003) administered self-report measures of trait rumination (ECQ-R), depressed mood, and sleep quality to a nonclinical sample. Results indicated significant bivariate correlations between trait rumination and various sleep indices, including SOL, such that higher levels of trait rumination were associated with worse sleep outcomes. Furthermore, rumination was significantly associated with overall sleep quality even after controlling for depressed mood. More recently, a comparison of good and poor sleepers (global PSQI scores > 6) revealed that the latter reported significantly higher scores on the RRS (Carney, Edinger, Meyer, Lindman, & Istre, 2006). These studies offer preliminary evidence that individuals with a ruminative response style are more likely to experience sleep difficulties than are individuals without such a cognitive vulnerability. However, due to the correlational nature of these data, neither study was able to establish a causal or temporal relationship between rumination and sleep disturbance. An equally plausible alternative explanation is that the inability to fall asleep after going to bed may trigger ruminative thinking. We are aware of only one sleep study in which state rumination was experimentally induced.

Guastella and Moulds (2007) assessed the impact of presleep rumination and distraction inductions on sleep quality in a sample of college students. Based on a median split of RRS scores, the sample was divided into two groups: high and low trait ruminators. Analyses revealed that high ruminators assigned to the rumination condition reported significantly worse sleep quality than did high ruminators in the distraction condition and low ruminators in either condition. In other words, ruminating prior to bed impaired sleep only among those predisposed to a ruminative response style. Guastella and Moulds reasoned that the low trait ruminators likely reverted back to their default nonruminative cognitive style following the rumination induction (2007; p. 1158). A more recent study on the association between poststressor rumination and objective sleep disruption reported similar findings (Zoccola et al., 2009). In this study, participants reported levels of in-lab state rumination after delivering a speech that was negatively evaluated by a panel of judges. The following night, sleep was monitored via actigraphy in participants' homes. Actiwatch data indicated that trait rumination (assessed by the ECQ-R) was related to significantly longer SOLs. On the other hand, though state rumination was not significantly associated with any sleep parameter, there was a significant interaction between trait and state rumination, such that high trait ruminators who engaged in more state rumination experienced the longest SOLs. Neither trait nor state rumination was related to TST or WASO.

Finally, Pillai, Steenburg, Ciesla, Roth, and Drake (2014) examined the effects of naturally occurring presleep rumination on self-reported and actigraphy-based sleep. This study adopted a week-long, daily sampling approach to overcome some of the limitations of prior studies, including first-night effects and low ecological validity. A sample of high trait ruminators (>1.5 standard deviations on the RRS) completed a short questionnaire after waking each morning for a period of 7 days. Participants reported the duration and quality of sleep they experienced the previous night, as well as levels of engagement in presleep rumination. Sleep was also assessed throughout this period in participants' home environments via wrist actigraphy. Analyses revealed that nightly levels of presleep rumination were associated with significantly longer actigraphy- and diary-based SOL. Notably, a 1-SD increase on the presleep rumination scale was associated with an approximately 7-min increase in actigraphy-based SOL, even after controlling for baseline sleep disturbance and depressive symptomatology. Other sleep parameters such as SE and TST were unrelated to rumination. Consistent with prior experimental data, this more naturalistic study suggests that individuals with a trait vulnerability to rumination do engage in presleep rumination, further indicating that this state phenomenon disrupts both actigraphy- and self-reported sleep.

In summary, though rumination is a reliable predictor of delayed sleep onset, its association with sleep maintenance is less robust (Zoccola et al., 2009; Pillai et al., 2014). With the exception of Carney et al. (2010), who found a significant association between trait rumination and diary-based SE and WASO, prior research has yet to establish an association between rumination and sleep maintenance difficulties. Arguably, null effects may simply reflect the methodological limitations of these studies. Three of the five studies reviewed here relied on samples of university students, a population notorious for erratic sleep schedules and poor sleep hygiene (Gaultney, 2011). Laboratory studies with a controlled sleep window are needed to test more definitively the association between rumination and sleep maintenance.

RUMINATION AND WORRY: COMMON AND DISTINGUISHING FEATURES

As alluded to earlier, rumination and worry emerged largely independently as respective cognitive pathways to depression and anxiety. However, the substantial comorbidity between depression and anxiety (Clark & Watson, 1991) has prompted researchers to examine whether rumination and worry are distinct or overlapping features of affective disorders. By definition, rumination involves dwelling on the past to understand and, potentially, alleviate negative mood states (Lyubomirsky & Nolen-Hoeksema, 1993), whereas worry is elicited by future threats and uncertainties (Dugas, Buhr, & Ladouceur, 2004). Thus, although rumination and worry both involve negatively valenced, repetitive thoughts (Brosschot et al., 2006; Watkins, 2008), they differ in terms of temporal focus (past vs. future) and perceived motivation (mood alleviation vs. uncertainty reduction). Empirical attempts to distinguish between rumination and worry within this conceptual framework have largely relied on assessing the specificity of these repetitive thought forms for various affective disorders. Despite some inconsistent findings in the past (for review, see Querstret & Cropley, 2013), the most robust thesis to emerge from this literature is that rumination and worry are significantly correlated and that each is uniquely associated (i.e., controlling for the other) with both depression and anxiety (Chelminsky & Zimmerman, 2003; McEvoy & Brans, 2013; McLaughlin et al., 2007).

A second group of studies has explored the factor structure of commonly used self-report measures of rumination and worry to identify shared or unique higher-order constructs. Segerstrom et al. (2000), for instance, used structural equation modeling to find a single repetitive thought factor underlying both rumination and worry, thus implying considerable overlap between these two mechanisms. In contrast, more recent investigations in both clinical (Goring & Papageorgiou, 2008; McEvoy & Brans, 2013) and nonclinical samples (Fresco, Frankel, Mennin, Turk, & Heimberg, 2002) have yielded some evidence of structural divergence. For instance, McEvoy et al. (2013) offer evidence for a four-factor model: reflective pondering, brooding, worrying, and general repetitive thought. This factor-analytic solution is especially noteworthy because the investigators removed disorder-specific items from the rumination and worry scales, in addition to controlling for the method variance associated with these instruments. However, with the exception of the reflective pondering factor, which was not predictive of anxiety, all other factors were significantly related to both depression and anxiety symptoms. Together, these data imply that, although rumination and worry may constitute phenomenologically distinct cognitive mechanisms, this distinction is less clinically meaningful in the context of affective disorders. Both forms of repetitive thought are predictive of affective disorders and warrant assessment and treatment in clinical settings. Not surprisingly, many recent reviews of this literature subsume rumination and worry under a common rubric, known variously as repetitive thought, repetitive negative thought, or perseverative thought (Mansell, Harvey, Watkins, & Shafran, 2008; Querstret & Cropley, 2013; Watkins, 2008).

With respect to sleep disturbance, however, we presently lack evidence to judge the respective merits of separating or combining the rumination and worry constructs. We are aware of only two studies that have examined the impact of both rumination and worry on sleep. As noted earlier, Carney et al. (2010) found that, although rumination was significantly related to sleep disturbance in insomnia patients, worry was not associated with sleep. A similar study in a sample of undergraduate students yielded nearly identical findings (Takano, Iijima, & Tanno, 2012). In this study, baseline trait rumination scores were associated with reduced subjective sleep quality at a 3-week follow-up, though no relationship between trait worry and sleep was found. However, there was a significant interaction between trait rumination and worry, such that rumination was associated with poor sleep quality among high trait worriers but not among low trait worriers. Thus, although worry did not exert a main effect on sleep, it amplified the effects of rumination.

Carney and colleagues reason that a potential explanation for these findings is that worry and sleep-related repetitive thought may be distinct constructs (2010, 2013). In a recent study, they administered the PSWQ and the DISRS to a nonclinical sample of college students. A factor-analysis of pooled items from these scales yielded a two-factor solution, such that items from the PSWQ and DISRS loaded onto separate factors. Though the authors cite this result as evidence for divergence between worry and sleep-related rumination, this factor-analytic solution may simply reflect the method variance in these scales (Harris & Bladen, 1994; McEvoy & Brans, 2013). Furthermore, though the correlation between PSWQ and the DISRS was not reported in this study, the correlation between the RRS and the DISRS was high ($r = 0.62$, $p < 0.01$). Data from other studies show that the RRS and the PSWQ are similarly correlated in both clinical (Carney et al., 2010: $r = 0.56$, $p < 0.01$) and nonclinical samples (Segerstrom et al., 2000: $r = 0.52$, $p < 0.01$; Takano et al., 2012: $r = 0.52$, $p < 0.01$). Finally, studies that rely on single measures of rumination and worry may only address the covariation between specific instruments and not the underlying constructs. Future studies must adopt multitrait-multimethod approaches before they can satisfactorily explain the relationships between these constructs (Williams & Brown, 1994). State measures of rumination and worry should also be emphasized to gain a more valid assay of the dynamic relationship between repetitive thought and sleep.

RUMINATION, WORRY, AND SLEEP: THEORETICAL MODELS

The inability to suppress mental activity while attempting to sleep is considered its own form of arousal called cognitive arousal (Harvey, Tang, & Browning, 2005). Therefore, the charge facing current research is twofold: (1) to explain why repetitive thought processes persist and (2) to show that these processes can override homeostatic and circadian sleep regulation. With respect to the durability of repetitive thought, transdiagnostic models suggest that repetitive thoughts are maintained by two mechanisms: faulty or maladaptive metacognitions about repetitive thought and ineffective strategies for suppressing repetitive thought (Clark, 2002). Perceptions about the function and meaning of repetitive thought vary. With respect to rumination, some data show that ruminators consciously prolong the ruminative response because they believe that rumination will help them gain insight into and eventually alleviate the target distress (Lyubomirsky & Nolen-Hoeksema, 1993). Similarly, worrying, as noted earlier, is often perceived as an adaptive coping strategy for planning or problem solving (Borkovec & Roemer, 1995). On the other hand, repetitive thoughts may be egodystonic and anxiety-provoking when, for instance, their content is related to trauma- or fear-specific schemas (Ehlers & Clark, 2000; Grisham & Williams, 2013). In such cases, counterproductive efforts to suppress repetitive thoughts are triggered, further prolonging repetitive thought processes (see the section on "Treatment Implications").

As for the interaction between repetitive thought and endogenous sleep regulation, research suggests that the attentional resources usurped by repetitive thought processes are on par with the demands of an executive or effortful cognitive task. Studies show that experimentally induced rumination degrades performance on effortful-attentional tasks such as random number generation (Watkins & Brown, 2002). Thus, it is conceivable that repetitive thought processes may be cognitively strenuous enough to preclude the relatively automatic inhibition of wakefulness that occurs around sleep onset (Borbely, 1982). However, none of the studies we reviewed assessed the association between repetitive thought and alertness, though such a relationship was typically inferred based on the association between repetitive thought and nocturnal SOL. Nocturnal SOL is a poor measure of arousal because considerable within- and between-person variability exists in this measure (Roehrs, Randall, Harris, Maan, & Roth, 2011).

The association between repetitive thought and other arousal indices, such as heart rate (HR) and HRV, offer some insight into this phenomenon. The cardiovascular health literature provides extensive and reliable evidence that state worry is associated with increased HR and reduced HRV (Aldao, Mennin, & McLaughlin, 2013; Hofmann et al., 2005). In a recent EMA study, Weise et al. (2013) showed that presleep worry is associated with increased skin conductance, elevated HR, and lower HRV, both during the presleep period and in the following sleep episode. Furthermore, these autonomic arousal indices were associated with significantly higher actigraphy-based nocturnal awakenings and reduced SE. Data on rumination and autonomic arousal are more equivocal. Although some studies report an association between state rumination and blunted HRV (Gerin, Davidson, Christenfeld, Goyal, & Schwartz, 2006; Key, Campbell, Bacon, & Gerin, 2008; Ottaviani, Shapiro, Davydov, Goldstein, & Mills, 2009), others report no association (Aldao et al., 2013). Further research on the autonomic phenotypes of rumination using standardized instruments is warranted to resolve these inconsistences.

Repetitive thought is also associated with markers of central nervous system (CNS) hyperarousal, such as hypothalamic-pituitary-adrenal axis (HPA) activity (Drake, Roehrs, & Roth, 2003; Rydstedt, Cropley, & Devereux, 2011). According to the sustained activation theory, the failure to downregulate physiological responses to stress exposure contributes to morbidity (Ursin & Eriksen, 2010). As repetitive thoughts offer a medium by which cognitive manifestations of stress endure, some researchers have examined whether repetitive thought inhibits CNS recovery. Cropley, Rydstedt, Devereux, and Middleton (2013) assessed the association between work-related rumination and salivary cortisol levels in a nonclinical sample of adult school teachers. Salivary cortisol (SC) is an easily procured measure of HPA activity, with empirically validated stress responsivity (Clow, Hucklebridge, Stalder, Evans, & Thorn, 2010). SC also exhibits a reliable diurnal pattern in healthy adults, such that its levels peak within a 30-45 min window of awakening, a phenomenon known as the CAR. SC levels subsequently decline over the remainder of the day until reaching a nadir late in the evening (Kudielka, Schommer, Hellhammer, & Kirschbaum, 2004). Participants in this study provided SC assays immediately before going to sleep, upon waking, and then 15, 30, and 45 min after waking. Participants also reported state levels of rumination before going to bed and again after waking in the morning. Analyses showed that rumination was associated with significant sleep disturbance, higher before-bed SC levels, and a blunted CAR in the morning. Notably, the association between repetitive thought and CAR dropped to insignificance when sleep disturbance was included in the model. The investigators speculated that sleep disturbance may have mediated the association between repetitive thought and CAR, though no statistical mediation analyses were performed. Other research groups report a similar association between repetitive thought and a blunted CAR (Kuehner, Holzhauer, & Huffziger, 2007; Rydstedt et al., 2011). Individuals with insomnia exhibit a remarkably similar pattern of elevated evening cortisol levels and a blunted CAR (Backhaus, Junghanns, & Hohagen, 2004; Vgontzas et al., 2001), implying an association between repetitive thought and the sleep-disruptive hyperarousal seen in insomnia.

In summary, there is now considerable evidence that rumination and worry are related to sleep disturbance, and several potential mechanisms have been proposed. Most of these findings are preliminary, however, and further investigation of the physiological correlates of cognitive arousal is warranted. Similarly, more laboratory-based PSG studies are needed because the association between repetitive thought and sleep architecture has not been examined. A recent study showed that a presleep negative affect-induction task was predictive of an increased latency to slow-wave sleep and an increased number of awakenings from rapid eye movement (REM) sleep (Vandekerckhove et al., 2011). Given the significant collinearity between affect and repetitive thought, it is important to ascertain whether repetitive thought is similarly related to sleep architecture. The unique neurobiology of REM sleep—limbic and forebrain levels of acetylcholine are significantly higher during REM sleep than during NREM sleep and quiet waking—has salient implications for the consolidation of emotional memories (for a review, see van der Helm & Walker, 2011). Furthermore, studies show that, although sleep loss impairs the encoding of neutral and positively valenced memories, negative memories appear relatively refractory (Walker & Stickgold, 2006). Thus, sleep loss leads to a selective disinhibition of negatively toned memories, which may then act as fodder for ruminative perseveration. Presently, these neurochemical mechanisms have largely been studied in the broad context of emotion regulation and not rumination or worry *per se*. We believe that future research along these lines, especially in diverse clinical samples, will further elucidate these phenomena, because disorder-specific research may only yield disorder-specific insight.

RUMINATION, WORRY, AND SLEEP: TREATMENT IMPLICATIONS

Early clinical trials of mindfulness-based techniques, such as mindfulness-based stress reduction and mindfulness-based cognitive therapy, show promise for alleviating both rumination (Jain et al., 2007; Kingston, Dooley, Bates, Lawlor, & Malone, 2007; Ramel, Goldin, Carmona, & McQuaid, 2004) and worry (Delgado et al., 2010; Yook et al., 2008). Rumination involves a perseverative analysis of discrepancies between current and desired mood states, often leading to self-blame (Lyubomirsky, Tucker, Caldwell, & Berg, 1999). Similarly, worry is triggered by the emotional discomfort associated with fear and uncertainty (Dugas et al., 2004). These aspects of repetitive thought are ideal targets for mindfulness-based interventions that emphasize a nonjudgmental, nonreactive acceptance of egodystonic thoughts and emotions. In a recent survey study, a nonclinical sample of university students completed the PSWQ and a trait measure of mindfulness (Fisak & von Lehe, 2012). Analyses revealed a significant inverse association between PSWQ scores and three factors of the mindfulness scale: nonreactivity to inner experience, nonjudgment of inner experience, and acting with awareness. Furthermore, a review of 19 clinical trials (which included a variety of treatment modalities, including cognitive behavior therapy, motivational interviewing, mindfulness-based meditation, etc.) concluded that interventions aimed at regulating one's emotional response to repetitive thought using mindfulness-based techniques are efficacious in alleviating rumination and worry (Querstret & Cropley, 2013).

Individuals with sleep difficulties typically engage in thought suppression to minimize the cognitive arousal triggered by rumination and worry (Ansfield, Wegner, & Bowser, 1996). Most thought suppression techniques are not only ineffective but are also associated with poor sleep outcomes, however. Consistent with Wegner's (1994) ironic process theory, engaging in thought suppression prior to bed results in a paradoxical increase in cognitive arousal (Harvey & Greenall, 2003). Thus, despite its intuitive appeal, thought suppression is not an effective antidote to repetitive thought. On the other hand, given its focus on acceptance and nonreactivity, mindfulness may represent a more suitable treatment (see Chapter 16 on mindfulness and sleep for additional information). Notably, although current treatments, such as cognitive behavior therapy for insomnia (CBTI), are effective in restructuring dysfunctional attitudes about sleep

(Roane, Dolan, Bramoweth, Rosenthal, & Taylor, 2012) and sleep-specific worry (Sunnhed & Jansson-Frojmark, 2014), their effects on rumination and more global, nonspecific worry have not been investigated. We therefore support recent efforts to supplement CBTI with mindfulness-based interventions (Ong, Shapiro, & Manber, 2008).

RUMINATION, WORRY, AND SLEEP: OTHER CLINICAL IMPLICATIONS

Though we emphasize the distinction between sleep disturbance and insomnia disorder throughout this chapter, research on acute sleep disturbance may nevertheless offer insight into premorbid vulnerabilities to insomnia disorder. A subset of normal sleepers exhibits heightened stress-related sleep disruption. Notably, this group exhibits this response in relation to a variety of stressors or challenges to the sleep system, including a first night in the sleep laboratory, caffeine administration, and circadian phase shifts (Bonnet & Arand, 2003; Drake, Jefferson, Roehrs, & Roth, 2006). Such sleep reactivity or the tendency to exhibit pronounced sleep disturbance in response to stress exposure has recently gained attention as a diathesis of insomnia disorder (Drake & Roth, 2006). The Ford Insomnia in Response to Stress Test (FIRST: Drake, Richardson, Roehrs, Scofield, & Roth, 2004), a nine-item self-report measure of sleep reactivity, assesses the likelihood of experiencing sleep disturbance in response to common psychosocial stressors. Studies suggest that sleep reactivity, as assessed by the FIRST, may constitute a trait vulnerability to insomnia disorder; it manifests premorbidly, shows within-person stability, and is predictive of insomnia onset (Drake, Roehrs, Richardson, & Roth, 2004; Drake, Scofield, & Roth, 2008).

Fernandez-Mendoza et al. (2010) found that trait rumination, as measured by the ECQ-R, was significantly associated with FIRST scores in a sample of good sleepers (Global PSQI<5), even after controlling for gender, depression, and anxiety. This study offers preliminary evidence that sleep reactivity and repetitive thought in the form of rumination are correlated vulnerabilities to insomnia disorder. The above data are cross-sectional, however, and the investigators did not rule out prior histories of insomnia disorder in their participants. Similarly, there were no worry measures in this study. To determine how trait repetitive thought relates to other vulnerabilities to insomnia disorder, such as sleep reactivity, future studies should assess levels of trait rumination and worry during the prodromal phase of the disorder. This research can help integrate the presently disparate literatures on cognitive and somatic/sleep-system vulnerabilities to insomnia.

The association between repetitive thought and sleep disturbance may also shed light on the comorbidity between insomnia disorder and depression. Studies suggest that 60-90% of individuals with depression experience clinically significant levels of insomnia (Kloss & Szuba, 2003). Clinical and basic science research has amassed a wealth of evidence for shared etiological pathways to insomnia disorder and depression in recent years, leading some to conclude that these disorders may represent alternate manifestations of the same or related diatheses (Pillai, Kalmbach, & Ciesla, 2011). For instance, neuroendocrinological studies suggest that the mechanisms responsible for the negative-feedback inhibition of stress-response systems, such as HPA activity, are impaired in both populations (Balbo, Leproult, & Van Cauter, 2010; Boyce & Ellis, 2005; McKay & Zakzanis, 2010; Roth, Roehrs, & Pies, 2007). We propose that repetitive thought may represent another common vulnerability to insomnia disorder and depression.

The depression literature suggests that individuals with depression are significantly more likely to engage in rumination in response to stress, and that this response style is associated with greater levels of depressive severity and relapse (Calmes & Roberts, 2007). Levels of pathological worry are similarly elevated in this population and are comparable to those observed in anxiety disorders (Chelminsky & Zimmerman, 2003). Thus, existing evidence suggests not only that repetitive thought characterizes depression, but also that this cognitive style is associated with sleep disturbance. Individuals with insomnia disorder also engage in repetitive thought, though the focus of cognitive perseveration in this population is typically on the daytime impairments caused by sleep loss (Carney et al., 2013). Hence, it stands to reason that sleep-focused and mood-focused repetitive thought may represent correlated vulnerabilities for insomnia disorder and depression, respectively. Although sleep-specific repetitive thought may trigger and maintain insomnia disorder, a more diffuse repetitive thought style will more likely precipitate depression or depression comorbid with insomnia disorder.

In a recent longitudinal study, a large community sample of adults ($n = 3496$) with no depression at baseline completed self-report measures of rumination and sleep disturbance (Batterham, Glozier, & Christensen, 2012). Follow-up data 4 years hence revealed that baseline sleep disturbance was a significant predictor of risk for depression (OR = 1.34; $p < 0.01$). This association was reduced to insignificance once rumination was included in the model, however, such that rumination was significantly associated with depression after controlling for sleep disturbance (OR = 1.12; $p < 0.01$). Importantly, rumination was measured using a shortened version of the RRS, which included depression-focused items (e.g., "think about how sad you feel"). Thus, the association between rumination and depression may have been artificially inflated due to scale collinearity. Similarly, sleep was not assessed using a standardized instrument, and there were no measures of worry. Despite these limitations, however, this study alludes to the possibility that repetitive thought may underlie the covariation between mood and sleep disturbance. More longitudinal studies of repetitive thought during the premorbid phase of depression and insomnia disorder are needed to test this hypothesis.

CONCLUSION

Our review points to a growing literature that supports the association between sleep disturbance and both rumination and worry. Presently, empirical evidence indicates that trait worriers are significantly more likely to experience sleep disturbance and that both experimentally induced and naturally occurring worry can impair subsequent sleep onset and maintenance. The rumination and sleep literature echoes these findings, with the exception that the association between rumination and objective indices of sleep maintenance has not been empirically substantiated. A number of methodological limitations limit the scope of current findings, however, including the use of single-night assays for sleep assessment, relying on subjective measures of sleep disturbance, correlational and cross-sectional designs, lack of standardization in the assessment of repetitive thought constructs, and an emphasis on disorder-specific models of repetitive thought and sleep. Consequently, there are several gaps in our current understanding of this relationship: (1) the mechanisms by which rumination and worry disturb sleep are poorly understood, (2) it is unknown whether repetitive thought has any impact on sleep architecture, and (3) the debate about whether rumination and worry are functionally distinct with respect to sleep remains unresolved. Targeting these neglected areas in future research will help elucidate an important link in the pathophysiology of sleep disturbance.

References

Aldao, A., Mennin, D. S., & McLaughlin, K. A. (2013). Differentiating worry and rumination: Evidence from heart rate variability during spontaneous regulation. *Cognitive Therapy and Research*, 37(3), 613–619.

Ansfield, M. E., Wegner, D. M., & Bowser, R. (1996). Ironic effects of sleep urgency. *Behaviour Research and Therapy*, 34(7), 523–531. http://dx.doi.org/10.1016/0005-7967(96)00031-9.

Armey, M. F., Fresco, D. M., Moore, M. T., Mennin, D. S., Turk, C. L., Heimberg, R. G., et al. (2009). Brooding and pondering: Isolating the active ingredients of depressive rumination with exploratory factor analysis and structural equation modeling. *Assessment*, 16(4), 315–327. http://dx.doi.org/10.1177/1073191109340388.

Backhaus, J., Junghanns, K., & Hohagen, F. (2004). Sleep disturbances are correlated with decreased morning awakening salivary cortisol. *Psychoneuroendocrinology*, 29(9), 1184–1191. http://dx.doi.org/10.1016/j.psyneuen.2004.01.010.

Bagby, R. M., & Parker, J. D. A. (2001). Relation of rumination and distraction with neuroticism and extraversion in a sample of patients with major depression. *Cognitive Therapy and Research*, 25(1), 91–102. http://dx.doi.org/10.1023/a:1026430900363.

REFERENCES

Bagby, R. M., Rector, N. A., Bacchiochi, J. R., & McBride, C. (2004). The stability of the response styles questionnaire rumination scale in a sample of patients with major depression. *Cognitive Therapy and Research, 28*(4), 527–538. http://dx.doi.org/10.1023/b:cotr.0000045562.17228.29.

Balbo, M., Leproult, R., & Van Cauter, E. (2010). Impact of sleep and its disturbances on hypothalamo-pituitary-adrenal axis activity. *International Journal of Endocrinology, 2010*, 759234. http://dx.doi.org/10.1155/2010/759234.

Batterham, P. J., Glozier, N., & Christensen, H. (2012). Sleep disturbance, personality and the onset of depression and anxiety: Prospective cohort study. *The Australian and New Zealand Journal of Psychiatry, 46*(11), 1089–1098. http://dx.doi.org/10.1177/0004867412457997.

Bonnet, M. H., & Arand, D. L. (2003). Situational insomnia: Consistency, predictors, and outcomes. *Sleep, 26*(8), 1029–1036.

Borbely, A. A. (1982). A two process model of sleep regulation. *Human Neurobiology, 1*(3), 195–204.

Borkovec, T. D. (1994). The nature, functions, and origins of worry. In G. C. L. Davey & F. Tallis (Eds.), *Worrying: Perspectives on theory, assessment and treatment* (pp. 5–33). Oxford, UK: John Wiley & Sons.

Borkovec, T. D., Robinson, E., Pruzinsky, T., & DePree, J. A. (1983). Preliminary exploration of worry: Some characteristics and processes. *Behaviour Research and Therapy, 21*(1), 9–16. http://dx.doi.org/10.1016/0005-7967(83)90121-3.

Borkovec, T. D., & Roemer, L. (1995). Perceived functions of worry among generalized anxiety disorder subjects: Distraction from more emotionally distressing topics? *Journal of Behavior Therapy and Experimental Psychiatry, 26*(1), 25–30. http://dx.doi.org/10.1016/0005-7916(94)00064-s.

Boyce, W., & Ellis, B. (2005). Biological sensitivity to context: I. An evolutionary-developmental theory of the origins and functions of stress reactivity. *Development and Psychopathology, 17*(2), 301.

Brosschot, J. F., Gerin, W., & Thayer, J. F. (2006). The perseverative cognition hypothesis: A review of worry, prolonged stress-related physiological activation, and health. *Journal of Psychosomatic Research, 60*(2), 113–124. http://dx.doi.org/10.1016/j.jpsychores.2005.06.074.

Brosschot, J. F., Van Dijk, E., & Thayer, J. F. (2007). Daily worry is related to low heart rate variability during waking and the subsequent nocturnal sleep period. *International Journal of Psychophysiology, 63*(1), 39–47. http://dx.doi.org/10.1016/j.ijpsycho.2006.07.016.

Buysse, D. J., Reynolds, C. F., Monk, T. H., & Berman, S. R. (1989). The Pittsburgh sleep quality index: A new instrument for psychiatric practice and research. *Psychiatry Research, 28*(2), 193–213. http://dx.doi.org/10.1016/0165-1781(89)90047-4.

Calmes, C. A., & Roberts, J. E. (2007). Repetitive thought and emotional distress: Rumination and worry as prospective predictors of depressive and anxious symptomatology. *Cognitive Therapy and Research, 31*(3), 343–356. http://dx.doi.org/10.1007/s10608-006-9026-9.

Cann, A., Calhoun, L. G., Tedeschi, R. G., Triplett, K. N., Vishnevsky, T., & Lindstrom, C. M. (2011). Assessing posttraumatic cognitive processes: The event related rumination inventory. *Anxiety, Stress & Coping: An International Journal, 24*(2), 137–156. http://dx.doi.org/10.1080/10615806.2010.529901.

Carney, C. E., Edinger, J. D., Meyer, B., Lindman, L., & Istre, T. (2006). Symptom-focused rumination and sleep disturbance. *Behavioral Sleep Medicine, 4*(4), 228–241. http://dx.doi.org/10.1207/s15402010bsm0404_3.

Carney, C. E., Harris, A. L., Falco, A., & Edinger, J. D. (2013). The relation between insomnia symptoms, mood, and rumination about insomnia symptoms. *Journal of Clinical Sleep Medicine, 9*(6), 567–575. http://dx.doi.org/10.5664/jcsm.2752.

Carney, C. E., Harris, A. L., Moss, T. G., & Edinger, J. D. (2010). Distinguishing rumination from worry in clinical insomnia. *Behaviour Research and Therapy, 48*(6), 540–546. http://dx.doi.org/10.1016/j.brat.2010.03.004.

Chelminsky, I., & Zimmerman, M. (2003). Pathological worry in depressed and anxious patients. *Journal of Anxiety Disorders, 17*, 533–546.

Clark, D. A. (2002). Unwanted mental intrusions in clinical disorders: An introduction. *Journal of Cognitive Psychotherapy, 16*(2), 123–126. http://dx.doi.org/10.1891/jcop.16.2.123.63995.

Clark, L. A., & Watson, D. (1991). Tripartite model of anxiety and depression: Psychometric evidence and taxonomic implications. *Journal of Abnormal Psychology, 100*(3), 316–336. http://dx.doi.org/10.1037/0021-843x.100.3.316.

Clow, A., Hucklebridge, F., Stalder, T., Evans, P., & Thorn, L. (2010). The cortisol awakening response: More than a measure of HPA axis function. *Neuroscience & Biobehavioral Reviews, 35*(1), 97–103. http://dx.doi.org/10.1016/j.neubiorev.2009.12.011.

Cropley, M., Rydstedt, L. W., Devereux, J. J., & Middleton, B. (2013). The relationship between work-related rumination and evening and morning salivary cortisol secretion. *Stress Health*, http://dx.doi.org/10.1002/smi.2538.

Delgado, L. C., Guerra, P., Perakakis, P., Vera, M. N., del Paso, G. R., & Vila, J. (2010). Treating chronic worry: Psychological and physiological effects of a training programme based on mindfulness. *Behaviour Research and Therapy, 48*(9), 873–882. http://dx.doi.org/10.1016/j.brat.2010.05.012.

Drake, C. L., Jefferson, C., Roehrs, T., & Roth, T. (2006). Stress-related sleep disturbance and polysomnographic response to caffeine. *Sleep Medicine, 7*(7), 567–572. http://dx.doi.org/10.1016/j.sleep.2006.03.019.

Drake, C. L., Richardson, G., Roehrs, T., Scofield, H., & Roth, T. (2004). Vulnerability to stress-related sleep disturbance and hyperarousal. *Sleep, 27*(2), 285–291.

Drake, C. L., Roehrs, T., Richardson, G., & Roth, T. (2004). *Vulnerability to chronic insomnia: A longitudinal population-based prospective study.* Paper presented at the Sleep, Philadelphia.

Drake, C. L., Roehrs, T., & Roth, T. (2003). Insomnia causes, consequences, and therapeutics: An overview. *Depression and Anxiety, 18*(4), 176.

Drake, C. L., & Roth, T. (2006). Predispostion in the evolution of insomnia: Evidence, potential mechanisms, and future directions. *Sleep Medicine Clinics, 1*, 333–349.

Drake, C. L., Scofield, H., & Roth, T. (2008). Vulnerability to insomnia: The role of familial aggregation. *Sleep Medicine, 9*(3), 297–302. http://dx.doi.org/10.1016/j.sleep.2007.04.012.

Dugas, M. J., Buhr, K., & Ladouceur, R. (2004). The role of intolerance of uncertainty in etiology and maintenance. In R. G. Heimberg, C. L. Turk, & D. S. Mennin (Eds.), *Generalized anxiety disorder: Advances in research and practice* (pp. 143–163). New York: Guilford Press.

Echiverri, A. M., Jaeger, J. J., Chen, J. A., Moore, S. A., & Zoellner, L. A. (2011). "Dwelling in the past": The role of rumination in the treatment of posttraumatic stress disorder. *Cognitive and Behavioral Practice, 18*(3), 338–349. http://dx.doi.org/10.1016/j.cbpra.2010.05.008.

Ehlers, A., & Clark, D. M. (2000). A cognitive model of posttraumatic stress disorder. *Behaviour Research and Therapy, 38*(4), 319–345. http://dx.doi.org/10.1016/S0005-7967(99)00123-0.

Espie, C. A. (2002). Insomnia: Conceptual issues in the development, persistence, and treatment of sleep disorder in adults. *Annual Review of Psychology, 53*, 215–243.

Fernandez-Mendoza, J., Vela-Bueno, A., Vgontzas, A. N., Ramos-Platon, M. J., Olavarrieta-Bernardino, S., Bixler, E. O., et al. (2010). Cognitive-emotional hyperarousal as a premorbid characteristic of individuals vulnerable to insomnia. *Psychosomatic Medicine, 72*(4), 397–403. http://dx.doi.org/10.1097/PSY.0b013e3181d75319.

Fichten, C. S., Libman, E., Creti, L., Amsel, R., Tagalakis, V., & Brender, W. (1998). Thoughts during awake times in older good and poor sleepers: The Self-Statement Test:60+. *Cognitive Therapy and Research, 22*(1), 1–20. http://dx.doi.org/10.1023/A:1018756618012.

Fisak, B., & von Lehe, A. C. (2012). The relation between the five facets of mindfulness and worry in a non-clinical sample. *Mindfulness, 3*(1), 15–21. http://dx.doi.org/10.1007/s12671-011-0075-0.

REFERENCES

Fresco, D. M., Frankel, A. N., Mennin, D. S., Turk, C. L., & Heimberg, R. G. (2002). Distinct and overlapping features of rumination and worry: The relationship of cognitive production to negative affective states. *Cognitive Therapy and Research, 26*(2), 179–188. http://dx.doi.org/10.1023/a:1014517718949.

Gaultney, J. F. (2011). The prevalence of sleep disorders in college students: Impact on academic performance. *Journal of American College of Health, 59*(2), 91–97.

Gerin, W., Davidson, K. W., Christenfeld, N. J., Goyal, T., & Schwartz, J. E. (2006). The role of angry rumination and distraction in blood pressure recovery from emotional arousal. *Psychosomatic Medicine, 68*(1), 64–72. http://dx.doi.org/10.1097/01.psy.0000195747.12404.aa.

Goring, H. J., & Papageorgiou, C. (2008). Rumination and worry: Factor analysis of self-report measures in depressed participants. *Cognitive Therapy and Research, 32*(4), 554–566. http://dx.doi.org/10.1007/s10608-007-9146-x.

Grisham, J. R., & Williams, A. D. (2013). Responding to intrusions in obsessive-compulsive disorder: The roles of neuropsychological functioning and beliefs about thoughts. *Journal of Behavior Therapy and Experimental Psychiatry, 44*(3), 343–350. http://dx.doi.org/10.1016/j.jbtep.2013.01.005.

Gross, R. T., & Borkovec, T. D. (1982). Effects of a cognitive intrusion manipulation on the sleep-onset latency of good sleepers. *Behavior Therapy, 13*(1), 112–116. http://dx.doi.org/10.1016/S0005-7894(82)80054-3.

Guastella, A. J., & Moulds, M. L. (2007). The impact of rumination on sleep quality following a stressful life event. *Personality and Individual Differences, 42*(6), 1151–1162. http://dx.doi.org/10.1016/j.paid.2006.04.028.

Hall, M., Buysse, D., Reynolds, C. F., Kupfer, D., & Baum, A. (1996). Stress-related intrusive thoughts disrupt sleep-onset and continuity. *Sleep Research, 25,* 163.

Hall, M., Buysse, D. J., Dew, M.A., Prigerson, H. G., Kupfer, D. J., Reynolds, 3rd. C. F. (1997). Intrusive thoughts and avoidance behaviors are associated with sleep disturbances in bereavement-related depression. *Depress Anxiety, 6*(3), 106–112.

Harris, M. M., & Bladen, A. (1994). Wording effects in the measurement of role conflict and role ambiguity: A multitrait-multimethod analysis. *Journal of Management, 20*(4), 887–901. http://dx.doi.org/10.1016/0149-2063(94)90034-5.

Harvey, A. G. (2002a). Identifying safety behaviors in insomnia. *The Journal of Nervous and Mental Disease, 190*(1), 16–21.

Harvey, A. G. (2002b). Trouble in bed: The role of pre-sleep worry and intrusions in the maintenance of insomnia. *Journal of Cognitive Psychotherapy, 16*(2), 161–177. http://dx.doi.org/10.1891/jcop.16.2.161.63992.

Harvey, A. G. (2005). Unwanted intrusive thoughts in insomnia. In D. A. Clark (Ed.), *Intrusive thoughts in clinical disorders: Theory, research, and treatment* (pp. 86–118). New York: Guilford Press.

Harvey, A. G., & Greenall, E. (2003). Catastrophic worry in primary insomnia. *Journal of Behavior Therapy and Experimental Psychiatry, 34*(1), 11–23.

Harvey, A. G., Tang, N. K., & Browning, L. (2005). Cognitive approaches to insomnia. *Clinical Psychology Review, 25*(5), 593–611. http://dx.doi.org/10.1016/j.cpr.2005.04.005.

Healey, E. S., Kales, A., Monroe, L. J., Bixler, E. O., Chamberlin, K., & Soldatos, C. R. (1981). Onset of insomnia: Role of life-stress events. *Psychosomatic Medicine, 43*(5), 439–451.

Hedberg, A. G. (1972). Review of 'State-Trait Anxiety Inventory'. *Professional Psychology, 3*(4), 389–390. http://dx.doi.org/10.1037/h0020743.

Hofmann, S. G., Moscovitch, D. A., Litz, B. T., Kim, H.-J., Davis, L. L., & Pizzagalli, D. A. (2005). The worried mind: Autonomic and prefrontal activation during worrying. *Emotion, 5*(4), 464–475. http://dx.doi.org/10.1037/1528-3542.5.4.464.

Jain, S., Shapiro, S. L., Swanick, S., Roesch, S. C., Mills, P. J., Bell, I., et al. (2007). A randomized controlled trial of mindfulness meditation versus relaxation training: Effects on distress, positive states of mind, rumination, and distraction. *Annals of Behavioral Medicine, 33*(1), 11–21. http://dx.doi.org/10.1207/s15324796abm3301_2.

Jansson, M., & Linton, S. J. (2006). The development of insomnia within the first year: A focus on worry. *British Journal of Health Psychology, 11*(3), 501–511. http://dx.doi.org/10.1348/135910705X57412.

Jansson-Fröjmark, M., Harvey, A. G., Norell-Clarke, A., & Linton, S. J. (2012). Associations between psychological factors and nighttime/daytime symptomatology in insomnia. *Cognitive Behaviour Therapy, 41*(4), 273–287. http://dx.doi.org/10.1080/16506073.2012.672454.

Kelly, W. E. (2002). Worry and sleep length revisited: Worry, sleep length, and sleep disturbance ascribed to worry. *The Journal of Genetic Psychology, 163*(3), 296–304. http://dx.doi.org/10.1080/00221320209598685.

Key, B. L., Campbell, T. S., Bacon, S. L., & Gerin, W. (2008). The influence of trait and state rumination on cardiovascular recovery from a negative emotional stressor. *Journal of Behavioral Medicine, 31*(3), 237–248. http://dx.doi.org/10.1007/s10865-008-9152-9.

Kingston, T., Dooley, B., Bates, A., Lawlor, E., & Malone, K. (2007). Mindfulness-based cognitive therapy for residual depressive symptoms. *Psychology and Psychotherapy: Theory, Research and Practice, 80*(2), 193–203. http://dx.doi.org/10.1348/147608306x116016.

Kloss, J., & Szuba, M. (2003). Insomnia in psychiatric disorders. In *Insomnia: Principles and management* (pp. 43–70). New York: Cambridge University Press.

Kudielka, B. M., Schommer, N. C., Hellhammer, D. H., & Kirschbaum, C. (2004). Acute HPA axis responses, heart rate, and mood changes to psychosocial stress (TSST) in humans at different times of day. *Psychoneuroendocrinology, 29*(8), 983–992. http://dx.doi.org/10.1016/j.psyneuen.2003.08.009.

Kuehner, C., Holzhauer, S., & Huffziger, S. (2007). Decreased cortisol response to awakening is associated with cognitive vulnerability to depression in a nonclinical sample of young adults. *Psychoneuroendocrinology, 32*(2), 199–209. http://dx.doi.org/10.1016/j.psyneuen.2006.12.007.

Kuisk, L. A., Bertelson, A. D., & Walsh, J. K. (1989). Presleep cognitive hyperarousal and affect as factors in objective and subjective insomnia. *Perceptual and Motor Skills, 69*(3 Pt. 2), 1219–1225.

Lavallee, L. F., & Campbell, J. D. (1995). Impact of personal goals on self-regulation processes elicited by daily negative events. *Journal of Personality and Social Psychology, 69*(2), 341–352. http://dx.doi.org/10.1037/0022-3514.69.2.341.

LeBlanc, M., Mérette, C., Savard, J., Ivers, H., Baillargeon, L., & Morin, C. M. (2009). Incidence and risk factors of insomnia in a population-based sample. *Sleep: Journal of Sleep and Sleep Disorders Research, 32*(8), 1027–1037.

Lichstein, K. L., & Fanning, J. (1990). Cognitive anxiety in insomnia: An analogue test. *Stress Medicine, 6*(1), 47–51. http://dx.doi.org/10.1002/smi.2460060110.

Lichstein, K. L., & Rosenthal, T. L. (1980). Insomniacs' perceptions of cognitive versus somatic determinants of sleep disturbance. *Journal of Abnormal Psychology, 89*(1), 105–107.

Llera, S. J., & Newman, M. G. (2010). Effects of worry on physiological and subjective reactivity to emotional stimuli in generalized anxiety disorder and nonanxious control participants. *Emotion, 10*(5), 640–650. http://dx.doi.org/10.1037/a0019351.

Lyubomirsky, S., & Nolen-Hoeksema S. (1993). Self-perpetuating properties of dysphoric rumination. *Journal of Personality and Social Psychology, 65*(2), 339–349.

Lyubomirsky, S., Tucker, K. L., Caldwell, N. D., & Berg, K. (1999). Why ruminators are poor problem solvers: Clues from the phenomenology of dysphoric rumination. *Journal of Personality and Social Psychology, 77*(5), 1041–1060.

Mansell, W., Harvey, A., Watkins, E. R., & Shafran, R. (2008). Cognitive behavioral processes across psychological disorders: A review of the utility and validity of the transdiagnostic approach. *International Journal of Cognitive Therapy, 1*(3), 181–191. http://dx.doi.org/10.1680/ijct.2008.1.3.181.

McEvoy, P. M., & Brans, S. (2013). Common versus unique variance across measures of worry and rumination: Predictive utility and mediational models for anxiety and depression. *Cognitive Therapy and Research, 37*(1), 183–196. http://dx.doi.org/10.1007/s10608-012-9448-5.

McKay, M., & Zakzanis, K. (2010). The impact of treatment on HPA axis activity in unipolar major depression. *Journal of Psychiatric Research, 44*(3), 192.

McLaughlin, K. A., Borkovec, T. D., & Sibrava, N. J. (2007). The effects of worry and rumination on affect states and cognitive activity. *Behavior Therapy, 38*(1), 23–38. http://dx.doi.org/10.1016/j.beth.2006.03.003.

Meyer, T. J., Miller, M. L., Metzger, R. L., & Borkovec, T. D. (1990). Development and validation of the Penn State Worry Questionnaire. *Behaviour Research and Therapy, 28*(6), 487–495. http://dx.doi.org/10.1016/0005-7967(90)90135-6.

Moberly, N. J., & Watkins, E. R. (2008). Ruminative self-focus, negative life events, and negative affect. *Behaviour Research and Therapy, 46*(9), 1034–1039. http://dx.doi.org/10.1016/j.brat.2008.06.004.

Nelson, J., & Harvey, A. G. (2003). An exploration of pre-sleep cognitive activity in insomnia: Imagery and verbal thought. *British Journal of Clinical Psychology, 42*(3), 271–288. http://dx.doi.org/10.1348/01446650360703384.

Nolen-Hoeksema, S. (1990). *Sex differences in depression.* Stanford, CA: Stanford University Press.

Nolen-Hoeksema, S. (1991). Responses to depression and their effects on the duration of depressive episodes. *Journal of Abnormal Psychology, 100*(4), 569–582. http://dx.doi.org/10.1037/0021-843X.100.4.569.

Nolen-Hoeksema, S., & Morrow, J. (1991). A prospective study of depression and posttraumatic stress symptoms after a natural disaster: The 1989 Loma Prieta earthquake. *Journal of Personality and Social Psychology, 61*(1), 115–121.

Nolen-Hoeksema, S., Morrow, J., & Fredrickson, B. L. (1993). Response styles and the duration of episodes of depressed mood. *Journal of Abnormal Psychology, 102*(1), 20–28.

Nolen-Hoeksema, S., Parker, L. E., & Larson, J. (1994). Ruminative coping with depressed mood following loss. *Journal of Personality and Social Psychology, 67*(1), 92–104.

Omvik, S., Pallesen, S., Bjorvatn, B., Thayer, J., & Nordhus, I. H. (2007). Night-time thoughts in high and low worriers: Reaction to caffeine-induced sleeplessness. *Behaviour Research and Therapy, 45*(4), 715–727. http://dx.doi.org/10.1016/j.brat.2006.06.006.

Ong, J. C., Shapiro, S. L., & Manber, R. (2008). Combining mindfulness meditation with cognitive-behavior therapy for insomnia: A treatment-development study. *Behavior Therapy, 39*(2), 171–182. http://dx.doi.org/10.1016/j.beth.2007.07.002.

Ottaviani, C., Shapiro, D., Davydov, D. M., Goldstein, I. B., & Mills, P. J. (2009). The autonomic phenotype of rumination. *International Journal of Psychophysiology, 72*(3), 267–275. http://dx.doi.org/10.1016/j.ijpsycho.2008.12.014.

Pillai, V., Kalmbach, D. A., & Ciesla, J. A. (2011). A meta-analysis of electroencephalographic sleep in depression: Evidence for genetic biomarkers. *Biological Psychiatry, 70*(10), 912–919.

Pillai, V., Steenburg, L., Ciesla, J. A., Roth, T., & Drake, C. L. (2014). A seven day actigraphy-based study of rumination and sleep disturbance among young adults with depressive symptoms. *Journal of Psychosomatic Research, 77*(1), 70–75.

Puterman, E., DeLongis, A., & Pomaki, G. (2010). Protecting us from ourselves: Social support as a buffer of trait and state rumination. *Journal of Social and Clinical Psychology, 29*(7), 797–820. http://dx.doi.org/10.1521/jscp.2010.29.7.797.

Querstret, D., & Cropley, M. (2013). Assessing treatments used to reduce rumination and/or worry: A systematic review. *Clinical Psychology Review, 33*(8), 996–1009. http://dx.doi.org/10.1016/j.cpr.2013.08.004.

Ramel, W., Goldin, P. R., Carmona, P. E., & McQuaid, J. R. (2004). The effects of mindfulness meditation on cognitive processes and affect in patients with past depression. *Cognitive Therapy and Research, 28*(4), 433–455. http://dx.doi.org/10.1023/b:cotr.0000045557.15923.96.

Roane, B. M., Dolan, D. C., Bramoweth, A. D., Rosenthal, L., & Taylor, D. J. (2012). Altering unhelpful beliefs about sleep with behavioral and cognitive therapies. *Cognitive Therapy and Research, 36*(2), 129–133. http://dx.doi.org/10.1007/s10608-011-9417-4.

Rodríguez-Muñoz, A., Notelaers, G., & Moreno-Jiménez, B. (2011). Workplace bullying and sleep quality: The mediating role of worry and need for recovery. *Behavioral Psychology/ Psicología Conductual: Revista Internacional Clínica y de la Salud*, *19*(2), 453–468.

Roehrs, T. A., Randall, S., Harris, E., Maan, R., & Roth, T. (2011). MSLT in primary insomnia: Stability and relation to nocturnal sleep. *Sleep*, *34*(12), 1647–1652. http://dx.doi.org/10.5665/sleep.1426.

Roger, D., & Najarian, B. (1989). The construction and validation of a new scale for measuring emotion control. *Personality and Individual Differences*, *10*(8), 845–853. http://dx.doi.org/10.1016/0191-8869(89)90020-2.

Roth, T., Roehrs, T., & Pies, R. (2007). Insomnia: Pathophysiology and implications for treatment. *Sleep Medicine Reviews*, *11*(1), 71–79. http://dx.doi.org/10.1016/j.smrv.2006.06.002.

Rydstedt, L. W., Cropley, M., & Devereux, J. (2011). Long-term impact of role stress and cognitive rumination upon morning and evening saliva cortisol secretion. *Ergonomics*, *54*(5), 430–435. http://dx.doi.org/10.1080/00140139.2011.558639.

Schlotz, W., Hellhammer, J., Schulz, P., & Stone, A. A. (2004). Perceived work overload and chronic worrying predict weekend-weekday differences in the cortisol awakening response. *Psychosomatic Medicine*, *66*(2), 207–214. http://dx.doi.org/10.1097/01.psy.0000116715.78238.56.

Segerstrom, S. C., Stanton, A. L., Alden, L. E., & Shortridge, B. E. (2003). A multidimensional structure for repetitive thought: What's on your mind, and how, and how much? *Journal of Personality and Social Psychology*, *85*(5), 909–921. http://dx.doi.org/10.1037/0022-3514.85.5.909.

Segerstrom, S. C., Tsao, J. I., Alden, L. E., & Craske, M. G. (2000). Worry and rumination: Repetitive thought as a concomitant and predictor of negative mood. *Cognitive Therapy and Research*, *24*(6), 671–688.

Sunnhed, R., & Jansson-Frojmark, M. (2014). Are changes in worry associated with treatment response in cognitive behavioral therapy for insomnia? *Cognitive Behaviour Therapy*, *43*(1), 1–11. http://dx.doi.org/10.1080/16506073.2013.846399.

Swanson, L. M., Pickett, S. M., Flynn, H., & Armitage, R. (2011). Relationships among depression, anxiety, and insomnia symptoms in perinatal women seeking mental health treatment. *Journal of Women's Health*, *20*(4), 553–558. http://dx.doi.org/10.1089/jwh.2010.2371.

Takano, K., Iijima, Y., & Tanno, Y. (2012). Repetitive thought and self-reported sleep disturbance. *Behavior Therapy*, *43*(4), 779–789.

Tang, N. K. Y., & Harvey, A. G. (2004a). Correcting distorted perception of sleep in insomnia: A novel behavioural experiment? *Behaviour Research and Therapy*, *42*(1), 27–39. http://dx.doi.org/10.1016/S0005-7967(03)00068-8.

Tang, N. K. Y., & Harvey, A. G. (2004b). Effects of cognitive arousal and physiological arousal on sleep perception. *Sleep: Journal of Sleep and Sleep Disorders Research*, *27*(1), 69–78.

Thomsen, D. K. (2006). The association between rumination and negative affect: A review. *Cognition and Emotion*, *20*(8), 1216–1235. http://dx.doi.org/10.1080/02699930500473533.

Thomsen, D. K., Mehlsen, M. Y., Christensen, S., & Zachariae, R. (2003). Rumination— relationship with negative mood and sleep quality. *Personality and Individual Differences*, *34*(7), 1293–1301. http://dx.doi.org/10.1016/s0191-8869(02)00120-4.

Treynor, W., Gonzalez, R., & Nolen-Hoeksema, S. (2003). Rumination Reconsidered: A psychometric analysis. *Cognitive Therapy and Research*, *27*(3), 247–259. http://dx.doi.org/10.1023/a:1023910315561.

Ursin, H., & Eriksen, H. R. (2010). Cognitive activation theory of stress (CATS). *Neuroscience & Biobehavioral Reviews*, *34*(6), 877–881. http://dx.doi.org/10.1016/j.neubiorev.2009.03.001.

van der Helm, E., & Walker, M. P. (2011). Sleep and emotional memory processing. *Sleep Medicine Clinics*, *6*(1), 31–43. http://dx.doi.org/10.1016/j.jsmc.2010.12.010.

Vandekerckhove, M., & Cluydts, R. (2010). The emotional brain and sleep: An intimate relationship. *Sleep Medicine Reviews*, *14*(4), 219–226.

Vandekerckhove, M., Weiss, R., Schotte, C., Exadaktylos, V., Haex, B., Verbraecken, J., et al. (2011). The role of presleep negative emotion in sleep physiology. *Psychophysiology, 48*(12), 1738–1744. http://dx.doi.org/10.1111/j.1469-8986.2011.01281.x.

Vgontzas, A. N., Bixler, E. O., Lin, H. M., Prolo, P., Mastorakos, G., Vela-Bueno, A., et al. (2001). Chronic insomnia is associated with nyctohemeral activation of the hypothalamic-pituitary-adrenal axis: Clinical implications. *The Journal of Clinical Endocrinology and Metabolism, 86*(8), 3787–3794. http://dx.doi.org/10.1210/jcem.86.8.7778.

Walker, M. P., & Stickgold, R. (2006). Sleep, memory, and plasticity. In S. T. Fiske, A. E. Kazdin, & D. L. Schacter (Eds.), *Annual review of psychology* (vol. 57, pp. 139–166). Palo Alto, CA, USA: Annual Reviews.

Watkins, E. R. (2008). Constructive and unconstructive repetitive thought. *Psychological Bulletin, 134*(2), 163–206. http://dx.doi.org/10.1037/0033-2909.134.2.163.

Watkins, E. R., & Brown, R. (2002). Rumination and executive function in depression: An experimental study. *Journal of Neurology, Neurosurgery & Psychiatry, 72*(3), 402.

Watts, F. N., Coyle, K., & East, M. P. (1994). The contribution of worry to insomnia. *British Journal of Clinical Psychology, 33*(2), 211–220.

Wegner, D. M. (1994). Ironic processes of mental control. *Psychological Review, 101*(1), 34–52. http://dx.doi.org/10.1037/0033-295x.101.1.34.

Weise, S., Ong, J., Tesler, N. A., Kim, S., & Roth, W. T. (2013). Worried sleep: 24-h monitoring in high and low worriers. *Biological Psychology, 94*(1), 61–70. http://dx.doi.org/10.1016/j.biopsycho.2013.04.009.

Weiss, D. S. (2007). The impact of event scale: Revised. In J. P. Wilson & C.S.-k. Tang (Eds.), *Cross-cultural assessment of psychological trauma and PTSD* (pp. 219–238). New York: Springer Science+Business Media.

Williams, L. J., & Brown, B. K. (1994). Method variance in organizational behavior and human resources research: Effects on correlations, path coefficients, and hypothesis testing. *Organizational Behavior and Human Decision Processes, 57*(2), 185–209. http://dx.doi.org/10.1006/obhd.1994.1011.

Wisco, B. E., & Nolen-Hoeksema, S. (2008). Ruminative response style. In K. S. Dobson & D. J. A. Dozois (Eds.), *Risk factors in depression* (pp. 221–236). San Diego, CA, USA: Elsevier.

Wood, J. V., Saltzberg, J. A., Neale, J. M., Stone, A. A., & Rachmiel, T. B. (1990). Self-focused attention, coping responses, and distressed mood in everyday life. *Journal of Personality and Social Psychology, 58*(6), 1027–1036. http://dx.doi.org/10.1037/0022-3514.58.6.1027.

Yook, K., Lee, S. H., Ryu, M., Kim, K. H., Choi, T. K., Suh, S. Y., et al. (2008). Usefulness of mindfulness-based cognitive therapy for treating insomnia in patients with anxiety disorders: A pilot study. *The Journal of Nervous and Mental Disease, 196*(6), 501–503. http://dx.doi.org/10.1097/NMD.0b013e31817762ac.

Zoccola, P. M., Dickerson, S. S., & Lam, S. (2009). Rumination predicts longer sleep onset latency after an acute psychosocial stressor. *Psychosomatic Medicine, 71*(7), 771–775. http://dx.doi.org/10.1097/PSY.0b013e3181ae58e8.

CHAPTER 11

Sleep, Sadness, and Depression

Rachel Manber and Christopher P. Fairholme

Department of Psychiatry and Behavioral Sciences, Stanford University, Stanford, California, USA

SLEEP, SADNESS, AND DEPRESSION

As one of the nine diagnostic symptoms of a depressive episode, poor sleep has been studied extensively in relation to depressive illnesses, such as major depressive disorder (MDD). This research has rarely isolated the symptom of sadness or depressed mood, however. Instead, it has focused on overall depressive symptom severity, often excluding sleep to avoid collinearity or comparing individuals with and without MDD. As a discrete emotion, sadness has also not been the focus of experimental research on the role of sleep in affective experiences. In general, research on sleep and emotion has conceptualized emotions using the following dimensions: appetitive (attracting engagement with the environment), defensive (compelling a response to threat) (Bradley, 2000), valence (positive-negative), and arousal (high-low) (Watson & Tellegen, 1985). As defined by these dimensional approaches, sadness is a low arousal negative emotion, as well as a low appetitive emotion. Largely, the experimental research reviewed in this book shows that poor sleep quality is associated with high negative and low positive emotions. In this research, positive affect and negative affect are measured using an average rating on a list of variously defined positive and negative emotions.

Researchers studying the relationship between sleep and depressive illness use a definition of sadness that differs from the one used by those researching sleep and emotions. Whereas the first group refers to sadness as a mood, defined as a longer-lasting affective experience, also referred to as depressed mood, the second group refers to sadness as an emotion, defined as an immediate and brief affective response to a stimulus. According to the *Diagnostic and Statistical Manual of Mental Disorders 5* (American Psychiatric Association, 2013), the symptom of depressed mood also encompasses feelings of emptiness and hopelessness. Specifically, depressed mood is experienced as pervasive (most of the day nearly every day), and it is one of two core symptoms of a depressive episode. The second core symptom of a depressive episode is anhedonia (low interest in or enjoyment from most activities that were previously interesting or enjoyable). Applying the dimensional classifications of emotions to the two core symptoms of depression, one can classify both depressed mood and anhedonia as low appetitive, negative valence, low arousal mood states.

In this chapter, we review data from both lines of research, with a focus on the bidirectional relationship between sleep and sadness. That is, we consider the following: (a) How do sleep abnormalities impact sadness in terms of depressive symptoms, diagnosis, and low positive affect? (b) How does sadness, when similarly dually defined, impact sleep? Our focus is on insufficient sleep, poor sleep quality (self-reported or objectively measured), and the distribution of the different sleep stages, including wakefulness, across the night. In terms of subjectively defined sleep difficulty, we concentrate on difficulties falling or staying asleep. We do not discuss nonrefreshing sleep or sleep disorders other than insomnia. Because not much of the existing research on sleep and affect has focused on sadness as a distinct emotion or mood, we have broadened our discussion to include research on sleep in relation to low positive affect, with low positive affectivity providing an index of low levels of happiness and other appetitive emotions. We discuss the relationship between sleep and sadness both as it is subjectively experienced and physiologically manifested.

HOW SLEEP IMPACTS SADNESS OR DEPRESSION

Evidence about how sleep quantity and quality impact sadness and related constructs comes from multiple lines of research studying both healthy and depressed samples. We first discuss evidence that poor sleep at one time point is associated with depression at a later time point among people who were not depressed at the initial assessment. This evidence includes data from longitudinal studies that examined poor sleep as a predictor of a depressive episode 1-3 years later among people who are not currently depressed; cross-sectional retrospective data derived from individuals who are currently depressed; and prospective data collected from previously depressed individuals in remission, related to the relative course of poor sleep and the recurrence of depression. We then review the literature on the more proximal impact of acute sleep loss on healthy individuals. Specifically, we consider next-day effects of experimentally induced sleep loss and naturally occurring fluctuation of sleep quantity or quality on sadness (low positive affect). We then shift our attention to the impact of sleep deprivation and poor sleep on individuals with depression. We discuss the impact of induced sleep deprivation on mood during a depressive episode and of poor sleep on the severity of the depressive illness and in response to its treatment.

HOW POOR AND INSUFFICIENT SLEEP IMPACT A FUTURE DEPRESSIVE EPISODE

A large and largely consistent body of literature indicates that poor sleep is a significant vulnerability risk for a future depressive episode. This research is summarized in a meta-analysis of 21 longitudinal epidemiological studies that have assessed sleep and depressive symptoms and controlled for baseline levels of depression. This meta-analysis concluded that, among individuals with no depression at baseline, those with poor sleep at baseline had a twofold risk of developing depression when reassessed 1 year or more later (average of 71±96 months), compared to those with no sleep difficulties at baseline (Baglioni et al., 2011). Although there is a convergence of evidence from longitudinal observations that poor sleep is a statistical risk for depressive episodes, the evidence does not imply causality. It is also not clear whether poor sleep represents a unique or shared vulnerability to depression. Individuals prone to experience poor sleep in response to stress might also be prone to developing a depressive episode under stress. Data from Rottenberg, Chambers, Allen, and Manber (2007) support the shared vulnerability hypothesis. These researchers found that impaired cardiac vagal control correlated with sad mood and middle-of-the-night sleep difficulty among patients with MDD, but not in those with other depression symptoms.

Consistent with the idea that poor sleep precedes depressive illness, a large epidemiological study found that the most common temporal pattern between sleep and mood disturbances among those with a first major depressive episode and insomnia was for insomnia symptoms to have emerged before MDD (41% of participants). In contrast, insomnia symptoms emerged at the same time as the MDD in 29% and after the depressive episode in 29% (Ohayon & Roth, 2003). The same study found a very different temporal relationship between poor sleep and anxiety disorders; in the vast majority of cases with anxiety disorder, insomnia emerged at the same time as or after the anxiety disorder (Ohayon & Roth, 2003). This study did not provide information about the timing of the emergence of poor sleep and a first depressive episode, however. Perlis, Giles, Buysse, Tu, et al. (1997) did provide such information in the context of recurrent depression. In this study, patients who remitted from a depressive disorder were followed longitudinally, and the authors periodically assessed them to ascertain the presence of depressive symptoms. The results indicated that poor sleep emerged a few weeks before the recurrence of MDD, and this finding suggests that poor sleep might be one of the first depressive symptoms to emerge, thus representing an early stage in the link between stress and depression.

THE IMPACT OF POOR SLEEP ON NEXT-DAY DEPRESSED MOOD IN NONDEPRESSED SAMPLES

The experimental data on the immediate impact of acute sleep loss on next-day sadness or low positive affect is relatively sparse. Haack and Mullington (2005) studied healthy participants in the laboratory for 12 consecutive days, with the participants randomized to 4- or 8-h nocturnal sleep periods, which in both conditions started at 11 p.m. Affect was assessed every 2h, using a list of 108 adjectives describing mood and pain symptoms, which were subsequently analyzed in four clusters, optimism-sociability, tiredness-fatigue, anger-aggression, and bodily discomfort. They found that a scale measuring optimism-sociability, which can be considered an appetitive affect, progressively declined over consecutive days of restricted sleep, with a total reduction of 15% of positive affect from baseline to the last day (Haack & Mullington, 2005). Talbot, McGlinchey, Kaplan, Dahl, and Harvey (2010) examined positive and negative affect among adolescents and adults following sleep deprivation and rest. Their study was unique because, in addition to composite scores on the positive and negative affect scales, they reported results for individual emotions. Regardless of age group, sleep deprivation led to lower positive affect compared to rested affect, as well as to lower ratings on the following distinct emotion concepts: interest, excitement, happy, strong, energetic, cheerful, active, proud, and delighted. The authors did not find an increase in sadness or feeling "miserable" following sleep deprivation, although there was a strong trend for higher ratings on the distinct emotion "miserable." Babson, Trainor, Feldner, and Blumenthal (2010) also found that total sleep deprivation is associated with next-day low positive affect, as defined by feeling slowed down, bored, withdrawn, and needing extra effort to get moving. Franzen, Siegle, and Buysse (2008) found that one night of sleep deprivation led to lower subjective reports of positive mood as well. Objective measures of emotional reactivity (based on pupil dilation while viewing positive, neutral, and negative visual stimuli) did not differ between the sleep-deprived and nonsleep-deprived participants when viewing positive images, however. In contrast, another study by Franzen, Buysse, Dahl, Thompson, and Siegle (2009) found that, after one night of sleep deprivation, participants exhibited greater pupil reactivity in response to, and in anticipation of, negative valence images compared to participants who were allowed to sleep for 8h. This latter finding is consistent with findings of exaggerated amygdala reactivity and reduced prefrontal connectivity in response to negative stimuli (for a review see Goldstein & Walker, 2014).

Another method for examining the impact of poor sleep on sadness and low positive affect has involved naturalistically examining day-to-day variations of these affective states in relation to sleep over a few consecutive nights. Using a sample of 25 middle-aged women, Berry and Webb (1985) examined the covariation of sleep with five emotional states (cheerful, energetic, angry, anxious, depressed) assessed via polysomnography and self-ratings within an hour before bedtime on four consecutive nights. Results demonstrated that, on two of the four nights, shorter latency to REM sleep was associated with a more cheerful and energetic mood on the following evening. The authors question the solidity of the results, however, because they tested a large number of correlations between multiple polysomnographically defined sleep parameters and mood states over multiple days (Berry & Webb, 1985). McCrae and colleagues (2008) have similarly used naturalistic observation of mood and sleep over time (14 days). In this study, sleep was measured both subjectively (with sleep diaries) and objectively (using actigraphy). The study found that subjective, but not objective, sleep parameters correlated with next-day mood, so that nights on which participants reported more time awake or lower sleep quality preceded days when the participants experienced less positive affect and more negative affect. Zohar, Tzischinsky, Epstein, and Lavie (2005) found that sleep loss intensified negative affect and diminished positive affect in medical residents whose mood and subjective sleep were naturalistically observed for 6 days. However, this study included both night and day shifts, raising the possibility that the observed effects might be impacted by circadian factors. Recently, de Wild-Hartmann and colleagues (2013) used naturalistic data to study the temporal relationship between sleep and mood. In this study 553 women recorded 10 ratings per day of four positive mood states (cheerful, content, energetic, and enthusiastic) and six negative mood states (insecure, lonely, anxious, low, guilty, and suspicious). Similar to the other two studies, this study found that lower sleep quality and longer sleep latency (self-rated daily in a sleep diary) were associated with reduced positive affect and increased negative affect the next day. Thus, experimental (sleep deprivation) and observational research suggests that, among nondepressed individuals, insufficient and poor sleep are associated with low positive affect the next day.

THE IMPACT OF SLEEP DEPRIVATION ON MOOD DURING A DEPRESSIVE EPISODE

In nondepressed samples, insufficient sleep is associated with future depressed states, but the impact of sleep curtailment on people with depressive illness is in the opposite direction. A large body of research consistently demonstrates that approximately 40-60% of depressed patients show a transient improvement of mood after one night of total sleep deprivation, but in approximately 80%, the depressed mood returns after the next night of sleep (see meta-analysis by Wu & Bunney, 1990). About half of the studied people experience the return of depressive symptoms even after a short nap the next day (Wiegand, Berger, Zulley, Lauer, & von Zerssen, 1987). Partial sleep deprivation in the second, but not in the first, half of the night also produces next-day mood elevation (for a review see Wehr, 1990), as do selective REM-sleep deprivation and selective non-REM-sleep deprivation (Vogel, Vogel, McAbee, & Thurmond, 1980). A consistent predictor of mood elevation following one night of total sleep deprivation among patients with depression is a pattern of diurnal variation of mood characterized by better mood in the second half of the day (Van den Hoofdakker, 1997). Researchers have proposed that depressed individuals with diurnal mood variation might be shifting more easily between mood states than those with a stable mood across the day (Ebert & Berger, 1998), but if differences in diurnal mood variation explained the transient mood elevation following sleep deprivation, one would expect that other diurnal patterns would also predict mood response to sleep deprivation, which is not the case. Others have proposed that sleep deprivation might exert its antidepressant effects through epigenetic processes that reset abnormal clock genes expressed in depression (Bunney & Bunney, 2013). This theory has not been tested yet, but it is consistent with findings that schedule advancement after total sleep deprivation extends the acute antidepressant effects of total sleep deprivation (for a review see Riemann, Berger, & Voderholzer, 2001).

THE IMPACT OF POOR SLEEP ON THE SEVERITY OF DEPRESSION

Poor sleep in depression contributes to the global severity of the depressive disorder. For example, depressed patients who experience sleep-continuity disturbance and early-morning awakening are more likely to have suicidal ideation than depressed patients without sleep disturbance (for a review see McCall & Black, 2013). This suggests that depressed patients who experience insomnia may have more severe depression not only because of their insomnia, which factors into measures of depression severity, but also because insomnia leads to and exacerbates other symptoms of depression. In addition to suicidality, insufficient sleep leads to daytime malaise and difficulties with attention and concentration, thus compromising the ability of patients with depression to cope. Difficulty coping could then contribute to helplessness and hopelessness. Thus, the contribution of poor sleep to the global severity of depression exceeds what would be expected if sleep disturbance were just a manifestation or consequence of depression.

THE IMPACT OF POOR SLEEP ON THE COURSE OF DEPRESSION TREATMENT

Poor sleep also has adverse effects on the course of standard antidepressant treatments. Patients with abnormal sleep profiles have significantly poorer clinical outcomes with respect to symptom ratings, attrition rates, and remission rates than do patients with more normal sleep profiles (Thase et al., 1997). Studies have repeatedly shown that poor subjective and objective (electroencephalographic) sleep is associated with slower and lower rates of remission from depression and a less stable treatment response (Buysse et al., 1997; Dew et al., 1997). Evidence also suggests that treatment that provides early improvement in sleep may hasten antidepressant treatment response (DiMascio et al., 1979; Manber et al., 2003; Thase et al., 2002). A study of individuals with chronic depression found that an antidepressant affecting sleep consolidation produced faster responses when compared with psychotherapy that did not address sleep, even though the two treatments showed equivalent antidepressant efficacy after 12 weeks (Manber et al., 2003; Thase et al., 2002).

In addition to its effects on treatment response, poor sleep is a common persistent residual symptom following successful treatment of depression. In fact, sleep disturbance and associated fatigue are present in approximately 44% and 38% of treatment completers, respectively, and thus, they are the most persistent residual symptoms following acute treatment of depression (Nierenberg et al., 1999). The next most common residual symptom found in the Nierenberg et al. study was anhedonia (27%). Residual insomnia has been shown to increase the risk for depressive relapse. For example, after treatment with a combination of nortriptyline and interpersonal psychotherapy, two-thirds of patients with persistent insomnia relapsed within 1 year after switching to pill placebo. In contrast, 90% of patients with good sleep at the end of the acute treatment remained well during the first year after discontinuing antidepressants (Reynolds et al., 1997). Incomplete remission and residual symptoms are associated with high rates of relapse and recurrence of depression (Buysse et al., 1996; Paykel et al., 1995; Simons, Murphy, Levine, & Wetzel, 1986). In one study, the depressive relapse rate among patients who ended treatment with residual symptoms was 76%, which was threefold greater than the rate for patients who achieved full remission (Paykel et al., 1995). A study of the long-term course of response to depression treatment found that responders who had residual symptoms experienced a major depressive episode sooner after remission (in the first 4 months) than did those without residual symptoms, who mostly relapsed after 12 months (Van Londen, Molenaar, Goekoop, Zwinderman, & Rooijmans, 1998). The high prevalence of sleep initiation and maintenance problems among depressed persons, as well as the possible adverse effects of poor sleep on depression severity and its therapeutic course, suggests that improving sleep early in the treatment process might produce a more rapid depressive treatment response, while reducing depressive relapse and therefore improving the long-term prognosis.

THE IMPACT OF SADNESS OR DEPRESSION ON SLEEP

We begin the second half of this chapter by discussing the impact of sadness and depression on sleep. This discussion is based on research on the association between the abnormal sleep architecture and poor sleep experienced during depression and the clinical course of depressive illness. We then discuss how depression and its core mood symptoms (sadness and anhedonia) impact sleep disorders such as insomnia, the delayed and advanced types of circadian rhythm sleep-wake, and sleep apnea. We conclude by considering the impact that low mood during the day has on sleep that night.

SUBJECTIVE SLEEP SYMPTOMS IN DEPRESSION

Most patients with MDD (60-84%) report some difficulty initiating and/ or maintaining sleep (Ford & Kamerow, 1989; Hamilton, 1989). It is not known how many people with MDD meet criteria for an insomnia disorder, which requires complaints of both poor sleep and a clinically meaningful impact, such as distress or impairment. Using weekly averages of sleep diary data in a large sample of patients with chronic depression, Manber and colleagues (2003) found that less than half of the participants had clinically meaningful sleep difficulties. Specifically, 45% of patients with chronic depression reported taking more than 30 min to fall asleep, 39% spent more than 30 min awake after sleep onset, and 25% woke up on average 60 min or more before planned or desired. Thus, although sleep complaints are common, clinically significant sleep complaints are less common.

OBJECTIVE SLEEP ABNORMALITIES IN DEPRESSION

Polysomnographically defined sleep abnormalities in depression have also been documented. Nearly 24 studies have documented that the distribution of sleep stages is altered in patients with MDD when compared to nonpsychiatric controls (Krystal, Thakur, & Roth, 2008). These abnormalities include an increase in percent-time spent awake, reduction in slow-wave sleep, and alterations in REM sleep. REM-sleep abnormalities in depression include faster entry into REM sleep, longer duration of the first REM episode, and greater REM density, which is defined as the number of eye movements during REM sleep (Palagini, Baglioni, Ciapparelli, Gemignani, & Riemann, 2013). Shorter REM latencies are associated with greater overall severity of depression (Bencca, 1994) and tend to be stable within individuals across time (Giles et al., 1989; Rush et al., 1986). This suggests that REM latency might be a vulnerability marker for depression and that individuals with this vulnerability also tend to have more severe depressive episodes. In contrast, REM density, which also correlates with depression severity and is more pronounced in recurrent depression (Thase et al., 1995), appears to be state-dependent because it normalizes when depression remits (Buysse, Kupfer, Frank, Monk, & Ritenour, 1992; Nofzinger et al., 1994). This suggests that REM density might be impacted by depression, but more research is needed in order to fully understand REM-sleep abnormalities that are state-dependent. Based on the above-discussed literature, some abnormal REM-sleep parameters might be more stable than others. Researchers have also suggested that the time course to normalization of REM-sleep abnormalities might depend on the time since depressive remission (Riemann et al., 2001). We hypothesize that the normalization of REM-sleep abnormalities that are not stable biological markers might also depend on the extent of recovery and the profile of residual depressive symptoms. The discussion has so far centered on objective sleep abnormalities in relation to depressive illness rather than sadness. One of the few studies that examined sleep in relation to specific depressive symptoms found that REM density might be relevant to depressed mood (Perlis, Giles, Buysse, Thase, et al., 1997). However, the pattern of the associations between REM density and low appetitive symptoms of depression was not consistent, suggesting that REM density might be most closely tied with depressed mood. Additional studies on the relationship between REM-sleep abnormalities and specific depressive symptoms are needed. In addition, studies examining the effects of experimentally induced sad mood on REM density might help to shed additional light on the nature of these relationships.

Based on findings that REM sleep is involved in emotional processing, particularly in the consolidation of autobiographical memories (for a review see Walker & van der Helm, 2009), researchers have hypothesized that REM-sleep abnormalities in depression might be markers of emotion dysregulation (Fosse, Stickgold, & Hobson, 2001). Goldstein and Walker (2014) propose that impaired discrimination of emotionally salient information in depression (Ritchey, Dolcos, Eddington, Strauman, & Cabeza, 2011) could also be related to REM-sleep abnormalities observed in patients with depression. They describe these abnormalities as representing exaggerated REM-sleep qualities and argue that this excess of REM sleep reduces noradrenaline availability, which, in turn, impacts amygdala sensitivity and produces impairments in emotional discrimination.

In addition to abnormalities in REM sleep, evidence shows a reduction in non-REM slow-wave sleep in depressed samples compared to nondepressed healthy samples (Kupfer et al., 1984). However, this polysomnographic difference has not been consistently reported. These inconsistent findings might be related to sample characteristics. For example, reduction in slow-wave activity appears to be present mostly in men (Armitage & Hoffmann, 2001; Reynolds et al., 1990) and mostly in young adults (Armitage, Hoffmann, Trivedi, & Rush, 2000). Comparing 76 outpatients with depression to 55 healthy control participants, Armitage and colleagues found that the amplitude of slow-wave activity was lower in the depressed group, especially among depressed men, and mostly among those between 20 and 30 years of age (Armitage et al., 2000). Differences in slow-wave activity between men and women were also twice as large among the depressed participants compared to the control participants (Armitage et al., 2000). Moreover, abnormal EEG patterns in depression are not universal. It has been estimated that 55% of outpatients with MDD do not have abnormal EEG patterns, as defined by a composite score combining sleep efficiency, REM latency, and REM density (Thase et al., 1997).

HOW SADNESS AND DEPRESSION IMPACT FUTURE SLEEP DISORDERS

Relatively few studies that examined poor sleep as a predictor of subsequent depression have also examined whether depression predicts subsequent poor sleep. In the studies that have examined this relationship, the results are largely mixed and suggest weak to no association. In one population-based study of 3000 individuals assessed at baseline and 1 year later (Jansson-Frojmark & Lindblom, 2008), insomnia at baseline was related to new episodes of depression a year later (OR=3.51, 4% of variance). In contrast, two other studies did not find a significant prediction of insomnia based on baseline depression. Buysse and colleagues (2008) examined the bidirectional association between insomnia and depression over a period of 20 years, using data collected in Zurich at six time points. Distinguishing between the experiences of these disorders, when alone and when coexisting, this study found that depression without comorbid insomnia was not longitudinally related to future insomnia. Skapinakis and colleagues (2013) also examined the bidirectional relationship between insomnia and depression. This study, which was conducted in the United Kingdom over a shorter time frame (18 months), found no significant association in either direction, although the relative magnitude of the odds ratios was large (OR=1.27, 95%

confidence interval [0.51-3.19] for the insomnia to depression association and OR=0.87, [0.56-1.35] for the reverse direction).

Considered proximally, sadness can lead to behaviors that can both contribute to and exacerbate existing sleep disorders, such as an insomnia disorder and circadian rhythm sleep-wake disorders. For example, patients with insomnia disorder who experience low mood and anhedonia often spend more time in bed (e.g., because nothing else feels good) and frequently have irregular sleep schedules (e.g., because of low intrinsic motivation to get out of bed coupled with variable extrinsic commitments in the morning). Extended time in bed while in a negative mood is detrimental to sleep because it can weaken the stimulus value of the bed as a cue for sleep. Irregular rise time is detrimental for sleep because it can lead to lower amplitude of the circadian curve, which reflects a weakening of the circadian process involved in the regulation of sleep. In these ways, behavioral responses to the experience of low mood could contribute to worse sleep. Indeed individuals with insomnia and high depressive symptom severity have more severe insomnia (Manber et al., 2011). In addition, the cognitive style and distortions common in depression could also contribute to worse sleep. For instance, the tendency of people with depression to engage in rumination and worry, particularly while in bed, can hinder their ability to fall or return to sleep. In addition, an attentional bias to negative emotional stimuli among individuals with depression (Joormann, 2004) could contribute to catastrophizing about the potential negative consequences of sleep loss, which could, in turn, increase arousal levels and further impair sleep. Consistent with this dynamic, studies have found that individuals with elevated scores on depression inventories have higher scores on the Dysfunctional Beliefs and Attitudes about Sleep questionnaire than do individuals with primary insomnia (Carney, Edinger, Manber, Garson, & Segal, 2007; Carney et al., 2010).

Low mood and depression can also contribute to the severity of a circadian sleep-wake disorder, delayed type (also known as delayed sleep phase syndrome). Two hallmark symptoms of this disorder are difficulty getting up in the morning at a conventional time and feeling most alert and engaged with the world late at night. Individuals with delayed-sleep-type circadian sleep disorder and depression often have great difficulty getting up in the morning, and they begin waking up or getting out of bed progressively later. They also tend to experience a diurnal mood pattern during which their depressed mood is less severe late at night and relatively worse in the morning. As a result, they begin to consistently delay their bedtimes, a practice that is likely negatively reinforced by the temporary improvement in mood. Thus, these individuals' sleep schedules become increasingly delayed, and they might find it more difficult to sleep at a time that supports their work or social schedules.

Low mood and depression can also contribute to the severity of a circadian sleep-wake disorder, advanced type (also known as advanced sleep phase syndrome). This disorder is characterized by a lifelong pattern of going to sleep and waking up earlier than most people, with concurrent difficulty maintaining wakefulness in the evening and sleep in the early morning hours. Two aspects of depression could lead to further exacerbation of this disorder. Low mood and anhedonia could lead to a person going to bed earlier than his or her already early habitual bedtime because, for example, nothing else feels good to do. Patients with advanced sleep phase syndrome who also experience early morning awakening as a symptom of depression might start waking up even earlier than before. Waking up earlier makes maintaining evening wakefulness even more difficult, and it can therefore lead to further advances in the sleep schedule. As a result, patients with circadian sleep-wake disorder-advanced type might have increasing difficulty maintaining a conventional sleep-wake schedule.

Another way in which sadness can contribute to or exacerbate sleep disorders is via weight gain. Depression can lead to weight gain because both low mood and anhedonia tend to lead to inactivity and/or overeating, which can result in weight gain. Weight gain has been consistently shown to exacerbate another sleep disorder, sleep apnea. Consistent with this connection, sleep apnea is more common among patients with depression (Ong, Gress, San Pedro-Salcedo, & Manber, 2009).

HOW MOOD IMPACTS NEXT-DAY SLEEP

Research examining day-to-day covariation of sleep and mood has largely focused on how sleep impacts next-day mood. One study has also examined the converse relationship. Specifically, de Wild-Hartmann and colleagues (2013) used naturalistic data to examine the bidirectional temporal relationship between sleep and mood. The researchers examined latency to sleep onset, the number of middle-of-the-night awakenings, total sleep time, and sleep quality. This study found little association between daytime affect and sleep that same night. The only significant finding was that daytime positive affect was associated with poorer self-rated same-night sleep quality. The authors have not interpreted this finding, possibly because they may have considered it spurious. However, in the context of this chapter, we take the opportunity to note that this study measured positive affect as the average of four positive valence emotions, of which three (cheerful, energetic, and enthusiastic) are also high arousal emotions and only one is a low arousal emotion (content). There is a consistent body of literature on the detrimental effects of high arousal on sleep. In fact, many researchers consider arousal downregulation to be a third sleep regulatory process (Sewitch, 1987) that is ancillary to the two process model of sleep (Borbely, 1982). Thus, the finding of de Wild-Hartmann and colleagues is consistent with the idea that high arousal emotions can be detrimental to sleep, independent of valence.

SUMMARY AND IMPLICATIONS FOR FUTURE RESEARCH

In general, researchers know more about the relationship between sleep and depressive disorder than they do about the connection between sleep and depressed affect, even if the latter is broadly defined to include low positive affect. This chapter presents evidence exploring the directionality of the association between sleep and depressed mood. Observational and experimental studies suggest that poor sleep quality and insufficient sleep lead to low positive affect among healthy, depression-free individuals. However, a separate line of research finds that the impact of sleep deprivation on depressed mood is in the opposite direction for individuals with depression; they respond to the challenge of sleep deprivation with transient mood elevation. This differential response to sleep deprivation among healthy and depressed individuals is intriguing but not well understood. One possible explanation could be related to how sleep deprivation impacts emotion regulation. Although less commonly reported in the systematic scientific literature, authors have reported nondepressed individuals responding to sleep deprivation with displays of giddiness and silly laughter (Bliss, 1959). This and the fact that depressed individuals with improved late-day mood are more likely to experience mood elevation in response to sleep deprivation suggest that people with and without depression exhibit individual differences in emotional response to sleep deprivation. Understanding causes of these individual differences could help researchers explain the paradoxical finding that those with and without depression respond differently to the challenge of sleep deprivation. One intriguing recent finding is that sleep deprivation enhances reactivity not only to negative stimuli but also to pleasure-evoking stimuli (Gujar, Yoo, Hu, & Walker, 2011). In this study, enhanced reactivity to positive stimuli was associated with altered connectivity in brain regions involved in affective and reward brain networks (Gujar et al., 2011). This has led the authors to speculate that the observed enhanced reactivity to pleasurable experiences might explain the reversal of anhedonia (defined by diminished ability to experience positive emotion in response to normally pleasurable events) among individuals with depression after a night of sleep deprivation. To better understand differential response to sleep deprivation among those with and without depression, researchers should simultaneously study the impact of sleep deprivation in both those with and without depression, thus removing variance due to the study procedures, the experimental manipulations of sleep, and methods for assessing mood (in terms of timing and measurement).

Studies on poor sleep as a predictor of the emergence or recurrence of a future depressive episode provide indirect evidence for the potential impact of poor sleep on mood. However, this inference is problematic both because (a) mood has not been analyzed separately, but only as it contributes to depression severity and diagnosis; and (b) alternative explanations to causality, such as common underlying vulnerability, need to be considered, despite prospective longitudinal methodology. This possibility is consistent with a study showing that, when compared to patients experiencing their first depressive episodes, those with recurrent depression have greater disturbances of sleep continuity and REM sleep and diminished slow-wave sleep both during a depressive episode and during early remission (Jindal et al., 2002). Yet, animal research suggests a potential causal link by which chronic sleep restriction leads to "a gradual and persistent desensitization of the 5-HT1A receptor system" (Roman, Walstra, Luiten, & Meerlo, 2005, p. 1509). If also true in humans, and given that depression is associated with altered serotonergic neurotransmission, desensitization of the serotonin system could explain the increased risk of depression among individuals with persistent sleep problems.

Fewer studies address the impact of low mood on sleep among individuals who are not depressed, and these studies generally find no association between the two, or they report weak evidence that a low arousal negative affect, such as sadness, predicts poor sleep the next night or that depression at baseline predicts a future episode of insomnia. On the other hand, people with depression experience greater sleep abnormalities than do people without depression, both in terms of difficulties initiating and maintaining sleep and in terms of REM-sleep abnormalities. Depression and sad mood might also contribute to the severity of other sleep disorders, such as insomnia, circadian rhythm sleep-wake disorders, and sleep apnea.

Research cannot yet fully explain the potentially complex relationship between sleep and sadness because very few studies have reported on sadness (or depressed mood) in isolation, and we found no research exclusively targeting sadness. Future research will need to separately study or compare the bidirectional relationship between sleep and sadness among people with and without depression. Given the paucity of research on sadness, we extended our review to include discussion of studies examining low positive affect. However, results from studies of the relationship between positive mood and sleep use checklists that include emotions with positive valence and varying degrees of arousal. For instance, excitement and contentment are positively valenced emotions with markedly different levels of arousal. Whereas low arousal positive mood is conducive to sleep, high arousal emotions are incompatible with sleep, independent of valence (positive and negative).

Another understudied area concerning the relationship between sleep and sadness is the relationship between hypersomnia and low arousal emotional states. There has been much confusion in the definition of hypersomnia. In the context of depression, DSM-5 defines hypersomnolence as prolonged nocturnal sleep or increased daytime sleep resulting in longer-than-normal sleep duration per 24-h day, but the term is often also used in reference to daytime sleepiness. For example, hypersomnolence disorder includes both sleepiness and long 24-h sleep duration in its definition (American Psychiatric Association, 2013). However, self-reported daytime sleepiness is not significantly correlated with long sleep duration (Ohayon, 2012), and they appear to be orthogonal when assessed in bipolar samples (Kaplan et al., 2014), suggesting that the two constructs might not be related. Although some research has addressed depressed mood and daytime sleepiness (e.g., daytime sleepiness has been found to predict a future occurrence of a depressive episode; Breslau, Roth, Rosenthal, & Andreski, 1996; Ford & Cooper-Patrick, 2001; Roberts, Roberts, & Chen, 2001), less is known about the relationship between depressed mood and long sleep duration. Moreover, it is not clear whether self-reported long sleep duration is actually associated with objectively measured sleep duration. A recent study used actigraphy to objectively measure sleep, ad the results suggest that, at least among euthymic patients with bipolar disorders, there is no significant correlation between the two (Kaplan et al., 2014). Hypersomnia is a manifestation of sleep disturbance in MDD and is therefore clinically relevant to depression. Yet, much remains unknown about the relationship between hypersomnia and depressed mood.

References

American Psychiatric Association. (2013). *Diagnostic and statistical manual of mental disorders* (5th ed.). Arlington, VA: American Psychiatric Publishing.

Armitage, R., & Hoffmann, R. F. (2001). Sleep EEG, depression and gender. *Sleep Medicine Review, 5*(3), 237–246. http://dx.doi.org/10.1053/smrv.2000.0144.

Armitage, R., Hoffmann, R., Trivedi, M., & Rush, A. J. (2000). Slow-wave activity in NREM sleep: Sex and age effects in depressed outpatients and healthy controls. *Psychiatry Research, 95*(3), 201–213.

Babson, K. A., Trainor, C. D., Feldner, M. T., & Blumenthal, H. (2010). A test of the effects of acute sleep deprivation on general and specific self-reported anxiety and depressive symptoms: An experimental extension. *Journal of Behavior Therapy and Experimental Psychiatry, 41*(3), 297–303. http://dx.doi.org/10.1016/j.jbtep.2010.02.008.

Baglioni, C., Battagliese, G., Feige, B., Spiegelhalder, K., Nissen, C., Voderholzer, U., et al. (2011). Insomnia as a predictor of depression: A meta-analytic evaluation of longitudinal epidemiological studies. *Journal of Affective Disorders, 135*(1–3), 10–19. http://dx.doi.org/10.1016/j.jad.2011.01.011.

Bencca, R. (1994). Mood disorders. In M. Kryger, M. Roth, & W. Dement (Eds.), *Principles and practice of sleep medicine* (pp. 899–913). Philadelphia: W.B. Saunders.

Berry, D. T., & Webb, W. B. (1985). Mood and sleep in aging women. *Journal of Personality and Social Psychology, 49*(6), 1724–1727.

Bliss, E. L., Clark, L. D., & West, C. D. (1959). Studies of sleep deprivation—Relationship to Schizophrenia. *Archives of Neurology and Psychiatry, 81*(3), 348–359.

Borbely, A. A. (1982). A two process model of sleep regulation. *Human Neurobiology, 1*(3), 195–204.

Bradley, M. M. (2000). Emotion and motivation. In J. Cacioppo, L. Tassinary, & G. Berntson (Eds.), *Handbook of psychophysiology* (pp. 602–642). New York: Cambridge University Press.

Breslau, N., Roth, T., Rosenthal, L., & Andreski, P. (1996). Sleep disturbance and psychiatric disorders: A longitudinal epidemiological study of young adults. *Biological Psychiatry, 39*(6), 411–418.

Bunney, B. G., & Bunney, W. E. (2013). Mechanisms of rapid antidepressant effects of sleep deprivation therapy: Clock genes and circadian rhythms. *Biological Psychiatry, 73*(12), 1164–1171. http://dx.doi.org/10.1016/j.biopsych.2012.07.020.

Buysse, D. J., Angst, J., Gamma, A., Ajdacic, V., Eich, D., & Rossler, W. (2008). Prevalence, course, and comorbidity of insomnia and depression in young adults. *Sleep, 31*(4), 473–480.

Buysse, D. J., Kupfer, D. J., Frank, E., Monk, T. H., & Ritenour, A. (1992). Electroencephalographic sleep studies in depressed outpatients treated with interpersonal psychotherapy: II. Longitudinal studies at baseline and recovery. *Psychiatry Research, 42*(1), 27–40.

Buysse, D. J., Reynolds, C. F., 3rd., Hoch, C. C., Houck, P. R., Kupfer, D. J., Mazumdar, S., et al. (1996). Longitudinal effects of nortriptyline on EEG sleep and the likelihood of recurrence in elderly depressed patients. *Neuropsychopharmacology, 14*(4), 243–252. http://dx.doi.org/10.1016/0893-133X(95)00114-S.

Buysse, D. J., Reynolds, C. F., 3rd., Houck, P. R., Perel, J. M., Frank, E., Begley, A. E., et al. (1997). Does lorazepam impair the antidepressant response to nortriptyline and psychotherapy? *Journal of Clinical Psychiatry, 58*(10), 426–432.

Carney, C. E., Edinger, J. D., Manber, R., Garson, C., & Segal, Z. V. (2007). Beliefs about sleep in disorders characterized by sleep and mood disturbance. *Journal of Psychosomatic Research, 62*(2), 179–188. http://dx.doi.org/10.1016/j.jpsychores.2006.08.006.

Carney, C. E., Edinger, J. D., Morin, C. M., Manber, R., Rybarczyk, B., Stepanski, E. J., et al. (2010). Examining maladaptive beliefs about sleep across insomnia patient groups. *Journal of Psychosomatic Research, 68*(1), 57–65. http://dx.doi.org/10.1016/j.jpsychores.2009.08.007.

de Wild-Hartmann, J. A., Wichers, M., van Bemmel, A. L., Derom, C., Thiery, E., Jacobs, N., et al. (2013). Day-to-day associations between subjective sleep and affect in regard to future depression in a female population-based sample. *British Journal of Psychiatry, 202*, 407–412. http://dx.doi.org/10.1192/bjp.bp.112.123794.

Dew, M. A., Reynolds, C. F., 3rd., Houck, P. R., Hall, M., Buysse, D. J., Frank, E., et al. (1997). Temporal profiles of the course of depression during treatment. Predictors of pathways toward recovery in the elderly. *Archives of General Psychiatry, 54*(11), 1016–1024.

DiMascio, A., Weissman, M. M., Prusoff, B. A., Neu, C., Zwilling, M., & Klerman, G. L. (1979). Differential symptom reduction by drugs and psychotherapy in acute depression. *Archives of General Psychiatry, 36*(13), 1450–1456.

Ebert, D., & Berger, M. (1998). Neurobiological similarities in antidepressant sleep deprivation and psychostimulant use: A psychostimulant theory of antidepressant sleep deprivation. *Psychopharmacology, 140*(1), 1–10.

Ford, D. E., & Cooper-Patrick, L. (2001). Sleep disturbances and mood disorders: An epidemiologic perspective. *Depression and Anxiety, 14*(1), 3–6.

Ford, D. E., & Kamerow, D. B. (1989). Epidemiologic study of sleep disturbances and psychiatric disorders. An opportunity for prevention? *JAMA, 262*(11), 1479–1484.

REFERENCES

Fosse, R., Stickgold, R., & Hobson, J. A. (2001). The mind in REM sleep: Reports of emotional experience. *Sleep*, *24*(8), 947–955.

Franzen, P. L., Buysse, D. J., Dahl, R. E., Thompson, W., & Siegle, G. J. (2009). Sleep deprivation alters pupillary reactivity to emotional stimuli in healthy young adults. *Biological Psychology*, *80*(3), 300–305. http://dx.doi.org/10.1016/j.biopsycho.2008.10.010.

Franzen, P. L., Siegle, G. J., & Buysse, D. J. (2008). Relationships between affect, vigilance, and sleepiness following sleep deprivation. *Journal of Sleep Research*, *17*(1), 34–41. http://dx.doi.org/10.1111/j.1365-2869.2008.00635.x.

Giles, D. E., Kupfer, D. J., Roffwarg, H. P., Rush, A. J., Biggs, M. M., & Etzel, B. A. (1989). Polysomnographic parameters in first-degree relatives of unipolar probands. *Psychiatry Research*, *27*(2), 127–136.

Goldstein, A. N., & Walker, M. P. (2014). The role of sleep in emotional brain function. *Annual Review of Clinical Psychology*, *10*, 679–708. http://dx.doi.org/10.1146/annurev-clinpsy-032813-153716.

Gujar, N., Yoo, S. S., Hu, P., & Walker, M. P. (2011). Sleep deprivation amplifies reactivity of brain reward networks, biasing the appraisal of positive emotional experiences. *Journal of Neuroscience*, *31*(12), 4466–4474. http://dx.doi.org/10.1523/JNEUROSCI.3220-10.2011.

Haack, M., & Mullington, J. M. (2005). Sustained sleep restriction reduces emotional and physical well-being. *Pain*, *119*(1–3), 56–64. http://dx.doi.org/10.1016/j.pain.2005.09.011.

Hamilton, M. (1989). Frequency of symptoms in melancholia (depressive illness). *British Journal of Psychiatry*, *154*, 201–206.

Jansson-Frojmark, M., & Lindblom, K. (2008). A bidirectional relationship between anxiety and depression, and insomnia? A prospective study in the general population. *Journal of Psychosomatic Research*, *64*(4), 443–449. http://dx.doi.org/10.1016/j.jpsychores.2007.10.016.

Jindal, R. D., Thase, M. E., Fasiczka, A. L., Friedman, E. S., Buysse, D. J., Frank, E., et al. (2002). Electroencephalographic sleep profiles in single-episode and recurrent unipolar forms of major depression: II. Comparison during remission. *Biological Psychiatry*, *51*(3), 230–236.

Joormann, J. (2004). Attentional bias in dysphoria: The role of inhibitory processes. *Cognition & Emotion*, *18*(1), 125–147. http://dx.doi.org/10.1080/02699930244000480.

Kaplan, K.A., Eidelman, P., Soehner, A.M., Gruber, J., Talbot, L.S., Gershon, A., & Harvey, A.G (2014). Hypersomnia in bipolar disorder: Clarifying a diagnostic dilemma. *Paper presented at The 28th Anniversary Meeting of the Associated Professional Sleep Societies*, Minneapolis, MN.

Krystal, A. D., Thakur, M., & Roth, T. (2008). Sleep disturbance in psychiatric disorders: Effects on function and quality of life in mood disorders, alcoholism, and schizophrenia. *Annals of Clinical Psychiatry*, *20*(1), 39–46. http://dx.doi.org/10.1080/10401230701844661.

Kupfer, D. J., Ulrich, R. F., Coble, P. A., Jarrett, D. B., Grochocinski, V., Doman, J., et al. (1984). Application of automated REM and slow wave sleep analysis: I. Normal and depressed subjects. *Psychiatry Research*, *13*(4), 325–334.

Manber, R., Bernert, R. A., Suh, S., Nowakowski, S., Siebern, A. T., & Ong, J. C. (2011). CBT for insomnia in patients with high and low depressive symptom severity: Adherence and clinical outcomes. *Journal of Clinical Sleep Medicine*, *7*(6), 645–652. http://dx.doi.org/10.5664/jcsm.1472.

Manber, R., Rush, A. J., Thase, M. E., Amow, B., Klein, D., Trivedi, M. H., et al. (2003). The effects of psychotherapy, nefazodone, and their combination on subjective assessment of disturbed sleep in chronic depression. *Sleep*, *26*(2), 130–136.

McCall, W. V., & Black, C. G. (2013). The link between suicide and insomnia: Theoretical mechanisms. *Current Psychiatry Reports*, *15*(9), 389. http://dx.doi.org/10.1007/s11920-013-0389-9.

McCrae, C. S., McNamara, J. P., Rowe, M. A., Dzierzewski, J. M., Dirk, J., Marsiske, M., et al. (2008). Sleep and affect in older adults: Using multilevel modeling to examine daily associations. *Journal of Sleep Research*, 17(1), 42–53. http://dx.doi.org/10.1111/j.1365-2869.2008.00621.x.

Nierenberg, A. A., Keefe, B. R., Leslie, V. C., Alpert, J. E., Pava, J. A., Worthington, J. J., 3rd., et al. (1999). Residual symptoms in depressed patients who respond acutely to fluoxetine. *Journal of Clinical Psychiatry*, 60(4), 221–225.

Nofzinger, E. A., Schwartz, R. M., Reynolds, C. F., 3rd., Thase, M. E., Jennings, J. R., Frank, E., et al. (1994). Affect intensity and phasic REM sleep in depressed men before and after treatment with cognitive-behavioral therapy. *Journal of Consulting and Clinical Psychology*, 62(1), 83–91.

Ohayon, M. M. (2012). Determining the level of sleepiness in the American population and its correlates. *Journal of Psychiatric Research*, 46(4), 422–427.

Ohayon, M. M., & Roth, T. (2003). Place of chronic insomnia in the course of depressive and anxiety disorders. *Journal of Psychiatric Research*, 37(1), 9–15.

Ong, J. C., Gress, J. L., San Pedro-Salcedo, M. G., & Manber, R. (2009). Frequency and predictors of obstructive sleep apnea among individuals with major depressive disorder and insomnia. *Journal of Psychosomatic Research*, 67(2), 135–141. http://dx.doi.org/10.1016/j.jpsychores.2009.03.011.

Palagini, L., Baglioni, C., Ciapparelli, A., Gemignani, A., & Riemann, D. (2013). REM sleep dysregulation in depression: State of the art. *Sleep Medicine Reviews*, 17(5), 377–390. http://dx.doi.org/10.1016/j.smrv.2012.11.001.

Paykel, E. S., Ramana, R., Cooper, Z., Hayhurst, H., Kerr, J., & Barocka, A. (1995). Residual symptoms after partial remission: An important outcome in depression. *Psychological Medicine*, 25(6), 1171–1180.

Perlis, M. L., Giles, D. E., Buysse, D. J., Thase, M. E., Tu, X., & Kupfer, D. J. (1997). Which depressive symptoms are related to which sleep electroencephalographic variables? *Biological Psychiatry*, 42(10), 904–913. http://dx.doi.org/10.1016/S0006-3223(96)00439-8.

Perlis, M. L., Giles, D. E., Buysse, D. J., Tu, X., & Kupfer, D. J. (1997). Self-reported sleep disturbance as a prodromal symptom in recurrent depression. *Journal of Affective Disorders*, 42(2–3), 209–212.

Reynolds, C. F., Frank, E., Houck, P. R., Mazumdar, S., Dew, M. A., Cornes, C., et al. (1997). Which elderly patients with remitted depression remain well with continued interpersonal psychotherapy after discontinuation of antidepressant medication? *American Journal of Psychiatry*, 154(7), 958–962.

Reynolds, C. F., Kupfer, D. J., Thase, M. E., Frank, E., Jarrett, D. B., Coble, P. A., et al. (1990). Sleep, gender, and depression: An analysis of gender effects on the electroencephalographic sleep of 302 depressed outpatients. *Biological Psychiatry*, 28(8), 673–684.

Riemann, D., Berger, M., & Voderholzer, U. (2001). Sleep and depression—Results from psychobiological studies: An overview. *Biological Psychology*, 57(1–3), 67–103.

Ritchey, M., Dolcos, F., Eddington, K. M., Strauman, T. J., & Cabeza, R. (2011). Neural correlates of emotional processing in depression: Changes with cognitive behavioral therapy and predictors of treatment response. *Journal of Psychiatric Research*, 45(5), 577–587. http://dx.doi.org/10.1016/j.jpsychires.2010.09.007.

Roberts, R., Roberts, C., & Chen, I. (2001). Functioning of adolescents with symptoms of disturbed sleep. *Journal of Youth & Adolescence*, 30, 1–18.

Roman, V., Walstra, I., Luiten, P. G., & Meerlo, P. (2005). Too little sleep gradually desensitizes the serotonin 1A receptor system. *Sleep*, 28(12), 1505–1510.

Rottenberg, J., Chambers, A. S., Allen, J. J., & Manber, R. (2007). Cardiac vagal control in the severity and course of depression: The importance of symptomatic heterogeneity. *Journal of Affective Disorders*, 103(1–3), 173–179. http://dx.doi.org/10.1016/j.jad.2007.01.028.

Rush, A. J., Erman, M. K., Giles, D. E., Schlesser, M. A., Carpenter, G., Vasavada, N., et al. (1986). Polysomnographic findings in recently drug-free and clinically remitted depressed patients. *Archives of General Psychiatry*, 43(9), 878–884.

Sewitch, D. E. (1987). Slow wave sleep deficiency insomnia: A problem in thermo-downregulation at sleep onset. *Psychophysiology, 24*(2), 200–215.

Simons, A. D., Murphy, G. E., Levine, J. L., & Wetzel, R. D. (1986). Cognitive therapy and pharmacotherapy for depression Sustained improvement over one year. *Archives of General Psychiatry, 43*(1), 43–48.

Skapinakis, P., Rai, D., Anagnostopoulos, F., Harrison, S., Araya, R., & Lewis, G. (2013). Sleep disturbances and depressive symptoms: An investigation of their longitudinal association in a representative sample of the UK general population. *Psychological Medicine, 43*(2), 329–339. http://dx.doi.org/10.1017/S0033291712001055.

Talbot, L. S., McGlinchey, E. L., Kaplan, K. A., Dahl, R. E., & Harvey, A. G. (2010). Sleep deprivation in adolescents and adults: Changes in affect. *Emotion, 10*(6), 831–841. http://dx.doi.org/10.1037/a0020138.

Thase, M. E., Buysse, D. J., Frank, E., Cherry, C. R., Cornes, C. L., Mallinger, A. G., et al. (1997). Which depressed patients will respond to interpersonal psychotherapy? The role of abnormal EEG sleep profiles. *American Journal of Psychiatry, 154*(4), 502–509.

Thase, M. E., Kupfer, D. J., Buysse, D. J., Frank, E., Simons, A. D., McEachran, A. B., et al. (1995). Electroencephalographic sleep profiles in single-episode and recurrent unipolar forms of major depression: I. Comparison during acute depressive states. *Biological Psychiatry, 38*(8), 506–515. http://dx.doi.org/10.1016/0006-3223(95)92242-A.

Thase, M. E., Rush, A. J., Manber, R., Kornstein, S. G., Klein, D. N., Markowitz, J. C., et al. (2002). Differential effects of nefazodone and cognitive behavioral analysis system of psychotherapy on insomnia associated with chronic forms of major depression. *Journal of Clinical Psychiatry, 63*(6), 493–500.

Van den Hoofdakker, R. H. (1997). Total sleep deprivation: Clinical and theoretical aspects. In A. Honig & H. van Praag (Eds.), *Depression: Neurobiological, psychopathological and therapeutic advances* (pp. 563–589). Chichester, England: John Wiley & Sons Ltd.

Van Londen, L., Molenaar, R. P., Goekoop, J. G., Zwinderman, A. H., & Rooijmans, H. G. (1998). Three- to 5-year prospective follow-up of outcome in major depression. *Psychological Medicine, 28*(3), 731–735.

Vogel, G. W., Vogel, F., McAbee, R. S., & Thurmond, A. J. (1980). Improvement of depression by REM sleep deprivation. New findings and a theory. *Archives of General Psychiatry, 37*(3), 247–253.

Walker, M. P., & van der Helm, E. (2009). Overnight therapy? The role of sleep in emotional brain processing. *Psychological Bulletin, 135*(5), 731–748. http://dx.doi.org/10.1037/a0016570.

Watson, D., & Tellegen, A. (1985). Toward a consensual structure of mood. *Psychological Bulletin, 98*(2), 219–235.

Wehr, T. A. (1990). Manipulations of sleep and phototherapy: Nonpharmacological alternatives in the treatment of depression. *Clinical Neuropharmacology, 13*(Suppl. 1), S54–S65.

Wiegand, M., Berger, M., Zulley, J., Lauer, C., & von Zerssen, D. (1987). The influence of daytime naps on the therapeutic effect of sleep deprivation. *Biological Psychiatry, 22*(3), 389–392.

Wu, J. C., & Bunney, W. E. (1990). The biological basis of an antidepressant response to sleep deprivation and relapse: Review and hypothesis. *American Journal of Psychiatry, 147*(1), 14–21.

Zohar, D., Tzischinsky, O., Epstein, R., & Lavie, P. (2005). The effects of sleep loss on medical residents' emotional reactions to work events: A cognitive-energy model. *Sleep, 28*(1), 47–54.

CHAPTER 12

The Interrelations Between Sleep, Anger, and Loss of Aggression Control

Jeanine Kamphuis* and Marike Lancel*,†

*Department of Forensic Psychiatry, Mental Health Services, Assen, The Netherlands
†Centre for Sleep and Psychiatry Assen, Mental Health Services, Assen, The Netherlands

INTRODUCTION

John, a 32-year-old man working as a consultant, recently became a father for the first time. His newborn currently keeps him and his wife awake most nights, resulting in too short and disrupted sleep. At work, a colleague makes a joke during a coffee break. Although normally a cheerful person up for a joke, John immediately snaps that the colleague should quit. His colleague looks at him and asks, "John, did you wake up on the wrong side of the bed?"

Most people have experienced how a night of disturbed sleep can negatively affect emotional reactivity the next day. The saying "waking up on the wrong side of the bed" shows that, even in normal daily conversation, we connect grumpiness and short-temperedness to poor sleep quality during the preceding night(s). Some might also recognize the potential effect of aggressive daytime interactions on nocturnal sleep (lying awake at night, still feeling tense after an argument). Despite all this, the relationship between sleep difficulties, anger, and aggression has not received a lot of scientific attention.

This chapter explores the interrelations between sleep, anger, and aggression. First, we consider different ways to define and investigate sleep, anger, and aggression. After that, we discuss several studies showing significant correlations between sleep disturbances and anger, hostility, irritability, and aggressive behavior. Then, we examine the causal directions of these interrelations from both sides; we look at evidence supporting the hypothesis that sleep loss may be a risk factor for daytime anger and loss of aggression control, and we look at evidence in support of the view that aggressive interactions and hostile personality traits can contribute to poor sleep. Thereafter, the chapter presents hypotheses on neurobiological mechanisms that may underlie the relationship between sleep, anger, and aggression, and we discuss risk factors that may impact the extent to which poor sleep affects emotional reactivity, considering the aggression-reducing effects of treating sleep difficulties and promoting proper sleep quality. Finally, we explore directions for future research.

ASPECTS OF SLEEP, ANGER, AND AGGRESSION

Sleep problems, anger, and aggression are broadly defined constructs. Consequently, there are many different ways of assessing sleep difficulties, anger, and aggression. Sleep can be defined as a recurring and reversible state of altered consciousness, inhibited sensory activity, and inhibition of nearly all voluntary muscles (adapted from MeSH, National Library of Medicine—Medical Subject Heading, 2014). When sleep is investigated, many different aspects can be explored, including subjective sleep duration as experienced by an individual, objective sleep duration measured with actigraphy or polysomnography, sleep quality (How does one experience his or her sleep? Was it refreshing? Did he or she awaken frequently and how did this affect his or her feelings? Was this the first time the person lay awake, or is it a chronic problem?), presence of sleep disorders (e.g., sleep-disordered breathing syndrome, restless leg syndrome, insomnia), and sleep stages measured by polysomnography. In many cases, people suffering from poor sleep have abnormalities in several of these areas, but not necessarily. Researchers do not yet know which aspects are most relevant to the association between sleep and aggression. The multiple ways of measuring sleep and defining sleep problems are important when interpreting the studies reviewed in this chapter.

Different aspects of anger and aggression can be defined as well. Anger is an emotion that involves a strong uncomfortable and emotional response to a perceived provocation (Novaco, 1986). Depending on various factors, the response to anger may differ between complete suppression to overt aggressive behavior. Such factors include the context of the situation at hand, one's coping strategies, and impulsivity (generally defined as a lack of behavioral inhibition or action without forethought) (Novaco, 1986). Whereas anger constitutes the emotional response to a situation, irritability can be viewed as the state in which anger is more easily evoked. In other words, "being irritable" means excessively responding to stimuli

(Ding, 2005). Hostility can be viewed as a set of cognitive variables containing cynicism (believing that others are selfishly motivated), mistrust (an overgeneralization that others will be hurtful and intentionally provoking), and denigration (evaluating others as dishonest, ugly, mean, and nonsocial) (Miller, Smith, Turner, & Guijarro, 1996). Anger can be a consequence of prior hostile cognitions (Eckhardt, Norlander, & Deffenbacher, 2004), and irritability, hostility, and anger may all contribute to aggressive behavior. Importantly, aggressive behavior in the animal kingdom is actually an important part of social communication. It is displayed to acquire and defend a territory, to obtain food water and shelter, and to maintain social status (Benus, Bohus, Koolhaas, & van Oortmerssen, 1991). Escalated aggressive behavior, which is displayed out of context and without inhibitory control, is called violence (de Boer, Caramaschi, Natarajan, & Koolhaas, 2009). For humans, this pathological form of aggression poses a social problem. Two subtypes of aggressive behavior are generally defined in humans: instrumental aggression, which is highly controlled and purposeful; and reactive aggression, which is more impulsive, unpredictable, and situationally dependent (Anderson & Bushman, 2002). With regard to ways of measuring anger and aggression, there are several options: self-report questionnaires measure subjective hostility, irritability, or anger; observations from the environment (health care workers, family, researchers) can give information about the extent to which an individual expresses hostility, anger, and aggression; and experimental tasks with clear endpoints may provide clues about how much aggression one is willing to engage in when responding to presented situations (for example, in computer games). In animal models, aggressive behavior can be monitored directly, but the researcher must consider whether this aggressive behavior is context-inappropriate or part of normal social communication between animals. In discussing the relationships between sleep, anger, and aggression, we are particularly interested in factors that increase the risk of context-inappropriate aggressive behavior, such as irritability, hostility, and anger. Although violent behavior is also a relevant topic for exploration, it is rarely included in studies as a primary endpoint. This chapter reviews studies measuring different aspects of anger and aggression.

CORRELATIONAL STUDIES SHOWING THE RELATIONSHIP BETWEEN SLEEP AND ANGER

Nonaggressive Populations

Sleep and anger have been correlated with each other in multiple studies (reviewed in Kamphuis, Meerlo, Koolhaas, & Lancel, 2012). For example, Schubert (1977) found that, in healthy German individuals, the scores on a personality inventory correlated with polysomnographically measured sleep. Higher scores on lack of aggression control and low frustration tolerance were also significantly associated with more superficial sleep (polysomnographically more stage N1 and N2) and more wakefulness. A group of US college students (total $n=117$) participated in a study performed by Pilcher, Ginter, and Sadowsky (1997). By means of self-report questionnaires, the authors assessed the students' sleep quality and mood states. Poor sleep quality (measured with the Pittsburgh sleep quality index (PSQI); Buysse, Reynolds, Monk, Berman, & Kupfer, 1989) significantly correlated with increased feelings of anger, tension, depression, fatigue, and confusion. In studying a sample of 1025 US students, Lund, Reider, Whiting, and Prichard (2010) found comparable results between sleep quality scores and mood states. In the study by Pilcher, a measure of sleep duration, based on daily sleep logs, was also included. Surprisingly, sleep duration did not correlate to any of the mood states. Granö, Vahtera, Virtanen, Keltikangas-Jarvinen, and Kivimaki (2008) found a slightly different result. They investigated a large Finnish hospital employee population (4832 women and 601 men) and subjected each participant to a set of self-report questionnaires at two time points separated by a 2-year interval. At time point 1, participants filled out a hostility self-report questionnaire to investigate their proneness to anger, irritability, and argumentativeness. At time point 2, the hostility questionnaire was filled out again, along with a self-report questionnaire on sleep disturbances and sleep duration in the prior 4 weeks. Results indicated that the more sleep disturbances a participant reported, the higher they scored on the hostility scale at time points 1 and 2. For sleep duration, the result was not that straightforward. Only for those participants who had a greater than 0.5 standard deviation difference between the two hostility scores (2 years apart), sleep duration was significantly relevant. More specifically, shorter sleep was associated with increasing hostility level (time point 1 < time point 2), whereas longer sleep was associated with decreasing hostility (time point 1 > time point 2). Other studies found similar relationships between hostility and sleep problems (Brissette & Cohen, 2002; Taylor, Fireman, & Levin, 2013).

In most of the above-mentioned studies, self-report questionnaires were used to investigate sleep and anger. Conflicting results have been observed using objective measures of sleep, however. For example, Tsuchiyama, Terao, Wang, Hoaki, and Goto (2013) combined the PSQI with actigraphy to measure sleep in 61 healthy Japanese adults. Higher self-rated hostility correlated significantly with higher scores on the PSQI (indicating a worse sleep quality), but not with any of the actigraphy parameters (total sleep time, sleep onset latency, and wake after sleep onset). In healthy populations, the validity and reliability of these actigraphy parameters are quite high: polysomnography-actigraphy correlations are generally above 0.80 (Sadeh & Acebo, 2002). Actigraphy was also used on

Israeli medical residents in order to investigate the effect of sleep duration on emotional reactivity (Zohar, Tzischinsky, Epstein, & Lavie, 2005). In these residents, a significant association was found between the amount of sleep loss they had suffered due to their work shifts in the hospital and negative emotional effects of disruptive events. Higher sleep loss correlated with more negative emotions. To date, physiological measures of anger or hostility have not been used in these types of studies.

Studies of children have produced evidence suggesting that sleep is not only associated with the internal emotion of anger or hostile tendencies, but also with actual overt problematic, aggressive behavior. Chervin, Dillon, Archbold, and Ruzicka (2003) recruited children between 2- and 14-years-old from two general hospitals and instructed their parents to fill out a questionnaire on pediatric sleep disorders and conduct. For children who were at high risk for sleep-disordered breathing problems, parents reported significantly more conduct problems, bullying, and other aggressive behaviors. Other studies in children have found similar positive associations between sleep-disordered breathing, but also other sleep problems (e.g., night-time awakenings) and parent-rated aggressive behavior (Gregory, van der Ende, Willis, & Verhulst, 2008; O'Brien et al., 2011; Reid, Hong, & Wade, 2009; Velten-Schurian, Hautzinger, Poets, & Schlarb, 2010). In summary, multiple studies have shown significant correlations between self-reported sleep difficulties and feelings of anger, hostile tendencies, and problematic aggressive behavior in nonaggressive populations.

Aggressive Populations

Few scientists have examined how sleep relates to anger, hostility, and aggression in so-called aggressive populations. Sleep disturbances may be partially responsible for the high rates of aggression in such groups (Kamphuis et al., 2012). One example an aggressive population is incarcerated offenders. In an adolescent prison population, self-reported shorter sleep duration and sleep disturbances were significantly associated with increased self-rated hostility (Ireland & Culpin, 2006). Another aggressive population consists of forensic psychiatric patients. These patients committed a crime or are at risk of committing a crime under the influence of mental disorders. In such a group, aggressive traits are typically present (Rosner, 2003). In Turkish military patients with antisocial personality disorder, poor sleep quality (measured with the PSQI) was significantly associated with higher self-reported aggression (Semiz et al., 2008). A similar result was found in a study of 96 Dutch forensic psychiatric inpatients (Kamphuis, Karsten, de Weerd, & Lancel, 2013). Participants completed two questionnaires on sleep: the PSQI, which is a measure of subjective sleep quality; and the sleep diagnosis list (SDL; Douglass et al., 1994; Sweere et al., 1998), which is used to measure chronic insomnia.

In addition, participants filled out the aggression questionnaire (AQ; Buss & Perry, 1992; Meesters, Muris, Bosma, Schouten, & Beuving, 1996) and the Barratt impulsiveness scale-11 (BIS-11; Patton, Stanford, & Barratt, 1995). In the AQ and BIS-11, participants have to score to what extent statements on aggression and impulsivity apply to them. To give a few examples: "I get in to fights more than other people" (AQ), "When somebody hits me, I will hit back" (AQ), "I act without thinking" (BIS-11), and "I spent more money than I make" (BIS-11). When people scored higher on the PSQI and on the SDL insomnia scale, with both results thus indicating worse sleep, they also had significantly higher scores on the AQ and BIS-11 (Figure 12.1). This indicates that the more sleep difficulties these forensic psychiatric inpatients have, the more tensed, irritable, and aggressive they experience themselves as being. This effect was found to be independent of medication effects, type of personality disorder, and history of substance abuse. Aggression parameters were not only measured with self-report questionnaires, but also by means of registered aggressive incidents within the hospital and clinician-ratings. The authors found that poor sleep quality and chronic insomnia were both significantly and independently associated with involvement in aggressive incidents and clinician-rated hostility. Thus, the more sleep problems a forensic patient reported, the higher the chance that patient had caused an aggressive incident and that a clinician had experienced the patient as hostile.

To summarize, few studies have investigated the potential contribution of sleep problems to aggression in specific "aggressive populations." Results suggest that sleep difficulties in these groups correlate not only with self-rated aspects of aggression, but also with actual aggressive behavior.

WHAT IS THE CAUSAL DIRECTION?

Correlational evidence does not give insight into the causal direction of the relationship between sleep and anger. Poor sleep may increase anger and, thereby, the risk of aggressive acts. On the other hand, aggressive behavior or even the stress of angry feelings may affect nocturnal sleep quality and duration. There is scientific support for both directions, indicating the complexity of the interrelations between sleep and anger. In this section, we explore evidence indicating that sleep loss may cause greater irritability, anger, and aggressive behavior and that anger and aggression may disrupt sleep.

Is Sleep Loss a Potential Risk Factor for Anger and Aggressive Behavior?

To answer that question, we investigate the direct influence of sleep disruption on emotion and anger control. Many people have experienced how poor sleep can affect their ability to control negative emotional responses,

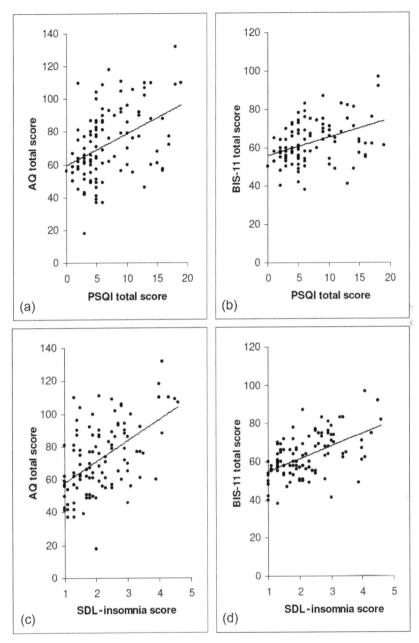

FIGURE 12.1 The relationship between sleep quality (PSQI) or chronic insomnia (SDL-insomnia) and self-rated aggression (AQ) and impulsivity (BIS-11) in 96 forensic psychiatric inpatients. Higher scores on the PSQI and SDL-insomnia indicate worse sleep quality and more severe chronic insomnia, respectively. Higher scores on the AQ and BIS-11 indicate higher levels of aggression and impulsivity, respectively. (a) The relation between sleep quality (PSQI) and self-reported aggression (AQ), $\beta = 0.40$, $p < 0.001$. (b) The relation between sleep quality (PSQI) and self-reported impulsivity (BIS-11), $\beta = 0.35$, $p < 0.001$. (c) The relation between chronic insomnia (SDL-insomnia) and self-reported aggression (AQ), $\beta = 0.54$, $p < 0.001$. (d) The relation between chronic insomnia (SDL-insomnia) and self-reported impulsivity (BIS-11), $\beta = 0.48$, $p < 0.001$. Correlations were corrected for several confounders, such as use of psychotropic medication, former drug abuse and personality disorders.

including angry and short-tempered remarks. Clinical observations within psychiatric hospitals suggest the same, even indicating that sleep problems contribute to complete loss of aggression control leading to escalated aggressive responses and acts. However, in order to positively answer the question of this paragraph, evidence with higher scientific quality is needed. Sleep deprivation studies may provide just such evidence.

Human Studies

Patrick and Gilbert conducted the first sleep deprivation study in humans in 1896. They kept three human beings awake for 90 hours and demonstrated a reduction in motor speed, decreased ability to memorize, diminished sensory acuity, and development of visual hallucinations (Patrick & Gilbert, 1896). No comments on anger or hostility were made. In 1964, John Ross performed an experiment, together with William Dement, in which they studied the psychiatric and neurological consequences of prolonged wakefulness in a more structured way: a 17-year-old high school boy, named Randy Gardner, volunteered to stay awake as long as possible in an attempt to break the world record for prolonged wakefulness. He remained awake for 11 days and was subjected to a series of tests every 6 hours. Already by the third day, the boy developed mood changes, and from the fourth day on, he became irritable and uncooperative. By the ninth day, he regularly did not finish his sentences, showed fragmented thoughts, and suffered from blurred vision (Ross, 1965). After that, many sleep deprivation studies have been performed.

Some studies specifically examined the effects of prolonged wakefulness on mood (Haack & Mullington, 2005; Kahn-Greene, Killgore, Kamimori, Balkin, & Killgore, 2007; Roth, Kramer, Lefton, & Thomas, 1976; Scott, McNaughton, & Polman, 2006). Results are not conclusive and seem to reflect an increased emotional reactivity rather than a specific effect on anger and aggression control. For example, after one night of sleep deprivation, healthy young men not only scored higher on the aggression subscale of a mood checklist, but also on the friendly subscale (Roth et al., 1976). In another study, the authors deprived 25 healthy military volunteers of sleep for 56 hours, and the participants completed an inventory on symptoms of psychopathology (Kahn-Greene et al., 2007). After sleep deprivation, some of the participants showed clinically relevant increased scores on different symptom domains. For instance, 25% of the participants had significantly more depressive symptoms, and 12.5% had increased paranoia scores, indicating that they felt easily mistreated and were more easily insulted. Only 8.3% of the sample had a relevant increase in antisocial symptomatology. It is not clear from the study whether or not there is an overlap between participants with increased scores on the different domains. Nevertheless, results suggest that total sleep deprivation can produce a broad spectrum of possible affective consequences. Increased anger, hostility, and aggression are possible outcomes, but certainly not the only ones.

The above-discussed studies have employed self-report mood and psychiatric symptom scales as outcome measures. Few studies aimed to investigate frustration tolerance and aggression control after sleep deprivation in a more objective manner. Two sleep-deprivation experiments using healthy university students employed a computer game against an unknown opponent to quantify aggression levels, more specifically by chosen noise levels blasted to the opponent following a win (Vohs, Glass, Todd Maddox, & Markman, 2011) and by counting the points stolen from the opponent (Cote, McCormick, Geniole, Renn, & MacAulay, 2013). The latter behavior had no advantage for the subject itself because the stolen points were not added up to the final score. Neither of these two studies showed an increase on aggression measures after one night of sleep deprivation. In fact, the study by Cote et al. (2013), which used the stolen points as a measure for reactive aggression, found decreased aggression levels and lowered testosterone levels in their male sleep-deprived participants. Other studies using objective outcome measures do suggest that people are less able to inhibit their aggressive responses when deprived of sleep, however. Kahn-Greene, Lipizzi, Conrad, Kamimori, and Killgore (2006) subjected human volunteers to 55 hours of sleep deprivation and measured frustration tolerance before and after the period of deprivation. Frustration tolerance was assessed with the Rosenzweig Picture Frustration Test. In this test the participant must write a response for a cartoon character that is confronted with various frustrating situations. For example, the character lends a book to a friend only to have it returned in a damaged state, or the character gets splashed with water from a puddle as a car passes through it (Figure 12.2). Sleep deprivation was associated with increased outward expression of aggressive responses, reduced willingness to take the blame, and greater tendency to assign blame to others. Authors explain this effect by connecting sleep loss to failing behavioral inhibition of aggressive impulses. Disrupted sleep may especially affect one's ability to inhibit behavior when faced with negative and frustrating circumstances. A study from Anderson and Platten (2011) showed that individuals were less able to inhibit their responses to negative emotional stimuli after 36 hours of sleep deprivation (Anderson & Platten, 2011). Interestingly, no effect of sleep deprivation was found for response inhibition to neutral and positive emotional stimuli.

Animal Studies

Scientists who employ gentle handling as a method to sleep deprive (e.g., Peñalva et al., 2003) animals commonly observe that sleep-deprived animals become less easy to handle and appear more irritable and aggressive. Licklider and Bunch (1946) first showed that sleep deprivation may lead to aggressive behavior in animals. He investigated the physiological effects of long-term selective rapid-eye-movement (REM) sleep deprivation, and he found quite striking effects on aggression. He subjected rats to

FIGURE 12.2 Examples from cartoons used in the Rosenzweig Picture Frustration Study (Rosenzweig, 1945; Rosenzweig, Clarke, Garfield, & Lehndorff, 1946).

enforced wakefulness by placing them together on top of a rotating drum surrounded by water. Rats died after 3-14 days, not directly due to sleep deprivation effects, but from fighting with each other. Licklider observed that his rats became so hyperreactive that even a slight physical contact lead to a vicious fight. This behavior was not always directed to the original aggressor, but occasionally several innocent rats would become involved in a fight, while the actual offender stood by and watched. In 1962, Webb performed a comparable study, in which he totally sleep deprived six rats by continuous forced locomotion and only paired them for 5 minutes each day in an observation cage (Webb, 1962). Rats started to exhibit aggressive behavior after 16 days of total sleep deprivation. Several studies confirmed these findings: most of these studies used selective REM-sleep deprivation by the flower pot method and demonstrated increased aggression in rats (de Paula & Hoshino, 2002; Hicks, Moore, Hayes, Phillips, & Hawkins, 1979; Marks & Wayner, 2005; Peder, Elomaa, & Johansson, 1986; Sloan, 1972) and mice (Benedetti, Fresi, Maccioni, & Smeraldi, 2008). The flower pot method, in which animals are placed on top of a flowerpot surrounded by water, has several disadvantages. As soon as animals enter REM-sleep, they fall into the water due to loss of muscle tone. This is presumed to be highly stressful and, thus, induces confounding factors to the experiment. Unfortunately, no studies address the impact of less stressful methods of sleep deprivation (e.g., forced locomotion and gentle stimulation) on aggressive behavior in animals. In order to investigate underlying mechanisms, exploring such methods seems to be an important direction for future studies.

Summary

Evidence from human and animal sleep deprivation studies supports, to some extent, the hypothesis that poor sleep may be a causal factor for irritability, anger, and increased aggression. Behavioral response inhibition to frustrating or negative emotional consequences may be particularly susceptible to sleep loss. Not all human studies investigating aggressive responses in frustrating situations have been able to show this effect of sleep deprivation, however. Future research should investigate whether this is due to the duration of sleep deprivation, selection of participants, or tests used to register aggressive responses.

Do Aggressive Tendencies and Actions Cause Poor Sleep?

The idea that aggression may cause sleep difficulties is not so hard to comprehend. An intense argument with your partner right before going to bed may make falling asleep more difficult. A fight induces a stress response in most people. Stress in the presleep period is thought to act on sleep quality through increased cognitive and somatic arousal (Krystal & Edinger, 2008). Yet, researchers have scarcely examined whether angry feelings and aggressive acts can directly disrupt sleep quality via this mechanism.

Human Studies

Brissette and Cohen (2002) asked 47 healthy adults in the United States to report the level of disagreement or conflict during their daily social interactions, in addition to positive and negative affect, for a period of 7 days. During this time, the subjects also reported on their sleep quality. Higher levels of conflict significantly correlated with sleep disturbances, but not with sleep duration, on the following night. However, on the days that participants reported more conflicts, they also reported more sleep problems on the previous night. This effect was partially mediated by negative affect during the day. These findings support a bidirectional relationship between sleep and anger. People with high hostility and anger traits (the disposition to experience angry feelings as a personality trait) have been found to exhibit altered and more pronounced stress responses (Suarez & Williams, 1989). This observation has led to the hypothesis that hostile or angry traits may predispose individuals to sleep problems. Correlational evidence supports this view (Granö et al., 2008; Taylor et al., 2013); however, as mentioned above, the evidence does not allow researchers to draw conclusions on causal direction. Adult males with high trait-anger, based on self-report questionnaire, showed cardiovascular hyperreactivity while being asleep compared to males with low trait-anger (Madigan, Dale, & Cross, 1997). In the study by Brissette and Cohen (2002), which was discussed earlier, the persons with high trait-hostility reported greater sleep disruption following conflict than their less hostile counterparts. Support for the hypothesis that hostile and aggressive traits may contribute to susceptibility for sleep problems comes from studies in aggressive populations. Among individuals with traits of antisocial personality disorder, characterized by irritability, hostility, aggressive behavior, impulsivity, lack of remorse, and deceptive acts, 58-80% suffer from poor sleep quality (Kamphuis et al., 2013; Semiz et al., 2008). This prevalence is extremely high compared to the rate in the general population. Although these data are based on self-report, polysomnographic findings support altered sleep in antisocial individuals. Lindberg et al. (2003) investigated the nighttime sleep of 19 male violent offenders with an antisocial personality disorder. They found more nighttime awakenings and lower sleep efficiency compared to healthy controls. These offenders also had a lower self-reported sleep quality. Results were not explained by alcoholism, sleep deprivation, or head injuries. With regard to sleep stages, individuals with antisocial personality disorder had more slow-wave sleep (SWS), compared to controls. Researchers do not yet know whether the high prevalence of disrupted sleep and differences in sleep architecture in individuals with hostile and aggressive traits is mediated by altered stress responses.

Animal Studies

Animal studies do show an effect of aggressive interactions on subsequent sleep. In these studies, animals (intruders) were placed in the territory of a bigger and older male conspecific (resident). In such a test, the resident generally attacked and defeated the intruder. Immediately following this defeat, losing rats (Meerlo, de Bruin, Strijkstra, & Daan, 2001; Meerlo, Pragt, & Daan, 1997) and mice (Lancel, Droste, Sommer, & Reul, 2003; Meerlo and Turek, 2001) showed increased amounts of non-REM sleep and/or elevated non-REM intensity, as reflected in elevated slow-wave activity (SWA) measured by electroencephalography. This effect was not seen after a sexual interaction, which is a nonaversive social interaction (Meerlo et al., 2001). Importantly, the experimental animals in the studies discussed in this paragraph were not the ones who came out as winners. They experienced a social defeat. To what extent they themselves displayed (defensive) aggressive behavior was not reported. Thus, whether we can conclude that confrontation with aggression or aggressive behavior itself affects sleep architecture is not clear from these results.

Summary

Involvement in conflicts can affect nighttime sleep. Hostile and antisocial individuals may represent a group in which these effects are particularly present. However, future studies are needed to elucidate the exact differences in sleep duration and quality between individuals with certain personality traits and how these individuals differ based on the effects daily stressors have on their nighttime sleep.

POSSIBLE NEUROBIOLOGICAL MECHANISMS UNDERLYING THE RELATIONSHIP BETWEEN SLEEP AND AGGRESSION

Several hypotheses address the effects of sleep loss on daytime aggression regulation, as well as the effects of aggressive interactions on sleep quality. Research into the interrelations between sleep disruptions, anger, frustration tolerance, and aggression control is relatively new. Therefore, the presented mechanisms should be seen as hypotheses generating directions for future studies, rather than proven pathophysiological pathways.

A Central Role for The Prefrontal Cortex (PFC)

The importance of the frontal lobes for anger and aggression control is well illustrated by the case of Phineas Gage. In 1848, the 25-year-old Phineas was working as a foreman at a railway construction site. Due a tragic chain of events, an iron bar tool, moving at high speed, penetrated his skull just under the left eye, emerged on top of the skull, and finally fell down on the ground some yards behind him. This accident completely damaged his middle left and frontal lobe (MacMillan, 2008). After this accident, his friends said he was "no longer Gage." His character had changed. Once a friendly, well-balanced, smart, and flexible man, he became stubborn, fitful, irreverent, irritable, and unable to postpone satisfaction. Gage's doctor stated, "The balance between his intellectual faculties and animal propensities seems to have been destroyed (Harlow, 1868)." Since the Gage incident, multiple human case studies, as well as lesion studies in animals, have shown that damage of the prefrontal area of the brain has several detrimental consequences for executive functions, including reduced behavioral inhibition and aggression control (reviewed in Giancola, 1995). Interestingly, imaging studies in highly aggressive individuals, such as criminals, psychopaths, and psychiatric patients with antisocial personality disorder, have shown anomalies in the prefrontal lobe, particularly in the orbitofrontal and dorsolateral PFC (Raine, Lencz, Bihrle, LaCasse, & Colletti, 2000; Yang & Raine, 2009). Some researchers have proposed that these deficits predispose an individual to antisocial behavior and, thus, aggressive behavior that no longer follows normal social rules. The PFC is considered to be the cortical region that controls such context-inappropriate aggression.

Sleep loss may negatively affect this prefrontal control function. After only 24 hours of sleep deprivation, the PFC shows reduced metabolic activity (Thomas et al., 2000). The behavioral, cognitive, and emotional consequences of sleep loss are comparable to what is seen after PFC lesions and anomalies (Dahl, 1996; Horne, 1993). For example, sleep-deprived individuals have problems initiating and maintaining goal-directed behavior (Dahl, 1996). They also show instability in emotional responses, reflecting increased emotional reactivity (Horne, 1993), diminished ability to understand the emotions of others (Kahn-Greene et al., 2006), and blunted recognition of facial emotional expressions (van der Helm, Gujar, & Walker, 2010). Yoo, Gujar, Hu, Jolesz, and Walker (2007) investigated the effect of 35 hours of sleep deprivation on the emotional responses of healthy volunteers, using functional magnetic resonance image. The participants were divided into a sleep-deprivation group and a sleep-control group. After the experimental night, all subjects performed an emotional stimulus-viewing task, involving the presentation of 100 pictures ranging from emotionally neutral to extremely aversive. The sleep-deprived group showed significantly greater amygdala activation in response to the negative pictures, compared to the control group. For the neutral images, no differences were found. Interestingly, the sleep-deprived subjects had a significantly weaker connectivity between the amygdala and the medial PFC and a greater connectivity from the amygdala to autonomic-activating centers in the brainstem, including the locus coeruleus. This can be interpreted as a failure of top-down, prefrontal control, associated with a hyperlimbic response by the amygdala to negative emotional stimuli. The authors hypothesized that sleep may be necessary in order to "reset" the correct brain reactivity to next-day emotional challenges (Yoo et al., 2007).

Based on this study, which directly supports sleep loss as a cause of reduced prefrontal inhibitory control to emotional responses, we can speculate about the causal mechanism of the relation between sleep disruption and aggression. In a similar manner, sleep loss may also contribute to failure of prefrontal inhibition of aggressive impulses in frustrating situations (Figure 12.3). This may lead to relatively uncontrolled, primitive, impulsive, aggressive reactions, and behavior.

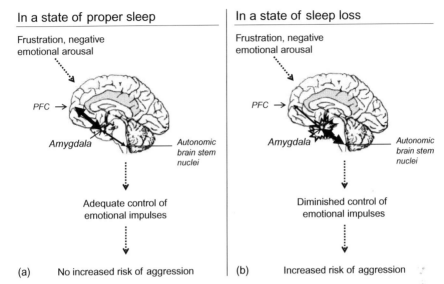

FIGURE 12.3 **Hypothesized causal mechanism between sleep loss and increased aggressiveness: loss of prefrontal cortical inhibition of emotional responses.** (a) In a state of proper sleep the prefrontal cortex (PFC) has a strong connectivity to the amygdala, thereby inhibiting its activity resulting in an adequate control of emotional impulses and context-appropriate, socially acceptable behaviour. (b) In a state of sleep loss the prefrontal cortex has a weaker connectivity to the amygdala, resulting in disinhibition and therefore greater activity of the amygdala in response to frustrating or emotional arousing situations. This contributes to an increased stimulation of the autonomic brain stimuli and thereby an increased risk of impulsive, emotional and aggressive responses.

Involvement of the Hypothalamic-Pituitary-Adrenal (HPA) Axis System

Researchers have suggested that differences in the functioning of the HPA-axis predispose individuals to both sleep disruptions and problematic aggressive behavior. Therefore, a disturbed stress system may mediate the relationship between sleep problems and irritable aggression. In fact, both may be outcomes of a dysfunctional HPA-axis. Due to the complexity of the HPA-axis system (e.g., differences in basal morning cortisol levels and cortisol levels in response to stressors), a discussion of the HPA-axis system's possible relationships to sleep and aggression is probably complex as well. First, we comment on the role of the HPA-axis system in aggressive behavior, and then we explore the relationship between the HPA-axis system and sleep.

Underarousal of the stress system may predispose individuals to criminal behavior (Raine, 1996). Many adults and children with conduct problems, antisocial individuals, and criminals have exhibited lower basal morning cortisol levels (McBurnett, Lahey, Rathouz, & Loeber, 2000; Oosterlaan, Geurts, Knol, & Sergeant, 2005). However, not all studies have been able to replicate these findings (Gerra, Zaimovic, Avanzini, & Chittolini, 1997; Schulz, Kirschbaum, Prüssner, & Hellhammer, 1998), and, with regard to stress responses, higher cortisol levels seem to be associated with higher aggression levels (Lopez-Duran, Olson, Hajal, Felt, & Vazquez, 2009). For example, children with extreme levels of conduct problems had higher cortisol responses to a psychological stress task than did other children with conduct problems (McBurnett et al., 2005). Persons with high trait-hostility have been found to have stronger cardiovascular reactivity to situations evoking suspiciousness and mistrust, compared to individuals with low trait-hostility (Weidner, Friend, Ficarrotto, & Mendell, 1989). A discussion of the interrelations between the HPA-axis system and aggression is complicated by the fact that the type of displayed aggression is often poorly defined. Hyperarousal and hyperreactivity of the HPA-axis system in response to a stressful situation seem to correlate better with reactive, defensive aggression than they do with proactive, instrumental aggression (Lopez-Duran et al., 2009; van Bokhoven et al., 2005). Reactive aggression reflects hypersensitivity to perceived threats (Vitaro, Brendgen, & Tremblay, 2002) and has been associated with hostile attributions (Schwartz et al., 1998). In contrast, proactive, instrumental aggression is goal-oriented, unprovoked, and often planned (Dodge & Coie, 1987; Vitaro et al., 2002). Antisocial or conduct-disordered individuals rarely display only one type of aggression; however, this mixed aggression profile may explain the contradicting results.

Sleep itself has a reciprocal relation with the HPA-axis. Sleep, particularly deep sleep (SWS), has an inhibitory effect on the HPA-axis, whereas HPA-axis activation produces arousal and wakefulness (Lancel, Müller-Preuss, Wigger, Landgraf, & Holsboer, 2002; Meerlo, Sgoifo, & Suchecki, 2008; Steiger, 2002; Vgontzas & Chrousos, 2002;). Insomnia is associated with hyperarousal of the HPA-axis system. Insomniacs have higher skin conductance, higher heart rate in response to stress, and higher levels of adrenocorticotropic hormone (ACTH) and cortisol during the evening and first part of the night, compared to noninsomniac controls (Vgontzas & Chrousos, 2002; Waters, Adams, Binks, & Varnado, 1993). An animal study with mice strains differing in their stress reactivity showed that mice with a lower responsiveness to stress had less REM sleep time and more SWA compared to mice with a high responsiveness to stress (Touma et al., 2009).

Given that hostile individuals presumably have increased and longer-lasting arousal after stress (Taylor et al., 2013), this HPA-dysfunction may mediate their heightened risk for poor sleep. This provides a model for how hostile tendencies can contribute to sleep problems, but it does not explain how sleep disturbances increase the risk for anger, hostility, and aggressive acts. Future studies are needed to explore the exact influence of the HPA-axis system on the relationship between sleep and aggression. For this research, researchers must clearly define the exact type of aggression that is being investigated.

Serotonin

The neurotransmitter serotonin or 5-hydroxytryptamine (5-HT) is involved in aggressive behavior and sleep. Therefore, serotoninergic activity may act as a neurobiological mechanism connecting sleep and anger or aggression. Low central serotonin may be a vulnerability factor for impulsive-aggressive behavior (Booij et al., 2010). In addition, increasing the serotonin availability in the synapses, by giving selective serotonin reuptake inhibitors, decreases anger, irritability, impulsivity, and aggression among violent offenders (Butler et al., 2010). Sleep deprivation and wakefulness are associated with higher serotonergic activity, compared to sleep (Peñalva et al., 2003; Portas, Bjorvatn, & Ursin, 2000). Interestingly, total sleep deprivation in rats increases 5-HT turnover in several brain areas, including the frontal cortex and hippocampus (Asikainen et al., 1997; Peñalva et al., 2003). Other rat studies indicated that chronic sleep restriction and disrupted sleep gradually lower central serotonin-receptor sensitivity (Novati et al., 2008; Roman, Walstra, Luiten, & Meerlo, 2005).

Although the findings for aggression and sleep seem compatible (sleep loss may induce increased use of 5-HT, potentially leading to a 5-HT deficiency, which is a risk factor for aggression), reality is not that simple. Growing insights into the role of serotonin in the development of aggression suggest that low 5-HT does not mediate current aggressive behavior, but it should be viewed as a risk factor for aggression that only acts contingent on other variables (Booij et al., 2010). An example of the complex interactions of serotonin with other variables comes from a study of healthy adult men (Kuepper et al., 2010). Trait-aggression was found to be significantly higher in men with decreased 5-HT availability in combination with high testosterone levels, but it was also higher in men with increased 5-HT availability and low testosterone. Speculatively, the potential causal relationship between sleep loss and aggressive behavior may be mediated by serotonergic activity only in the presence or absence of other variables. Thus, serotonin may be an important link connecting sleep problems to anger, irritability, and aggression, but no conclusive remarks can be made. It is an interesting direction for future studies.

GROUPS AT RISK

Not everybody responds to poor sleep with anger or short-temperedness. In even fewer people, this greater irritability results in loss of impulse control and escalated aggression. Thus, interindividual differences likely contribute to the variability in effects of sleep loss on aggression. Based on the hypothesis that sleep loss negatively influences the inhibitory function of the PFC, we can speculate that individuals with poor baseline PFC inhibitory functioning are more vulnerable to the effects of sleep disruptions on aggression control. Many psychiatric disorders have been associated with structural and functional impairments in the PFC compared to controls. For example, patients with antisocial personality disorder who committed violent offences have exhibited decreased glucose uptake in the PFC (Raine, Buchsbaum, & Lacasse, 1997). The size of the dorsolateral PFC is also inversely correlated with impulsivity in patients with borderline personality disorder (Sala et al., 2011). Patients with schizophrenia with a weaker connectivity between the PFC and the amygdala have higher self-reported aggression (Hoptman et al., 2010). Finally, men with a substance abuse disorder were shown to have a smaller grey matter volume in the orbitofrontal cortex, ventromedial PFC, and premotor cortex, compared to men without a substance abuse disorder (Schiffer et al., 2011). It is easy to imagine that, when these groups start to suffer from sleep problems, their PFC function deteriorates further. This increases the individual's risk for crossing the aggression control threshold, leading to escalated, context-inappropriate aggressive behavior. In a clinical sample of forensic psychiatric patients who often suffered from antisocial or borderline personality disorder, psychotic disorder, or substance abuse disorder, researchers found that sleep quality and insomnia were significant predictors for aggressive incidents within the hospital (Kamphuis, Dijk, Spreen, & Lancel, 2014).

Interindividual differences in serotonergic function may also predict consequences of sleep loss and/or the emergence of sleep problems after aggressive interactions. Sleep deprivation most certainly affects serotonergic activity in the brain, but the individual variations of this effect is unclear (Kamphuis et al., 2012). Differences in central serotonin function are associated with variability in personality, affect, and temperament. Such differences also likely influence the interaction between sleep and aggression. The current body of knowledge does not allow speculating on which constellation of serotonergic functioning increases the risk for negative emotional consequences due to loss of sleep, however. This inability to draw conclusions is especially due to the complex role serotonin plays in aggression. The same applies to interindividual differences in HPA-axis functioning. Taken together, based on the theory that sleep problems may weaken the top-down prefrontal inhibition of aggressive impulses, we hypothesize that individuals with a low impulse control are at increased risk for escalated aggression as a result of disturbed sleep. For this reason, psychiatric patients with various types of disorders, but especially antisocial personality disorders, and prisoners may represent groups at high risk.

CLINICAL IMPLICATIONS

From a clinical perspective, it is important to identify the individuals who are highly susceptible for the causal interrelations between sleep and anger, irritability, and aggression. As stated earlier, psychiatric patients and prison populations may be at high risk for the emotional consequences of sleep loss. If so, attending staff should pay close attention to their sleep, adequately treating sleep disorders and promoting proper sleep quality. A limited number of studies report on how the treatment of sleep disturbances affects anger- and aggression-regulation capacities. For example, children with enlarged tonsils may suffer from obstructive sleep apnea syndrome (OSAS), causing frequent nighttime awakenings and disturbed sleep. Some of these children with OSAS show problematic, aggressive behavior during the day (Mitchell & Kelly, 2005, 2006; Mulvaney et al., 2006), with the behaviors sometimes being so serious that psychiatric admission is necessary (Pakyurek, Gutkovich, & Weintraub, 2002). After treating them with adenotonsillectomy, sleep quality improves, and daytime parent- and teacher-rated aggressive and conduct behavior diminishes as well (Mitchell & Kelly, 2005, 2006; Mulvaney et al., 2006; Pakyurek et al., 2002). In sex offenders suffering from OSAS, comparable effects have been found: treatment of OSAS with continuous positive airway pressure improved sleep, but also significantly reduced scores on an AQ (Booth, Fedoroff, Curry, & Douglass, 2006). Finally, a study in adolescents recently completing a program for substance abuse disorders also supports the positive effect of sleep improvement on aggressive tendencies (Haynes et al., 2006). Adolescents from this group who suffered from sleep problems followed a 6-week behavioral sleep therapy. The adolescents reported significantly fewer aggressive ideations and actions with improved total sleep time. These studies support the hypothesis that treatment of sleep complaints can potentially reduce anger and aggression.

In this chapter, we have also explored the possibility that aggressive interactions and hostile traits can affect nighttime sleep. From this point of view, one can argue that, by reducing daytime aggression or feelings of hostility, the individual can experience improved sleep quality. No clinical studies or case reports directly support this conclusion, however. General relaxation and mindfulness techniques seem to improve sleep quality (Caldwell, Harrison, Adams, Quin, & Greeson, 2010; Winbush, Gross, & Kreitzer, 2007), which may be mediated by an effect on irritability (Caldwell et al., 2010), but additional research in this area is needed. To conclude, clinicians and their clients might benefit from focusing on the treatment of sleep difficulties in order to achieve better daytime anger and aggression control.

CONCLUSIONS AND FUTURE DIRECTIONS

The interrelations between sleep, anger, and aggression are complex, but evidence supports causal bidirectional effects. By decreasing behavioral inhibitory capacities, sleep loss may be a risk factor for increased anger and loss of aggression control in frustrating situations. Individuals with low baseline behavioral inhibition, as reflected by poor impulse and emotion control, may be particularly at risk for these detrimental consequences of sleep difficulties. Paying attention to sleep complaints, treating sleep disturbances, and promoting proper sleep quality seem to have aggression-reducing effects; however, studies in well-defined risk groups are needed. Researchers are still debating the extent to which aggressive interactions and aggressive or hostile traits affect nighttime sleep, and there is also a need for future studies in this direction. In addition, studies investigating the neurobiological mechanisms underlying the relationship between sleep and aggression are of particular interest.

References

Anderson, C. A., & Bushman, B. J. (2002). Human aggression. *Annual Review of Psychology, 53*(1), 27–51.
Anderson, C., & Platten, C. R. (2011). Sleep deprivation lowers inhibition and enhances impulsivity to negative stimuli. *Behavioural Brain Research, 217*, 463–466.
Asikainen, M., Toppila, J., Alanko, L., Ward, D. J., Stenberg, D., & Porkka-Heiskanen, T. (1997). Sleep deprivation increases brain serotonin turnover in the rat. *Neuroreport, 8*(7), 1577–1582.
Benedetti, F., Fresi, F., Maccioni, P., & Smeraldi, E. (2008). Behavioural sensitization to repeated sleep deprivation in a mice model of mania. *Behavioural Brain Research, 187*(2), 221–227.
Benus, R. F., Bohus, B., Koolhaas, J. M., & van Oortmerssen, G. A. (1991). Heritable variation for aggression as a reflection of individual coping strategies. *Experientia, 47*(10), 1008–1019.
Booij, L., Tremblay, R. E., Leyton, M., Séguin, J. R., Vitaro, F., Gravel, P., et al. (2010). Brain serotonin synthesis in adult males characterized by physical aggression during childhood: A 21-year longitudinal study. *PLoS One, 5*(6), e11255, 22.
Booth, B. D., Fedoroff, J. P., Curry, S. D., & Douglass, A. B. (2006). Sleep apnea as a possible factor contributing to aggression in sex offenders. *Journal of Forensic Science, 51*, 1178–1181.
Brissette, I., & Cohen, S. (2002). The contribution of individual differences in hostility to the associations between daily interpersonal conflict, affect and sleep. *Personality and Social Psychology Bulletin, 28*, 1265–1274.
Buss, A. H., & Perry, M. (1992). The aggression questionnaire. *Journal of Personality and Social Psychology, 63*, 452–459.
Butler, T., Schofield, P. W., Greenberg, D., Allnutt, S. H., Indig, D., Carr, V., et al. (2010). Reducing impulsivity in repeat violent offenders: An open label trial of a selective serotonin reuptake inhibitor. *The Australian and New Zealand Journal of Psychiatry, 44*(12), 1137–1143.
Buysse, D. J., Reynolds, C. F., 3rd., Monk, T. H., Berman, S. R., & Kupfer, D. J. (1989). The Pittsburgh sleep quality index: A new instrument for psychiatric practice and research. *Psychiatry Research, 28*, 193–213.
Caldwell, K., Harrison, M., Adams, M., Quin, R. H., & Greeson, J. (2010). Developing mindfulness in college students through movement based courses: Effects on self-regulatory self-efficacy, mood, stress, and sleep quality. *Journal of American College Health, 58*(5), 433–442.
Chervin, R. D., Dillon, J. E., Archbold, K. H., & Ruzicka, D. L. (2003). Conduct problems and symptoms of sleep disorders in children. *Journal of the American Academy of Child and Adolescent Psychiatry, 42*, 201–208.

Cote, K. A., McCormick, C. M., Geniole, S. N., Renn, R. P., & MacAulay, S. D. (2013). Sleep deprivation lowers reactive aggression and testosterone in men. *Biological Psychology, 92*(2), 249–256.

Dahl, R. E. (1996). The regulation of sleep and arousal: Development and psychopathology. *Development and Psychopathology, 8*, 3–27.

de Boer, S. F., Caramaschi, D., Natarajan, D., & Koolhaas, J. M. (2009). The vicious cycle towards violence: Focus on the negative feedback mechanisms of brain serotonin neurotransmission. *Frontiers in Behavioral Neuroscience, 3*, 52.

de Paula, H. M., & Hoshino, K. (2002). Correlation between the fighting rates of REM sleep-deprived rats and susceptibility to the 'wild running' of audiogenic seizures. *Brain Research, 926*(1-2), 80–85.

Ding, C. (2005). Applications of multidimensional scaling profile analysis in developmental research: An example using adolescent irritability patterns. *International Journal of Behavioral Development, 29*, 185–197.

Dodge, K. A., & Coie, J. D. (1987). Social-information-processing factors in reactive and proactive aggression in children's peer groups. *Journal of Personality and Social Psychology, 53*(6), 1146–1158.

Douglass, A. B., Bornstein, R., Nino-Murcia, G., Keenan, S., Miles, L., Zarcone, V. P., Jr., et al. (1994). The sleep disorders questionnaire. I: Creation and multivariate structure of SDQ. *Sleep, 17*, 160–167.

Eckhardt, C., Norlander, B., & Deffenbacher, J. (2004). The assessment of anger and hostility: A critical review. *Aggression and Violent Behavior, 9*(1), 17–43.

Gerra, G., Zaimovic, A., Avanzini, P., & Chittolini, B. (1997). Neurotransmitter-neuroendocrine responses to experimentally induced aggression in humans: Influence of personality variable. *Psychiatry Research, 66*(1), 33–43.

Giancola, P. (1995). Evidence for dorsolateral and orbital prefrontal cortical involvement in the expression of aggressive behavior. *Aggressive Behavior, 21*, 431–450.

Granö, N., Vahtera, J., Virtanen, M., Keltikangas-Jarvinen, L., & Kivimaki, M. (2008). Association of hostility with sleep duration and sleep disturbances in an employee population. *International Journal of Behavioral Medicine, 15*, 73–80.

Gregory, A. M., van der Ende, J., Willis, T. A., & Verhulst, F. C. (2008). Parent-reported sleep problems during development and self-reported anxiety/depression, attention problems and aggressive behavior in later life. *Archives of Pediatrics & Adolescent Medicine, 162*(4), 330–335.

Haack, M., & Mullington, J. M. (2005). Sustained sleep restriction reduces emotional and physical well-being. *Pain, 119*(1-3), 56–64.

Harlow, J. M. (1868). Recovery from the passage of an iron bar through the head. Originally published in *Publications of the Massachusetts Medical Society, 2*, 327–347.

Haynes, P. L., Bootzin, R. R., Smith, L., Cousins, J., Cameron, M., & Stevens, S. (2006). Sleep and aggression in substance-abusing adolescents: Results from an integrative behavioral sleep-treatment pilot program. *Sleep, 29*, 512–520.

Hicks, R. A., Moore, J. D., Hayes, C., Phillips, N., & Hawkins, J. (1979). REM sleep deprivation increases aggressiveness in male rats. *Physiology & Behavior, 22*(6), 1097–1100.

Hoptman, M. J., D'Angelo, D., Catalano, D., Mauro, C. J., Shehzad, Z. E., Clare Kelly, A. M., et al. (2010). Amygdalofrontal functional disconnectivity and aggression in schizophrenia. *Schizophrenia Bulletin, 36*(5), 1020–1028.

Horne, J. A. (1993). Human sleep, sleep loss and behaviour. Implications for the prefrontal cortex and psychiatric disorder. *The British Journal of Psychiatry: The Journal of Mental Science, 162*, 413–419.

Ireland, J. L., & Culpin, V. (2006). The relationship between sleeping problems and aggression, anger, and impulsivity in a population of juvenile and young offenders. *The Journal of Adolescent Health: Official Publication of the Society for Adolescent Medicine, 38*, 649–655.

Kahn-Greene, E. T., Killgore, D. B., Kamimori, G. H., Balkin, T. J., & Killgore, W. D. (2007). The effects of sleep deprivation on symptoms of psychopathology in healthy adults. *Sleep Medicine, 8*(3), 215–221.

Kahn-Greene, E. T., Lipizzi, E. L., Conrad, A. K., Kamimori, G. H., & Killgore, W. D. S. (2006). Sleep deprivation adversely affects interpersonal responses to frustration. *Personality and Individual Differences, 41*, 1433–1443.

Kamphuis, J., Dijk, D. J., Spreen, M., & Lancel, M. (2014). The relation between poor sleep, impulsivity and aggression in forensic psychiatric patients. *Physiology & Behavior, 123*, 168–173.

Kamphuis, J., Karsten, J., de Weerd, A., & Lancel, M. (2013). Sleep disturbances in a clinical forensic psychiatric population. *Sleep Medicine, 14*(11), 1164–1169.

Kamphuis, J., Meerlo, P., Koolhaas, J. M., & Lancel, M. (2012). Poor sleep as a potential causal factor in aggression and violence. *Sleep Medicine, 13*, 327–334.

Krystal, A. D., & Edinger, J. D. (2008). Measuring sleep quality. *Sleep Medicine, 9*(1), S10–S17.

Kuepper, Y., Alexander, N., Osinsky, R., Mueller, E., Schmitz, A., Netter, P., et al. (2010). Aggression-interactions of serotonin and testosterone in healthy men and women. *Behavioural Brain Research, 206*(1), 93–100.

Lancel, M., Droste, S. K., Sommer, S., & Reul, J.M.H.M. (2003). Influence of regular voluntary exercise on spontaneous and social stress-affected sleep in mice. *The European Journal of Neuroscience, 17*, 2171–2179.

Lancel, M., Müller-Preuss, P., Wigger, A., Landgraf, R., & Holsboer, F. (2002). The CRH1 receptor antagonist R121919 attenuates stress-elicited sleep disturbances in rats, particularly in those with high innate anxiety. *Journal of Psychiatric Research, 36*, 197–208.

Licklider, J. C. R., & Bunch, M. E. (1946). Effects of enforced wakefulness upon the growth and the maze learning performance of white rats. *Journal of Comparative Psychology, 39*, 339–350.

Lindberg, N., Tani, P., Appelberg, B., Naukkarinen, H., Stenberg, D., Rímon, R., et al. (2003). Sleep among habitually violent offenders with antisocial personality disorder. *Neuropsychobiology, 47*(198), 205.

Lopez-Duran, N. L., Olson, S. L., Hajal, N. J., Felt, B. T., & Vazquez, D. M. (2009). Hypothalamic pituitary adrenal axis functioning in reactive and proactive aggression in children. *Journal of Abnormal Child Psychology, 37*, 169–182.

Lund, H. G., Reider, B. D., Whiting, A. B., & Prichard, J. R. (2010). Sleep patterns and predictors of disturbed sleep in a large population of college students. *The Journal of Adolescent Health: Official Publication of the Society for Adolescent Medicine, 46*(2), 124–132.

MacMillan, M. (2008). Phineas Gage—Unravelling the myth. *The Psychologist, 21*(9), 828–831.

Madigan, M. F., Jr., Dale, J. A., & Cross, J. D. (1997). No respite during sleep: Heart-rate hyperreactivity to rapid eye movement sleep in angry men classified as type A. *Perceptual and Motor Skills, 85*, 1451–1454.

Marks, C. A., & Wayner, M. J. (2005). Effects of sleep disruption on rat dentate granule cell LTP in vivo. *Brain Research Bulletin, 66*(2), 114–119.

McBurnett, K., Lahey, B. B., Rathouz, P. J., & Loeber, R. (2000). Low salivary cortisol and persistent aggression in boys referred for disruptive behavior. *Archives of General Psychiatry, 57*(1), 38–43.

McBurnett, K., Raine, A., Stouthamer-Loeber, M., Loeber, R., Kumar, A. M., Kuman, M., et al. (2005). Mood and hormone responses to psychological challenge in adolescent males with conduct problems. *Biological Psychiatry, 57*(10), 1109–1116.

Meerlo, P., de Bruin, E. A., Strijkstra, A. M., & Daan, S. (2001). A social conflict increases EEG slow-wave activity during subsequent sleep. *Physiology & Behavior, 73*, 331–335.

Meerlo, P., Pragt, B., & Daan, S. (1997). Social stress induces high intensity sleep in rats. *Neuroscience Letters, 225*, 41–44.

Meerlo, P., Sgoifo, A., & Sucheki, D. (2008). Restricted and disrupted sleep: Effects on autonomic function, neuroendocrine stress systems and stress responsivity. *Sleep Medicine Reviews, 12*(3), 197–210.

Meerlo, P., & Turek, F. W. (2001). Effects of social stimuli on sleep in mice: Non-rapid-eye-movement (NREM) sleep is promoted by aggressive interaction but not by sexual interaction. *Brain Research, 907*(1-2), 84–92.

Meesters, C., Muris, P., Bosma, H., Schouten, E., & Beuving, S. (1996). Psychometric evaluation of the Dutch version of the aggression questionnaire. *Behaviour Research and Therapy, 34*, 839–843.

Miller, T. Q., Smith, T. W., Turner, C. W., & Guijarro, M. L. (1996). Meta-analytic review of research on hostility and physical health. *Psychological Bulletin, 119*(2), 322–348.

Mitchell, R. B., & Kelly, J. (2005). Child behavior after adenotonsillectomy for obstructive sleep apnea syndrome. *The Laryngoscope, 115*, 2051–2055.

Mitchell, R. B., & Kelly, J. (2006). Long-term changes in behavior after adenotonsillectomy for obstructive sleep apnea syndrome in children. *Otolaryngology-Head and Neck Surgery: Official Journal of American Academy of Otolaryngology-Head and Neck Surgery, 134*(3), 374–378.

Mulvaney, S. A., Goodwin, J. L., Morgan, W. J., Rosen, G. R., Quan, S. F., & Kaemingk, K. L. (2006). Behavior problems associated with sleep disordered breathing in school-aged children-the Tucson children's assessment of sleep apnea study. *Journal of Pediatric Psychology, 31*(3), 322–330.

National Library of Medicine—Medical Subject Heading (2014) MeSH, MeSH Descriptor Data, http://www.nlm.nih.gov/cgi/mesh/2014/MB_cgi.

Novaco, R. W. (1986). Anger as a clinical and social problem. In R. Blanchard, & C. Blanchard (Eds.), *Advances in the study of aggression* (Vol. II). New York: Academic Press.

Novati, A., Roman, V., Cetin, T., Hagewoud, R., Den Boer, J. A., Luiten, P. G. M., et al. (2008). Chronically restricted sleep leads to depression-like changes in neurotransmitter receptor sensitivity and neuroendocrine stress reactivity in rats. *Sleep, 31*, 1579–1585.

O'Brien, L. M., Lucas, N. H., Felt, B. T., Hoban, T. F., Ruzicka, D. L., Jordan, R., et al. (2011). Aggressive behavior, bullying, snoring, and sleepiness in schoolchildren. *Sleep Medicine, 12*, 652–658.

Oosterlaan, J., Geurts, H. M., Knol, D. L., & Sergeant, J. A. (2005). Low basal salivary cortisol is associated with teacher-reported symptoms of conduct disorder. *Psychiatry Research, 134*(1), 1–10.

Pakyurek, M., Gutkovich, Z., & Weintraub, S. (2002). Reduced aggression in two inpatient children with the treatment of their sleep disorder. *Journal of the American Academy of Child and Adolescent Psychiatry, 41*, 1025.

Patrick, G. T. W., & Gilbert, J. A. (1896). Studies from the psychological laboratory of the University of Iowa: On the effects of loss of sleep. *Psychological Review, 3*, 469–483.

Patton, J. H., Stanford, M. S., & Barratt, E. S. (1995). Factor structure of the Barratt impulsiveness scale. *Journal of Clinical Psychology, 51*, 768–774.

Peder, M., Elomaa, E., & Johansson, G. (1986). Increased aggression after rapid eye movement sleep deprivation in Wistar rats is not influenced by reduction of dimensions of enclosure. *Behavioral and Neural Biology, 45*(3), 287–291.

Peñalva, R. G., Lancel, M., Flachskamm, C., Reul, J.M.H.M., Holsboer, F., & Linthorst, A. C. E. (2003). Effect of sleep and sleep deprivation on serotonergic neurotransmission in the hippocampus: A combined *in vivo* microdialysis/EEG study in rats. *The European Journal of Neuroscience, 17*, 1896–1906.

Pilcher, J. J., Ginter, D. R., & Sadowsky, B. (1997). Sleep quality versus sleep quantity: Relationships between sleep and measures of health, well-being and sleepiness in college students. *Journal of Psychosomatic Research, 42*, 583–596.

Portas, C. M., Bjorvatn, B., & Ursin, R. (2000). Serotonin and the sleep/wake cycle: Special emphasis on microdialysis studies. *Progress in Neurobiology, 60*, 13–35.

Raine, A. (1996). Autonomic nervous system factors underlying disinhibited, antisocial, and violent behavior biosocial perspectives and treatment implications. *Annals of the New York Academy of Sciences, 794*, 46–59.

Raine, A., Buchsbaum, M., & Lacasse, L. (1997). Brain abnormalities in murderers indicated by positron emission tomography. *Biological Psychiatry, 42*(6), 495–508.

Raine, A., Lencz, T., Bihrle, S., LaCasse, L., & Colletti, P. (2000). Reduced prefrontal gray matter volume and reduced autonomic activity in antisocial personality disorder. *Archives of General Psychiatry, 57,* 119–127, discussion 128-129.

Reid, G. J., Hong, R. Y., & Wade, T. J. (2009). The relation between common sleep problems and emotional and behavioral problems among 2- and 3-year-olds in the context of known risk factors for psychopathology. *Journal of Sleep Research, 18*(1), 49–59.

Roman, V., Walstra, I., Luiten, P. G. M., & Meerlo, P. (2005). Too little sleep gradually desensitizes the 5-HT 1A receptor system in rats. *Sleep, 28,* 1505–1510.

Rosenzweig, S. (1945). The picture association method and its application in a study of reaction to frustration. *Journal of Personality, 14*(1), 3–23.

Rosenzweig, S., Clarke, H. J., Garfield, M. S., & Lehndorff, A. (1946). Scoring samples for the Rosenzweig Picture-Frustration Study. *Journal of Psychology: Interdisciplinary and Applied, 21,* 45–72.

Rosner, R. (2003). *Principles and practice of forensic psychiatry* (2 ed.). Boca Raton: CRC Press.

Ross, J. J. (1965). Neurological findings after prolonged sleep deprivation. *Archives of Neurology, 12,* 399–403.

Roth, A. T., Kramer, M., Lefton, W. L., Thomas. (1976). The effects of sleep deprivation on mood. *Psychiatric Journal of the University of Ottawa, 1,* 136–139.

Sadeh, A., & Acebo, C. (2002). The role of actigraphy in sleep medicine. *Sleep Medicine Reviews, 6*(2), 113–124.

Sala, M., Caverzasi, E., Lazzaretti, M., Morandotti, N., De Vidovich, G., Marraffini, E., et al. (2011). Dorsolateral prefrontal cortex and hippocampus sustain impulsivity and aggressiveness in borderline personality disorder. *Journal of Affective Disorders, 131*(1-3), 417–421.

Schiffer, B., Müller, B. W., Scherbaum, N., Hodgins, S., Forsting, M., Wiltfang, J., et al. (2011). Disentangling structural brain alterations associated with violent behavior from those associated with substance use disorders. *Archives of General Psychiatry, 68*(10), 1039–1049.

Schubert, F. (1977). Personality traits and polygraphic sleep parameters: Correlations between personality factors and polygraphically recorded sleep in healthy subjects. *Waking & Sleeping, 1,* 165–170.

Schulz, P., Kirschbaum, C., Prüssner, J., & Hellhammer, D. (1998). Increased free cortisol secretion after awakening in chronically stressed individuals due to work overload. *Stress Medicine, 14*(2), 91–97.

Schwartz, D., Dodge, K. A., Coie, J. D., Hubbard, J. A., Cillessen, A. H., Lamerise, E. A., et al. (1998). Social-cognitive and behavioral correlates of aggression and victimization in boys' play groups. *Journal of Abnormal Child Psychology, 26*(6), 431–440.

Scott, J. P., McNaughton, L. R., & Polman, R. C. (2006). Effects of sleep deprivation and exercise on cognitive, motor performance and mood. *Physiology & Behavior, 87*(2), 396–408.

Semiz, U. B., Algul, A., Basoglu, C., Ates, M. A., Ebrinc, S., Cetin, M., et al. (2008). The relationship between subjective sleep quality and aggression in male subjects with antisocial personality disorder. *Türk Psikiyatri Dergisi, 19*(4), 373–381.

Sloan, M. A. (1972). The effects of deprivation of rapid eye movement (REM) sleep on maze learning and aggression in the albino rat. *Journal of Psychiatric Research, 9*(2), 101–111.

Steiger, A. (2002). Sleep and the hypothalamo-pituitary-adrenocortical system. *Sleep Medicine Reviews, 6,* 125–138.

Suarez, E. C., & Williams, R. B. (1989). Situational determinants of cardiovascular and emotional reactivity in high and low hostile men. *Psychosomatic Medicine, 51*(4), 404–418.

Sweere, Y., Kerkhof, G. A., De Weerd, A. W., Kamphuisen, H. A., Kemp, B., & Schimsheimer, R. J. (1998). The validity of the Dutch sleep disorders questionnaire (SDQ). *Journal of Psychosomatic Research, 45,* 549–555.

Taylor, N. D., Fireman, G. D., & Levin, R. (2013). Trait hostility, perceived stress, and sleep quality in a sample of normal sleepers. *Sleep Disorders, 2013,* 1–8.

Thomas, M., Sing, H., Belenky, G., Holcomb, H., Mayberg, H., Dannals, R., et al. (2000). Neural basis of alertness and cognitive performance impairments during sleepiness. I. Effects of 24 h of sleep deprivation on waking human regional brain activity. *Journal of Sleep Research, 9*(4), 335–352.

Touma, C., Fenzl, T., Ruschel, J., Palme, R., Holsboer, F., Kimura, M., et al. (2009). Rhythmicity in mice selected for extremes in stress reactivity: Behavioural, endocrine and sleep changes resembling endophenotypes of major depression. *PLoS One, 4*(1), e4325.

Tsuchiyama, K., Terao, T., Wang, Y., Hoaki, N., & Goto, S. (2013). Relationship between hostility and subjective sleep quality. *Psychiatry Research, 209*(3), 545–548.

van Bokhoven, I., van Goozen, S. M., van Engeland, H., Schaal, B., Arseneault, L., Séguin, J. R., et al. (2005). Salivary cortisol and aggression in a population-based longitudinal study of adolescent males. *Journal of Neural Transmission, 112*(8), 1083–1096.

van der Helm, E., Gujar, N., & Walker, M. P. (2010). Sleep deprivation impairs the accurate recognition of human emotions. *Sleep, 33*(3), 335–342.

Velten-Schurian, K., Hautzinger, M., Poets, C. F., & Schlarb, A. A. (2010). Association between sleep patterns and daytime functioning in children with insomnia: The contribution of parent-reported frequency of night waking and wake time after sleep onset. *Sleep Medicine, 11*, 281–288.

Vgontzas, A. N., & Chrousos, G. P. (2002). Sleep, the hypothalamic-pituitary-adrenal axis, and cytokines: Multiple interactions and disturbances in sleep disorders. *Endocrinology and Metabolism Clinics of North America, 31*(1), 15–36.

Vitaro, F., Brendgen, M., & Tremblay, R. E. (2002). Reactively and proactively aggressive children: Antecedent and subsequent characteristics. *Journal of Child Psychology and Psychiatry, and Allied Disciplines, 43*(4), 495–505.

Vohs, K. D., Glass, B. D., Todd Maddox, W. T., & Markman, A. B. (2011). Ego depletion is not just fatigue: Evidence from a total sleep deprivation experiment. *Social Psychological and Personality Science, 2*(2), 166–173.

Waters, W. F., Adams, S. G., Jr., Binks, P., & Varnado, P. (1993). Attention, stress and negative emotion in persistent sleep-onset and sleep-maintenance insomnia. *Sleep, 16*(2), 128–136.

Webb, W. B. (1962). Some effects of prolonged sleep deprivation on the hooded rat. *Journal of Comparative and Physiological Psychology, 55*, 791–793.

Weidner, G., Friend, R., Ficarrotto, T. J., & Mendell, N. R. (1989). Hostility and cardiovascular reactivity to stress in women and men. *Psychosomatic Medicine, 51*, 36–45.

Winbush, N. Y., Gross, C. R., & Kreitzer, M. J. (2007). The effects of mindfulness-based stress reduction on sleep disturbance: A systematic review. *Explore (NY), 3*(6), 585–591.

Yang, Y., & Raine, A. (2009). Prefrontal structural and functional brain imaging findings in antisocial, violent, and psychopathic individuals: A meta-analysis. *Psychiatry Research: Neuroimaging, 174*(2), 81–88.

Yoo, S. S., Gujar, N., Hu, P., Jolesz, F. A., & Walker, M. P. (2007). The human emotional brain without sleep—A prefrontal amygdala disconnect. *Current Biology, 17*, R877–R878.

Zohar, D., Tzischinsky, O., Epstein, R., & Lavie, P. (2005). The effects of sleep loss on medical residents' emotional reactions to work events: A cognitive-energy model. *Sleep, 28*(1), 47–54.

SECTION 2

SLEEP AND POSITIVE AFFECT

CHAPTER 13

Positive Affect as Resilience and Vulnerability in Sleep

Anthony D. Ong[*], Emily D. Bastarache[*], and Andrew Steptoe[†]

[*]Department of Human Development, Cornell University, Ithaca, New York, USA
[†]Department of Epidemiology and Public Health, University College, London, UK

Extensive research has documented the importance of sleep for promoting restorative processes and protecting against impairments in a range of neurobehavioral functions, including emotion regulation, immune control, and learning or memory consolidation (Baglioni, Spiegelhalder, Lombardo, & Riemann, 2010; Irwin, Carrillo, & Olmstead, 2010; Payne & Kensinger, 2010; Xie et al., 2013). Moreover, deficits in fundamental aspects of sleep, including reduced sleep efficiency and increased sleep disturbances, can have profound health effects that contribute to increased risks for adult morbidity and all-cause mortality (Dew et al., 2003; Kripke, Garfinkel, Wingard, Klauber, & Marler, 2002; Li et al., 2014; Reid et al., 2006). Given the significant role of sleep in psychiatric and health morbidities, it is important to advance our understanding of the key factors that contribute to individual differences in sleep quality.

Growing evidence suggests that positive affect (PA) is an important factor affecting individuals' overall sleep. Adults who report high levels of PA have been shown to exhibit improved sleep patterns (Fosse, Stickgold, & Hobson, 2002; Steptoe, O'Donnell, Marmot, & Wardle, 2008). In contrast, those who experience difficulties in regulating PA report greater sleep disturbances (Ong et al., 2013; Talbot, Hairston, Eidelman, Gruber, & Harvey, 2009). In this chapter, we focus on what is known about the relationship between PA and sleep, emphasizing the major approaches, empirical findings, and methodological gaps that currently exist in the literature. We review recent evidence that indicates the effects of PA on sleep outcomes are bivalent in character, exerting both risk-protective and risk-augmenting effects. We then summarize evidence of potential pathways underlying the association between PA and sleep. We conclude with a discussion of priority recommendations for future research.

POSITIVE AFFECT AS PROMOTIVE INFLUENCE AND RESILIENCE

Although sleep loss adversely affects health and emotional well-being, PA appears to be associated with good or restorative sleep. However, perhaps due to the historic press to understand the etiological significance of negative emotions (Fredrickson, 2013; Kahn, Sheppes, & Sadeh, 2013), researchers have only recently begun to examine the extent to which positive affective states contribute to adaptive sleep patterns. A small number of studies have investigated the role of PA as a *promotive factor*, increasing the likelihood of optimal sleep, as well as a *protective factor*, decreasing vulnerability in the face of risk (Sameroff, Gutman, & Peck, 2003). We review the evidence of promotive and protective effects below.

PROMOTIVE EFFECTS

We define PA as pleasant feeling states such as joy, contentment, love, and happiness that motivate adaptive behavior and engagement with the environment (Pressman & Cohen, 2005). As documented in previous selective reviews (Baglioni et al., 2010; Boehm & Kubzansky, 2012; Kahn et al., 2013), PA directly predicts sleep outcomes. In the following section, we summarize findings from previous investigations, as well as more recent work, chronicling the direct contribution of PA to sleep. Findings are organized according to studies that have conceptualized PA as a trait (i.e., stable, enduring disposition), a state (i.e., transient feelings), or a discrete emotional category (i.e., experiences that are distinct in their phenomenology and activation level). Of the 21 studies identified, 8 conceptualized PA as a trait, 7 assessed PA on a daily basis, and 6 measured discrete positive emotions (i.e., vigor, gratitude, and love).

Trait Positive Affect

Eight previous studies have investigated the unique effects of trait PA on sleep (see Table 13.1). Most of these studies assessed trait PA using

TABLE 13.1 Effects of Trait Positive Affect on Sleep

PA Measure	Sleep Measure	NA Adjustment	Authors/Year	Findings
$PANAS_{(10)}$	$PSQI_{(19)}$, $SSI_{(12)}$	$PANAS_{(10)}$	Gray and Watson (2002)	PA uniquely predicted better sleep quality (i.e., lower PSQI and SSI scores)
$PA_{(17)}$	Sleep Quality$_{(30)}$	$NA_{(20)}$	Fortunato and Harsh (2006)	PA had a unique direct effect on sleep quality and moderated the effect of interpersonal conflict
$DISS_{(5)}$[a]	$PghSD_{(2)}$		Buysse et al. (2007)	PA associated with better sleep efficiency in insomnia patients
$PA_{(3)}$[a]	$JSPS_{(4)}$	$GHQ_{(30)}$	Steptoe et al. (2008)[a]	PA uniquely predicted good sleep and attenuated the effects of psychological risk factors
Happiness$_{(4)}$	$JSPS_{(3)}$	$PANAS_{(10)}$	Jackowska, Dockray, Hendrickx, and Steptoe (2011)	PA uniquely predicted with better sleep efficiency
$PANAS_{(10)}$	$PSQI_{(19)}$	$BDI-II_{(20)}$, $STAI_{(20)}$ $STAXI_{(10)}$	Stewart, Rand, Hawkins, and Stines (2011)	PA not uniquely associated with sleep quality
$PA_{(6)}$	Sleep Quality$_{(2)}$, Actigraphy	$NA_{(6)}$	Ong et al. (2013)	PA uniquely predicted morning rest and overall sleep
$CES-D_{(4)}$	$PSQI_{(19)}$	$CES-D_{(16)}$	Fredman, Gordon, Heeren, and Stuver (2014)	PA uniquely predicted better sleep quality (i.e., lower PSQI scores) among caregivers

Subscripted number in parenthesis = number of items; PA, positive affect; NA, negative affect; BDI, Beck depression inventory; CES-D, Center for Epidemiological Studies-Depression scale (PA subscale); DISS, daytime symptoms in insomnia scale; GHQ, general health questionnaire; JPSP, Jenkins sleep problems scale; PANAS, positive and negative affect schedule; PSQI, Pittsburgh sleep quality index; PghSD, Pittsburgh sleep diary; SSI, subjective sleep inefficiency; STAI, state-trait anxiety inventory; STAXI, state-trait anger expression inventory.
[a] Trait PA derived from ecological momentary assessment.

single-administration, paper-and-pencil questionnaires; however, two studies (Buysse et al., 2007; Steptoe et al., 2008) repeatedly measured PA using ecological momentary assessment approaches (Shiffman, Stone, & Hufford, 2008). All but one study (i.e., Buysse et al., 2007) controlled for potential confounding factors, such as negative affect (NA) or psychological distress symptoms. Overall, findings indicate that higher levels of PA are uniquely promotive of better sleep quality in both clinical and nonclinical samples of adults. Moreover, these associations appear to be independent of NA, suggesting that high trait PA may have a salutary health effect that is distinct from the effect associated with low NA. Of note, one study did not find any effect of PA on sleep quality (i.e., Stewart, Rand, Hawkins, & Stines, 2011).

Daily Positive Affect

Evidence linking daily PA and sleep has also emerged from studies involving children, adolescence, and adults. Of the seven studies identified (see Table 13.2), all controlled for the influence of NA. Two studies reported bidirectional relations between daytime PA and nighttime sleep quality (Cousins et al., 2011; van Zundert, , van Roekel, Engels, & Scholte, in press). Of the two studies reporting a unique association between daily PA and sleep, one study reported a negative relation between daily happiness and irregular sleep duration among ethnically diverse adolescents (Fuligni & Hardway, 2006), while another study of Canadian college students reported that daily PA was associated with sleep quality but not sleep quantity (Galambos & Dalton, 2009). Two studies failed to find statistically significant associations between daily PA and sleep (Brissette & Cohen, 2002; Wrzus, Wagner, & Riediger, 2014). Of note, these studies focused on the absence rather than the presence of good sleep (i.e., adequate or restorative sleep). Finally, one study (i.e., de Wild-Hartmann et al., 2013) reported a negative relation between prior daytime PA and subsequent sleep quality.

Discrete Emotions

A number of studies have examined the effects of specific positive affective states (i.e., vigor, gratitude, and love) on sleep outcomes (see Table 13.3). For example, an early cross-sectional study by Bardwell et al. (1999) found that vigor was positively associated with sleep duration in patients with and without sleep apnea. In comparison, reciprocal inverse relations between vigor and insomnia were reported in a recent longitudinal study of working adults (Armon, Melamed, & Vinokur, 2014). In terms of the effects of gratitude on improved sleep, a positive association has been found in a sample of healthy adults (Wood, Joseph, Lloyd, & Atkins, 2009)

TABLE 13.2 Effects of Daily Positive Affect on Sleep

PA Measure	Sleep Measure	NA Adjustment	Authors/Year	Findings
$POMS_{(9)}$	Sleep Disturbance$_{(2)}$	$POMS_{(12)}$	Brissette and Cohen (2002)	PA not uniquely associated with sleep disturbance
Happiness$_{(3)}$	Sleep Time$_{(2)}$	$POMS_{(10)}$	Fuligni and Hardway (2006)	Happiness positively associated with more sleep and negatively associated with irregular sleep duration
$PANAS_{(10)}$	Sleep Quantity & Quality$_{(2)}$	$PANAS_{(10)}$	Galambos and Dalton (2009)	PA uniquely associated with sleep quality (but not sleep quantity)
PANAS-C$_{(4)}$[a]	Actigraphy	PANAS-C$_{(4)}$[a]	Cousins et al. (2011)	Bidirectional relations between daytime PA and nighttime sleep. Higher daytime PA associated with *more* time in bed for youth with depression and *less* time in bed for youth with anxiety.
$PA_{(4)}$[a]	Sleep Quantity & Quality$_{(5)}$	$NA_{(6)}$[a]	de Wild-Hartmann et al. (2013)	Sleep quality and quantity (period/latency, number of awakenings) predicted subsequent daytime PA, whereas prior daytime PA was negatively associated with subsequent sleep quality
$PA_{(5)}$[a]	Sleep Quality$_{(1)}$	$NA_{(5)}$[a]	van Zundert, van Roekel, Engels, and Scholte, in press	Bidirectional relations between daytime PA and nighttime sleep quality. Poorer sleep quality predicted and was predicted by low PA
$PA_{(6)}$[a]	Sleep Duration$_{(1)}$	$NA_{(6)}$[a]	Wrzus, Wagner, and Riediger (2014)	Affect balance not uniquely associated with sleep duration

Subscripted number in parenthesis = number of items; PANAS-C, positive and negative affect schedule for children; POMS, profile of mood state.
[a] *Daily PA derived from ecological momentary assessment.*

TABLE 13.3 Effects of Discrete Positive Affective States on Sleep

PA Measure	Sleep Measure	NA Adjustment	Authors/Year	Findings
VIGOR				
POMS[7]	Sleep Duration[1]		Bardwell, Berry, Ancoli-Israel, and Dimsdale (1999)	Vigor was positively correlated with sleep time in patients with and without sleep apnea
Vigor Measure[14]	AIS-5[5]	Neuroticism[8]	Armon, Melamed, and Vinokur (2014)	Bidirectional relations between vigor and insomnia symptoms. Vigor and insomnia are negatively associated with each over time.
GRATITUDE				
GQ-6[6]	PSQI[19]	Neuroticism[4]	Wood, Joseph, Lloyd, and Atkins (2009)	Gratitude uniquely associated with sleep quality (i.e., lower PSQI scores)
GQ-6[6]	PSQI[19]	HADS[14]	Ng and Wong (2012)	Gratitude uniquely associated with sleep quality (i.e., lower PSQI scores) in chronic pain patients
LOVE				
Romantic Love[6]	Sleep Log[5]		Brand, Luethi, von Planta, Hatzinger, and Holsboer-Trachsler (2007)	Adolescents in romantic love reported fewer hours of sleep, better sleep quality, lower daytime sleepiness, and heightened daytime concentration
Marital Quality[1]	Sleep Disturbance[4]	Anxiety[4], Depression[20]	Troxel, Buysse, Hall, and Matthews (2009)	Women high in marital happiness reported fewer sleep disturbances

Subscripted number in parenthesis = number of items; AIS-5, Athens insomnia scale; GQ, gratitude questionnaire; HADS, hospital anxiety and depression scale; POMS, profile of mood states; PSQI, Pittsburgh sleep quality index.

and chronic pain patients (Ng & Wong, 2012). Finally, early-stage intense romantic love (Brand, Luethi, von Planta, Hatzinger, & Holsboer-Trachsler, 2007) and current marital happiness (Troxel, Buysse, Hall, & Matthews, 2009) have both been linked to improved sleep quality and fewer sleep disturbances.

PROTECTIVE EFFECTS

Minimal research to date has focused on the potential stress-buffering or protective effects of PA (for a discussion, see Folkman, 2008; Folkman & Moskowitz, 2000; Fredrickson & Cohn, 2008). As demonstrated in the prior work, PA can buffer the effects of stress in at least two ways. First, PA may *indirectly* influence sleep by modifying the effects of stress. In an illustrative study, Steptoe et al. (2008) found that trait PA, as measured by ecological momentary assessment, partly mediated the association between psychosocial risk factors (e.g., financial strains, poor social relationships, and psychological distress) and poor sleep (i.e., restless sleep and trouble falling asleep) in a sample of middle-aged and older adults (age range 58-72). Second, PA may act as a *moderator*, either accentuating or attenuating the impact of risk factors on sleep. For example, Fortunato and Harsh (2006) demonstrated a mitigating effect of interpersonal conflicts on sleep quality (i.e., falling asleep and reinitiating sleep) among individuals high in trait PA. Similarly, a recent study by Fredman et al. (2014) found that PA was associated with fewer sleep problems in caregivers (but not noncaregivers), thus suggesting a protective effect. Taken together, existing data provide preliminary evidence for a protective function of PA and suggest that this is an important direction for future research.

MECHANISMS

To the best of our knowledge, no data addresses the mechanisms accounting for the presumptive beneficial effects of PA on sleep (Kahn et al., 2013). However, prior reviews of the literature (e.g., Boehm & Kubzansky, 2012; Harvey, Murray, Chandler, & Soehner, 2011; Kahn et al., 2013; Pressman & Cohen, 2005) suggest a number of variables that could be on the pathway to restorative or good sleep. These include stress hormones and inflammatory markers (e.g., cortisol, interleukin-6), neurobiological processes (e.g., nocturnal heart rate variability (HRV), circadian and serotonergic function), emotional brain networks and REM sleep, health behaviors (e.g., physical activity), and social relationships. These hypothesized mechanisms have yet to be empirically investigated.

SUMMARY

Using a variety of approaches to conceptualize PA (i.e., traits, daily processes, discrete emotions), studies have documented that PA directly predicts sleep outcomes. Overall, the pattern of findings suggests that trait-like measures provide the most consistent evidence of an association between PA and good sleep. At present, less evidence exists for the protective effects of PA. Finally, a critical direction for future research is to elucidate the mechanisms by which PA contributes to adaptive sleep outcomes.

POSITIVE AFFECT AS RISK AND VULNERABILITY

Diathesis-stress/dual-risk models (Monroe & Simons, 1991; Sameroff, 1983) center on the prediction that some individuals, due to an inherent personal characteristic or diathesis, are disproportionately more vulnerable to adverse environmental conditions. From this perspective, *risk factors* are those that increase the odds of maladjustment, whereas *vulnerability factors* encompass variables that exacerbate the effects of risks (Zuckerman, 1999). In the following sections, we review research that establishes associations between sleep disturbance and various mood disorders associated with PA dysregulation. We then examine a number of potential mechanisms that may underlie these associations. Overall, the available evidence supports a complex and transactional relationship (Kahn et al., 2013), with both PA and sleep emerging as significant risks and vulnerability factors.

MOOD DISORDERS AND SLEEP

Anxiety and Depression

Epidemiological studies of specific psychiatric outcomes such as anxiety and depression have identified sleep problems and changes in normal sleep patterns among participants (Breslau, Roth, Rosenthal, & Andreski, 1996; Jansson-Frojmark & Lindblom, 2008; Neckelmann, Mykletun, & Dahl, 2007). Indeed, sleep difficulties have been considered hallmark symptoms as well as distinct *Diagnostic Statistical Manual 5* (DSM-5) diagnoses of both generalized anxiety disorder and major depressive disorder (American Psychiatric Association, 2013). Baglioni et al. (2010) summarized bidirectional links between *insomnia*—defined as "difficulties in initiating/maintaining sleep and/or non-restorative sleep accompanied by decreased daytime functioning, persisting for at least four weeks" (p. 227)—and depression and anxiety. Nineteen of the 21 longitudinal studies reviewed found that insomnia symptoms predicted an increased risk for future depression, and five studies reported a prospective relation between insomnia and anxiety (Baglioni et al., 2010). Moreover, some evidence indicates that the presence of insomnia is associated with higher risk for future depression than anxiety (Jansson-Frojmark & Lindblom, 2008; e.g., Morphy, Dunn, Lewis, Boardman, & Croft, 2007). Other studies have suggested a potential causal relationship between insomnia and depression in patients who have no previous history of depression (Breslau et al., 1996; Ford & Kamerow, 1989; Ohayon & Roth, 2003). Additionally, insomnia and depression may be linked through specific impairments in sleep and affect. For example, a study by Koffel and Watson (2009) found that, when compared with nighttime symptoms of insomnia (e.g., long sleep latency, minutes awake at night), daytime symptoms (e.g., sleepiness, motivation reduction) showed stronger relations to depression than anxiety and were associated with heightened NA and diminished PA. Thus, insomnia (especially daytime symptoms of sleepiness) appears to predict the onset of depression; however, the affective mechanisms (heightened NA, low PA) underlying these relationships have, to date, received limited attention.

Increasing evidence suggests that, despite symptom overlap, etiological differences between anxiety and depression may be reflected in anhedonia or chronically low levels of PA (Mineka, Watson, & Clark, 1998; Watson, 2000; Watson & Naragon-Gainey, 2010). However, studies examining the link between sleep and low PA have produced mixed results. For example, Bower, Bylsma, Morris, and Rottenberg (2010) found that daytime dysfunction and poor subjective sleep quality (as measured by the Pittsburgh Sleep Index [PSQI]) were unique risk factors for low ambulatory PA, even after accounting for the effects of depression status and anxiety symptoms. Poor sleep has also been implicated as a vulnerability factor for low PA. For example, Zohar, Tzischinsky, Epstein, and Lavie (2005) found that sleep loss resulted in a dampening effect on ambulatory PA in response to daily positive events. Experimental studies have also shown that poor sleep (i.e., sleep duration and quality) can suppress positive affective responses in healthy adolescents (Dagys et al., 2012; Talbot, McGlinchey, Kaplan, Dahl, & Harvey, 2010) and adults (Franzen, Siegle, & Buysse, 2008; Paterson et al., 2011; Williams, Cribbet, Rau, Gunn, & Czajkowski, 2013), a conclusion confirmed by daily diary investigations of chronic pain patients (Hamilton, Catley, & Karlson, 2007; Hamilton et al., 2008). Nevertheless, other studies have yielded contrasting findings concerning relations between low PA and sleep disturbance. For example, Watson and Naragon-Gainey (2010) recently reviewed a broad range of evidence indicating that, although low PA shows considerable specificity to core affective deficits in depression (e.g., depressed mood/dysphoria, lassitude), it is only weakly related to symptoms that are characteristic of sleep dysfunction (e.g., insomnia, motor disturbance, and appetite loss). Overall, inconsistencies in previous studies may be partly explained by differences in the measurement of sleep and affect (e.g., behavior/physiological measures vs. subjective reports), as well as discrepancy in assessment methods (e.g., retrospective vs. experience sampling).

Bipolar Disorder

Clinicians have identified sleep disturbance as a core diagnostic feature of bipolar disorder (American Psychiatric Association, 2013). Comprehensive reviews conducted by Harvey and colleagues (Harvey, 2008; Kaplan & Harvey, 2009) have found that a majority of patients report a reduced need for sleep during episodes of mania, as well as varying rates of insomnia and hypersomnia during bipolar depression. Sleep abnormalities, including greater fragmentation of rapid eye movement (REM) sleep and longer sleep onset latency (SOL: length of time it takes to transition from full wakefulness to sleep) are also present during the period between episodes (Jones, Hare, & Evershed, 2005; Millar, Espie, & Scott, 2004; Talbot et al., 2009). Furthermore, evidence across a range of studies (for reviews, see Harvey, 2008; Jackson, Cavanagh, & Scott, 2003) indicates that sleep disturbance contributes to the onset and maintenance of manic symptoms among individuals with bipolar disorder.

Data from comparative studies suggests that *hypersomnia*—defined by "a combination of prolonged nighttime sleep episodes, increased nighttime wakefulness, frequent daytime napping, and excessive daytime sleepiness (EDS)"(Kaplan & Harvey, 2009, p. 276)—may be a feature of bipolar disorder that differentiates it from major depressive disorder. For example, Kaplan and Harvey (2009) summarize data indicating that hypersomnia is more prevalent in bipolar depression than it is in major depressive disorder, and it is uniquely predictive of future bipolar episodes (Akiskal & Benazzi, 2005; Benazzi, 2006; Bowden, 2005; Hantouche, 2005). Moreover, a recent study by Kaplan, Gruber, Eidelman, Talbot, and Harvey (2011) found that the presence of hypersomnia during the interepisode period predicted subsequent depressive symptoms, adding to growing evidence that interepisode sleep disturbance is an important contributor to relapse in bipolar patients (Harvey, 2008; Jackson et al., 2003).

MECHANISMS

Minimal data address the mechanisms contributing to sleep and positive affective functioning in depression and/or bipolar disorder. Integrative reviews of the literature (Harvey, 2008; Harvey et al., 2011; Kaplan & Harvey, 2009) have implicated circadian rhythm disruption and increased sensitivity to light exposure as potential neurobiological processes contributing to sleep disturbance across depression and bipolar patients, although the effects of these putative mechanisms on positive affective responses have yet to be investigated. Recent work also suggests that amplifying positive emotional states (via self-focused rumination) increases PA across both major depressive and bipolar disorders, but attempts to dampen PA paradoxically intensify these states in bipolar individuals (Gilbert, Nolen-Hoeksema, & Gruber, 2013). The implications of these PA regulation strategies for reducing sleep disturbances (i.e., insomnia and hypersomnia) remain unclear, however.

Several studies have demonstrated links between various dynamic aspects of PA regulation (e.g., reactivity and variability) and functional impairments in sleep and diminished well-being. For example, Ong et al. (2013) recently reported on a daily diary study of PA and sleep. Greater positive affective reactivity in response to daily events was associated with poorer sleep efficiency, as measured by a small wrist sensor (a sleep actigraph). Other studies have found associations between heightened variability in PA and increased depression and anxiety symptoms (Gruber, Kogan, Quoidbach, & Mauss, 2013) and low self-esteem (Kuppens, Allen, & Sheeber, 2010). Given that mood lability and reactivity represent central features across psychiatric (e.g., bipolar disorder) and sleep disturbances (e.g., insomnia), researchers should examine PA variability and reactivity as potential mechanisms that contribute to the development of sleep impairments in individuals with depression and bipolar disorder.

SUMMARY

Although ample empirical evidence links sleep disturbances to psychiatric outcomes such as depression and bipolar disorder, the majority of longitudinal studies have focused on unidirectional effects, with sleep problems predicting later affective difficulties. Moreover, although evidence links sleep to PA, the specific direction and nature (risk vs. vulnerability) of the effects are not always clear, and this ambiguity presents an important direction for future research. Similarly, limited research to date has focused on PA dysregulation as a potential mechanism underpinning the association between sleep disturbance and the presence of psychiatric disorder.

CONCLUSIONS

Existing evidence demonstrates important links between PA and sleep. The data reviewed show that disturbances in PA (e.g., chronically low/elevated PA) are strongly linked to various forms of psychopathology (e.g., depression, bipolar disorder) and sleep impairment (e.g., insomnia, hypersomnia). More limited empirical data exists on the impact of PA on sleep. To date, much of this research has focused on the role of PA in nighttime and daytime impairments in sleep. Less attention has been given to whether PA is associated with good or restorative sleep. Overall, the limitations in the existing data provide an important impetus for future work. Below, we highlight several critical but, as yet, unresolved issues.

First, as noted earlier, several authors have suggested that a bidirectional relationship exists between sleep and emotions (Baglioni et al., 2010; Kahn et al., 2013) and between sleep and PA in particular (Ong et al., 2013; Steptoe et al., 2008). However, with a few exceptions (e.g., Armon et al., 2014; Cousins et al., 2011), studies have rarely examined the reciprocal or bidirectional links between PA and sleep. Similarly, controlled experimental studies investigating the effect of PA on sleep outcomes are especially scarce (for an exception, see Talbot et al., 2009). Thus, prospective and experimental studies addressing the causal relationship between PA and sleep are urgently needed.

Second, the vast majority of studies have used subjective measures of sleep. It remains unclear whether standard subjective assessments (e.g., PSQI, sleep diaries) and objective methods (e.g., actigraph, polysomnographic monitoring) are equivalent or whether they assess different underlying processes with potentially differing sleep etiologies (cf. Kahn et al., 2013; Kaplan & Harvey, 2009). As has been noted by others (Boehm & Kubzansky, 2012; Ong, 2010; e.g., Pressman & Cohen, 2005), the measurement of PA also raises fundamental (but understudied) questions. For example, is the association between PA and sleep moderated by affective arousal? We could identify only one study (i.e., Schwerdtfeger, Friedrich-Mai, & Gerteis, in press) suggesting that deactivated PA (e.g., relaxed, even-tempered, and content) throughout the day was associated with beneficial cardiac function (elevated HRV and diminished heart rate [HR]) during sleep, thus supporting the hypothesis that low arousal PA throughout the day may foster good nighttime sleep (see also Kok & Fredrickson, 2010). Additional research in this area is needed to determine what level of adaptive PA arousal is required to detect an association with sleep outcomes.

Relatedly, although several studies suggest a link between sleep and low PA (e.g., Bower et al., 2010), little work to date has examined (within nonclinical samples) the effects of excessive or extreme levels of high PA on sleep. Recent reviews (Grant & Schwartz, 2011; Gruber, Mauss, & Tamir, 2011) suggest that, at very high levels, PA may confer detrimental outcomes. For example, Friedman and colleagues found that extremely cheerful people were more likely to engage in risky health behaviors (Martin et al., 2002) that increased their risk of early mortality (Friedman et al., 1993). Such investigations could confirm and extend previous clinical observations and experimental data concerning the link between PA deficits and sleep quality (Harvey, 2008; Kaplan & Harvey, 2009).

Finally, as noted earlier, a dearth of evidence exists on the mechanisms contributing to good sleep. Although a number of candidate pathways linking PA and restorative processes have been identified (Boehm & Kubzansky, 2012; Pressman & Cohen, 2005), the evidence base to date includes few formal tests of mechanistic hypotheses. Indeed, we could identify no studies that have directly examined the mechanisms by which PA influences adaptive sleep patterns. To the extent that progress can be made on these issues, research on sleep and affect may begin to create theoretically informed links to other neighboring fields currently attempting to probe the adaptive significance of positive affect.

References

Akiskal, H., & Benazzi, F. (2005). Atypical depression: A variant of bipolar II or a bridge between unipolar and bipolar II? *Journal of Affective Disorders, 84*, 209–217.

American Psychiatric Association. (2013). *Diagnostic and statistical manual of mental disorders* (5th ed.). Arlington: American Psychiatric Publishing.

Armon, G., Melamed, S., & Vinokur, A. (2014). The reciprocal relationship between vigor and insomnia: A three-wave prospective study of employed adults. *Journal of Behavioral Medicine, 37*, 664–674.

Baglioni, C., Spiegelhalder, K., Lombardo, C., & Riemann, D. (2010). Sleep and emotions: A focus on insomnia. *Sleep Medicine Reviews, 14*, 227–238.

Bardwell, W. A., Berry, C. C., Ancoli-Israel, S., & Dimsdale, J. E. (1999). Psychological correlates of sleep apnea. *Journal of Psychosomatic Research, 47*, 583–596.

Benazzi, F. (2006). Symptoms of depression as possible markers of bipolar II disorder. *Progress in Neuro-Psychopharmacology and Biological Psychiatry, 30*, 471–477.

Boehm, J. K., & Kubzansky, L. D. (2012). The heart's content: The association between positive psychological well-being and cardiovascular health. *Psychological Bulletin, 138*, 655–691.

Bowden, C. (2005). A different depression: Clinical distinctions between bipolar and unipolar depression. *Journal of Affective Disorders, 84*, 117–125.

Bower, B., Bylsma, L. M., Morris, B. H., & Rottenberg, J. (2010). Poor reported sleep quality predicts low positive affect in daily life among healthy and mood-disordered persons. *Journal of Sleep Research, 19*(2), 323–332.

Brand, S., Luetlii, M., von Planta, A., Hatzinger, M., & Holsboer-Trachsler, E. (2007). Romantic love, hypomania, and sleep pattern in adolescents. *Journal of Adolescent Health, 41*, 69–76.

Breslau, N., Roth, T., Rosenthal, L., & Andreski, P. (1996). Sleep disturbance and psychiatric disorders: A longitudinal epidemiological study of young adults. *Biological Psychiatry, 39*, 411–418.

Brissette, I., & Cohen, S. (2002). The contribution of individual differences in hostility to the associations between daily interpersonal conflict, affect, and sleep. *Personality and Social Psychology Bulletin, 28*(9), 1265–1274.

Buysse, D., Thompson, W., Scott, J., Franzen, P., Germain, A., Hall, M., et al. (2007). Daytime symptoms in primary insomnia: A prospective analysis using ecological momentary assessment. *Sleep Medicine Reviews, 8*, 198–208.

Cousins, J., Whalen, D., Dahl, R., Forbes, E., Olino, T., Ryan, N., et al. (2011). The bidirectional association between daytime affect and night time sleep in youth with anxiety and depression. *Journal of Pediatric Psychology, 36*, 969–979.

Dagys, N., McGlinchey, E., Talbot, L., Kaplan, K., Dahl, R., & Harvey, A. (2012). Double trouble? The effects of sleep deprivation and chronotype on adolescent affect. *Journal of Child Psychology and Psychiatry, 53*, 660–667.

de Wild-Hartmann, J. A., Wichers, M., van Bemmel, A., Derom, C., Thiery, E., Jacobs, N., et al. (2013). Day-to-day associations between subjective sleep and affect in regard to future depression in a female population-based sample. *British Journal of Psychiatry, 202*, 407–412.

Dew, M. A., Hoch, C. C., Buysse, D. J., Monk, T. H., Begley, A. E., Houck, P. R., et al. (2003). Healthy older adults' sleep predicts all-cause mortality at 4 to 19 years of follow-up. *Psychosomatic Medicine, 65*, 63–73.

Folkman, S. (2008). The case of positive emotions in the stress process. *Anxiety, Stress, and Coping, 21*, 3–14.

Folkman, S., & Moskowitz, J. T. (2000). Positive affect and the other side of coping. *American Psychologist, 55*(6), 647–654.

Ford, D. E., & Kamerow, D. B. (1989). Epidemiologic study of sleep disturbances and psychiatric disorders: An opportunity for prevention? *JAMA, 262*, 1479–1484.

Fortunato, V. J., & Harsh, J. (2006). Stress and sleep quality: The moderating role of negative affectivity. *Personality and Individual Differences, 41*, 825–836.

Fosse, R., Stickgold, R., & Hobson, A. (2002). Emotional experience during rapid-eye-movement sleep in narcolepsy. *Sleep, 25*, 724–732.

Franzen, P. L., Siegle, G. J., & Buysse, D. J. (2008). Relationships between affect, vigilance, and sleepiness following sleep deprivation. *Journal of Sleep Research, 17*(1), 34–41.

Fredman, L., Gordon, S., Heeren, T., & Stuver, S. (2014). Positive affect is associated with fewer problems in older caregivers but not noncaregivers. *The Gerontologist, 54*, 559–569.

Fredrickson, B. L. (2013). Positive emotions broaden and build. In E. A. Plant, & P. G. Devine (Eds.), *Advances in experimental social psychology* (vol. 47, pp. 1–53). Burlington: Academic Press.

Fredrickson, B. L., & Cohn, M. A. (2008). Positive emotions. In M. Lewis, J. M. Haviland-Jones, & L. F. Barrett (Eds.), *Handbook of emotions* (3rd ed., pp. 777–796). New York: Guilford Press.

Friedman, H. S., Tucker, J. S., Tomlinson-Keasey, C., Schwartz, J. E., Wingard, D. L., & Criqui, M. H. (1993). Does childhood personality predict longevity? *Journal of Personality and Social Psychology, 65*, 176–185.

Fuligni, A. J., & Hardway, C. (2006). Daily variation in adolescents' sleep, activities, and psychological well-being. *Journal of Research on Adolescence, 16*, 353–378.

Galambos, N., & Dalton, A. (2009). Losing sleep over it: Daily variation in sleep quantity and quality in Candian students' first semester of university. *Journal of Research on Adolescence, 19*, 741–761.

Gilbert, K. E., Nolen-Hoeksema, S., & Gruber, J. (2013). Positive emotion dysregulation across mood disorders: How amplifying versus dampening predicts emotional reactivity and illness course. *Behaviour Research and Therapy, 51*(11), 736–741.

Grant, A. M., & Schwartz, B. (2011). Too much of a good thing: The challenge and opportunity of the inverted U. *Perspectives on Psychological Science, 6*, 61–76.

Gray, E. K., & Watson, D. (2002). General and specific traits of personality and their relation to sleep and academic performance. *Journal of Personality, 70*(2), 177–206.

Gruber, J., Kogan, A., Quoidbach, J., & Mauss, I. B. (2013). Happiness is best kept stable: Positive emotion variability is associated with poorer psychological health. *Emotion, 13*, 1–6. http://dx.doi.org/10.1037/a0030262.

Gruber, J., Mauss, I., & Tamir, M. (2011). A dark side of happiness? How, when and why happiness is not always good. *Perspectives on Psychological Science, 6*, 222–233.

Hamilton, N. A., Affeck, G., Tennen, H., Karlson, C., Luxton, D., Preacher, K. J., et al. (2008). Fibromyalgia: The role of sleep in affect and in negative event reactivity and recovery. *Health Psychology, 27*, 490–494.

Hamilton, N. A., Catley, D., & Karlson, C. (2007). Sleep and the affective response to stress and pain. *Health Psychology, 26*(3), 288.

Hantouche, A. (2005). Bipolar II vs. unipolar depression: Psychopathologic differenation by dimensional measures. *Journal of Affective Disorders, 84,* 127–132.

Harvey, A. (2008). Sleep and ciradian rhythms in bipolar disorder: Seeking synchrony, harmony, and regulation. *American Journal of Psychiatry, 165,* 820–829.

Harvey, A., Murray, G., Chandler, R., & Soehner, A. (2011). Sleep disturbance as transdiagnostic: Consideration of neurobiological mechanisms. *Clinical Psychology Review, 31,* 225–235.

Irwin, M. R., Carrillo, C., & Olmstead, R. (2010). Sleep loss activates cellular markers of inflammation: Sex differences. *Brain, Behavior, and Immunity, 24,* 54–57.

Jackowska, M., Dockray, S., Hendrickx, H., & Steptoe, A. (2011). Psychological factors and sleep efficiecy: Discrepancies between subjective and objective evaluations of sleep. *Psychosomatic Medicine, 73,* 810–816.

Jackson, A., Cavanagh, J., & Scott, J. (2003). A systematic review of manic and depressive prodromes. *Journal of Affective Disorders, 74,* 209–217.

Jansson-Frojmark, M., & Lindblom, K. (2008). A bidirectional relationship between anxiety anddepression, and insomnia? Aprospective study in the general population. *Journal of Psychosomatic Research, 64,* 443–449.

Jones, S., Hare, D., & Evershed, K. (2005). Actigraphic assessment of circadian activity and sleep patterns in bipolar disorder. *Bipolar Disorders, 7,* 176–186.

Kahn, M., Sheppes, G., & Sadeh, A. (2013). Sleep and emotions: Bidirectional links and underlying mechanisms. *International Journal of Psychophysiology, 89*(2), 218–228.

Kaplan, K., Gruber, J., Eidelman, P., Talbot, L., & Harvey, A. (2011). Hypersomnia in interepisode bipolar disorder: Does it have prognostic significance? *Journal of Affective Disorders, 132*(3), 438–444.

Kaplan, K., & Harvey, A. (2009). Hypersomnia across mood disorders: A review and synthesis. *Sleep Medicine Reviews, 13,* 275–285.

Koffel, E., & Watson, D. (2009). The two-factor structure of sleep complaints and its relation to depression and anxiety. *Journal of Abnormal Psychology, 118*(1), 183.

Kok, B. E., & Fredrickson, B. L. (2010). Upward spirals of the heart: Autonomic flexibility, as indexed by vagal tone, reciprocally and prospectively predicts positive emotions and social connectedness. *Biological Psychology, 85,* 432–436. http://dx.doi.org/10.1016/j.biopsycho.2010.09.005.

Kripke, D. F., Garfinkel, L., Wingard, D. L., Klauber, M. R., & Marler, M. R. (2002). Mortality associated with sleep duration and insomnia. *Archives of General Psychiatry, 59,* 131–136.

Kuppens, P., Allen, N., & Sheeber, L. (2010). Emotional inertia and psychological maladjustment. *Psychological Science, 21,* 984–991.

Li, Y., Zhang, X., Winkelman, J., Redline, S., Hu, F., Stampfer, M., et al. (2014). The association between insomnia symptoms and mortality: A prospective study of US men. *Circulation, 129,* 737–746.

Martin, L. R., Friedman, H. S., Tucker, J. S., Tomlinson-Keasey, C., Criqui, M. H., & Schwartz, J. E. (2002). A life course perspective on childhood cheerfulness and its relation to mortality risk. *Personality and Social Psychology Bulletin, 28,* 1155–1165.

Millar, A., Espie, C., & Scott, J. (2004). The sleep of remitted bipolar outpatients: A controlled naturalistic study using actigraphy. *Journal of Affective Disorders, 80,* 145–153.

Mineka, S., Watson, D., & Clark, L. A. (1998). Comorbidity of anxiety and unipolar mood disorders. *Annual Review of Psychology, 49,* 377–412.

Monroe, S. M., & Simons, A. D. (1991). Diathesis-stress theories in the context of life stress research: Implications for the depressive disorders. *Psychological Bulletin, 110,* 406–425.

Morphy, H., Dunn, K., Lewis, M., Boardman, H., & Croft, P. (2007). Epidemiology of insomnia: A longitudinal study in a UK population. *Sleep, 30,* 274–280.

Neckelmann, D., Mykletun, A., & Dahl, A. (2007). Chronic insomnia as a risk factor for developing anxiety and depression. *Sleep Medicine Reviews, 30*, 873–880.

Ng, M., & Wong, W. (2012). The differential effects of gratitude and sleep on psychological distress in patients with chronic pain. *Journal of Health Psychology, 18*, 263–271.

Ohayon, M. M., & Roth, T. (2003). Place of chronic insomnia in the course of depressive and anxiety disorders. *Journal of Psychiatric Research, 37*, 9–15.

Ong, A. D. (2010). Pathways linking positive emotion and health in later life. *Current Directions in Psychological Science, 19*, 358–362.

Ong, A. D., Exner-Cortens, D., Riffin, C., Steptoe, A., Zautra, A. J., & Almeida, D. M. (2013). Linking stable and dynamic features of positive affect to sleep. *Annals of Behavioral Medicine, 46*, 52–61. http://dx.doi.org/10.1007/s12160-013-9484-8.

Paterson, J. L., Dorrian, J., Ferguson, S. A., Jay, S. M., Lamond, N., Murphy, P. J., et al. (2011). Changes in structural aspects of mood during 39-66 h of sleep loss using matched controls. *Applied Ergonomics, 42*, 196–201.

Payne, J. D., & Kensinger, E. A. (2010). Sleep's role in the consolidation of emotional episodic memories. *Current Directions in Psychological Science, 19*(5), 290–295.

Pressman, S. D., & Cohen, S. (2005). Does positive affect influence health? *Psychological Bulletin, 131*, 925–971.

Reid, K. J., Marinovick, Z., Finkel, S., Statsinger, J., Golden, R., Harter, K., et al. (2006). Sleep: A marker of physical and mental health in the elderly. *American Journal of Geriatric Psychiatry, 14*, 860–866.

Sameroff, A. J. (1983). Developmental systems: Contexts and evolution. In P. Mussen (Ed.), *Handbook of child psychology* (vol. 1, pp. 237–294). New York: Wiley.

Sameroff, A. J., Gutman, L., & Peck, S. (2003). Adaptation among youth facing multiple risks: Prospective research findings. In S. S. Luthar (Ed.), *Resilience and vulnerability: Adaptation in the context of childhood adversities* (pp. 364–391). New York: Cambridge University Press.

Schwerdtfeger, A., Friedrich-Mai, P., & Gerteis, A. (2014). Daily positive affect and nocturnal cardiac activation. *International Journal of Behavioral Medicine*, in press.

Shiffman, S., Stone, A. A., & Hufford, M. R. (2008). Ecological momentary assessment. *Annual Review of Clinical Psychology, 4*, 1–32.

Steptoe, A., O'Donnell, K., Marmot, M., & Wardle, J. (2008). Positive affect, psychological well-being, and good sleep. *Journal of Psychosomatic Research, 64*(4), 409–415. http://dx.doi.org/10.1016/j.jpsychores.2007.11.008.

Stewart, J. C., Rand, K. L., Hawkins, M. A., & Stines, J. A. (2011). Associations of the shared and unique aspects of positive and negative emotional factors with sleep quality. *Personality and Individual Differences, 50*(5), 609–614.

Talbot, L., Hairston, I., Eidelman, P., Gruber, J., & Harvey, A. (2009). The effect of mood on sleep onset latency and REM sleep in interepisode bipolar disorder. *Journal of Abnormal Psychology, 118*(3), 448.

Talbot, L., McGlinchey, E., Kaplan, K., Dahl, R., & Harvey, A. (2010). Sleep deprivation in adolescents and adults: Changes in affect. *Emotion, 10*, 831–841.

Troxel, W., Buysse, D., Hall, M., & Matthews, K. A. (2009). Marital happiness and sleep disturbances in a multi-ethnic sample of middle-aged women. *Behavioral Sleep Medicine, 7*, 2–19.

van Zundert, R., van Roekel, E., Engels, R., & Scholte, R. (2013). Reciprocal associations between adolescents' night-time sleep and daytime affect and the role of gender and depressive symptoms. *Journal of Youth and Adolescence*, in press.

Watson, D. (2000). *Mood and temperament*. New York: Guilford Press.

Watson, D., & Naragon-Gainey, K. (2010). On the specificity of positive emotional dysfunction in psychopathology: Evidence from the mood and anxiety disorders and schizophrenia/schizotypy. *Clinical Psychology Review, 30*, 839–848.

Williams, P. G., Cribbet, M. R., Rau, H. K., Gunn, H. E., & Czajkowski, L. A. (2013). The effects of poor sleep on cognitive, affective, and physiological responses to a laboratory stressor. *Annals of Behavioral Medicine, 46*(1), 40–51.

Wood, A. M., Joseph, S., Lloyd, J., & Atkins, S. (2009). Gratitude influences sleep through the mechanism of pre-sleep cognitions. *Journal of Psychosomatic Research, 66*(1), 43–48.

Wrzus, C., Wagner, G. G., & Riediger, M. (2014). Feeling good when sleeping in? Day-to-day associations between sleep duration and affective well-being differ from youth to old age. *Emotion, 14*, 624–628.

Xie, L., Kang, K., Xu, Q., Chen, M., Liao, Y., Thiyagarajan, M., et al. (2013). Sleep drives metabolite clearance from the adult brain. *Science, 342*, 373–377.

Zohar, D., Tzischinsky, O., Epstein, R., & Lavie, P. (2005). The effects of sleep loss on medical residents' emotional reactions to work events: A cognitive-energy model. *Sleep Medicine Reviews, 28*, 47–54.

Zuckerman, M. (1999). *Vulnerability to psychopathology: A biosocial model*. Washington, DC: American Psychological Association.

CHAPTER 14

Sleep and Biological Rhythms in Mania

Rébecca Robillard and Ian B. Hickie

Clinical Research Unit, Brain and Mind Research Institute, University of Sydney, Sydney, Australia

MANIA AS AN ACTIVATION SWITCH DYSFUNCTION

Motor Activation, Deregulation of Energy Levels, and the Experience of a Decreased Need for Sleep

The diagnostic criteria for an episode of mania include a period of at least 1 week of elevated, expansive, or irritable mood co-occurring with increased energy levels or increased engagement in goal-directed activities, as well as a constellation of other possible symptoms that represent a departure from one's usual behavior, including a reduction in sleep duration (American Psychiatric Association, 2000). Although sleep reduction is not necessary for the diagnosis of a manic episode, factor analyses in large-scale studies have suggested that it represents a core element of the phenomenology of mania and a significant predictor of recurrence (Kessler, Rubinow, Holmes, Abelson, & Zhao, 1997).

The *Diagnostic Statistical Manual 5* (DSM-5; American Psychiatric Association [APA], 2013) refers to individuals with mania experiencing a "decreased need for sleep," giving the example of feeling rested after sleeping for about 3 hours. As opposed to the diurnal fatigue characteristic of those suffering from insomnia, sleep reduction in the context of a manic episode is accompanied by a subjective feeling of increased energy across the day, suggesting that less sleep may be needed to function. However, in some cases of mania, reduced sleep is linked to a subjective experience of feeling "tired but wired," which is reminiscent of the classical state of insomniacs at bedtime. Furthermore, this view of a decreased need for sleep in mania is difficult to reconcile with the fact that other daytime symptoms characteristic of mania, such as irritable mood, increased distractibility, and altered judgment (often leading to higher risk-taking behavior), can all be triggered by sleep loss. Therefore, this high-energy/reduced-sleep feature of mania might reflect a dysfunctional state in which sleep needs are unmet rather than a period during which sleep needs are reduced. For instance, as we discuss in later sections of this chapter, evidence suggests that sleep and circadian disturbances may be predisposing and aggravating factors mediating affective and cognitive difficulties in mania-related disorders (Boland & Alloy, 2013; McKenna & Eyler, 2012).

Periods of increased psychomotor activity and reduced sleep duration during mania are often followed by physiological exhaustion and increased sleep durations beyond pre-episode levels. This suggests that sleep and energy homeostasis have been driven to extremes, and that the body's need for sleep has been ignored. Accordingly, although sleep loss can trigger manic episodes, subsequent recovery sleep has been shown to attenuate or terminate manic episodes in most cases (e.g., Wehr, Goodwin, Wirz-Justice, Breitmaier, & Craig, 1982). The restoration of the sleep-wake cycle has long been a therapeutic priority for manic states, and it has been facilitated by sedative medication, forcing sleep to resume after periods of overactivation. Prior to the common use of sedative agents, authors reported manic episodes with prolonged sleeplessness eventually leading to death (Bell, 1849). These extreme and dramatic examples imply that, despite sustained subjective energy, sleep loss in the context of manic episodes can be expressed independently of the vital physiological need for sleep.

The concept of sleep need is closely related to the regulation of sleep pressure, or sleep homeostasis. Through the homeostatic sleep process, sleep pressure (reflective of the physiological sleep need) increases progressively with the hours of wakefulness and decreases across the subsequent sleep episode. This can be monitored by changes in the electroencephalogram, with slow-wave activity increasing progressively with the time spent awake and decreasing exponentially across sleep. Furthermore, slow-wave sleep and slow-wave activity are significantly enhanced after sleep deprivation, which can be seen as a mechanism for restoring sleep homeostasis by increasing the amount of deep recuperative sleep in response to increased sleep pressure. From this perspective, changes in the dynamics of slow-wave activity across the day and night could possibly characterize the so-called reduced "sleep need" in manic states. To our knowledge, no study has specifically assessed this question. In a case study, Kupfer and Heninger Heninger (1972) observed that an elderly patient with recurring manic episodes failed to show the normal REM-sleep rebound following REM-sleep deprivation (i.e., a normal increase in the amount of REM sleep in the sleep episode following selective REM-sleep deprivation). Although this may suggest that the restorative equilibrium of some aspects of sleep may be altered in manic states, the finding is likely to have been strongly influenced by the effects of age on sleep regulation. Therefore, studies specifically investigating the build-up and dissipation of sleep pressure in the context of mania could help to better understand the possible variations in sleep need associated with mania.

The Typical Cyclic Nature of Mania Symptoms

Prospectively, the rate of recurrence of manic episodes has been estimated to be 35% over 1.75 years (Kora et al., 2008). In fact, episodes of mania often occur in a cyclical pattern, alternating with periods characterized by either euthymic or depressive mood over hours, days, or months. Long periods of low energy and depression are typical of bipolar disorder. Some depressive symptoms frequently persist during manic episodes, but they may be less salient than manic symptoms. From this perspective, the changes in motor and mental activation occurring during mania could represent the core cyclical component of affective illnesses such as bipolar disorder.

Interestingly, in a considerable subset of patients, the periodicity of mood switches appears to have some degree of intraindividual stability (Bunney, Murphy, Goodwin, & Borge, 1972; Sitaram, Gillin, & Bunney, 1978). These switches may be influenced by individual profiles of endogenous biological rhythms. In healthy individuals, subjective mood follows a circadian pattern generally defined by mood being at its best in the morning, worsening across the day, and reaching its lowest point around the time when the core body temperature reaches its nadir (Boivin et al., 1997; Wood & Magnello, 1992). Some evidence suggests that this pattern is more pronounced and sometimes even reversed (i.e., worse mood upon awakening and improvement across the day) in individuals with affective disorders (Gordijn, Beersma, Bouhuys, Reinink, & Van den Hoofdakker, 1994; Hall, Sing, & Romanoski, 1991; Tolle & Goetze, 1987). Early case studies revealed that rapid (within 24 hours) switches into manic states frequently occur in the later portion of the night or early morning and that these night time mood switches appear to be predictive of more severe subsequent sleep disturbances and mania symptoms (Gillin, Mazure, Post, Jimerson, & Bunney, 1977; Kupfer & Heninger, 1972; Sitaram et al., 1978). Interestingly, the circadian pattern of mood switches appears to vary according to the individual's length of mania cycles; those with a 48-hour period would commonly switch into mania during the night, and those with a 24-hour period would more often switch during the day (Wilk & Hegerl, 2010). In several cases of rapidly cycling mania, mood switches often coincided with periods of two or more sleepless nights alternating with sleeping nights (i.e., extended wakefulness or 48-hour sleep-wake cycles), during which mania symptoms worsened (Wehr et al., 1982). This relationship between shifts in mood and sleep, as well as other potential abnormalities in endogenous rhythms, might induce desynchrony between some circadian rhythms and the external 24-hour cycle, with the rhythms of other functions remaining entrained (Halberg, 1968; Kripke, Mullaney, Atkinson, & Wolf, 1978). This pattern could then lead to successive misalignments and realignments of these circadian rhythms as their respective cycles progress, and it could also possibly lead to recurrent changes in mood. Subsequent findings have indicated that the period length and internal synchrony across circadian rhythms may be abnormal during periods of mood disorders (Hasler, Buysse, Kupfer, & Germain, 2010; Lewy, 2009; Robillard, Naismith, Rogers, Scott, et al., 2013).

Mania-related illnesses also tend to follow infradian patterns, with marked changes in mood, sleep, social activity, and weight across seasons (Baek et al., 2011; Choi et al., 2011; Shand, Scott, Anderson, & Eagles, 2011). For instance, bipolar disorder is often associated with extended sleep duration and a higher likelihood of depression during winter (Faedda et al., 1993; Simonsen, Shand, Scott, & Eagles, 2011). Historically, there have been reports of increased hospitalizations for mania in autumn and spring. More recent findings confirm the occurrence of hypomania/mania symptoms is greater during fall (Akhter et al., 2013), and other observations appear to indicate a yearly biphasic rhythm with peaks in winter (December) and summer (notably in June) (Yang et al., 2013). Interestingly, mania symptoms seem to be more sensitive to seasonality in males than females and in younger compared to older persons (Yang et al., 2013). However, the pathophysiological mechanisms underlying these cyclical variations could also be influenced by exogenous rhythms, including the likely sensitivity of mood to changes in light-dark, activity-rest, and social/occupation cycles across days and seasons.

Longitudinal studies have identified key changes in sleep at the turning point of mood switches. In most cases, the day preceding the onset of a manic/hypomanic episode is characterized by a notable and often sudden reduction in sleep duration (e.g., 0-5 hours of sleep) and a decrease in the amount of REM sleep (Bunney, Murphy, Goodwin, & Borge, 1972; Kupfer & Heninger, 1972; Sitaram et al., 1978; Wehr et al., 1982). Most of these sleep changes typically persist and worsen across the episode and are often accompanied by increases in noradrenaline and dopamine levels (Bunney, Goodwin, Murphy, House, & Gordon, 1972; Post et al., 1977). Conversely, the transition from mania to depression episodes has been associated with increased REM sleep and total sleep duration, as well as decreased noradrenaline (Bunney, Goodwin, Murphy, House, & Gordon, 1972). Although induced sleep deprivation can reduce depressive symptoms, several experimental studies have demonstrated that sleep deprivation can also precipitate mania symptoms in a significant proportion of individuals with affective disorders (ranging between 5 and 29% across various studies) (Kasper & Wehr, 1992; Wu & Bunney, 1990; Figure 14.1).

These findings are in line with the observation that mania onset is often attributed to external factors likely to induce sleep disturbances, such as changing medications, recreational drug use, shift-work, having a new child, stressful life events, and traveling across time zones (Davenport & Adland, 1982; Peet & Peters, 1995; Ranga, Krishnan, Swartz, Larson, & Santoliquido, 1984; Wehr, Sack, & Rosenthal, 1987). For example, an observational study conducted near one of the busiest airports in the world highlighted an association between the onset of hypomanic episodes and west-to-east overnight flights (Jauhar & Weller, 1982). Moreover, the fact that restoring and realigning the sleep-wake cycle via benzodiazepines and chronobiotic agents often reduces symptoms of bipolar disorder further supports the potential causative role of sleep loss and circadian desynchrony in mania-related illnesses (Berk, Ichim, & Brook, 1999; Calabrese, Guelfi, Perdrizet-Chevallier, & Agomelatine Bipolar Study Group, 2007; Fornaro et al., 2013; McElroy et al., 2011; Meehan et al., 2001).

These observations reinforce the hypothesis that the cyclic recurrence of mania may be closely linked to the dysregulation of biological rhythms and that changes in the sleep-wake cycle could be one of the mechanisms involved in the emergence and maintenance of manic states. However, researchers must conduct further longitudinal studies using broader samples in order to ascertain whether consistent circadian and seasonal patterns of switches in and out of manic states are present in specific patient subgroups.

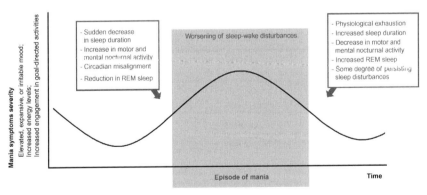

FIGURE 14.1 Schematic representation of sleep and circadian changes before, during, and after a mania episode.

SLEEP AND CIRCADIAN DISTURBANCES IN MANIA-RELATED ILLNESSES

Bipolar Disorder

Bipolar disorder encompasses the most common forms of recurring manic illness. The main symptoms of bipolar disorder include the occurrence of at least one episode of mania or hypomania, typically alternating with episodes of depression, fatigue, and low energy. In some syndromes, mania alternates with periods of low mood and energy, but, nevertheless, mania is considered to be distinct and independent from depression (Hickie, 2014; Merikangas et al., 2014). Individuals with bipolar disorder commonly exhibit several abnormalities of the sleep-wake cycle across these various phases of illness, as well as during the euthymic phases between episodes (for a recent review, see Robillard, Naismith, & Hickie, 2013).

When compared to healthy controls, persons in the manic phase of bipolar disorder appear to have longer sleep onset latency, shorter sleep duration, and lower sleep efficiency (Hudson, Lipinski, Frankenburg, Grochocinski, & Kupfer, 1988; Linkowski, Kerkhofs, Rielaert, & Mendlewicz, 1986). Furthermore, several studies have observed lower slow-wave sleep and REM sleep together with increased stage 1 sleep in hypomanic/manic patients, as compared to either controls or values typically found in healthy persons (Akiskal, 1984; Hudson et al., 1988, 1992; Kupfer, Wyatt, & Snyder, 1970; Mendels & Hawkins, 1971). Some (Hudson et al., 1988), but not all (Hartman, 1968; Mendels & Hawkins, 1971) studies also observed shorter REM latencies and increased REM density during manic phases. An early longitudinal study followed six persons with bipolar disorder across illness phases for a 10- to 26-month period, and the study revealed that, during manic/hypomanic phases, sleep duration and the proportion of REM were reduced, while REM latency was quite variable (Hartman, 1968). In fact, during manic episodes, they observed that, on most nights, participants obtained 4-5 hours of sleep and that several nights encompassed less than 15 minutes of REM sleep. This pattern contrasts with the increased amounts of REM sleep often seen during depressive episodes, as compared to euthymic phases.

Stage 4 sleep (the deeper component of slow-wave sleep) appeared to be highly variable and somewhat reduced during mania in some (Hartman, 1968) but not all (Hudson et al., 1988) studies. This variation may be modulated by recurring rebounds of slow-wave sleep following nights of severe sleep reduction during mania. Importantly, most of these sleep changes seem to be associated with higher symptom severity (Hudson et al., 1988, 1992; Linkowski et al., 1986). Nevertheless, sleep-wake disturbances are found to persist during euthymic phases in about 70% of persons with bipolar disorder (Harvey, Schmidt, Scarna, Semler, & Goodwin, 2005). These disturbances are often characterized by unstable sleep patterns, poorer subjective sleep quality and sleep efficiency, prolonged sleep latency, middle insomnia, shorter sleep duration, and extended time in bed (Brill, Penagaluri, Roberts, Gao, & El-Mallakh, 2011; Giglio et al., 2010; Gruber et al., 2011; Rocha, Neves, & Correa, 2013). Notably, although unstable sleep patterns during euthymic phases of bipolar disorder have been associated with subsequent manic and depressive symptoms, shorter sleep duration appears to be more specifically predictive of manic/hypomanic symptoms (Gruber et al., 2011; Leibenluft, Albert, Rosenthal, & Wehr, 1996).

In many instances, the difficult morning waking frequently observed in persons with bipolar disorder (Ashman et al., 1999; Robillard, Naismith, Rogers, Ip, et al., 2013) may indicate delayed circadian rhythms. However, to date, few studies have assessed changes in circadian markers during the manic phase of bipolar disorder. Using prospective measures of core body temperature in six patients with bipolar disorder in different phases of the illness, a group of researchers recently found a shorter circadian period, increased amplitude, and similar rates of phase advance and delay during manic episodes, as compared to euthymic phases and healthy controls (Fukuyama & Uchimura, 2011). In another interesting study, researchers monitored sleep and core body temperature in three women with bipolar disorder before, during, and after 3 to 4 weeks of isolation from external time cues (Wehr et al., 1985). In two of these three participants, the normal expected delay in the sleep-wake cycle during temporal isolation was accompanied by an abnormal shortening of the sleep-wake period, resulting in a misalignment between the sleep and temperature cycles and in the onset of a manic episode. Despite the small sample size of this study, its findings highlight the potential role of external synchronizers (i.e., zeitgeber) in maintaining some degree of circadian alignment, as well as possibly preventing or minimizing manic symptomatology. The possibility that affective illnesses can be linked to decreased sensitivity to various zeitgebers, which could lead to some degree of temporal isolation, requires further exploration. Nevertheless, because bipolar disorder is commonly associated with later sleep-wake preferences (i.e., chronotypes) and because persons in manic phases are likely to alter their light-dark and activity-rest cycles due to their drastically shortened sleep durations, self-modulated changes in zeitgebers represent another potential factor contributing to the maintenance of mania.

Adolescence and Early-Onset Bipolar Disorder

In bipolar disorder, earlier onset of illness has been associated with more severe symptoms, poorer functional outcomes, and worse prognosis (Berk et al., 2007; Geller Axelson & Luby, 2003; Perlis et al., 2004). The common overlap of mania symptoms with other developmental or psychiatric disorders, as well as the limitations inherent in children or adolescents and parental reports of symptoms, contribute to the diagnostic challenges and difficulties in characterizing early-onset bipolar disorder (Angold, Erkanli, Costello, & Rutter, 1996; Biederman & Jellinek, 1998). As in the adult world, the combination of sleep disturbances with high daytime energy levels and the cyclic nature of symptoms are considered to constitute core features of early-onset bipolar disorder (Geller et al., 1998), and this characterization can potentially assist diagnostic decisions.

Similar to results found in adults, early-onset bipolar disorder is associated with sleep initiation difficulties, poor sleep consolidation, low amounts of stage 4 sleep and REM sleep, and increased stage 1 sleep (Lofthouse, Fristad, Splaingard, & Kelleher, 2007; Mehl et al., 2006; Rao et al., 2002). However, some studies failed to show clear alterations in REM features (Rao et al., 2002), suggesting that REM-sleep disturbances, which are more commonly seen in adults with bipolar depression, are possibly attenuated in younger persons. Compared to older persons with bipolar disorder, younger individuals present with longer and more variable sleep duration (Gruber et al., 2011). However, this difference in sleep duration could be influenced by parental regulation of sleep schedules. A high proportion of youth with bipolar disorder also experience frequent nightmares, significant difficulties sleeping away from main attachment figures, nocturnal enuresis (often amplified by medication), increased fear of the dark, and sleep-related anxiety (Baroni, Hernandez, Grant, & Faedda, 2012; Lofthouse et al., 2007; Mehl et al., 2006). Importantly, in children and adolescents with bipolar disorder, sleep difficulties are associated with worse evening mood, excessive guilt, and self-harm (Roybal et al., 2011), and delayed sleep-wake patterns have been found to be predictive of a more severe and chronic course of illness (Staton, 2008).

Sleep and endogenous circadian rhythms undergo a significant delay during normal adolescence and young adulthood (Crowley, Acebo, & Carskadon, 2007). Affective disorders appear to interact with these normal age-related changes, often leading to more pronounced sleep and activity phase delays, as well as lower chronotypic 'morningness' preference and evening tiredness in younger individuals compared to older ones (Chung et al., 2012; Robillard et al., 2014). Furthermore, compared to young persons with unipolar depression, those with the bipolar subtype show more pronounced phase shifts toward later sleep schedules and later evening melatonin onset

(Robillard, Naismith, Rogers, Ip, et al., 2013; Robillard, Naismith, Rogers, Scott, et al., 2013). Strikingly, about 22% of young persons with bipolar disorder undergo a reversal of the sleep-wake cycle (Baroni et al., 2012).

Cyclothymia

Cyclothymia is characterized by recurrent subthreshold episodes of hypomania and mild to moderate depression alternating over at least 2 years (1 year in children). These mood switches often emerge during adolescence and typically occur rapidly and unpredictably, leading to significant instability and psychological distress (Akiskal, 2001; Brieger & Marneros, 1998). Clinical observations suggest that sleep disturbances frequently occur in conjunction with cyclothymia (Akiskal, Khani, & Scott-Strauss, 1979). A large-scale web-based survey indicated that "cyclothymic temperament" has been associated with prolonged sleep latency, later sleep schedule, frequent nighttime awakenings, poor sleep quality, and high rates of daily and occasional use of sleep medications (Ottoni, Lorenzi, & Lara, 2011). Recently, sleep disturbances have also been found to be worse in young persons with cyclothymia compared to clinical controls and generally similar, albeit slightly lower, to those with bipolar disorder (Van Meter, Youngstrom, Youngstrom, Feeny, & Findling, 2011).

Although it has been proposed that the core pathophysiology of cyclothymia is closely related to internal desynchrony among biological rhythms (Papousek, 1975), limited empirical data is currently available. A recent case study of a middle-age cyclothymic female noted a combination of higher "energetic mood" toward the end of the day and sleep duration of about 3 hours (Totterdell & Kellett, 2008). In many cases, cyclothymia is also subject to seasonal changes, with more prominent depressive episodes during winter (Akiskal, 2001).

Because of its more subtle symptoms, cyclothymia may go undetected or misrepresented in clinical settings. Nevertheless, recent epidemiological findings estimate that approximately 40% of the population with a history of major depressive disorder also reports experiencing subthreshold mania symptoms at some stage of their life (Angst et al., 2010). Thus, researchers may have to reach beyond formal clinical settings when investigating the interplay between sleep or circadian rhythms and low-severity mania.

Obsessive-Compulsive Symptoms

Obsessive-compulsive features are relatively common during manic episodes, and the recurrent drives to perform specific behaviors with related physical activation characteristic of obsessive-compulsive disorders

(OCD) somewhat resemble manic states. Not surprisingly, the comorbidity between OCD and mania-related illnesses such as bipolar disorder is high (Kruger, Cooke, Hasey, Jorna, & Persad, 1995; Masi et al., 2004; Perugi et al., 1997).

The sleep of persons suffering from OCD is characterized by reduced sleep duration and efficiency (Hohagen et al., 1994; Insel et al., 1982). In fact, recent findings suggest that over 90% of young persons with OCD experience sleep-related problems, including nightmares, sleeping difficulties, and being overtired (Storch et al., 2008). Although some studies observed no clear abnormality in sleep architecture (Hohagen et al., 1994; Walsleben et al., 1990), others indicated a profile somewhat similar to that of mania, with shortened REM latency and lower amounts of stage 4 sleep (Insel et al., 1982).

As appears to be the case for mania, some cyclic patterns have also been found for OCD symptoms. Early observations suggest that compulsive ritual behaviors follow a circadian pattern, decreasing across the day (Janet, 1908). Some endogenous circadian rhythms also appear to be disrupted; the circadian rhythmicity of body temperature appears to be attenuated (Millet et al., 1998). Furthermore, although the circadian variations of cortisol seem to be preserved in OCD, some studies observed a significant delay (by about 2 hours compared to controls) of the nocturnal peak of melatonin (Monteleone, Catapano, Del Buono, & Maj, 1994). Compared to healthy controls, persons with OCD have also shown significantly reduced melatonin and elevated cortisol across the 24-hour cycle (Catapano, Monteleone, Fuschino, Maj, & Kemali, 1992; Monteleone et al., 1994). Some of these endocrine abnormalities have seem to vary with the severity of OCD symptoms (Monteleone et al., 1994).

Attention Deficit Hyperactivity Disorder

The hyperactive symptomatology of attention deficit hyperactivity disorder (ADHD) resembles the motor activation, impulsive behavior, and distractibility that occur during manic episodes. ADHD also shares common characteristics with some aspects of the sleep and circadian profiles typically seen in manic states. Approximately 55-83% of persons with ADHD report sleep complaints (Corkum, Tannock, & Moldofsky, 1998; Philipsen, Hornyak, & Riemann, 2006). Objective sleep assessments have identified delayed and unstable sleep-wake patterns, lower sleep efficiency, and lower amounts of REM sleep (Gruber, Sadeh, & Raviv, 2000; Gruber et al., 2009; Rybak, McNeely, Mackenzie, Jain, & Levitan, 2007; Van Veen, Kooij, Boonstra, Gordijn, & Van Someren, 2010). As is the case for mania, sleep disturbances are thought to contribute to the pathophysiology of ADHD (Corkum et al., 1998; Golan, Shahar, Ravid, & Pillar, 2004; Yuen & Pelayo, 1999). Researchers have also proposed that circadian abnormalities in melatonin rhythms underlie some of the sleep difficulties (Van der Heijden, Smits, Van Someren, & Gunning, 2005; Van Veen et al., 2010).

The circadian profile linked to ADHD is characterized by delayed evening melatonin onset and morning cortisol peak, reduced melatonin amplitude, and disorganized circadian rhythmicity of melatonin (Baird, Coogan, Siddiqui, Donev, & Thome, 2012; Van der Heijden et al., 2005; Van Veen et al., 2010). Interestingly, the severity of ADHD symptoms reportedly correlates with a shorter period of the activity-rest cycle and with later chronotypes (Baird et al., 2012; Rybak et al., 2007). Similarly, some evidence suggests that abnormal timing of the cortisol rhythm is more prevalent in those with more severe hyperactivity symptoms (Kaneko, Hoshino, Hashimoto, Okano, & Kumashiro, 1993). Furthermore, mood, physical activity, and energy, all overlapping features of hyperactivity and mania, appear to be especially sensitive to seasonal changes in individuals with ADHD (Levitan, Jain, & Katzman, 1999; Rybak et al., 2007).

Despite the similarities in the symptomatology and sleep and circadian profiles of ADHD and mania, one of the most reliable distinguishing features is the subjective feeling of reduced need for sleep reported in 72% of young persons with bipolar spectrum disorders, compared to 6% in those with ADHD (Geller et al., 2002; Kowatch, Youngstrom, Danielyan, & Findling, 2005).

THE IMPACT OF SLEEP DISTURBANCES ON AFFECTIVE REGULATION, EXECUTIVE FUNCTIONS, AND RISK-TAKING BEHAVIOR

Affective Regulation

Observational and experimental studies have confirmed that sleep plays a critical role in affective regulation (for a recent review, see Walker & van der Helm, 2009). For example, the learning of contextual fear and adverse consequence avoidance, two important affective components of behavioral inhibition and risk-taking decisions, often affects subsequent sleep architecture (Smith, 1985; Smith, Young, & Young, 1980). Also, the consolidation of such learning is impaired after sleep loss, especially after REM-sleep deprivation (Beaulieu & Godbout, 2000; Walker & Stickgold, 2004). Accordingly, reductions in REM sleep during manic episodes have been hypothesized to induce changes in monoamine regulation leading to mood alterations in mania (Hartman, 1968).

Several experimental studies have focused on the impact of sleep loss on negative emotions, using tasks and scales specifically targeting depressed mood. However, authors have recently proposed that sleep loss leads to higher reactivity to positive as well as negative affective experiences, both from a neural and behavioral perspective (Gujar, Yoo, Hu, & Walker, 2011). More specifically, sleep deprivation enhances positive appraisal of emotional stimuli, while proportionally amplifying the reactivity of the mesolimbic system. This is in line with other findings showing that sleep deprivation enhances dopaminergic activity in brain networks involved in affective processing and reward (Volkow et al., 2008, 2009). Importantly, a series of studies conducted on persons with mania have also identified functional abnormalities in these networks in relation to various affective processing (Altshuler et al., 2005; Bermpohl et al., 2009, 2010; Elliott et al., 2004). Therefore, the sleep disturbances emerging during manic states may contribute to the pathogenesis of affective processing abnormalities. In addition, sleep deprivation studies have highlighted other affective changes compatible with manic states. For instance, sleep loss is known to augment irritability and frustration (Horne, 1985; Killgore, 2010). Furthermore, emotional lability following sleep deprivation can lead to transient euphoria, overexcitement, and disproportionate positive emotional reactions (Dahl, 1996; Horne, 1985, 1993).

In clinical populations, studies have reported manic episodes triggered by induced sleep deprivation (Barbini, Bertelli, Colombo, & Smeraldi, 1996; Wehr, 1992). These converging lines of evidence suggest that mania-related reductions in total sleep duration and REM sleep could possibly contribute to abnormalities in affective regulation processes, notably via functional alterations in the mesolimbic system.

Cognition

In addition to its incapacitating mood variations, mania is accompanied by significant cognitive dysfunctions. A few studies overcame the considerable challenges of cognitive testing during manic episodes and found widespread impairment across multiple cognitive domains, including sustained attention, working memory, verbal learning and memory, visuospatial functions, and executive functions such as verbal fluency, planning, problem solving, and cognitive flexibility, with impulsive/disinhibited and perseverative behavior (Basso, Lowery, Neel, Purdie, & Bornstein, 2002; Clark, Iversen, & Goodwin, 2001; Malhi et al., 2007; Martinez-Aran et al., 2004; Murphy et al., 1999). Furthermore, mania has been associated with positive affect bias on an attention-shifting task (Murphy et al., 1999). Aside from these observations based on standardized objective neuropsychological measures, the functional impacts of mania-related cognitive dysfunctions, especially in the executive domain, are also seen in the form of perseverative, risk-taking, and reward-seeking behaviors. There appears to be some variability in executive dysfunctions associated with mania, however, and some authors have proposed that these dysfunctions may affect a subgroup of persons with mania (Clark et al., 2001). Similarly, one could postulate that executive and other cognitive difficulties may be notably more prominent in individuals with worse sleep disturbances and/or at periods when these sleep disturbances are especially pronounced.

Indeed, most of the cognitive dysfunctions characteristic of manic states are known to be sensitive to sleep loss. Experimental studies conducted in healthy persons demonstrated that sleep deprivation alters sustained attention, working memory, and verbal learning (Drummond et al., 2000; Lim & Dinges, 2010). Although researchers have not yet reached a consensus on the effects of sleep loss on executive functions, some studies observed alterations in verbal fluency, planning and cognitive flexibility, and perseverative and impulsive behavior (Couyoumdjian et al., 2010; Drummond, Paulus, & Tapert, 2006; Harrison & Horne, 1999; Jones & Harrison, 2001; Nilsson et al., 2005). Sleep deprivation has also been shown to increase one's inclination toward immediate gratification and to take risks to obtain potential gains, while altering one's response to losses after making risky decisions (Killgore, Balkin, & Wesensten, 2006; Venkatraman, Chuah, Huettel, & Chee, 2007). Beyond the laboratory environment, sleep disturbances and delayed sleep-schedules in adolescents have been linked with self-reported risk-taking behavior (O'Brien & Mindell, 2005) and difficulties in emotional regulation thought to contribute to further sleep disturbances by increasing arousal (Dahl, 2002).

THERAPEUTIC IMPLICATIONS

Sleep and Circadian Disturbances as Predictors of Manic Episodes

For most psychiatric disorders, including mania-related illnesses, early detection optimizes preventative and treatment outcomes (Berk et al., 2007; McGorry, Hickie, Yung, Pantelis, & Jackson, 2006; Miklowitz & Chang, 2008). Sleep-wake disturbances tend to appear several years before the onset of mania-related illnesses, often during puberty (Ritter, Marx, Bauer, Leopold, & Pfennig, 2011). Also, it is estimated that sleep-wake disturbances precede the emergence of manic episodes in 77% of persons suffering from mania (Jackson, Cavanagh, & Scott, 2003). Thus, many consider changes in the sleep-wake cycle to be the most frequent prodromal sign of the emergence and recurrence of manic episodes.

Increasing evidence supports the use of sleep-wake and circadian characteristics to assist in the early identification of those at risk for illnesses involving manic episodes. For instance, many individuals with a personal history of depression and subthreshold mania symptoms or with a family history of bipolar disorder, depression, or schizoaffective disorder and subthreshold mood symptoms present similar sleep-wake patterns to those with established bipolar disorder (Ritter et al., 2012). This includes prolonged sleep onset latency, increased and variable nighttime activity, recurring insomnia and hypersomnia, difficulties awakening and getting out of bed in the morning, and heightened sensitivity to shifts in circadian rhythms. Also, in those with a family history of bipolar disorder, sleep disorders increase the likelihood of developing bipolar disorder (Duffy, Alda, Hajek, Sherry, & Grof, 2010). Interestingly, persons at risk for bipolar disorder also appear to display stronger interdependence of mood and sleep variations compared to healthy controls (Duffy et al., 2010; Ritter et al., 2012), which may reflect an increased sensitivity to the impacts of sleep loss on affective regulation (described in the previous section). Interestingly, a multivariate predictive model designed to predict the risks for bipolar disorder based on a series of potential phenomenological and biological markers identified sleep disturbances, circadian and homeostatic deregulation, and polymorphisms in circadian genes as relevant predictors (Brietzke et al., 2012). Although further longitudinal studies are required, some evidence indicates that sleep and circadian profiles change across the course of the illness. For instance, in a young cohort of persons with various neuropsychiatric disorders, including mania-related illnesses, studies have recently shown that sleep consolidation problems are more characteristic of earlier stages of illness, but sleep-phase delay and lower evening melatonin concentrations are more prominent in later stages (Naismith et al., 2012; Scott et al., 2014).

Aside from their potential use as markers of illness onset and general progression, sleep-wake patterns also constitute a barometer for episode recurrence. For individuals with chronic mania, sudden sleep reductions on the days preceding the episodes provide a concrete objective warning sign that the individual and clinician can use to prepare for an eventual relapse (Jackson et al., 2003). Similarly, it may be relevant to integrate seasonality in screening and clinical management plans. For instance, specific interventions to stabilize the sleep-wake cycle at the onset of spring may be particularly relevant in the prevention of mania. Such therapeutic strategies assisting patients to recognize early signs of emerging episodes have been found to minimize the impacts of mania on socio-occupational functioning and to prolong the time between episodes (Perry, Tarrier, Morriss, McCarthy, & Limb, 1999).

Sleep- and Circadian-Based Interventions for the Management of Mania Symptoms

Considering that sleep and circadian disturbances can influence affective regulation and mania-related cognitive and behavioral difficulties, and that abnormal sleep and circadian profiles may represent risk factors for mania-related illnesses, there is a great potential for relevant therapeutic applications. Studies have recently reviewed the application of sleep- and circadian-based interventions for a broad range of affective disorders (Benedetti, 2012; Hickie, Naismith, Robillard, Scott, & Hermens, 2013). The following section focuses on interventions especially relevant for mania symptoms.

Early interventions for manic episodes often focus on sedation to prevent physical exhaustion and restoration of the normal sleep-wake cycle. For instance, benzodiazepines can be a useful adjunctive treatment in addition to lithium carbonate to hasten sleep and circadian readjustments in the acute phases of mania (Modell, Lenox, & Weiner, 1985; Post et al., 1996). However, benzodiazepines also bear elevated risks for tolerance and abuse, which is especially undesirable in the context of mania. Furthermore, even after receiving sedative and antipsychotic medication, it is estimated that 90% of hospitalized manic patients sleep less than 7 hours and that more than 80% still present sleep initiation difficulties and/or early awakenings (Winokur, Clayton, & Reich, 1969). Other pharmacological agents impacting sleep and circadian systems with low potential for abuse, tolerance, or daytime sedation therefore represent attractive alternatives (Hickie & Rogers, 2011). For instance, the chronobiotic impacts of melatonin or melatonin agonists, such as Agomelatine and Ramelton, have promising effects on both sleep and depressive symptoms in bipolar disorder (Fornaro et al., 2013; McElroy et al., 2011; Norris et al., 2013), but little is known about the potential effects of such medications on mania symptoms. A series of studies observed frank improvements in sleep duration and manic symptoms after exogenous melatonin intake in patients with both mania and treatment-resistant insomnia (Bersani & Garavini, 2000).

Most long-term treatments for mania, such as lithium carbonate, stabilize sleep-wake and circadian cycles. One of the first polysomnographic studies conducted during mania/hypomania showed that lithium carbonate decreased the percentage of REM sleep and REM activity, increased slow-wave sleep, and extended REM latency (Kupfer, Wyatt, Greenspan, Scott, & Snyder, 1970). Consistent with a normalization of the sleep profile associated with manic states, these changes seemed to co-occur with symptom improvements, and participants most often reverted to their initial sleep profiles if they reduced or ceased lithium intake, suggesting that the mood-stabilizing effects of lithium are strongly related to changes in sleep. More recently, studies have demonstrated that lithium also inhibits glycogen synthase kinase 3-b and Per2, which results in the lengthening of the abnormally short circadian period characteristic of bipolar disorder (Li, Lu, Beesley, Loudon, & Meng, 2012).

Cognitive behavioral therapy for insomnia can be useful for affective illnesses such as depression. However, some of the key techniques essential for this therapy to be effective may not be appropriate in the context of mania. For example, although the increased sleep pressure resulting from sleep-restriction therapy (i.e., limiting the length of time spent in bed to the time usually spent asleep) and stimulus-control therapy (i.e., getting out of bed whenever the last 15 minutes or so have been spent awake) can restore homeostatic sleep mechanisms and even induce antidepressant effects, these treatments may also trigger manic episodes. Conversely, in persons with rapidly cycling mania, increasing the length of time spent in bed can have some beneficial effects (Wehr et al., 1998; Wirz-Justice, Quinto, Cajochen, Werth, & Hock, 1999). Implementing and maintaining a bedtime routine can also help to reduce nighttime arousal and facilitate the regularization of sleep schedule in persons with bipolar disorder (Harvey, Mullin, & Hinshaw, 2006).

Cognitive behavioral interventions for mania often focus on enforcing regular schedules for sleep, exercise, and social activities (e.g., Frank et al., 2005; Lam et al., 2003; Shen, Alloy, Abramson, & Sylvia, 2008). Unstable social rhythms have been found to predict the recurrence of manic episodes in bipolar disorder (Shen et al., 2008). Consequently, interpersonal and social rhythm therapy, a circadian-based intervention developed for the specific context of bipolar disorder, aims to stabilize these social circadian rhythms primarily by establishing daily routines for sleep, wake, eating, and purposeful activities. Used as an adjunctive treatment, this has been shown to prolong asymptomatic periods, improve social and occupational functioning, and reduce depressive and, to some degree, mania symptoms (Bouwkamp et al., 2013; Frank et al., 2005; Hlastala, Kotler, McClellan, & McCauley, 2010).

Manipulations of the light/dark cycle constitute other nonpharmacological interventions susceptible to improve mania symptomatology by restoring sleep and circadian rhythms. Indeed, the unstable sleep-wake schedules characteristic of manic states are likely to generate chaotic light exposure patterns that could contribute to further alterations in circadian entrainment and mood instability. According to self-report measures, mood appears to be more sensitive to light in individuals with bipolar disorder as compared to healthy controls (Ritter et al., 2012). Phototherapy, or bright light exposure, is known to alleviate depressive symptoms, but it can also induce hypomania/mania (Schwitzer, Neudorfer, Blecha, & Fleischhacker, 1990). A few studies of individuals with bipolar disorder observed positive outcomes of light therapy on depressive symptoms (Deltito, Moline, Pollak, Martin, & Maremmani, 1991; Papatheodorou & Kutcher, 1995; Wirz-Justice et al., 1999). Light could possibly have potential mood-stabilizing properties in the context of mania provided that its chronobiotic effects are used carefully to realign circadian rhythms. For instance, preliminary results support the use of increased light early in the day and darkness in the evening to stabilize sleep and mood in bipolar disorder (Phelps, 2008; Wirz-Justice et al., 1999).

Overall, several intervention pathways focusing on the restoration of sleep and circadian rhythms may have significant advantages as adjunctive therapies for manic states. For instance, in support of sleep enhancing interventions, longer sleep durations in early stages of hospitalization during manic episodes have been associated with prompt therapeutic response and early discharge (Nowlin-Finch, Altshuler, Szuba, & Mintz, 1994). Longer sleep during hospitalization has also been shown to improve cooperative attitudes during hospitalization (Barbini et al., 1996). In addition to possible improvement in cognition, this could lead to a better disposition to adhere to treatment and engage in psychotherapy.

CONCLUSION

Due to its unique combination of reduced sleep duration and enhanced daytime energy and its typical cyclical nature, mania is closely related to sleep and chronobiological systems. Several pathophysiological mechanisms at the affective, cognitive, and behavioral levels may explain the potentially bidirectional relationship between mania and sleep/circadian disturbances. Furthermore, sleep and biological rhythm disturbances often precede the occurrence of manic episodes, and increasing evidence suggests that they can contribute to the onset, maintenance, and cyclic recurrence of manic episodes. Consequently, sleep and chronobiological changes represent relevant predictive biomarkers and therapeutic targets for mania.

References

Akhter, A., Fiedorowicz, J. G., Zhang, T., Potash, J. B., Cavanaugh, J., Solomon, D. A., et al. (2013). Seasonal variation of manic and depressive symptoms in bipolar disorder. *Bipolar Disorder*. http://dx.doi.org/10.1111/bdi.12072.

Akiskal, H. S. (1984). Characterologic manifestations of affective disorders: Toward a new conceptualization. *Integrative Psychiatry*, 2(3), 83–88.

Akiskal, H. S. (2001). Dysthymia and cyclothymia in psychiatric practice a century after Kraepelin. *Journal of Affective Disorders*, 62(1–2), 17–31.

Akiskal, H., Khani, M., & Scott-Strauss, A. (1979). Cyclothymic temperamental disorders. *Psychiatric Clinics of North America*, 2, 527–554.

Altshuler, L., Bookheimer, S., Proenza, M. A., Townsend, J., Sabb, F., Firestine, A., et al. (2005). Increased amygdala activation during mania: A functional magnetic resonance imaging study. *American Journal of Psychiatry*, 162(6), 1211–1213. http://dx.doi.org/10.1176/appi.ajp.162.6.1211.

American Psychiatric Association. (2013). *Diagnostic and Statistical Manual of Mental Disorders* (5th ed.). Arlington, VA: American Psychiatric Publishing.

Angold, A., Erkanli, A., Costello, E. J., & Rutter, M. (1996). Precision, reliability and accuracy in the dating of symptom onsets in child and adolescent psychopathology. *Journal of Child Psychology and Psychiatry, 37*(6), 657–664.

Angst, J., Cui, L., Swendsen, J., Rothen, S., Cravchik, A., Kessler, R. C., et al. (2010). Major depressive disorder with subthreshold bipolarity in the National Comorbidity Survey Replication. *American Journal of Psychiatry, 167*(10), 1194–1201. http://dx.doi.org/10.1176/appi.ajp.2010.09071011.

Ashman, S. B., Monk, T. H., Kupfer, D. J., Clark, C. H., Myers, F. S., & Leibenluft, Frank E. (1999). Relationship between social rhythms and mood in patients with rapid cycling bipolar disorder. *Psychiatry Research, 86*(1), 1–8.

Baek, J. H., Park, D. Y., Choi, J., Kim, J. S., Choi, J. S., Ha, K., et al. (2011). Differences between bipolar I and bipolar II disorders in clinical features, comorbidity, and family history. *Journal of Affective Disorders, 131*(1–3), 59–67. http://dx.doi.org/10.1016/j.jad.2010.11.020.

Baird, A. L., Coogan, A. N., Siddiqui, A., Donev, R. M., & Thome, J. (2012). Adult attention-deficit hyperactivity disorder is associated with alterations in circadian rhythms at the behavioural, endocrine and molecular levels. *Molecular Psychiatry, 17*(10), 988–995. http://dx.doi.org/10.1038/mp.2011.149.

Barbini, B., Bertelli, S., Colombo, C., & Smeraldi, E. (1996). Sleep loss, a possible factor in augmenting manic episode. *Psychiatry Research, 65*(2), 121–125.

Baroni, A., Hernandez, M., Grant, M. C., & Faedda, G. L. (2012). Sleep disturbances in pediatric bipolar disorder: A comparison between bipolar I and bipolar NOS. *Frontiers in Psychiatry, 3*, 22. http://dx.doi.org/10.3389/fpsyt.2012.00022.

Basso, M. R., Lowery, N., Neel, J., Purdie, R., & Bornstein, R. A. (2002). Neuropsychological impairment among manic, depressed, and mixed-episode inpatients with bipolar disorder. *Neuropsychology, 16*(1), 84–91.

Beaulieu, I., & Godbout, R. (2000). Spatial learning on the Morris Water Maze Test after a short-term paradoxical sleep deprivation in the rat. *Brain and Cognition, 43*(1–3), 27–31.

Bell, L. (1849). On a form of disease resembling some advanced states of mania and fever. *The American Journal of Insanity, 6*, 97–127.

Benedetti, F. (2012). Antidepressant chronotherapeutics for bipolar depression. *Dialogues in Clinical NeuroSciences, 14*(4), 401–411.

Berk, M., Dodd, S., Callaly, P., Berk, L., Fitzgerald, P., de Castella, A. R., et al. (2007). History of illness prior to a diagnosis of bipolar disorder or schizoaffective disorder. *Journal of Affective Disorders, 103*(1–3), 181–186. http://dx.doi.org/10.1016/j.jad.2007.01.027.

Berk, M., Ichim, L., & Brook, S. (1999). Olanzapine compared to lithium in mania: A double-blind randomized controlled trial. *International Clinical Psychopharmacology, 14*(6), 339–343.

Bermpohl, F., Dalanay, U., Kahnt, T., Sajonz, B., Heimann, H., Ricken, R., et al. (2009). A preliminary study of increased amygdala activation to positive affective stimuli in mania. *Bipolar Disorder, 11*(1), 70–75. http://dx.doi.org/10.1111/j.1399-5618.2008.00648.x.

Bermpohl, F., Kahnt, T., Dalanay, U., Hagele, C., Sajonz, B., Wegner, T., et al. (2010). Altered representation of expected value in the orbitofrontal cortex in mania. *Human Brain Mapping, 31*(7), 958–969. http://dx.doi.org/10.1002/hbm.20909.

Bersani, G., & Garavini, A. (2000). Melatonin add-on in manic patients with treatment resistant insomnia. *Progress in Neuro-Psychopharmacology and Biological Psychiatry, 24*(2), 185–191.

Biederman, J., & Jellinek, M. S. (1998). Resolved: Mania is mistaken for ADHD in prepubertal children. *Journal of the American Academy of Child & Adolescent Psychiatry, 37*(10), 1091–1099.

Boivin, D. B., Czeisler, C. A., Dijk, D. J., Duffy, J. F., Folkard, S., Minors, D. S., et al. (1997). Complex interaction of the sleep-wake cycle and circadian phase modulates mood in healthy subjects. *Archives of General Psychiatry, 54*(2), 145–152.

Boland, E. M., & Alloy, L. B. (2013). Sleep disturbance and cognitive deficits in Bipolar disorder: Toward an integrated examination of disorder maintenance and functional impairment. *Clinical Psychology Review, 33*(1), 33–44. http://dx.doi.org/10.1016/j.cpr.2012.10.001.

Bouwkamp, C. G., de Kruiff, M. E., van Troost, T. M., Snippe, D., Blom, M. J., de Winter, R. F., et al. (2013). Interpersonal and social rhythm group therapy for patients with Bipolar disorder. *International Journal of Group Psychotherapy, 63*(1), 97–115. http://dx.doi.org/10.1521/ijgp.2013.63.1.97.

Brieger, P., & Marneros, A. (1998). Dysthymia and cyclothymia–serious consequences of rarely diagnosed disorders. *Versicherungsmedizin, 50*(6), 215–218.

Brietzke, E., Mansur, R. B., Soczynska, J. K., Kapczinski, F., Bressan, R. A., & McIntyre, R. S. (2012). Towards a multifactorial approach for prediction of Bipolar disorder in at risk populations. *Journal of Affective Disorders, 140*(1), 82–91. http://dx.doi.org/10.1016/j.jad.2012.02.016.

Brill, S., Penagaluri, P., Roberts, R. J., Gao, Y., & El-Mallakh, R. S. (2011). Sleep disturbances in euthymic bipolar patients. *Annals of Clinical Psychiatry, 23*(2), 113–116.

Bunney, W. E., Jr., Goodwin, F. K., Murphy, D. L., House, K. M., & Gordon, E. K. (1972). The "switch process" in manic-depressive illness. II. Relationship to catecholamines, REM sleep, and drugs. *Archives of General Psychiatry, 27*(3), 304–309.

Bunney, W. E., Jr., Murphy, D. L., Goodwin, F. K., & Borge, G. F. (1972). The "switch process" in manic-depressive illness. I. A systematic study of sequential behavioral changes. *Archives of General Psychiatry, 27*(3), 295–302.

Calabrese, J. R., Guelfi, J. D., Perdrizet-Chevallier, C., & Agomelatine Bipolar Study Group. (2007). Agomelatine adjunctive therapy for acute bipolar depression: Preliminary open data. *Bipolar Disorder, 9*(6), 628–635. http://dx.doi.org/10.1111/j.1399-5618.2007.00507.x.

Catapano, F., Monteleone, P., Fuschino, A., Maj, M., & Kemali, D. (1992). Melatonin and cortisol secretion in patients with primary obsessive-compulsive disorder. *Psychiatry Research, 44*(3), 217–225.

Choi, J., Baek, J. H., Noh, J., Kim, J. S., Choi, J. S., Ha, K., et al. (2011). Association of seasonality and premenstrual symptoms in bipolar I and bipolar II disorders. *Journal of Affective Disorders, 129*(1–3), 313–316. http://dx.doi.org/10.1016/j.jad.2010.07.030.

Chung, J. K., Lee, K. Y., Kim, S. H., Kim, E. J., Jeong, S. H., Jung, H. Y., et al. (2012). Circadian rhythm characteristics in mood disorders: Comparison among bipolar I disorder, bipolar II disorder and recurrent major depressive disorder. *Clinical Psychopharmacology and Neuroscience, 10*(2), 110–116. http://dx.doi.org/10.9758/cpn.2012.10.2.110.

Clark, L., Iversen, S. D., & Goodwin, G. M. (2001). A neuropsychological investigation of prefrontal cortex involvement in acute mania. *American Journal of Psychiatry, 158*(10), 1605–1611.

Corkum, P., Tannock, R., & Moldofsky, H. (1998). Sleep disturbances in children with attention-deficit/hyperactivity disorder. *Journal of the American Academy of Child & Adolescent Psychiatry, 37*(6), 637–646. http://dx.doi.org/10.1097/00004583-199806000-00014.

Couyoumdjian, A., Sdoia, S., Tempesta, D., Curcio, G., Rastellini, E., De Gennaro, L., et al. (2010). The effects of sleep and sleep deprivation on task-switching performance. *Journal of Sleep Research, 19*(1 Pt 1), 64–70. http://dx.doi.org/10.1111/j.1365-2869.2009.00774.x.

Crowley, S. J., Acebo, C., & Carskadon, M. A. (2007). Sleep, circadian rhythms, and delayed phase in adolescence. *Sleep Medicine, 8*(6), 602–612. http://dx.doi.org/10.1016/j.sleep.2006.12.002.

Dahl, R. E. (1996). The impact of inadequate sleep on children's daytime cognitive function. *Seminars in Pediatric Neurology, 3*(1), 44–50.

Dahl, R. E. (2002). The regulation of sleep-arousal, affect, and attention in adolescence: Some questions and speculations. In M. A. Carskadon (Ed.), *Adolescent sleep patterns: Biological, social and psychological influences* (pp. 269–284). Cambridge: Cambridge University Press.

Davenport, Y. B., & Adland, M. L. (1982). Postpartum psychoses in female and male bipolar manic-depressive patients. *American Journal of Orthopsychiatry, 52*(2), 288–297.

Deltito, J. A., Moline, M., Pollak, C., Martin, L. Y., & Maremmani, I. (1991). Effects of phototherapy on non-seasonal unipolar and bipolar depressive spectrum disorders. *Journal of Affective Disorders, 23*(4), 231–237.

Drummond, S. P., Brown, G. G., Gillin, J. C., Stricker, J. L., Wong, E. C., & Buxton, R. B. (2000). Altered brain response to verbal learning following sleep deprivation. *Nature, 403*(6770), 655–657. http://dx.doi.org/10.1038/35001068.

Drummond, S. P., Paulus, M. P., & Tapert, S. F. (2006). Effects of two nights sleep deprivation and two nights recovery sleep on response inhibition. *Journal of Sleep Research, 15*(3), 261–265. http://dx.doi.org/10.1111/j.1365-2869.2006.00535.x.

Duffy, A., Alda, M., Hajek, T., Sherry, S. B., & Grof, P. (2010). Early stages in the development of bipolar disorder. *Journal of Affective Disorders, 121*(1–2), 127–135. http://dx.doi.org/10.1016/j.jad.2009.05.022.

Elliott, R., Ogilvie, A., Rubinsztein, J. S., Calderon, G., Dolan, R. J., & Sahakian, B. J. (2004). Abnormal ventral frontal response during performance of an affective go/no go task in patients with mania. *Biological Psychiatry, 55*(12), 1163–1170. http://dx.doi.org/10.1016/j.biopsych.2004.03.007.

Faedda, G. L., Tondo, L., Teicher, M. H., Baldessarini, R. J., Gelbard, H. A., & Floris, G. F. (1993). Seasonal mood disorders. Patterns of seasonal recurrence in mania and depression. *Archives of General Psychiatry, 50*(1), 17–23.

Fornaro, M., McCarthy, M. J., De Berardis, D., De Pasquale, C., Tabaton, M., Martino, M., et al. (2013). Adjunctive agomelatine therapy in the treatment of acute bipolar II depression: A preliminary open label study. *Neuropsychiatric Disease and Treatment, 9*, 243–251. http://dx.doi.org/10.2147/NDT.S41557.

Frank, E., Kupfer, D. J., Thase, M. E., Mallinger, A. G., Swartz, H. A., Fagiolini, A. M., et al. (2005). Two-year outcomes for interpersonal and social rhythm therapy in individuals with bipolar I disorder. *Archives of General Psychiatry, 62*(9), 996–1004. http://dx.doi.org/10.1001/archpsyc.62.9.996.

Fukuyama, H., & Uchimura, N. (2011). Changes in rectal temperature rhythm on manic states of bipolar patients. 久留米大学文学部紀要, 社会福祉学科編 (Minutes of Kurume University Department of Literature, social welfare discipline), *10*, 53–58.

Geller Axelson, D., & Luby, J. (2003). Phenomenology and longitudinal course of children with a prepubertal and early adolescent bipolar disorder phenotype. In B.G.M.P. DelBello (Ed.), *Bipolar disorderer in childhood and early adolescence* (pp. 25–50). New York, NY: Guilford Press.

Geller, B., Williams, M., Zimerman, B., Frazier, J., Beringer, L., Warner, K. L., et al. (1998). Prepubertal and early adolescent bipolarity differentiate from ADHD by manic symptoms, grandiose delusions, ultra-rapid or ultradian cycling. *Journal of Affective Disorders, 51*(2), 81–91.

Geller, B., Zimerman, B., Williams, M., Delbello, M. P., Bolhofner, K., Craney, J. L., et al. (2002). DSM-IV mania symptoms in a prepubertal and early adolescent Bipolar disorderer phenotype compared to attention-deficit hyperactive and normal controls. *Journal of Child and Adolescent Psychopharmacology, 12*(1), 11–25. http://dx.doi.org/10.1089/10445460252943533.

Giglio, L. M., Magalhaes, P. V., Andersen, M. L., Walz, J. C., Jakobson, L., & Kapczinski, F. (2010). Circadian preference in Bipolar disorderer. *Sleep and Breathing, 14*(2), 153–155. http://dx.doi.org/10.1007/s11325-009-0301-3.

Gillin, J. C., Mazure, C., Post, R. M., Jimerson, D., & Bunney, W. E., Jr. (1977). An EEG sleep study of a bipolar (manic-depressive) patient with a nocturnal switch process. *Biological Psychiatry, 12*(6), 711–718.

Golan, N., Shahar, E., Ravid, S., & Pillar, G. (2004). Sleep disorders and daytime sleepiness in children with attention-deficit/hyperactivity disorder. *Sleep, 27*(2), 261–266.

Gordijn, M. C., Beersma, D. G., Bouhuys, A. L., Reinink, E., & Van den Hoofdakker, R. H. (1994). A longitudinal study of diurnal mood variation in depression; characteristics and significance. *Journal of Affective Disorders, 31*(4), 261–273.

Gruber, J., Miklowitz, D. J., Harvey, A. G., Frank, E., Kupfer, D., Thase, M. E., et al. (2011). Sleep matters: Sleep functioning and course of illness in Bipolar disorderer. *Journal of Affective Disorders, 134*(1–3), 416–420. http://dx.doi.org/10.1016/j.jad.2011.05.016.

Gruber, R., Sadeh, A., & Raviv, A. (2000). Instability of sleep patterns in children with attention-deficit/hyperactivity disorder. *Journal of the American Academy of Child and Adolescent Psychiatry, 39*(4), 495–501. http://dx.doi.org/10.1097/00004583-200004000-00019.

Gruber, R., Xi, T., Frenette, S., Robert, M., Vannasinh, P., & Carrier, J. (2009). Sleep disturbances in prepubertal children with attention deficit hyperactivity disorder: A home polysomnography study. *Sleep, 32*(3), 343–350.

Gujar, N., Yoo, S. S., Hu, P., & Walker, M. P. (2011). Sleep deprivation amplifies reactivity of brain reward networks, biasing the appraisal of positive emotional experiences. *Journal of Neuroscience, 31*(12), 4466–4474. http://dx.doi.org/10.1523/JNEUROSCI.3220-10.2011.

Halberg, F. (1968). Physiological considerations underlying rhythmometry, with special reference to emotional illness. Paper presented at the Symposium on Biological Cycles and Psychiatry, Geneva.

Hall, D. P., Jr., Sing, H. C., & Romanoski, A. J. (1991). Identification and characterization of greater mood variance in depression. *American Journal of Psychiatry, 148*(10), 1341–1345.

Harrison, Y., & Horne, J. A. (1999). One night of sleep loss impairs innovative thinking and flexible decision making. *Organizational Behavior and Human Decision Processes, 78*(2), 128–145. http://dx.doi.org/10.1006/obhd.1999.2827.

Hartman, E. (1968). Longitudinal studies of sleep and dream patterns in manic-depressive patients. *Archives of General Psychiatry, 19*(3), 312–329.

Harvey, A. G., Mullin, B. C., & Hinshaw, S. P. (2006). Sleep and circadian rhythms in children and adolescents with Bipolar disorderer. *Development and Psychopathology, 18*(4), 1147–1168. http://dx.doi.org/10.1017/S095457940606055X.

Harvey, A. G., Schmidt, D. A., Scarna, A., Semler, C. N., & Goodwin, G. M. (2005). Sleep-related functioning in euthymic patients with Bipolar disorderer, patients with insomnia, and subjects without sleep problems. *American Journal of Psychiatry, 162*(1), 50–57. http://dx.doi.org/10.1176/appi.ajp.162.1.50.

Hasler, B. P., Buysse, D. J., Kupfer, D. J., & Germain, A. (2010). Phase relationships between core body temperature, melatonin, and sleep are associated with depression severity: Further evidence for circadian misalignment in non-seasonal depression. *Psychiatry Research, 178*(1), 205–207. http://dx.doi.org/10.1016/j.psychres.2010.04.027.

Hickie, I. B. (2014). Evidence for separate inheritance of mania and depression challenges current concepts of bipolar mood disorder. *Molecular Psychiatry, 19*(2), 153–155. http://dx.doi.org/10.1038/mp.2013.173.

Hickie, I. B., Naismith, S. L., Robillard, R., Scott, E. M., & Hermens, D. F. (2013). Manipulating the sleep-wake cycle and circadian rhythms to improve clinical management of major depression. *BMC Medicine, 11*, 79. http://dx.doi.org/10.1186/1741-7015-11-79.

Hickie, I. B., & Rogers, N. L. (2011). Novel melatonin-based therapies: Potential advances in the treatment of major depression. *Lancet, 378*(9791), 621–631. http://dx.doi.org/10.1016/S0140-6736(11)60095-0.

Hlastala, S. A., Kotler, J. S., McClellan, J. M., & McCauley, E. A. (2010). Interpersonal and social rhythm therapy for adolescents with Bipolar disorderer: Treatment development and results from an open trial. *Depression and Anxiety, 27*(5), 457–464. http://dx.doi.org/10.1002/da.20668.

Hohagen, F., Lis, S., Krieger, S., Winkelmann, G., Riemann, D., Fritsch-Montero, R., et al. (1994). Sleep EEG of patients with obsessive-compulsive disorder. *European Archives of Psychiatry and Clinical Neuroscience, 243*(5), 273–278.

Horne, J. A. (1985). Sleep function, with particular reference to sleep deprivation. *Annals of Clinical Research*, 17(5), 199–208.

Horne, J. A. (1993). Human sleep, sleep loss and behaviour. Implications for the prefrontal cortex and psychiatric disorder. *British Journal of Psychiatry*, 162, 413–419.

Hudson, J. I., Lipinski, J. F., Frankenburg, F. R., Grochocinski, V. J., & Kupfer, D. J. (1988). Electroencephalographic sleep in mania. *Archives of General Psychiatry*, 45(3), 267–273.

Hudson, J. I., Lipinski, J. F., Keck, P. E., Jr., Aizley, H. G., Lukas, S. E., Rothschild, A. J., et al. (1992). Polysomnographic characteristics of young manic patients. Comparison with unipolar depressed patients and normal control subjects. *Archives of General Psychiatry*, 49(5), 378–383.

Insel, T. R., Gillin, J. C., Moore, A., Mendelson, W. B., Loewenstein, R. J., & Murphy, D. L. (1982). The sleep of patients with obsessive-compulsive disorder. *Archives of General Psychiatry*, 39(12), 1372–1377.

Jackson, A., Cavanagh, J., & Scott, J. (2003). A systematic review of manic and depressive prodromes. *Journal of Affective Disorders*, 74(3), 209–217.

Janet, P. (1908). *Les obsessions et la psychasthenie*. Bailliere, Paris.

Jauhar, P., & Weller, M. P. (1982). Psychiatric morbidity and time zone changes: A study of patients from Heathrow airport. *British Journal of Psychiatry*, 140, 231–235.

Jones, K., & Harrison, Y. (2001). Frontal lobe function, sleep loss and fragmented sleep. *Sleep Medicine Reviews*, 5(6), 463–475. http://dx.doi.org/10.1053/smrv.2001.0203.

Kaneko, M., Hoshino, Y., Hashimoto, S., Okano, T., & Kumashiro, H. (1993). Hypothalamic-pituitary-adrenal axis function in children with attention-deficit hyperactivity disorder. *Journal of Autism and Developmental Disorders*, 23(1), 59–65.

Kasper, S., & Wehr, T. A. (1992). The role of sleep and wakefulness in the genesis of depression and mania. *Encephale*, 18(Spec No. 1), 45–50.

Kessler, R. C., Rubinow, D. R., Holmes, C., Abelson, J. M., & Zhao, S. (1997). The epidemiology of DSM-III-R bipolar I disorder in a general population survey. *Psychological Medicine*, 27(5), 1079–1089.

Killgore, W. D. (2010). Effects of sleep deprivation on cognition. *Progress in Brain Research*, 185, 105–129. http://dx.doi.org/10.1016/B978-0-444-53702-7.00007-5.

Killgore, W. D., Balkin, T. J., & Wesensten, N. J. (2006). Impaired decision making following 49 h of sleep deprivation. *Journal of Sleep Research*, 15(1), 7–13. http://dx.doi.org/10.1111/j.1365-2869.2006.00487.x.

Kora, K., Saylan, M., Akkaya, C., Karamustafalioglu, N., Tomruk, N., Yasan, A., et al. (2008). Predictive factors for time to remission and recurrence in patients treated for acute mania: Health outcomes of manic episodes (HOME) study. *The Primary Care Companion—Journal of Clinical Psychiatry*, 10(2), 114–119.

Kowatch, R. A., Youngstrom, E. A., Danielyan, A., & Findling, R. L. (2005). Review and meta-analysis of the phenomenology and clinical characteristics of mania in children and adolescents. *Bipolar Disorder*, 7(6), 483–496. http://dx.doi.org/10.1111/j.1399-5618.2005.00261.x.

Kripke, D. F., Mullaney, D. J., Atkinson, M., & Wolf, S. (1978). Circadian rhythm disorders in manic-depressives. *Biological Psychiatry*, 13(3), 335–351.

Kruger, S., Cooke, R. G., Hasey, G. M., Jorna, T., & Persad, E. (1995). Comorbidity of obsessive compulsive disorder in Bipolar disorderer. *Journal of Affective Disorders*, 34(2), 117–120.

Kupfer, D. J., & Heninger, G. R. (1972). REM activity as a correlate of mood changes throughout the night. Electroencephalographic sleep patterns in a patient with a 48-hour cyclic mood disorder. *Archives of General Psychiatry*, 27(3), 368–373.

Kupfer, D. J., Wyatt, R. J., Greenspan, K., Scott, J., & Snyder, F. (1970). Lithium carbonate and sleep in affective illness. *Archives of General Psychiatry*, 23(1), 35–40.

Kupfer, D. J., Wyatt, R. J., & Snyder, F. (1970). Comparison between electroencephalographic and systematic nursing observations of sleep in psychiatric patients. *Journal of Nervous and Mental Disease*, 151(6), 361–368.

Lam, D. H., Watkins, E. R., Hayward, P., Bright, J., Wright, K., Kerr, N., et al. (2003). A randomized controlled study of cognitive therapy for relapse prevention for bipolar affective disorder: Outcome of the first year. *Archives of General Psychiatry, 60*(2), 145–152.

Leibenluft, E., Albert, P. S., Rosenthal, N. E., & Wehr, T. A. (1996). Relationship between sleep and mood in patients with rapid-cycling Bipolar disorderer. *Psychiatry Research, 63*(2–3), 161–168.

Levitan, R. D., Jain, U. R., & Katzman, M. A. (1999). Seasonal affective symptoms in adults with residual attention-deficit hyperactivity disorder. *Comprehensive Psychiatry, 40*(4), 261–267.

Lewy, A. J. (2009). Circadian misalignment in mood disturbances. *Current Psychiatry Reports, 11*(6), 459–465.

Li, J., Lu, W. Q., Beesley, S., Loudon, A. S., & Meng, Q. J. (2012). Lithium impacts on the amplitude and period of the molecular circadian clockwork. *PloS One, 7*(3), e33292. http://dx.doi.org/10.1371/journal.pone.0033292.

Lim, J., & Dinges, D. F. (2010). A meta-analysis of the impact of short-term sleep deprivation on cognitive variables. *Psychological Bulletin, 136*(3), 375–389. http://dx.doi.org/10.1037/a0018883.

Linkowski, P., Kerkhofs, M., Rielaert, C., & Mendlewicz, J. (1986). Sleep during mania in manic-depressive males. *European Archives of Psychiatry and Neurological Sciences, 235*(6), 339–341.

Lofthouse, N., Fristad, M., Splaingard, M., & Kelleher, K. (2007). Parent and child reports of sleep problems associated with early-onset bipolar spectrum disorders. *Journal of Family Psychology, 21*(1), 114–123. http://dx.doi.org/10.1037/0893-3200.21.1.114.

Malhi, G. S., Ivanovski, B., Hadzi-Pavlovic, D., Mitchell, P. B., Vieta, E., & Sachdev, P. (2007). Neuropsychological deficits and functional impairment in bipolar depression, hypomania and euthymia. *Bipolar Disorder, 9*(1–2), 114–125. http://dx.doi.org/10.1111/j.1399-5618.2007.00324.x.

Martinez-Aran, A., Vieta, E., Reinares, M., Colom, F., Torrent, C., Sanchez-Moreno, J., et al. (2004). Cognitive function across manic or hypomanic, depressed, and euthymic states in Bipolar disorderer. *American Journal of Psychiatry, 161*(2), 262–270.

Masi, G., Perugi, G., Toni, C., Millepiedi, S., Mucci, M., Bertini, N., et al. (2004). Obsessive-compulsive bipolar comorbidity: Focus on children and adolescents. *Journal of Affective Disorders, 78*(3), 175–183. http://dx.doi.org/10.1016/S0165-0327(03)00107-1.

McElroy, S. L., Winstanley, E. L., Martens, B., Patel, N. C., Mori, N., Moeller, D., et al. (2011). A randomized, placebo-controlled study of adjunctive ramelteon in ambulatory bipolar I disorder with manic symptoms and sleep disturbance. *International Clinical Psychopharmacology, 26*(1), 48–53. http://dx.doi.org/10.1097/YIC.0b013e3283400d35.

McGorry, P. D., Hickie, I. B., Yung, A. R., Pantelis, C., & Jackson, H. J. (2006). Clinical staging of psychiatric disorders: A heuristic framework for choosing earlier, safer and more effective interventions. *Australian & New Zealand Journal of Psychiatry, 40*(8), 616–622. http://dx.doi.org/10.1111/j.1440-1614.2006.01860.x.

McKenna, B. S., & Eyler, L. T. (2012). Overlapping prefrontal systems involved in cognitive and emotional processing in euthymic Bipolar disorderer and following sleep deprivation: A review of functional neuroimaging studies. *Clinical Psychology Review, 32*(7), 650–663. http://dx.doi.org/10.1016/j.cpr.2012.07.003.

Meehan, K., Zhang, F., David, S., Tohen, M., Janicak, P., Small, J., et al. (2001). A double-blind, randomized comparison of the efficacy and safety of intramuscular injections of olanzapine, lorazepam, or placebo in treating acutely agitated patients diagnosed with bipolar mania. *Journal of Clinical Psychopharmacology, 21*(4), 389–397.

Mehl, R. C., O'Brien, L. M., Jones, J. H., Dreisbach, J. K., Mervis, C. B., & Gozal, D. (2006). Correlates of sleep and pediatric Bipolar disorderer. *Sleep, 29*(2), 193–197.

Mendels, J., & Hawkins, D. R. (1971). Longitudinal sleep study in hypomania. *Archives of General Psychiatry, 25*(3).

Merikangas, K. R., Cui, L., Heaton, L., Nakamura, E., Roca, C., Ding, J., et al. (2014). Independence of familial transmission of mania and depression: Results of the NIMH family study of affective spectrum disorders. *Molecular Psychiatry, 19*(2), 272. http://dx.doi.org/10.1038/mp.2013.181.

Miklowitz, D. J., & Chang, K. D. (2008). Prevention of Bipolar disorderer in at-risk children: Theoretical assumptions and empirical foundations. *Development and Psychopathology, 20*(3), 881–897. http://dx.doi.org/10.1017/S0954579408000424.

Millet, B., Touitou, Y., Poirier, M. F., Bourdel, M. C., Hantouche, E., Bogdan, A., et al. (1998). Plasma melatonin and cortisol in patients with obsessive-compulsive disorder: Relationship with axillary temperature, physical activity, and clinical symptoms. *Biological Psychiatry, 44*(9), 874–881.

Modell, J. G., Lenox, R. H., & Weiner, S. (1985). Inpatient clinical trial of lorazepam for the management of manic agitation. *Journal of Clinical Psychopharmacology, 5*(2), 109–113.

Monteleone, P., Catapano, F., Del Buono, G., & Maj, M. (1994). Circadian rhythms of melatonin, cortisol and prolactin in patients with obsessive-compulsive disorder. *Acta Psychiatrica Scandinavica, 89*(6), 411–415.

Murphy, F. C., Sahakian, B. J., Rubinsztein, J. S., Michael, A., Rogers, R. D., Robbins, T. W., et al. (1999). Emotional bias and inhibitory control processes in mania and depression. *Psychological Medicine, 29*(6), 1307–1321.

Naismith, S. L., Hermens, D. F., Ip, T. K., Bolitho, S., Scott, E., Rogers, N. L., et al. (2012). Circadian profiles in young people during the early stages of affective disorder. *Translational Psychiatry, 2*, e123. http://dx.doi.org/10.1038/tp.2012.47.

Nilsson, J. P., Soderstrom, M., Karlsson, A. U., Lekander, M., Akerstedt, T., Lindroth, N. E., et al. (2005). Less effective executive functioning after one night's sleep deprivation. *Journal of Sleep Research, 14*(1), 1–6. http://dx.doi.org/10.1111/j.1365-2869.2005.00442.x.

Norris, E. R., Karen, B., Correll, J. R., Zemanek, K. J., Lerman, J., Primelo, R. A., et al. (2013). A double-blind, randomized, placebo-controlled trial of adjunctive ramelteon for the treatment of insomnia and mood stability in patients with euthymic Bipolar disorderer. *Journal of Affective Disorders, 144*(1–2), 141–147. http://dx.doi.org/10.1016/j.jad.2012.06.023.

Nowlin-Finch, N. L., Altshuler, L. L., Szuba, M. P., & Mintz, J. (1994). Rapid resolution of first episodes of mania: Sleep related? *Journal of Clinical Psychiatry, 55*(1), 26–29.

O'Brien, E. M., & Mindell, J. A. (2005). Sleep and risk-taking behavior in adolescents. *Behavioral Sleep Medicine, 3*(3), 113–133. http://dx.doi.org/10.1207/s15402010bsm0303_1.

Ottoni, G. L., Lorenzi, T. M., & Lara, D. R. (2011). Association of temperament with subjective sleep patterns. *Journal of Affective Disorders, 128*(1–2), 120–127. http://dx.doi.org/10.1016/j.jad.2010.06.014.

Papatheodorou, G., & Kutcher, S. (1995). The effect of adjunctive light therapy on ameliorating breakthrough depressive symptoms in adolescent-onset Bipolar disorderer. *Journal of Psychiatry and Neuroscience, 20*(3), 226–232.

Papousek, M. (1975). Chronobiological aspects of cyclothymia (author's transl). *Fortschritte der Neurologie, Psychiatrie, und ihrer Grenzgebiete, 43*(8), 381–440.

Peet, M., & Peters, S. (1995). Drug-induced mania. *Drug Safety, 12*(2), 146–153.

Perlis, R. H., Miyahara, S., Marangell, L. B., Wisniewski, S. R., Ostacher, M., DelBello, M. P., et al. (2004). Long-term implications of early onset in Bipolar disorderer: Data from the first 1000 participants in the systematic treatment enhancement program for Bipolar disorderer (STEP-BD). *Biological Psychiatry, 55*(9), 875–881. http://dx.doi.org/10.1016/j.biopsych.2004.01.022.

Perry, A., Tarrier, N., Morriss, R., McCarthy, E., & Limb, K. (1999). Randomised controlled trial of efficacy of teaching patients with Bipolar disorderer to identify early symptoms of relapse and obtain treatment. *BMJ, 318*(7177), 149–153.

Perugi, G., Akiskal, H. S., Pfanner, C., Presta, S., Gemignani, A., Milanfranchi, A., et al. (1997). The clinical impact of bipolar and unipolar affective comorbidity on obsessive-compulsive disorder. *Journal of Affective Disorders, 46*(1), 15–23.

Phelps, J. (2008). Dark therapy for Bipolar disorderer using amber lenses for blue light blockade. *Medical Hypotheses, 70*(2), 224–229. http://dx.doi.org/10.1016/j.mehy.2007.05.026.

Philipsen, A., Hornyak, M., & Riemann, D. (2006). Sleep and sleep disorders in adults with attention deficit/hyperactivity disorder. *Sleep Medicine Reviews, 10*(6), 399–405. http://dx.doi.org/10.1016/j.smrv.2006.05.002.

Post, R. M., Ketter, T. A., Pazzaglia, P. J., Denicoff, K., George, M. S., Callahan, A., et al. (1996). Rational polypharmacy in the bipolar affective disorders. *Epilepsy Research. Supplement, 11*, 153–180.

Post, R. M., Stoddard, F. J., Gillin, J. C., Buchsbaum, M. S., Runkle, D. C., Black, K. E., et al. (1977). Alterations in motor activity, sleep, and biochemistry in a cycling manic-depressive patient. *Archives of General Psychiatry, 34*(4), 470–477.

Ranga, K., Krishnan, R., Swartz, M. S., Larson, M. J., & Santoliquido, G. (1984). Funeral mania in recurrent bipolar affective disorders: Reports of three cases. *Journal of Clinical Psychiatry, 45*(7), 310–311.

Rao, U., Dahl, R. E., Ryan, N. D., Birmaher, B., Williamson, D. E., Rao, R., et al. (2002). Heterogeneity in EEG sleep findings in adolescent depression: Unipolar versus bipolar clinical course. *Journal of Affective Disorders, 70*(3), 273–280.

Ritter, P. S., Marx, C., Bauer, M., Leopold, K., & Pfennig, A. (2011). The role of disturbed sleep in the early recognition of Bipolar disorderer: A systematic review. *Bipolar Disorder, 13*(3), 227–237. http://dx.doi.org/10.1111/j.1399-5618.2011.00917.x.

Ritter, P. S., Marx, C., Lewtschenko, N., Pfeiffer, S., Leopold, K., Bauer, M., et al. (2012). The characteristics of sleep in patients with manifest Bipolar disorderer, subjects at high risk of developing the disease and healthy controls. *Journal of Neural Transmission, 119*(10), 1173–1184. http://dx.doi.org/10.1007/s00702-012-0883-y.

Robillard, R., Naismith, S. L., & Hickie, I. B. (2013). Recent advances in sleep-wake cycle and biological rhythms in Bipolar disorderer. *Current Psychiatry Reports, 15*(10), 402. http://dx.doi.org/10.1007/s11920-013-0402-3.

Robillard, R., Naismith, S. L., Rogers, N. L., Ip, T. K. C., Hermens, D. F., Scott, E. M., et al. (2013). Delayed sleep phase in young people with unipolar or bipolar affective disorders. *Journal of Affective Disorders, 145*(2), 260–263. http://dx.doi.org/10.1016/j.jad.2012.06.006.

Robillard, R., Naismith, S. L., Rogers, N. L., Scott, E. M., Ip, T. K. C., Hermens, D. F., et al. (2013). Sleep-wake cycle and melatonin rhythms in adolescents and young adults with mood disorders: Comparison of unipolar and bipolar phenotypes. *European Psychiatry, 28.7*, 412–416. http://dx.doi.org/10.1016/j.eurpsy.2013.04.001.

Robillard, R., Naismith, S. L., Smith, K., Rogers, N. L., White, D., Terpening, Z., et al. (2014). Sleep-wake cycle in young and older persons with a lifetime history of mood disorders. *PloS One, 9*(2), http://dx.doi.org/10.1371/journal.pone.0087763.

Rocha, P. M., Neves, F. S., & Correa, H. (2013). Significant sleep disturbances in euthymic bipolar patients. *Comprehensive Psychiatry.* http://dx.doi.org/10.1016/j.comppsych.2013.04.006.

Roybal, D. J., Chang, K. D., Chen, M. C., Howe, M. E., Gotlib, I. H., & Singh, M. K. (2011). Characterization and factors associated with sleep quality in adolescents with bipolar I disorder. *Child Psychiatry & Human Development, 42*(6), 724–740. http://dx.doi.org/10.1007/s10578-011-0239-0.

Rybak, Y. E., McNeely, H. E., Mackenzie, B. E., Jain, U. R., & Levitan, R. D. (2007). Seasonality and circadian preference in adult attention-deficit/hyperactivity disorder: Clinical and neuropsychological correlates. *Comprehensive Psychiatry, 48*(6), 562–571. http://dx.doi.org/10.1016/j.comppsych.2007.05.008.

Schwitzer, J., Neudorfer, C., Blecha, H. G., & Fleischhacker, W. W. (1990). Mania as a side effect of phototherapy. *Biological Psychiatry, 28*(6), 532–534.

Scott, E., Robillard, R., Hermens, D. F., Naismith, S. L., Rogers, N. L., Ip, T. K., et al. (2014). Dysregulated sleep-wake cycles in young people are associated with emerging stages of major mental disorders. *Early Intervention in Psychiatry.* http://dx.doi.org/10.1111/eip.12143, article first published online: 28 Apr 2014; Early View (Online Version of Record published before inclusion in an issue).

Shand, A. J., Scott, N. W., Anderson, S. M., & Eagles, J. M. (2011). The seasonality of bipolar affective disorder: Comparison with a primary care sample using the Seasonal Pattern Assessment Questionnaire. *Journal of Affective Disorders, 132*(1–2), 289–292. http://dx.doi.org/10.1016/j.jad.2011.02.015.

Shen, G. H., Alloy, L. B., Abramson, L. Y., & Sylvia, L. G. (2008). Social rhythm regularity and the onset of affective episodes in bipolar spectrum individuals. *Bipolar Disorder, 10*(4), 520–529. http://dx.doi.org/10.1111/j.1399-5618.2008.00583.x.

Simonsen, H., Shand, A. J., Scott, N. W., & Eagles, J. M. (2011). Seasonal symptoms in bipolar and primary care patients. *Journal of Affective Disorders, 132*(1–2), 200–208. http://dx.doi.org/10.1016/j.jad.2011.02.018.

Sitaram, N., Gillin, J. C., & Bunney, W. E., Jr. (1978). The switch process in manic-depressive illness. Circadian variation in time of switch and sleep and manic ratings before and after switch. *Acta Psychiatrica Scandinavica, 58*(3), 267–278.

Smith, C. (1985). Sleep states and learning: A review of the animal literature. *Neuroscience & Biobehavioral Reviews, 9*(2), 157–168.

Smith, C., Young, J., & Young, W. (1980). Prolonged increases in paradoxical sleep during and after avoidance-task acquisition. *Sleep, 3*(1), 67–81.

Staton, D. (2008). The impairment of pediatric bipolar sleep: Hypotheses regarding a core defect and phenotype-specific sleep disturbances. *Journal of Affective Disorders, 108*(3), 199–206. http://dx.doi.org/10.1016/j.jad.2007.10.007.

Storch, E. A., Murphy, T. K., Lack, C. W., Geffken, G. R., Jacob, M. L., & Goodman, W. K. (2008). Sleep-related problems in pediatric obsessive-compulsive disorder. *Journal of Anxiety Disorders, 22*(5), 877–885. http://dx.doi.org/10.1016/j.janxdis.2007.09.003.

Tolle, R., & Goetze, U. (1987). On the daily rhythm of depression symptomatology. *Psychopathology, 20*(5–6), 237–249.

Totterdell, P., & Kellett, S. (2008). Restructuring mood in cyclothymia using cognitive behavior therapy: An intensive time-sampling study. *Journal of Clinical Psychology, 64*(4), 501–518. http://dx.doi.org/10.1002/jclp.20444.

Van der Heijden, K. B., Smits, M. G., Van Someren, E. J., & Gunning, W. B. (2005). Idiopathic chronic sleep onset insomnia in attention-deficit/hyperactivity disorder: A circadian rhythm sleep disorder. *Chronobiology International, 22*(3), 559–570. http://dx.doi.org/10.1081/CBI-200062410.

Van Meter, A., Youngstrom, E. A., Youngstrom, J. K., Feeny, N. C., & Findling, R. L. (2011). Examining the validity of cyclothymic disorder in a youth sample. *Journal of Affective Disorders, 132*(1–2), 55–63. http://dx.doi.org/10.1016/j.jad.2011.02.004.

Van Veen, M. M., Kooij, J. J., Boonstra, A. M., Gordijn, M. C., & Van Someren, E. J. (2010). Delayed circadian rhythm in adults with attention-deficit/hyperactivity disorder and chronic sleep-onset insomnia. *Biological Psychiatry, 67*(11), 1091–1096. http://dx.doi.org/10.1016/j.biopsych.2009.12.032.

Venkatraman, V., Chuah, Y. M., Huettel, S. A., & Chee, M. W. (2007). Sleep deprivation elevates expectation of gains and attenuates response to losses following risky decisions. *Sleep, 30*(5), 603–609.

Volkow, N. D., Tomasi, D., Wang, G. J., Telang, F., Fowler, J. S., Wang, R. L., et al. (2009). Hyperstimulation of striatal D2 receptors with sleep deprivation: Implications for cognitive impairment. *NeuroImage, 45*(4), 1232–1240. http://dx.doi.org/10.1016/j.neuroimage.2009.01.003.

Volkow, N. D., Wang, G. J., Telang, F., Fowler, J. S., Logan, J., Wong, C., et al. (2008). Sleep deprivation decreases binding of [11C]raclopride to dopamine D2/D3 receptors in the human brain. *Journal of Neuroscience, 28*(34), 8454–8461. http://dx.doi.org/10.1523/JNEUROSCI.1443-08.2008.

Walker, M. P., & Stickgold, R. (2004). Sleep-dependent learning and memory consolidation. *Neuron, 44*(1), 121–133. http://dx.doi.org/10.1016/j.neuron.2004.08.031.

Walker, M. P., & van der Helm, E. (2009). Overnight therapy? The role of sleep in emotional brain processing. *Psychological Bulletin, 135*(5), 731–748. http://dx.doi.org/10.1037/a0016570.

Walsleben, J., Robinson, D., Lemus, C., Hackshaw, R., Norman, R., & Alvir, J. (1990). Polysomnographic aspects of obsessive-compulsive disorders. *Sleep Research, 19*.

Wehr, T. A. (1992). Improvement of depression and triggering of mania by sleep deprivation. *Journal of the American Medical Association, 267*(4), 548–551.

Wehr, T. A., Goodwin, F. K., Wirz-Justice, A., Breitmaier, J., & Craig, C. (1982). 48-hour sleep-wake cycles in manic-depressive illness: Naturalistic observations and sleep deprivation experiments. *Archives of General Psychiatry, 39*(5), 559–565.

Wehr, T. A., Sack, D. A., Duncan, W. C., Mendelson, W. B., Rosenthal, N. E., Gillin, J. C., et al. (1985). Sleep and circadian rhythms in affective patients isolated from external time cues. *Psychiatry Research, 15*(4), 327–339.

Wehr, T. A., Sack, D. A., & Rosenthal, N. E. (1987). Sleep reduction as a final common pathway in the genesis of mania. *American Journal of Psychiatry, 144*(2), 201–204.

Wehr, T. A., Turner, E. H., Shimada, J. M., Lowe, C. H., Barker, C., & Leibenluft, E. (1998). Treatment of rapidly cycling bipolar patient by using extended bed rest and darkness to stabilize the timing and duration of sleep. *Biological Psychiatry, 43*(11), 822–828.

Wilk, K., & Hegerl, U. (2010). Time of mood switches in ultra-rapid cycling disorder: A brief review. *Psychiatry Research, 180*(1), 1–4. http://dx.doi.org/10.1016/j.psychres.2009.08.011.

Winokur, G., Clayton, P., & Reich, T. (1969). *Manic depressive illness*. St. Louis: CV Mosby Co.

Wirz-Justice, A., Quinto, C., Cajochen, C., Werth, E., & Hock, C. (1999). A rapid-cycling bipolar patient treated with long nights, bedrest, and light. *Biological Psychiatry, 45*(8), 1075–1077.

Wood, C., & Magnello, M. E. (1992). Diurnal changes in perceptions of energy and mood. *Journal of the Royal Society of Medicine, 85*(4), 191–194.

Wu, J. C., & Bunney, W. E. (1990). The biological basis of an antidepressant response to sleep deprivation and relapse: Review and hypothesis. *American Journal of Psychiatry, 147*(1), 14–21.

Yang, A. C., Yang, C. H., Hong, C. J., Liou, Y. J., Shia, B. C., Peng, C. K., et al. (2013). Effects of age, sex, index admission, and predominant polarity on the seasonality of acute admissions for Bipolar disorderer: A population-based study. *Chronobiology International, 30*(4), 478–485. http://dx.doi.org/10.3109/07420528.2012.741172.

Yuen, K. M., & Pelayo, R. (1999). Sleep disorders and attention-deficit/hyperactivity disorder. *Journal of the American Medical Association, 281*(9), 797.

CHAPTER 15

Physical Activity, Sleep, and Biobehavioral Synergies for Health

Matthew P. Buman and Shawn D. Youngstedt*,†,‡*

*School of Nutrition and Health Promotion, Arizona State University, Phoenix, Arizona, USA
†College of Nursing and Health Innovation, Arizona State University, Phoenix, Arizona, USA
‡Phoenix VA Healthcare System, Phoenix, Arizona, USA

Sleep is essential for health and has important functions in metabolism, emotion, memory, immunity, and learning (Siegel, 2005). Poor sleep has been associated with morbidity and mortality (Cappuccio, D'Elia, Strazzullo, & Miller, 2010b; Mokdad, Stroup, & Giles, 2003), and studies have directly linked it to leading causes of death, including heart disease, diabetes, and stoke (Cappuccio, D'Elia, Strazzullo, & Miller, 2010a; Gangwisch et al., 2006, 2007; Jiménez-Conde, Ois, Rodríguez-Campello, Gomis, & Roquer, 2007). Poor sleep is also strongly associated with depression and anxiety (Perlis, Giles, Buysse, Tu, & Kupfer, 1997), memory and learning outcomes (Stickgold, 2013), and reduced quality of life (Institute of Medicine (US) Committee on Sleep Medicine and Research, 2006). Sleep even plays an important role in accidental deaths, with 2.5% of all fatal motor vehicle crashes due to drowsy driving (National Highway Traffic Safety Administration, 2012).

Poor sleep, insufficient sleep, and sleep disorders are also highly prevalent in the population. About 30% of the adults report no more than 6 h of sleep per night (Centers for Disease Control and Prevention (CDC), 2012; Gangwisch et al., 2007). Persistent insomnia is present in about 10% of the adult population (Ancoli-Israel & Roth, 1999), and more occasional or transient forms of insomnia are likely much higher but difficult to define precisely (Roth & Roehrs, 2003). Obstructive sleep apnea (OSA) and restless leg syndrome (RLS) are experienced by 20% (Young, Peppard, & Gottlieb, 2002) and 10% (Phillips et al., 2000) of adults, respectively. The importance of sleep health and its impact on public health continue to become more apparent (Perry, Patil, & Presley-Cantrell, 2013).

Long-term, sustainable treatments for poor sleep are needed. Pharmacological and nonpharmacological treatments are widely used. Pharmacological treatments have only short-term efficacy, have common side effects, and are associated with considerable risk for morbidity and mortality (Kripke, Langer, & Kline, 2012; Mallon, Broman, & Hetta, 2009). Nonpharmacological treatments such as cognitive-behavioral therapy for insomnia are efficacious (Morin et al., 1999) and widely used, but these approaches are costly and time-intensive to deliver and thus limit widespread dissemination (Richardson, 2000). Researchers suggest and sleep organizations recognize that regular exercise is as an important component and potential alternative treatment for poor sleep (Buman & King, 2010; Schutte-Rodin, Broch, Buysse, Dorsey, & Sateia, 2008; Youngstedt, O'Connor, & Dishman, 1997). Exercise is an appealing alternative for a number of reasons. First, the lay public generally believes that regular exercise improves sleep (Hasan, Urponen, Vuori, & Partinen, 1988; National Sleep Foundation, 2013; Vuori, Urponen, Hasan, & Partinen, 1988), suggesting the acceptability of the approach. Second, when performed appropriately, exercise is relatively simple, inexpensive, and safe for the large majority of the population (Youngstedt, 2005). Third, exercise is compelling given its well-established concomitant benefits for longevity and disease prevention independent of sleep (Physical Activity Guidelines Advisory Committee, 2008), potentially providing a broader impact on overall health than do other pharmacological or nonpharmacological treatments for poor sleep.

The first purpose of this chapter is to review the evidence supporting the relationship between exercise and sleep, including epidemiologic studies, acute laboratory-based studies, and randomized controlled exercise-training studies, with special considerations for temporal relationships, exercise modalities, and common sleep disorders. Our second purpose is to examine emerging trends in the field, including the relationship between sedentary behavior and sleep and a 24-h approach that captures the synergies between sleep, sedentary behavior, and exercise for chronic disease prevention.

EXERCISE AND SLEEP: OVERVIEW OF PRIMARY FINDINGS

The study of the relationship between exercise and sleep has a rich history spanning more than 40 years of original investigation. The observational evidence is consistent in supporting the association between self-reported exercise and better sleep. Exercise has been associated with greater ease of falling asleep, greater perceived deepness of sleep, higher morning alertness, and better overall sleep quality (Mizuno et al., 2004; Vuori et al., 1988; Youngstedt & Kline, 2006). A number of studies have also shown exercise to be associated with lower prevalence of sleep disorders, including insomnia (Sherrill, Kotchou, & Quan, 1998) and OSA (Quan et al., 2007).

Acute laboratory-based, randomized, controlled experimental studies have generally found more modest relationships between exercise and sleep. These studies typically measured a single bout of exercise and examined a subsequent night or consecutive nights of sleep, using objective methods (i.e., polysomnography). Meta-analyses of these studies have found moderate relationships with total sleep time, slow-wave sleep, delayed rapid eye movement (REM) onset, and reduced REM duration (Driver & Taylor, 2000; Youngstedt et al., 1997). However, an important limitation of the vast majority of this research is that subjects were typically young and lacked sleep complaints, thus limiting the potential benefit of exercise given the relatively high sleep quality of the subjects (Youngstedt, 2003).

Randomized controlled intervention studies have explored more thoroughly the chronic effects of exercise. These studies have typically examined exercise doses similar to US national guidelines of 150 min per week of moderate-to-vigorous physical activity (Physical Activity Guidelines Advisory Committee, 2008). A recent review identified at least 14 randomized controlled trials of exercise with a sleep outcome that ranged in length from 4 to 52 weeks (Buman & King, 2010). Findings from these studies generally favored the idea that exercise improved sleep relative to control treatments. Notably, the observed effects were strongest among self-reported sleep metrics, including sleep onset latency and overall sleep quality (King, Baumann, O'Sullivan, Wilcox, & Castro, 2002; King, Oman, Brassington, Bliwise, & Haskell, 1997), and the most pronounced improvements were among older adults (Reid et al., 2010; Singh et al., 2005) and poorer sleepers (Tworoger et al., 2003).

TEMPORAL EXERCISE-SLEEP RELATIONSHIPS

Recent studies have examined potential bidirectional effects of the exercise-sleep relationship. These studies have primarily used multilevel, intensive, repeated measure designs that temporally order exercise and sleep occasions and then examine the relative strength of the relationships between exercise and better sleep and between sleep and more exercise. The results of these studies have been mixed, but they generally support a bidirectional relationship between exercise and sleep. For example, Dzierzewski et al. (2014) reported that exercise positively predicted better subjective sleep quality on the subsequent night and, in a reciprocal model, that better sleep quality predicted higher levels of exercise on the subsequent day. Likewise, Lambiase, Gabriel, Kuller, and Matthews (2013) reported that objectively measured total activity and moderate-to-vigorous physical activity were associated with great sleep efficiency on the subsequent night, and the reciprocal relationship was also true. Finally, in contrast to the previous two studies, Baron, Reid, and Zee (2012) reported self-reported exercise did not predict subjective or objective sleep variables on the subsequent night; however, longer sleep onset latency predicted lower levels of exercise on the subsequent day. The inconsistencies in these studies are likely due to a number of factors, including the diverse populations, varying sample sizes, and inconsistent measurement methods (i.e., objective vs. subjective) for both exercise and sleep. However, these studies offer an important future direction for research because they will continue to illuminate the temporal nature of the exercise-sleep relationship, while disentangling the relative acute and chronic effects that exercise may have on sleep.

EXERCISE MODALITIES AND SLEEP

In studies of exercise and sleep, aerobic exercise has been the most common mode of exercise studied. However, resistance exercise, yoga, and tai chi have also been associated with better overall sleep. Older studies had suggested that the improvement in sleep following exercise training were mediated by increases in cardiorespiratory fitness (Griffin & Trinder, 1978; Horne, 1981; Shapiro, Bortz, Mitchell, Bartel, & Jooste, 1981). This hypothesis suggests that the greatest sleep improvements following exercise training should be observed via aerobic exercise. Other evidence suggests that the relationship between exercise and sleep may be independent of cardiorespiratory fitness gains (Meintjes, Driver, & Shapiro, 1989; Youngstedt et al., 1997). Randomized controlled trials of moderate-to-vigorous intensity aerobic exercise consistent within national guidelines led to improvements in sleep quality, and these changes were correlated with improvements in cardiorespiratory fitness (King et al., 1997, 2008; Tworoger et al., 2003). For example, Tworoger et al. (2003) found that the individuals with greater fitness improvements were more likely to report greater total sleep time.

A small number of studies have also examined the effects of resistance training on sleep quality (Ferris, Williams, Shen, O'Keefe, & Hale, 2005). The primary mechanism hypothesized to drive this effect has been growth hormone secretion. Growth hormone secretion is particularly pronounced during slow-wave sleep (Van Cauter et al., 2004), suggesting that resistance exercise may particularly promote this restorative form of sleep. However, current experimental studies have not identified the dose of resistance training that optimally promotes growth hormone secretion during the subsequent night of sleep (Breus, O'Connor, & Ragan, 2000; Tuckow et al., 2006). Singh and colleagues have conducted two short-term (10 and 8 weeks) randomized controlled trials (Singh, Clements, & Fiatorone, 1997; Singh et al., 2005) among depressed older adults using high-intensity progressive resistance training, however, and they found favorable results for subjective sleep quality. A low intensity condition of resistance exercise was added to their second trial (Singh et al., 2005), with significant but less favorable results compared to the high-intensity resistance-training condition.

Yoga has emerged as an important treatment modality for a host of chronic conditions due to its strong antidepressive and antianxiolytic effects (Cramer, Lauche, Langhorst, & Dobos, 2013). There have been a number of controlled and uncontrolled studies of the effects of yoga on sleep quality (Elavsky & McAuley, 2007; Khalsa, 2004; Manjunath & Telles, 2005). In general, these studies have not produced positive results in subjective sleep outcomes. However, it should be noted that these studies primarily employed relaxed forms of yoga that focused on breathing exercises. More intense forms of yoga may be the ones that elicit physiological adaptations more similar to those that occur with aerobic or resistance training. Alternatively, subjects in the yoga studies might not have had substantially poor sleep. A recent review of yoga as a treatment for insomnia among cancer patients and survivors concluded that, despite methodological limitations, yoga is an effective treatment in this population (Mustian, 2013). This conclusion was largely based on a 4-week gentle hatha and restorative yoga intervention in 410 cancer survivors, and the authors observed improvements in subjective insomnia symptoms and actigraphy-derived wakefulness after sleep onset and efficiency (Mustian et al., 2013).

PUTATIVE BIOLOGICAL MECHANISMS OF EXERCISE EFFECTS ON SLEEP

The mechanism(s) by which exercise promotes sleep are still unclear, partly because little research with animal models has addressed this topic. Classic theories that sleep has an energy conservation (Berger & Phillips, 1987) or body restitution function (Oswald, 1980) predict that exercise would have a unique sleep-promoting function because exercise can have unique energy-expenditure and body-deterioration consequences. Interestingly, some evidence supports exceptionally long sleep following extreme sporting events, such as marathon and ultramarathon competitions (Driver et al., 1994; Shapiro, Griesel, Bartel, & Jooste, 1975), as well as during weeks involving extremely high training loads experienced by runners in training camp (Aritake, 2014). These patterns could partly reflect a sleep rebound from the night(s) preceding the competition, on which sleep was not assessed. Athletes commonly have profound insomnia preceding such competitions, exacerbated by early start times for the competitions. Moreover, in the Driver et al. (1994), sleep was allowed ad lib after, but not before the extreme exercise. Nonetheless, the long sleep durations are greater than is typically observed.

Conceivably, exercise might elicit the accumulation of somnogenic substances. For example, adenosine is one substance that accumulates in response to exercise, and it is associated with sleep. Conversely, the blockage of adenosine receptors with caffeine disturbs sleep. The presence of an adenosine mechanism mediating slow-wave sleep enhancement after exercise has been indicated by a study by Youngstedt et al., which showed a threefold greater increase in SWS following exercise combined with placebo compared with caffeine treatment (Youngstedt, O'Connor, Crabbe, & Dishman, 2000).

The thermogenic hypothesis that sleep serves a temperature-regulating function (McGinty & Szymusiak, 1990) predicts the unique benefit of exercise, which elevates temperature more readily than any other stimulus. Classic studies by Horne and colleagues were supportive of this hypothesis (Horne & Reid, 1985; Horne & Staff, 1983). Most compellingly, Horne and Reid (1985) found that the increase in SWS following exercise was blunted when temperature elevation during exercise was suppressed by having the participants run while wearing wet clothes in front of a high-powered fan. However, some factors raise questions about the veracity of this hypothesis with respect to exercise and sleep. First, many studies have shown that acute exercise has a minimal effect on SWS (Youngstedt et al., 1997). Second, the increase in SWS following exercise was accompanied by a similar decrease in REM sleep, and it is not clear whether SWS or REM should be considered deeper, more restful sleep (Rechtschaffen, 1997). Indeed, some research has found a negative correlation between SWS after exercise and self-reported sleep quality (Driver et al., 1994).

Conceivably, a thermoregulation function could contribute to sleep improvement following exercise training (Youngstedt, 2005). The decline in body temperature from lights out to the nighttime temperature nadir could be considered a thermolytic challenge, and failure to downregulate temperature has been noted in insomniacs (Monroe, 1967) for whom it may contribute to sleeping problems. Exercise training promotes heat acclimation and efficient temperature downregulation, which could offset this problem.

Exercise could promote sleep via decreases in anxiety and physiological hyperarousal as well. Anxiety disturbs sleep, essentially by definition, and insomnia has been increasingly linked to hyperarousal (Buysse, 2013). The benefits of acute exercise for self-reported anxiety, as well as psychophysiological indices of anxiety, such as blood pressure, muscle tension, and improved heart rate variability, are well established. Likewise, exercise training typically results in chronic improvements in these measures.

Well-established antidepressive effects of exercise could also moderate sleep improvements. One of the most robust effects of acute exercise on sleep is a reduction in REM sleep (Youngstedt et al., 1997), which could mediate these effects. The effects of exercise on the circadian system could also promote sleep. In addition, phase-shifting effects and/or reduced day-to-day fluctuation of circadian timing could mediate sleep improvements.

EXERCISE AND OBSTRUCTIVE SLEEP APNEA

The prevalence of OSA among adults is approximately 10-20%. OSA is associated with increased mortality, as well as an increased prevalence of cardiovascular disease, hypertension, diabetes, obesity, and accidents. The most common treatment for OSA, positive airway pressure (PAP), is associated with many negative side effects (e.g., dry mouth) and low adherence, and many patients conclude that its burdens exceed its benefits. Indeed, the overall benefits of PAP are not impressive, as PAP often does not reverse associated morbidity (e.g., hypertension and impaired glucose tolerance). Moreover, curiously, PAP commonly leads to weight gain, a particularly disconcerting side effect for a condition so closely linked with obesity.

Thus, alternative or adjuvant treatments are needed. Compelling evidence suggests that exercise training is an efficacious treatment for OSA. For example, in a randomized controlled trial of 42 overweight adults with OSA, Kline et al. (2011) found a 24% reduction in OSA severity following 12 weeks of moderate-intensity aerobic and resistance exercise training. The mechanisms of this effect are unestablished. However, it is clear that exercise can elicit significant reductions in OSA independent of changes in body weight or composition, though there seems to be additive benefits for OSA for interventions that involve both exercise and weight loss (Iftikhar, Kline, & Youngstedt, 2014). Although improvements in OSA with exercise training are modest compared to the effects of PAP, exercise can elicit chronic improvements in OSA, which remain apparent following non-exercise days, whereas the effects of PAP only occur so long as it is used. Moreover, in contrast with PAP, exercise has unparalleled health benefits, which make it an attractive alternative or adjuvant treatment.

Moreover, the extant literature suggests that exercise training elicits greater improvement in sleepiness than PAP does (Iftikhar et al., 2014; Marshall et al., 2006).

EXERCISE AND RESTLESS LEG SYNDROME

RLS is associated with excruciating "creepy-crawly" sensations, an overwhelming urge to move the legs, severe sleep disruption, and depression. The symptoms of RLS show a circadian pattern so that they reach their peak during the evening. RLS has a clear genetic link (Schormair & Winkelmann, 2011), and it is also associated with iron deficiency. RLS is typically treated with dopamine agonists, which have negative side effects and paradoxically sometimes exacerbate symptoms (Romenets & Postuma, 2013). There are multiple rationales for using exercise as a treatment for RLS. Moving the legs is virtually the only means of relief from acute symptoms. Moreover, RLS patients report that daytime exercise is associated with reduced symptoms of RLS, but that excessively vigorous exercise can sometimes exacerbate symptoms. Epidemiologic research confirms that regular exercise is associated with reduced RLS prevalence (Phillips et al., 2000). Most compelling, two randomized controlled trials have found comparable efficacy of exercise training and dopamine agonists (de Mello, Esteves, & Tufik, 2004; Giannaki et al., 2013).

SEDENTARY BEHAVIOR AND SLEEP

Sedentary behavior is emerging as a novel risk factor for cardiovascular disease, diabetes, mental illness, and all-cause mortality (Katzmarzyk, Church, Craig, & Bouchard, 2009; Thorp, Owen, Neuhaus, & Dunstan, 2011; Warren et al., 2010). Sedentary behavior is defined as sitting or reclining with low energy expenditure (Sedentary Behaviour Research Network, 2012), and the detrimental consequences of sedentary behavior are largely independent of the robust benefit physical activity has on chronic health disease risk (Owen, Healy, Matthews, & Dunstan, 2010). Surprisingly, little research has explored whether sedentary behavior is linked to sleep quality independent of physical activity. In the American Time Use Survey, common sedentary pursuits (i.e., work commute time, television viewing) were associated with shortened sleep duration (Basner et al., 2007). In youth, video gaming and computer use have been linked to shortened sleep duration and increased sleep onset latency (Foti, Eaton, Lowry, & McKnight-Ely, 2011; Higuchi, Motohashi, Liu, & Maeda, 2005). Most recently, two published abstracts have indicated that delayed sleep timing was associated with more self-reported minutes sitting and lower levels of free-living physical activity (Baron, 2014; Shechter & St-Onge, 2014). Baron (2014) suggested that individuals with delayed sleep timing patterns also reported less routinized exercise patterns and more difficulty making time for exercise. These initial links suggest additional preliminary work needs to be conducted to include both self-reported and objective measures of sedentary behavior and sleep. If additional cross-sectional studies continue to observe promising relationships, acute experimental work examining mechanisms of this relationship would be warranted.

SYNERGIES AMONG SLEEP, SEDENTARY BEHAVIOR, AND EXERCISE

The health-promoting benefits of regular exercise (e.g., brisk walking, swimming, hard physical labor) have been well documented (Physical Activity Guidelines Advisory Committee, 2008). More recently, even light-intensity activities (i.e., lifestyle behaviors such as leisurely walking or household chores) appear to have additional health benefits (Buman et al., 2010; Healy et al., 2007). As noted previously, studies have connected sedentary time to reduced health outcomes, even after accounting for exercise (Healy, Matthews, Dunstan, Winkler, & Owen, 2011; Owen et al., 2010). Sleep quality and duration also have unique associations with health outcomes (Cappuccio et al., 2010b).

Whereas suboptimal sleep duration, prolonged sedentary behavior, and lack of exercise produce distinct consequences, time is finite for individuals. Thus, increasing time spent in one behavior inevitably requires decreasing time in another behavior. Researchers must then understand not only the health impact of a given health behavior, but also the role of the behavior that it displaces (Mekary, Willett, Hu, & Ding, 2009). This is important because, although the benefits of exercise are well recognized, individuals cannot feasibly exercise during all waking hours. Understanding the potential benefits of replacing sedentary time with other behaviors (i.e., light-intensity activity or sleep), above and beyond time spent in exercise, is an important task not currently being undertaken in prevention efforts. The isotemporal substitution paradigm (Mekary et al., 2009), which is historically rooted in nutritional epidemiology, adequately accounts for the fixed-time nature of these behaviors, and it provides insights on how varying distributions of sleep, sedentary behavior, and exercise may impact health outcomes. A recent study by Buman et al. (2014) has applied the isotemporal substitution paradigm in the context of obesity, diabetes risk, and other cardiometabolic risk factors, using a population-based sample. Findings from this study demonstrated that reallocating 30 min per day of sedentary time for exercise, light physical activity, or sleep produced clinically and statistically meaningful decreases in obesity and fasting insulin, along with increases in HDL cholesterol and decreases in triglycerides.

A 24-H APPROACH TO CHRONIC DISEASE PREVENTION

In light of the observed relationships between exercise and sleep, emerging relationships between sedentary behavior and sleep, and the fixed-time nature of these behaviors, we propose a 24-h approach to chronic disease prevention (Figure 15.1). Central to this approach is the idea that behaviors cannot be viewed in isolation and changes in one behavior inevitably necessitate spontaneous and/or volitional changes in another. We propose that individuals place a priority on the interplay of the full 24-h spectrum of health behaviors, including sleep, sedentary behavior, and exercise, as well as how these behaviors can be collectively optimized to produce the maximum health benefit.

This approach depends on measurement strategies appropriate for capturing the full 24-h spectrum of behaviors. Historically, advances in measurement tools have evolved relatively independently in the sleep and exercise fields. More recently, integration of technologies and the availability of commercial sensors have triggered a shift toward 24-h concurrent assessments of sleep, sedentary behavior, and exercise, with the use of wrist accelerometry. Recent studies have demonstrated cross-validation

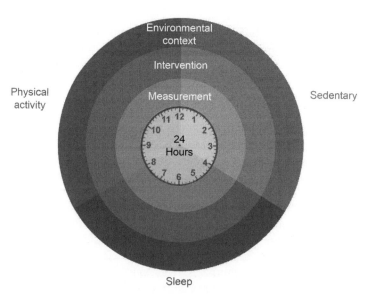

FIGURE 15.1 A 24-h approach to chronic disease prevention.

of sensors for measuring exercise and sleep constructs (Cellini, Buman, McDevitt, Ricker, & Mednick, 2013; Lambiase, Gabriel, Chang, Kuller, & Matthews, 2014). Ongoing, large-scale population-based surveillance projects (National Health and Nutrition Examination Survey and United Kingdom Biobank) have recently moved objective data collection devices for exercise and sedentary behavior metrics to the wrist position to allow the collection of sleep patterns and to increase compliance. Self-reported exercise, sedentary behavior, and sleep measures can already be easily combined to get a more comprehensive view of the full 24 h. Additional studies are needed to validate existing and emerging consumer- and research-grade sensors, however. To achieve the goal of long-term continuous monitoring, studying the feasibility of measurement approaches is perhaps just as important as validation. Understanding the factors that influence the user experience, including device placement, sleep/wake transitions, and behavioral feedback, will be critical.

In light of the relationship between exercise and sleep, researchers must quantify the effects that an intervention targeting a single behavior may have on related behaviors. An increase of 30 min per day of exercise, in line with national recommendations, may or may not have a maximal benefit for health, depending on scheduling and the type of activity displaced. For some, increased exercise may have a null or even detrimental effect if it occurs in place of a critical sleep period. We suggest that future clinical interventions should seek to identify the optimal combinations of changes in sleep, sedentary behavior, and exercise for maximal health benefit. Additional research should also address change synergies within these behaviors, especially regarding temporal sequencing of behaviors. Finally, targeted interventions for sleep (i.e., sleeping pills, cognitive-behavioral interventions) should also explore how changes in sleep metrics may positively or negatively impact changes in sedentary and exercise behaviors.

Experimental evidence suggests that neighborhood characteristics may contribute to obesity and diabetes (Ludwig et al., 2011). The social (e.g., social connectivity, proximity to others) and built (e.g., crime, street connectivity) environments are key predictors of exercise (King et al., 2011). Perceived sleep quality appears to be a mediator of the relationship between neighborhood quality (e.g., perceptions of crime, litter, and pleasantness) and self-rated health. Taken together, these results suggest that the environment (both social and physical features) appears to play a role in health across the 24 h. For sleep, this includes the microenvironment (i.e., bedroom), and established sleep hygiene recommendations exist for optimal sleep conditions. However, the world beyond the bedroom, which includes macroenvironmental features such as ambient noise and temperature, actual and perceived crime, and social capital, also should be accounted for. Researchers have not fully developed validated scales and objective measures of these metrics within the sleep context. Furthermore, only a few studies have described environments likely to reduce sedentary behavior, including the use of height-adjustable desks in the workplace (Alkhajah et al., 2012) and television and electronics restrictions in the home (Otten, Jones, Littenberg, & Harvey-Berino, 2009). Collectively, the field must take a broader contextual view of health behaviors that accounts for their interrelationships and dependence on environmental factors so that we might harness the interplay of the behaviors across the 24 h.

SUMMARY

Exercise and sleep exhibit a moderate-strength relationship, and this relationship seems to be stronger among poorer sleepers. Given the challenges with pharmacological and other nonpharmacological treatments for poor sleep, exercise and related approaches are promising, and they are likely to broadly impact the individual, enhancing physical and emotional health. The use of exercise as a stand-alone or adjunct treatment for sleep disorders, including OSA and RLS, is an exciting new area of research. As a construct independent from physical activity, sedentary behavior may have unique effects on aspects of sleep. Collectively, despite the fact that sleep, sedentary behavior, and exercise have unique impacts on health outcomes, these behaviors are linked by virtue of their time-dependent nature, and therefore, they require a more comprehensive, 24-h perspective in order capture their full impact on health.

References

Alkhajah, T. A., Reeves, M. M., Eakin, E. G., Winkler, E. A. H., Owen, N., & Healy, G. N. (2012). Sit–stand workstations: A pilot intervention to reduce office sitting time. *American Journal of Preventive Medicine, 43*(3), 298–303.

Ancoli-Israel, S., & Roth, T. (1999). Characteristics of insomnia in the United States: Results of the 1991 national sleep foundation survey. I. *Sleep, 22*, S347–S353.

Aritake, S. (2014). *Current research and clinical insights into sleep, recovery and performance in elite athletes.* Paper presented at the SLEEP, Minneapolis, MN.

Baron K. G., (Ed.). (2014 June 2). Early to bed, early to rise makes easier to exercise: The role of sleep timing in physical activity and sedentary behavior. *Sleep*, Minneapolis, MN.

Baron, K. G., Reid, K. J., & Zee, P. C. (2012). Exercise to improve sleep in insomnia: Exploration of the bidirectional effects. *Journal of Clinical Sleep Medicine, 9*(8), 819–824.

Basner, M., Fomberstein, K. M., Razavi, F. M., Banks, S., William, J. H., Rosa, R. R., et al. (2007). American time use survey: Sleep time and its relationship to waking activities. *Sleep, 30*(9), 1085–1095.

Berger, R., & Phillips, N. (1987). Comparative aspects of energy metabolism, body temperature and sleep. *Acta Physiologica Scandinavica Supplement, 574*, 21–27.

Breus, M. J., O'Connor, P. J., & Ragan, S. T. (2000). Muscle pain induced by novel eccentric exercise does not disturb the sleep of normal young men. *The Journal of Pain, 1*(1), 67–76.

Buman, M. P., Hekler, E. B., Haskell, W. L., Pruitt, L., Conway, T. L., Cain, K. L., et al. (2010). Objective light-intensity physical activity associations with rated health in older adults. *American Journal of Epidemiology, 172*(10), 1155–1165.

Buman, M. P., & King, A. C. (2010). Exercise as a treatment to enhance sleep. *American Journal of Lifestyle Medicine*, *4*(6), 500–514.

Buman, M. P., Winkler, E. A., Kurka, J. M., Hekler, E. B., Baldwin, C. M., Owen, N., et al. (2014). Reallocating time to sleep, sedentary behaviors, or active behaviors: Associations with cardiovascular disease risk biomarkers, NHANES 2005–2006. *American Journal of Epidemiology*, *179*(3), 323–334.

Buysse, D. J. (2013). Insomnia. *Journal of the American Medical Association*, *309*(7), 706.

Cappuccio, F., D'Elia, L., Strazzullo, P., & Miller, M. A. (2010a). Quantity and quality of sleep and incidence of type 2 diabetes: A systematic review and meta-analysis. *Diabetes Care*, *33*(2), 414–420.

Cappuccio, F. P., D'Elia, L., Strazzullo, P., & Miller, M. A. (2010b). Sleep duration and all-cause mortality: A systematic review and meta-analysis of prospective studies. *Sleep*, *33*(5), 585–592.

Cellini, N., Buman, M. P., McDevitt, E. A., Ricker, A. A., & Mednick, S. C. (2013). Direct comparison of two actigraphy devices with polysomnographically recorded naps in healthy young adults. *Chronobiology International*, *30*(5), 691–698.

Centers for Disease Control and Prevention (CDC). (2012). Short sleep duration among workers—United States, 2010. *Morbidity and Mortality Weekly Report*, *61*(16), 281–285.

Cramer, H., Lauche, R., Langhorst, J., & Dobos, G. (2013). Yoga for depression: A systematic review and meta-analysis. *Depression and Anxiety*, *30*(11), 1068–1083.

de Mello, M. T., Esteves, A. M., & Tufik, S. (2004). Comparison between dopaminergic agents and physical exercise as treatment for periodic limb movements in patients with spinal cord injury. *Spinal Cord*, *42*(4), 218–221.

Driver, H. S., Rogers, G. G., Mitchell, D., Borrow, S. J., Allen, M., Luus, H. G., et al. (1994). Prolonged endurance exercise and sleep disruption. *Medicine and Science in Sports and Exercise*, *26*(7), 903–907.

Driver, H. S., & Taylor, S. R. (2000). Exercise and sleep. *Sleep Medicine Reviews*, *4*(4), 387–402.

Dzierzewski, J. M., Buman, M. P., Giacobbi, P. R., Roberts, B. L., Aiken Morgan, A. T., Marsiske, M., et al. (2014). Exercise and sleep in community dwelling older adults: Evidence for a reciprocal relationship. *Journal of Sleep Research*, *23*(1), 61–68.

Elavsky, S., & McAuley, E. (2007). Lack of perceived sleep improvement after 4-month structured exercise programs. *Menopause*, *14*(3), 535–540.

Ferris, L. T., Williams, J. S., Shen, C. L., O'Keefe, K. A., & Hale, K. B. (2005). Resistance training improves sleep quality in older adults a pilot study. *Journal of Sports Science & Medicine*, *4*(3), 354–360.

Foti, K. E., Eaton, D. K., Lowry, R., & McKnight-Ely, L. R. (2011). Sufficient sleep, physical activity, and sedentary behaviors. *American Journal of Preventive Medicine*, *41*(6), 596–602.

Foundation, National Sleep (2013). *2013 sleep in America poll: Exercise and sleep*. Arlington, VA: National Sleep Foundation.

Gangwisch, J. E., Heymsfield, S. B., Boden Albala, B., Buijs, R. M., Kreier, F., Pickering, T. G., et al. (2006). Short sleep duration as a risk factor for hypertension analyses of the first National Health and Nutrition Examination Survey. *Hypertension*, *47*(5), 833–839.

Gangwisch, J. E., Heymsfield, S. B., Boden-Albala, B., Buijs, R. M., Kreier, F., Pickering, T. G., et al. (2007). Sleep duration as a risk factor for diabetes incidence in a large US sample. *Sleep*, *30*(12), 1667–1673.

Giannaki, C. D., Sakkas, G. K., Karatzaferi, C., Hadjigeorgiou, G. M., Lavdas, E., Kyriakides, T., et al. (2013). Effect of exercise training and dopamine agonists in patients with uremic restless legs syndrome: A six-month randomized, partially double-blind, placebo-controlled comparative study. *BMC Nephrology*, *14*(1), 194.

Griffin, S. J., & Trinder, J. (1978). Physical fitness, exercise, and human sleep. *Psychophysiology*, *15*(5), 447–450.

Hasan, J., Urponen, H., Vuori, I., & Partinen, M. (1988). Exercise habits and sleep in a middle-aged finnish population. *Acta Physiologica Scandinavica Supplementum*, *574*, 33–35.

Healy, G. N., Dunstan, D. W., Salmon, J., Cerin, E., Shaw, J. E., Zimmet, P. Z., et al. (2007). Objectively measured light-intensity physical activity is independently associated with 2-h plasma glucose. *Diabetes Care, 30*(6), 1384–1389.

Healy, G. N., Matthews, C. E., Dunstan, D. W., Winkler, E. A. H., & Owen, N. (2011). Sedentary time and cardio-metabolic biomarkers in US adults: NHANES 2003–06. *European Heart Journal, 32*(5), 590–597.

Higuchi, S., Motohashi, Y., Liu, Y., & Maeda, A. (2005). Effects of playing a computer game using a bright display on presleep physiological variables, sleep latency, slow wave sleep and rem sleep. *Journal of Sleep Research, 14*(3), 267–273.

Horne, J. (1981). The effects of exercise upon sleep: A critical review. *Biological Psychology, 12*(4), 241–290.

Horne, J., & Reid, A. (1985). Night-time sleep eeg changes following body heating in a warm bath. *Electroencephalography and Clinical Neurophysiology, 60*(2), 154–157.

Horne, J., & Staff, L. (1983). Exercise and sleep: Body-heating effects. *Sleep, 6*(1), 36–46.

Iftikhar, I. H., Kline, C. E., & Youngstedt, S. D. (2014). Effects of exercise training on sleep apnea: A meta-analysis. *Lung, 192*(1), 175–184.

Institute of Medicine (US) Committee on Sleep Medicine and Research. (2006). In H. R. Colten & B. M. Altevogt (Eds.), *Sleep disorders and sleep deprivation: An unmet public health problem*. Washington, DC: National Academy of Sciences.

Jiménez-Conde, J., Ois, A., Rodríguez-Campello, A., Gomis, M., & Roquer, J. (2007). Does sleep protect against ischemic stroke? Less frequent ischemic strokes but more severe ones. *Journal of Neurology, 254*(6), 782–788.

Katzmarzyk, P. T., Church, T. S., Craig, C. L., & Bouchard, C. (2009). Sitting time and mortality from all causes, cardiovascular disease, and cancer. *Medicine and Science in Sports and Exercise, 41*(5), 998–1005.

Khalsa, S. B. S. (2004). Treatment of chronic insomnia with yoga: A preliminary study with sleep–wake diaries. *Applied Psychophysiology and Biofeedback, 29*(4), 269–278.

King, A. C., Baumann, K., O'Sullivan, P., Wilcox, S., & Castro, C. (2002). Effects of moderate-intensity exercise on physiological, behavioral, and emotional responses to family caregiving a randomized controlled trial. *The Journals of Gerontology Series A: Biological Sciences and Medical Sciences, 57*(1), M26–M36.

King, A. C., Oman, R. F., Brassington, G. S., Bliwise, D. L., & Haskell, W. L. (1997). Moderate-intensity exercise and self-rated quality of sleep in older adults: A randomized controlled trial. *Journal of the American Medical Association, 277*(1), 32–37.

King, A. C., Pruitt, L. A., Woo, S., Castro, C. M., Ahn, D. K., Vitiello, M. V., et al. (2008). Effects of moderate-intensity exercise on polysomnographic and subjective sleep quality in older adults with mild to moderate sleep complaints. *The Journals of Gerontology Series A: Biological Sciences and Medical Sciences, 63*(9), 997–1004.

King, A. C., Sallis, J. F., Frank, L. D., Saelens, B. E., Cain, K., Conway, T. L., et al. (2011). Aging in neighborhoods differing in walkability and income: Associations with physical activity and obesity in older adults. *Social Science and Medicine, 73*(10), 1525–1533.

Kline, C. E., Crowley, E. P., Ewing, G. B., Burch, J. B., Blair, S. N., Durstine, J. L., et al. (2011). The effect of exercise training on obstructive sleep apnea and sleep quality: A randomized controlled trial. *Sleep, 34*(12), 1631–1640.

Kripke, D. F., Langer, R. D., & Kline, L. E. (2012). Hypnotics' association with mortality or cancer: A matched cohort study. *BMJ Open, 2*(1), 1–8.

Lambiase, M. J., Gabriel, K. P., Chang, Y.-F., Kuller, L. H., & Matthews, K. A. (2014). Utility of actiwatch sleep monitor to assess waking movement behavior in older women. *Medicine and Science in Sports and Exercise*, http://dx.doi.org/10.1249/mss.0000000000000361.

Lambiase, M. J., Gabriel, K. P., Kuller, L. H., & Matthews, K. A. (2013). Temporal relationships between physical activity and sleep in older women. *Medicine and Science in Sports and Exercise, 45*(12), 2362–2368.

Ludwig, J., Sanbonmatsu, L., Gennetian, L., Adam, E., Duncan, G. J., Katz, L. F., et al. (2011). Neighborhoods, obesity, and diabetes—A randomized social experiment. *New England Journal of Medicine, 365*(16), 1509–1519.

Mallon, L., Broman, J.-E., & Hetta, J. (2009). Is usage of hypnotics associated with mortality? *Sleep Medicine, 10*(3), 279–286.

Manjunath, N., & Telles, S. (2005). Influence of yoga & ayurveda on self-rated sleep in a geriatric population. *Indian Journal of Medical Research, 121*(5), 683–690.

Marshall, N. S., Barnes, M., Travier, N., Campbell, A. J., Pierce, R. J., McEvoy, R. D., et al. (2006). Continuous positive airway pressure reduces daytime sleepiness in mild to moderate obstructive sleep apnoea: A meta-analysis. *Thorax, 61*(5), 430–434.

McGinty, D., & Szymusiak, R. (1990). Keeping cool: A hypothesis about the mechanisms and functions of slow-wave sleep. *Trends in Neurosciences, 13*(12), 480–487.

Meintjes, A., Driver, H., & Shapiro, C. (1989). Improved physical fitness failed to alter the eeg patterns of sleep in young women. *European Journal of Applied Physiology and Occupational Physiology, 59*(1–2), 123–127.

Mekary, R. A., Willett, W. C., Hu, F. B., & Ding, E. L. (2009). Isotemporal substitution paradigm for physical activity epidemiology and weight change. *American Journal of Epidemiology, 170*(4), 519–527.

Mizuno, K., Kunii, M., Seita, T., Ono, S., Komada, Y., & Shirakawa, S. (2004). Effects of habitual exercise on sleep habits and sleep health in middle-aged and older japanese women. *Japanese Journal of Physical Fitness and Sports Medicine, 53*(5), 527–536.

Mokdad, A. H., Stroup, D. F., & Giles, W. H. (2003). Public health surveillance for behavioral risk factors in a changing environment recommendations from the behavioral risk factor surveillance team. *Morbidity and Mortality Weekly Report, 52*(RR09), 1–12.

Monroe, L. J. (1967). Psychological and physiological differences between good and poor sleepers. *Journal of Abnormal Psychology, 72*(3), 255–264.

Morin, C. M., Hauri, P. J., Espie, C. A., Spielman, A. J., Buysse, D. J., & Bootzin, R. R. (1999). Nonpharmacologic treatment of chronic insomnia. An american academy of sleep medicine review. *Sleep, 22*(8), 1134–1156.

Mustian, K. M. (2013). Yoga as treatment for insomnia among cancer patients and survivors: A systematic review. *European Medical Journal, 1*, 106–115.

Mustian, K. M., Sprod, L. K., Janelsins, M., Peppone, L. J., Palesh, O. G., Chandwani, K., et al. (2013). Multicenter, randomized controlled trial of yoga for sleep quality among cancer survivors. *Journal of Clinical Oncology, 31*(26), 3233–3241.

National Highway Traffic Safety Administration. (2012). *Traffic safety facts crash stats: Drowsy driving*. Washington, DC: US Department of Transportation.

Network, Sedentary Behaviour Research. (2012). Letter to the editor: Standardized use of the terms "sedentary" and "sedentary behaviours". *Applied Physiology, Nutrition, and Metabolism, 37*(3), 540–542.

Oswald, I. (1980). Sleep as a restorative process: Human clues. *Progress in Brain Research, 53*, 279–288.

Otten, J. J., Jones, K. E., Littenberg, B., & Harvey-Berino, J. (2009). Effects of television viewing reduction on energy intake and expenditure in overweight and obese adults: A randomized controlled trial. *Archives of Internal Medicine, 169*(22), 2109–2115.

Owen, N., Healy, G. N., Matthews, C. E., & Dunstan, D. W. (2010). Too much sitting: The population health science of sedentary behavior. *Exercise and Sport Sciences Reviews, 38*(3), 105–113.

Perlis, M. L., Giles, D. E., Buysse, D. J., Tu, X., & Kupfer, D. J. (1997). Self-reported sleep disturbance as a prodromal symptom in recurrent depression. *Journal of Affective Disorders, 42*(2), 209–212.

Perry, G. S., Patil, S. P., & Presley-Cantrell, L. R. (2013). Raising awareness of sleep as a healthy behavior. *Preventing Chronic Disease, 10*, E133.

Phillips, B., Young, T., Finn, L., Asher, K., Hening, W. A., & Purvis, C. (2000). Epidemiology of restless legs symptoms in adults. *Archives of Internal Medicine, 160*(14), 2137–2141.

Physical Activity Guidelines Advisory Committee. (2008). *Physical activity guidelines advisory committee report, 2008.* Washington, DC: US Department of Health and Human Services.

Quan, S. F., O'Connor, G. T., Quan, J. S., Redline, S., Resnick, H. E., Shahar, E., et al. (2007). Association of physical activity with sleep-disordered breathing. *Sleep and Breathing, 11*(3), 149–157.

Rechtschaffen, A. (1997). Current perspectives on the function of sleep. *Perspectives in Biology and Medicine, 41*(3), 359–390.

Reid, K. J., Baron, K. G., Lu, B., Naylor, E., Wolfe, L., & Zee, P. C. (2010). Aerobic exercise improves self-reported sleep and quality of life in older adults with insomnia. *Sleep Medicine, 11*(9), 934–940.

Richardson, G. S. (2000). Managing insomnia in the primary care setting: Raising the issues. *Sleep, 23*(Suppl.), S9–S12, discussion S13-15.

Romenets, S. R., & Postuma, R. B. (2013). Treatment of restless legs syndrome. *Current Treatment Options in Neurology, 15*(4), 396–409.

Roth, T., & Roehrs, T. (2003). Insomnia: Epidemiology, characteristics, and consequences. *Clinical Cornerstone, 5*(3), 5–15.

Schormair, B., & Winkelmann, J. (2011). Genetics of restless legs syndrome: Mendelian, complex, and everything in between. *Sleep Medicine Clinics, 6*(2), 203–215.

Schutte-Rodin, S., Broch, L., Buysse, D., Dorsey, C., & Sateia, M. (2008). Clinical guideline for the evaluation and management of chronic insomnia in adults. *Journal of Clinical Sleep Medicine, 4*(5), 487–504.

Shapiro, C. M., Bortz, R., Mitchell, D., Bartel, P., & Jooste, P. (1981). Slow-wave sleep: A recovery period after exercise. *Science, 214*(4526), 1253–1254.

Shapiro, C., Griesel, R., Bartel, P., & Jooste, P. (1975). Sleep patterns after graded exercise. *Journal of Applied Physiology, 39*(2), 187–190.

Shechter, A., & St-Onge, M. (2014). *Delayed sleep timing is associated with low levels of free-living physical activity.* Paper presented at the SLEEP, Minneapolis, MN.

Sherrill, D. L., Kotchou, K., & Quan, S. F. (1998). Association of physical activity and human sleep disorders. *Archives of Internal Medicine, 158*(17), 1894–1898.

Siegel, J. M. (2005). Clues to the functions of mammalian sleep. *Nature, 437*(7063), 1264–1271.

Singh, N. A., Clements, K. M., & Fiatorone, M. A. (1997). Sleep, sleep deprivation, and daytime activities a randomized controlled trial of the effect of exercise on sleep. *Sleep, 20*(2), 95–101.

Singh, N. A., Stavrinos, T. M., Scarbek, Y., Galambos, G., Liber, C., & Singh, M. A. F. (2005). A randomized controlled trial of high versus low intensity weight training versus general practitioner care for clinical depression in older adults. *The Journals of Gerontology Series A: Biological Sciences and Medical Sciences, 60*(6), 768–776.

Stickgold, R. (2013). Parsing the role of sleep in memory processing. *Current Opinion in Neurobiology, 23*(5), 847–853.

Thorp, A. A., Owen, N., Neuhaus, M., & Dunstan, D. W. (2011). Sedentary behaviors and subsequent health outcomes in adults: A systematic review of longitudinal studies, 1996–2011. *American Journal of Preventive Medicine, 41*(2), 207–215.

Tuckow, A. P., Rarick, K. R., Kraemer, W. J., Marx, J. O., Hymer, W. C., & Nindl, B. C. (2006). Nocturnal growth hormone secretory dynamics are altered after resistance exercise: Deconvolution analysis of 12-hour immunofunctional and immunoreactive isoforms. *American Journal of Physiology-Regulatory, Integrative and Comparative Physiology, 291*(6), R1749–R1755.

Tworoger, S. S., Yasui, Y., Vitiello, M. V., Schwartz, R. S., Ulrich, C. M., Aiello, E. J., et al. (2003). Effects of a yearlong moderate-intensity exercise and a stretching intervention on sleep quality in postmenopausal women. *Sleep, 26*(7), 830–838.

Van Cauter, E., Latta, F., Nedeltcheva, A., Spiegel, K., Leproult, R., Vandenbril, C., et al. (2004). Reciprocal interactions between the gh axis and sleep. *Growth Hormone and IGF Research, 14*, 10–17.

Vuori, I., Urponen, H., Hasan, J., & Partinen, M. (1988). Epidemiology of exercise effects on sleep. *Acta Physiologica Scandinavica. Supplementum, 574*, 3–7.

Warren, T. Y., Vaughan, B., Hooker, S. P., Sui, X., Church, T. S., & Blair, S. N. (2010). Sedentary behaviors increase risk of cardiovascular disease mortality in men. *Medicine and Science in Sports and Exercise, 42*(5), 879–885.

Young, T., Peppard, P. E., & Gottlieb, D. J. (2002). Epidemiology of obstructive sleep apnea: A population health perspective. *American Journal of Respiratory and Critical Care Medicine, 165*(9), 1217–1239.

Youngstedt, S. D. (2003). Ceiling and floor effects in sleep research. *Sleep Medicine Reviews, 7*(4), 351–365.

Youngstedt, S. D. (2005). Effects of exercise on sleep. *Clinics in Sports Medicine, 24*(2), 355–365.

Youngstedt, S. D., & Kline, C. E. (2006). Epidemiology of exercise and sleep. *Sleep and Biological Rhythms, 4*(3), 215–221.

Youngstedt, S. D., O'Connor, P. J., Crabbe, J. B., & Dishman, R. K. (2000). The influence of acute exercise on sleep following high caffeine intake. *Physiology and Behavior, 68*(4), 563–570.

Youngstedt, S. D., O'Connor, P. J., & Dishman, R. K. (1997). The effects of acute exercise on sleep: A quantitative synthesis. *Sleep, 20*(3), 203–214.

CHAPTER 16

Mindfulness, Affect, and Sleep: Current Perspectives and Future Directions

Sheila N. Garland[*], Willoughby Britton[†], Noemi Agagianian[‡], Roberta E. Goldman[†], Linda E. Carlson[§], and Jason C. Ong[¶]

[*]Department of Family Medicine and Community Health, Perelman School of Medicine, University of Pennsylvania, Philadelphia, Pennsylvania, USA
[†]Department of Psychiatry and Human Behavior, Brown University Medical School, Providence, Rhode Island, USA
[‡]Department of Psychology, Haverford College, Haverford, Pennsylvania, USA
[§]Department of Oncology, University of Calgary, Calgary, Alberta, Canada
[¶]Department of Behavioral Sciences, Rush University Medical Center, Chicago, Illinois, USA

MINDFULNESS, AFFECT, AND SLEEP: CURRENT PERSPECTIVES AND FUTURE DIRECTIONS

The world is becoming increasingly complex as new technology continually introduces new, exciting, and stimulating ways to live. Fortunately, the human brain has an amazing capacity to process information, learning and adapting to this new technology. At the same time, however, the modern lifestyle challenges the biological and emotional capacity of the brain, as we are regularly inundated by immense amounts of stimuli, pressured to stay awake at times that are incongruent with biological clocks, and trained to remain vigilant for social media updates 24 h a day. Given these increased demands dividing our attention, one might argue that the constant distractions of technology have led to a "mindless" way of living.

Therefore, one possible antidote to the stress of modern society involves the practice of becoming more mindful of our thoughts, feelings, and actions in the present moment.

This chapter provides an introduction to the concept of mindfulness and examines the impact of mindfulness on sleep and affect. First, we define mindfulness and describe the most frequently utilized mindfulness-based interventions (MBIs). Second, we discuss the evidence that MBIs influence sleep and affect. Third, we examine the theoretical models that explain the salutary effects of mindfulness on affect and sleep. Fourth, we consider future directions of research and healthcare implications.

MINDFULNESS-BASED INTERVENTIONS

In the modern context, mindfulness has been defined as the quality of bringing attention to one's experience of the present moment without judgment or attachment to outcome (Kabat-Zinn, 1990). Meditation is the most commonly recommended method of developing mindfulness. The key to mindfulness meditation (MM) is to *intentionally* pay *attention* to present experience with an *attitude* that is accepting and nonjudging (Shapiro, Carlson, Astin, & Freedman, 2006). Mindfully being present contrasts with a layered analysis of experience and thinking that is either past- or future-focused. Regular practice of MM is thought to lead to an overall diffusion of mindfulness throughout everyday life. A key goal of the practice of MM is the development of an awareness of impermanence and constant change. Practicing nonattachment to outcomes and letting go of needing things to be a certain way can relieve the suffering caused by grasping at specific desired outcomes. This corresponds to reduced emotional distress and cultivation of positive emotions (Germer & Neff, 2013).

The practice of MM has been incorporated into clinical interventions as a potential pathway to more effectively decrease perceived stress and to regulate emotional reactivity associated with a broad range of clinical and nonclinical conditions. The first formal Westernized MBI, mindfulness-based stress reduction (MBSR) was developed in 1979 by Jon Kabat-Zinn and colleagues at the University of Massachusetts Medical Center. MBSR incorporates elements of MM training with stress reduction techniques in an 8-week group format (Kabat-Zinn, 1990). The program focuses on several core mindfulness practices, including the body scan; sitting, breathing, and walking meditations; and gentle Hatha yoga (see description of practices in Table 16.1). In addition to these mindfulness practices, a key element of MBSR is experiential learning through body awareness and group discussions. Since that time, hundreds of research

TABLE 16.1 Formal Mindfulness Meditations in MBSR

Meditation Practice	Quiet/Movement	Brief Description
Body scan	Quiet	Bringing awareness to a part of the body and letting go in a sequence throughout the body
Breathing meditation	Quiet	Observing the breath and awareness of the body while breathing
Sitting meditation	Quiet	Observing thoughts, sensations, and feelings arising in the mind and body
Hatha yoga	Movement	Light stretching and gentle yoga postures that bring awareness to the body and its movements
Walking meditation	Movement	Slow deliberate walk to bring awareness to the body in motion

studies have evaluated the efficacy of MBSR for treating myriad physical and psychological symptoms and disorders, and many review articles and meta-analyses are now available (Bohlmeijer, Prenger, Taal, & Cuijpers, 2010; Chiesa & Serretti, 2009; Zainal, Booth, & Huppert, 2013).

In the late 1990s, components of MBSR were combined with cognitive therapy to create mindfulness-based cognitive therapy (MBCT), an 8-week group program to reduce the rate of relapse in patients in remission from major depression (Ma & Teasdale, 2004; Teasdale et al., 2000). Evidence suggests that people who have recovered from depression remain vulnerable to relapse because of previously learned associations between dysphoric mood and patterns of negative, self-devaluative, and hopeless thinking (Marchetti, Koster, Sonuga-Barke, & De Raedt, 2012). As such, subsequent low mood is likely to reactivate old thought patterns and trigger an escalating, self-perpetuating cycle of ruminative cognitive-affective processing. Each subsequent depressive episode is thought to strengthen these associations and less of a catalyst is required to provoke relapse (Post, 2007). MBCT is based on the theory that the risk of depression recurrence can be minimized if people can learn to be more aware of, and disengage from, negative thought patterns and feelings in times of potential relapse or recurrence. Unlike cognitive therapy, which teaches people to change the content of their thoughts, MBCT teaches patients to change one's awareness and relationship to these thoughts. A recent meta-analysis concluded that MBCT: (1) was associated with a 34% reduction in the risk of relapse/recurrence compared to usual or placebo controls, (2) is as effective as maintenance antidepressant medication, and (3) appears to be even more beneficial for patients with three or more depressive episodes (Piet & Hougaard, 2011).

The use of MBIs has grown rapidly in the primary or adjunctive treatment of a wide range of mental health conditions, including anxiety (Evans et al., 2008; Kim et al., 2009), unipolar depression (Bondolfi et al., 2010; Kuyken et al., 2008), bipolar disorder (Williams et al., 2008), suicidal behavior (Williams, Duggan, Crane, & Fennell, 2006), psychosis (Johnson et al., 2011; Newman Taylor, Harper, & Chadwick, 2009), personality disorders (Soler et al., 2012), eating disorders (Wanden-Berghe, Sanz-Valero, & Wanden-Berghe, 2011), addiction (Hsu, Grow, & Marlatt, 2008), autism (Spek, van Ham, & Nyklicek, 2013), and posttraumatic stress disorder (Omidi, Mohammadi, Zargar, & Akbari, 2013). Clinicians have also used MBIs to treat the psychological and physical effects associated with several chronic diseases, including cancer (Carlson et al., 2013; Carlson & Speca, 2010), HIV/AIDS (Gonzalez-Garcia et al., 2014), fibromyalgia (Schmidt et al., 2011), heart disease (Hughes et al., 2013; Parswani, Sharma, & Iyengar, 2013), diabetes (van Son et al., 2013), brain injury (Johansson, Bjuhr, & Ronnback, 2012), solid organ transplant (Gross et al., 2010), and irritable bowel syndrome (Zernicke et al., 2013). More recently, online adaptations of MBIs have been developed and tested (Zernicke et al., 2013). The consistency of results across conditions suggests that MBIs may target the same underlying processes, regardless of the particular disease presentation (Carlson, 2012).

EVIDENCE FOR THE EFFICACY OF MBIs ON SLEEP

Interest in the application of MBIs for treating insomnia and disturbed sleep related to a physical or psychological condition has steadily increased. Treatment studies examining the impact of MBIs on sleep are largely heterogeneous with respect to study design, population, outcomes, and other factors, making interpretation and aggregation difficult. In summarizing this literature, we discuss important aspects of these studies with respect to the population under study, outcome measures, and comparison groups. Further details for each study are provided in Table 16.2.

STUDY CHARACTERISTICS

The sample characteristics were quite variable across studies that examined the use of MBIs to improve sleep. Of the studies identified, the majority were in clinical samples: seven were conducted in patients with

TABLE 16.2 A Descriptive Summary of Mindfulness-Based Interventions for Sleep

Date	First Author	Sample/sex Population	Design/Intervention	Assessment Times	Intervention Details	Sleep and Psychological Outcome Measures	Main Results (p-value)
1994	Goldenberg	121/Female & male Fibromyalgia	Nonrandomized Tx—MM+CBT Com—Waitlist	T1—Baseline T2—Post-program	Group 10 sessionss, weekly 120 min	Visual analog scale of sleep quality SCL-90-R	No significant diff between groups ↓ overall distress (<0.0001)
2003	Carlson	59/Female & male Breast & prostate cancer	Single group/MBSR	T1—Baseline T2—Post-program	Group 8 sessionss, weekly 90 min 3-h retreat	Self reported SQ EORTC, POMS, SOSI	↑ SQ (<0.05) ↑ QL (0.05) and ↓ symptoms of stress (0.01) No significant change in mood disturbance
2003	Shapiro	63/Female Breast cancer	RCT Tx—MBSR C—Free choice TAU	T1—Baseline T2—Post-program T3—3 months T4—9 months	Group 6 sessionss, weekly 120 min 6-h retreat	SD—SE, SQ, feeling rested Single unit-weighted factor scale of psychosocial distress	No significant time or group effect for SE. Informal mindfulness practice was significantly associated with feeling more rested Baseline distress was significantly associated with ↓ SQ and restfulness
2004	Gross	20/Female & male Solid organ transplant	Single group/MBSR	T1—Baseline T2—Post-program T3—3 months	Group 8 sessionss, weekly 120 min	PSQI CESD, SF-12, STAI	↓ sleep disturbance at T2 (0.01) and T3 (0.002) Depression ↓ at T2 (0.006) but not T3 (0.19); anxiety ↓ at T3 (0.04); QL was not significant ↑ at either T

(Continued)

TABLE 16.2 A Descriptive Summary of Mindfulness-Based Interventions for Sleep—Cont'd

Date	First Author	Sample/sex Population	Design/Intervention	Assessment Times	Intervention Details	Sleep and Psychological Outcome Measures	Main Results (p-value)
2004	Roth	86/Female & male Inner city heterogeneous medical patients	Nonrandomized Tx—MBSR Com—No intervention	T1—Baseline T2—Post-Program	Group 8 sessionss, weekly 120min	Number of nights of good sleep in past 4 weeks SF-36	No significant group×time interaction effect for sleep quality ↑ general health (0.033), vitality (0.05), and social functioning (0.014), ↓ role limitations caused by physical health problems (<0.001), and emotional problems (0.050)
2005	Carlson	63/Female & male All cancers	Single group/MBSR	T1—Baseline T2—Post-program	Group 8 sessions, weekly 90min 3-h retreat	PSQI POMS, SOSI	↓ sleep disturbance (<0.001) ↓ stress (<0.001), mood disturbance (<0.001) and fatigue (<0.001), ↓ stress was associated with ↓ sleep disturbance (0.005)
2005	Kreitzer	20/Female & male Solid Organ Transplant Recipients	Single group/MBSR	T1—Baseline T2—Post-program T3—3months T4—6months	Group 8 sessions, weekly 150min	PSQI CES-D, STAI, IIRS, TRS	↓ sleep disturbance over time (<0.001) ↓ depression (0.015). No significant ↓ anxiety or transplant-related stressors

Year	Author	N/Sex/Population	Design	Time points	Intervention	Measures	Results
2006	Carmody	15/Female Menopause-related hot flashes	Single group/MBSR	T1—Baseline T2—Post-program T3—4 weeks	Group 8 sessions, weekly 150 min 7-h retreat	WHIIRS SCL-90-R, PSS, TMS, MENQOL HFRDIS	↓ insomnia symptoms (<0.01) ↓ psychological symptoms (<0.05) and stress (<0.05), ↑ mindfulness (<0.01) and QL (<0.01) ↓ Hot flash frequency (<0.01) and severity (<0.01)
2006	Heidenreich	14/Female & male Primary insomnia	Single group/MBCT	T1—Baseline T2—Post-program	Group 8 sessions, Weekly 120 min 7-h retreat	SD—TST, SL TCQ-I, FEPS-II	↑ TST (<0.01), SL (<0.05) ↓ insomnia-related reappraisal (<0.05), worry (<0.05), and focus on sleep (<0.01)
2008	Klatt	48/Female & male University faculty and staff	RCT Tx—Low dose MBSR C—Waitlist	T1—Baseline T2—Post-program	Group 6 sessions, weekly 60 min	PSQI PSS, MAAS	↓ total sleep disturbance in the T (0.002) and C groups (0.007) ↓ stress (0.003) and ↑ mindfulness (0.02) for the T group only
2008	Yook	19/Female & male Anxiety disorder patients	Single Group/MBCT	T1—Baseline T2—Post-program	Group 8 sessions, Weekly 120 min	PSQI HAM-A, HAM-D, PSWQ, RRS	↑ SQ (0.01) and SE (<0.01), ↓ SL (0.02), use of sleep medication (0.04), and daytime dysfunction (0.01). No change in sleep duration. PSQI total effect size = 1.32 ↓ anxiety (<0.01), depression (<0.01), worry (<0.01), rumination (<0.01)

(Continued)

TABLE 16.2 A Descriptive Summary of Mindfulness-Based Interventions for Sleep—Cont'd

Date	First Author	Sample/sex Population	Design/Intervention	Assessment Times	Intervention Details	Sleep and Psychological Outcome Measures	Main Results (p-value)
2008	Ong	30/Female & male Insomnia	Single group/MBT-I	T1—Baseline T2—Post-program	Group 6 sessions, weekly 90-120min	SD—SL, WASO, TST, SE ISI, GSES, PSAS, DBAS KIMS, PANAS	↓ SL (0.001), ↓ WASO (0.002), TST (ns), ↑ SE (0.001) ↓ insomnia severity (0.001), sleep effort (0.001), pre-sleep arousal (0.001), dysfunctional sleep beliefs (0.001) No significant change in mindfulness or affect (positive or negative)
2009	Ong	21/Female & male Insomnia		T3—6 months T4—12 months		SD—TWT, TST, SE ISI, GSES, PSAS KIMS, PANAS	↓ TWT (0.005), TST (ns), ↑ SE (0.005) ↓ insomnia severity (0.005), sleep effort (0.005), pre-sleep arousal (0.01) No significant change in mindfulness or affect

Year	Author	N/Sex/Population	Design	Time points	Intervention	Measures	Results
2010	Britton	26/Female & male Partially remitted unipolar depression	RCT Tx—MBCT C—Waitlist	T1—Baseline T2—Post-program	Group 8 sessions, Weekly 180 min 6-h retreat	SD—SL, WASO, NWAK, TST, SE PSG—NWAK, SWS, SL, WASO, TST, SE BDI	SD—↓ SL (0.002), WASO (0.006), NWAK (0.006) and ↑ TST (0.006) and SE (0.001) in both groups PSG—T had ↑ NWAK and ↓ SWS, no significant change in SL, WASO, TST, or SE ↓ depression in T (0.02) Meditation practice was correlated with ↑ PSG arousals, awakenings, stage 1, and ↓ SWS
2010	Britton	55/Female & male Adolescents in an outpatient substance abuse program	Single group/MBSR & CBT-I	T1—Baseline T2—Post-program T3—3 months T4—12 months	Group 6 sessions, weekly 90 min	SD—SL, WASO, NWAK, TST, SE GMDI, SEI, SPI, PSWQ	↓ in distress (0.001) and sleepiness (0.05) Meditation practice was significantly associated with ↑ TST and self-efficacy TST was significantly associated with ↓ worry, substance use and ↑ self-efficacy
2010	Gross	138/Female & male Solid organ transplant	RCT Tx—MBSR C1—Health education C2—Waitlist	T1—Baseline T2—Post-program T3—6 months T4—12 months	Group 8 sessions, weekly 150 min 6-h retreat	PSQI STAI, CESD, SF-36	↓ sleep disturbance maintained at T4 (0.02) for Tx group ↓ anxiety (0.02) at T4 for Tx group No significant effect for depression or QL at T4

(Continued)

TABLE 16.2 A Descriptive Summary of Mindfulness-Based Interventions for Sleep—Cont'd

Date	First Author	Sample/sex Population	Design/Intervention	Assessment Times	Intervention Details	Sleep and Psychological Outcome Measures	Main Results (p-value)
2011	Carmody	110/Female Menopause-related hot flashes	RCT Tx—MBSR C—Waitlist	T1—Baseline T2—Post-program T3—3 months	MBSR 8 sessions, weekly 150 min 7-h retreat	WHIIRS MENQOL HADS-A, PSS	↓ SQ (0.009) pre-post program in Tx versus C ↑ in QOL (0.022), ↓ anxiety (0.005), and perceived stress (0.001) at T2. No change at T3
2011	Gross	30/Female & male Insomnia	RCT Tx—MBSR C—Pharmacotherapy	T1—Baseline T2—Post-program T3—3 months	Group 8 sessions, weekly 150 min 6-h retreat	SD & actigraphy—SL, WASO, NWAK, TST, SE PSQI, ISL, DBAS, SSES STAI, CESD, SF-12	SD—greater ↓ SL (0.001) and ↑ TST (0.01) and SE (0.01) in Tx group at T3 than C group Actigraphy—↓ SL (0.04) in the Tx group; ↑ TST (0.05) and SE (0.05) for the C group ↓ sleep disturbance, insomnia severity in both groups (<0.001) at T3. Only the Tx group ↑ sleep self-efficacy and ↓ dysfunctional sleep beliefs No significant effect for depression, anxiety or QL

Year	Author	N/Sex/Condition	Design	Time points	Format	Measures	Results
2011	Schmidt	177/Female Fibromyalgia	RCT Tx—MBSR C1-PMR+stretching C2—Waitlist	T1—Baseline T2—Post-program T3—2 months	Group 8 sessions, weekly 150 min 7-h retreat	PSQI CES-D, STAI, PLC FMI	↓ sleep disturbance for C1 (0.015) and Tx (0.004) at T3 ↓ CES-D (0.012) and STAI (0.003) for Tx. All groups ↑ QL ↑ mindfulness in Tx only
2012	Lengacher	82/Female Breast cancer	RCT Tx—MBSR C—Usual care	T1—Baseline T2—Post-program	Group 6 sessions, weekly 120 min	MDASI	↓ in sleep disturbance (0.001) and fatigue (0.009) ↑ mood in the Tx group
2012	Britton	23/Female & male Antidepressant medication users Unipolar depression	RCT Tx—MBCT C—Waitlist	T1—Baseline T2—Post-program	Group 8 sessions, weekly 180 min 7-h retreat	SD—TWT, TST, SE, SQ PSG—TWT, TST, SE, SWS, stage 1 sleep BDI	↓ TWT (0.046) and ↑ SE (0.007) for Tx group No difference in TST or SQ ↓ TWT (0.046) and ↑ SE (0.09) for Tx group No diff. in TST, SWS, or stage 1 sleep ↓ depression symptoms for Tx group (0.07)

(Continued)

TABLE 16.2 A Descriptive Summary of Mindfulness-Based Interventions for Sleep—Cont'd

Date	First Author	Sample/sex Population	Design/Intervention	Assessment Times	Intervention Details	Sleep and Psychological Outcome Measures	Main Results (p-value)
2013	Andersen	336/Female Breast cancer	RCT Tx—MBSR C—Usual care	T1—Baseline T2—Post-program T3—6 months T4—12 months	Group 8 sessionss, weekly 120 min 5-h retreat	MOS-SS GSI from SCL-90-R Hot Flush Score	↓ sleep disturbance (0.03) in Tx versus C at T2, but not T3 or T4. ↑ sleep adequacy at T2 (0.06) and ↓ somnolence at T3 (0.02) No significant effects were found for psychological distress and hot flushes as modifiers of effect of MBSR on sleep quality
2013	Bei	10/Female 9th grade (ages 13-15) girls with PSQI ≥9	Single group/sleep information, CBT and MM	T1—Baseline T2—Post-program	Group 6 sessions, weekly 90 min	Actigraphy—TST, SL, WASO, SE, SQ PSQI SCAS	20 min ↑ TST ($d=0.46$); 17 min ↓ SL ($d=0.53$). No effect for WASO or SE ↓ sleep disturbance ($d=0.51$). SD—19 min ↓ SL ($d=0.80$) and SQ ($d=0.84$). No effect for TST and SE No significant change in anxiety

Year	Author	N/Sex Population	Design	Timepoints	Format	Measures	Results
2013	Nakamura	57/Female & male All cancers	RCT Tx1—MBB Tx2—MM C—Sleep education	T1—Baseline T2—Post-program T3—2 months	Group 3 sessions, weekly 120 min	MOS-SS FACT-G, PSS, CESD, IES, FFMQ, PANAS	↓ sleep disturbance at T3 for the MBB (0.02) and MM groups (0.04) compared to C No significant between group diff in QL, perceived stress, depression, distress, mindfulness, and affect (positive or negative)
2013	Gonzalez-Garcia	39/Female & male HIV	RCT Tx—MBCT C—Usual care	T1—Baseline T2—Post-program T3—3 months	Group 8 sessions, weekly 150 min	Sleep item of NHP NHP total, PSS-10, BDI-II, BAI	↓ sleep disturbance at T2 (0.03, $d = -0.74$) and T3 (0.05, $d = -0.60$) ↑ QL at T2 ($<0.001, d = -1.4$) and T3 ($<0.001, d = -1.9$). ↓ stress in Tx versus C at T2 ($<0.001, d = -1.6$) and T3 ($<0.001, d = -2.1$), ↓ depression in Tx versus C at T2 ($<0.001, d = -1.3$) and T3 ($<0.001, d = -1.4$), ↓ anxiety in Tx versus C at T2 ($<0.001, d = 1.1$) and T3 ($<0.001, d = -1.3$)
2013	Salmoirago-Blotcher	174/Female & male Heterogeneous physical and mental health conditions	Single group/MBSR	T1—Baseline T2—Post-program	Group 7 sessionss, weekly 150 min 7-h retreat	Sleep quality item of HBQ FFMQ	↑ sleep quality (0.01) Mindfulness scores were positively associated with SQ (<0.01) but not other health behaviors

(Continued)

TABLE 16.2 A Descriptive Summary of Mindfulness-Based Interventions for Sleep—Cont'd

Date	First Author	Sample/sex Population	Design/Intervention	Assessment Times	Intervention Details	Sleep and Psychological Outcome Measures	Main Results (p-value)
2014	Garland	111/Female & male Nonmetastatic cancer and insomnia	Noninferiority RCT T1—CBT-I T2—MBSR	T1—Baseline T2—Post-program T3—3 months	CBT-I Group 8 sessions, weekly 90 min MBSR Group 8 sessions, weekly 90 min 6-h retreat	SD & Actigraphy—SL, WASO, TST, SE ISI, PSQI, DBAS CSOSI, POMS-SF	SD—CBT-I ↓ SL (0.004) and ↑ SE (0.008) compared to MBSR at T3. Both groups ↓ WASO (<0.001) and ↑ TST (<0.001) at T3 Actigraphy—↓ SL (0.003) in CBT-I and ↑ TST (0.004) in MBSR at T3. No group effect for WASO or SE MBSR was inferior to CBT-I for ↓ insomnia severity at T2 (0.35), but noninferior at T3 (0.02). Compared to MBSR, CBT-I ↓ sleep disturbance (<0.001) and dysfunctional sleep beliefs (<0.001) Both groups↓ stress (<0.001) and mood disturbance (<0.001)

BAI, Beck Anxiety Inventory; BDI, Beck Depression Inventory; C, Control Group; CBT-I, cognitive behavior therapy for insomnia; CESD, Center for Epidemiologic Studies Depression Scale; Com, Comparison Group; CSOSI, Calgary Symptoms of Stress Inventory; EORTC, European Organization for Research and Treatment of Cancer Quality of Life Questionnaire; FACIT-B, functional assessment of cancer treatment-breast; FACT-G, functional assessment of cancer therapy-general; FEPS-II, questionnaire to assess central personality variables in insomnia; FFMQ, five facet mindfulness questionnaire; FIQ, fibromyalgia impact questionnaire; FMI, Freiburg Mindfulness Inventory; GCQ, Giessen Compliant Questionnaire; GMDI, general mental distress index; GSES, Glasgow Sleep Effort Scale; GSI, Global Severity Index; HADS-A, Hospital Anxiety and Depression Scale; HAM-A, Hamilton Anxiety Rating Scale; HAM-D, Hamilton Depression Rating Scale; HBQ, Health Behaviors Questionnaire; HFRDIS, Hot Flash-Related Daily Interference Scale; IES, Impact of Events Scale; IIRS, Illness Intrusiveness Rating Scale; ISI, Insomnia Severity Index; KIMS, Kentucky Inventory of Mindfulness; K-SADS, Schedule of Affective Disorders and Schizophrenia Children's Version; MAAS, Mindful Attention Awareness Scale; MBB, mind-body bridging; MBSR, mindfulness-based stress reduction; MBCT, mindfulness-based cognitive therapy; MDASI, M.D. Anderson Symptom Inventory; MENQOL, Menopause-Related Quality of Life; MOS-SS, Medical Outcomes Study-Sleep Scale; MM, Mindfulness meditation; NHP, Nottingham Health Profile; NWAK, Number of Awakenings; PANAS, Positive and Negative Affect Scale; PENN, Penn State Worry Questionnaire; PLC, Quality of Life Profile for the Chronically Ill; PMR, Progressive Muscle Relaxation; PMS-SF, Profile of Mood Stress-Short Form; POMS, Profile of Mood States; PSG, Polysomnograp'ıy; PSQI, Pittsburgh Sleep Quality Index; PSAS, Pre Sleep Arousal Scale; PSS, Perceived Stress Scale; PSWQ, Penn State Worry Questionnaire; QL, Quality of Life; RRS, Ruminative Response Scale; SCAS, Spence Children's Anxiety Scale (SCAS); SCI, Shapiro Control Inventory; SCS, Self Compassion Scale; SCL-90-R, Hopkins Symptom Checklist-90-revised; SD, sleep diary; SEI, self efficacy index; SOC, Sense of Coherence; SF-12, Short Form Health Survey; SOSI, Symptoms of Stress Inventory; SPI, substance problem incex; SQ, Sleep Quality; SSES, Sleep Self Efficacy Scale; STAI, State Trait Anxiety Inventory; SWS, slow wave sleep; TCQ-I, Thoughts Control Questionnaire for Insomnia; TMS, Toronto Mindfulness Scale; TWT, Total Wake Time; TRS, Transplant Related Stressors; WHIIRS, Women's Health Initiative Insomnia Rating Scale.

cancer (Andersen et al., 2013; Carlson & Garland, 2005; Garland et al., 2014; Lengacher et al., 2012; Nakamura, Lipschitz, Kuhn, Kinney, & Donaldson, 2013; Shapiro, Bootzin, Figueredo, Lopez, & Schwartz, 2003), four in individuals with primary insomnia disorder (Bei et al., 2013; Gross et al., 2011; Heidenreich, Tuin, Pflug, Michal, & Michalak, 2006; Ong, Shapiro, & Manber, 2008; Ong, Shapiro, & Manber, 2009), three in patients undergoing solid organ transplants (Gross et al., 2004, 2010; Kreitzer, Gross, Ye, Russas, & Treesak, 2005), two in fibromyalgia patients (Goldenberg et al., 1994; Schmidt et al., 2011), two in patients with partially remitted unipolar depression (Britton, Haynes, Fridel, & Bootzin, 2010; Britton, Shahar, Szepsenwol, & Jacobs, 2012), two in individuals with insomnia secondary to menopause (Carmody, Crawford, & Churchill, 2006; Carmody et al., 2011) or generalized anxiety disorder (Yook et al., 2008), and one in adolescent substance abuse outpatients (Britton, Bootzin, et al., 2010; Britton, Haynes, et al., 2010). In addition, three studies were conducted in nonclinical community samples (Klatt, Buckworth, & Malarkey, 2009; Roth & Robbins, 2004; Salmoirago-Blotcher, Crawford, Carmody, Rosenthal, & Ockene, 2011).

OUTCOME MEASURES FOR SLEEP

Sleep can be measured by a variety of methods, ranging from subjective measures, such as daily sleep diaries, and validated questionnaires to objective measures, such as actigraphy and polysomnography (PSG). A brief overview of these methods is presented here (please see Chapter 4 for more detailed information). Studies most commonly used subjective measures, which have the advantage of nightly or repeated at-home measurement. Daily sleep diaries, a standard outcome measure in many studies (Carney et al., 2012), require that participants monitor their previous night's sleep for several days or weeks. Sleep diary variables include bedtime, wake time, sleep latency (SL), wake after sleep onset (WASO), number of nocturnal awakenings (NWAK), perceived total sleep time (TST), sleep efficiency (SE; time asleep/time spent in bed×100), and a sleep quality rating. Sleep diaries are considered superior to retrospective aggregate scales in both resistance to recall bias and sensitivity to change over time (Carney et al., 2012).

Wrist actigraphy is carried out using a watch-like device that is worn on the arm and estimates sleep using an accelerometer to detect movement. Actigraphy is also widely used and well validated in many samples and conditions and correlates well with PSG on most sleep parameters (0.15-0.92, average=0.71) (Morgenthaler et al., 2007). Laboratory-based PSG is considered the gold standard measurement, as well as the basis of the physiological definition of sleep (Rechtschaffen & Kales, 1969). PSG measurement allows for the collection of multiple physiological indicators of sleep, including electroencephalography (EEG), eye movements (EOG), and muscle tone (EMG), but it often requires that patients sleep in an unfamiliar environment, which can negatively impact generalizability.

In general, MM shows stronger sleep-promoting effects for subjectively measured sleep than for objectively measured sleep. For example, in adolescent girls participating in a mindfulness-based multicomponent sleep intervention, Bei et al. (2013) reported a 19-min reduction in SL ($d = 0.80$) and improved sleep quality ratings ($d = 0.84$), using sleep diaries, while improvements in actigraphy translated into a reduction of 17-min in the time it took to fall asleep ($d = 0.53$), 20 min more sleep per night ($d = 0.46$), and a 3.4% increase in SE ($d = 0.51$). Similarly, Britton, Haynes, Fridel, and Bootzin (2012) and Britton, Shahar, Szepsenwol, and Jacobs (2012) found improvements on both subjective and objective measurements of sleep, with large effect sizes (mean $d = 0.92$) for diaries and medium effect sizes (mean $d = 0.57$) for PSG in antidepressant users participating in MBCT. Furthermore, for subjectively measured sleep measures, aggregate retrospective sleep questionnaires appear to be more amenable to change than prospective daily sleep diaries and actigraphic sleep measurements. In individuals with insomnia disorder participating in MBSR, Gross et al. (2011) found that sleep questionnaires yielded the highest improvements with large to very large effects sizes ($d = 0.80$-2.4), followed by sleep diaries with small to large effects ($d = 0.23$-0.80), while objective, actigraphy-based improvements were small ($d = 0.10$-0.31). Although the causes of the variations in measurement are yet to be determined, the differential improvements observed between subjective and objective sleep outcomes for MBIs may be explained by improvements in affect influencing perception of sleep quality not captured in objective measures.

Some studies that used both subjective and objective measures of sleep have reported conflicting findings. Following MBSR in patients with insomnia disorder, Gross et al. (2011) found an increase in diary-based TST ($d = +0.21$) but a *decrease* in actigraphy-based TST in MBSR ($d = -0.21$). Britton, Bootzin, et al. (2010) and Britton, Haynes, et al. (2010) also found opposite effects depending on whether sleep was measured subjectively or objectively in patients with partially remitted depression being treated with MBCT. Diary-based measures of sleep, including SL, NWAK, minutes of WASO, and SE improved significantly pre-to-post with large to very large pre-to-post effect sizes ($d = 1.4$-2.0). However, the objective PSG sleep profiles indicated increased awakenings and stage 1 sleep ($d = 0.87$) and a significant decrease in slow-wave sleep ($d = 1.40$) relative to controls. The time (minutes) spent in MM practice was positively correlated with several indices of increased cortical arousal, including the amount of stage 1 sleep ($r = 0.80$, $p < 0.005$), number of arousals ($r = 0.59$, $p < 0.05$), number of awakenings ($r = 0.57$, $p < 0.05$), and TST ($r = 0.80$, $p < 0.005$), so that more time spent meditating was associated with greater cortical arousal and less sleep. Although this dose-related increase in cortical arousal may appear counterintuitive, increased cortical arousal, wakefulness, and decreases in sleep propensity are common in meditators, and from a traditional Buddhist perspective, these states are an expected and desired outcome of the practice (Britton, Lindahl, Cahn, Davis, & Goldman, 2014).

Some evidence also seems to show that sleep quality ratings and retrospective questionnaires are influenced more by meditation than are specific sleep parameters, such as sleep duration (Andersen et al., 2013; Shapiro et al., 2003). For example, in women with breast cancer, Andersen et al. (2013) found improved "sleep disturbance" and "sleep quality" but no change in sleep duration for MBSR, when it was compared to treatment as usual. Also in women with breast cancer, Shapiro et al. (2003) found significant improvements in sleep quality ratings, but no change in sleep parameters (i.e., sleep duration, SE) for MBSR. One possible explanation for the discrepancy is that the self-report ratings can be influenced by self-perception of sleep and affect and that improving affect or changing the perception of sleep could improve self-reported ratings of sleep quality.

Overall, these studies suggest that the effects of MBIs on sleep might depend on how sleep is measured. Specifically, effect sizes for MBIs are larger when using retrospective, subjective measures of sleep quality as opposed to specific sleep parameters or objective measures of sleep. The discrepancy between subjective and objective measures of sleep has been well documented in sleep research, especially among people with insomnia (Edinger et al., 2000; Silva et al., 2007; Zhang & Zhao, 2007), and a similar trend appears to be emerging in studies using meditation.

COMPARISON CONDITIONS

A number of randomized controlled trials (RCTs) have found positive effects of MM training on self-reported sleep quality in comparison with passive controls (waitlist or treatment as usual) (Britton, Haynes, et al., 2012; Britton, Shahar, et al., 2012; Carmody et al., 2011; Gross et al., 2011, 2010; Klatt et al., 2009; Lengacher et al., 2012; Nakamura et al., 2013). However, other RCTs have found no difference between the meditation training and passive control conditions (Britton, Bootzin, et al., 2010; Britton, Haynes, et al., 2010; Goldenberg et al., 1994; Klatt et al., 2009; Roth & Robbins, 2004; Schmidt et al., 2011; Shapiro et al., 2003). The lack of differences between intervention and control was due to a lack of improvement in either group (Shapiro et al., 2003) or comparable improvement in both groups (Britton, Haynes, et al., 2012; Britton, Shahar, et al., 2012; Klatt et al., 2009; Roth & Robbins, 2004).

In patients with partially remitted depression, Britton, Bootzin, et al. (2010) and Britton, Haynes, et al., (2010) found that both the MBCT and waitlist controls improved significantly on diary sleep measures, which the authors attributed to the beneficial effect of monitoring sleep daily for 10 weeks (i.e., the Hawthorne effect; McCarney et al., 2007). Similarly, in working adults, Klatt et al. (2009) reported that low dose MBSR ($d = 0.76$) and a waitlist control group ($d = 0.68$) improved significantly with a medium effect size on the PSQI global sleep disturbance score, even though one group received no treatment.

When MBIs are compared to active controls, the effects are mixed. When compared to a "health education" control, matched MBIs have been statistically superior in two studies (Gross et al., 2010; Nakamura et al., 2013). Gross et al. (2010) found significantly greater self-reported improvements in sleep with a medium effect size ($d = 0.51$) following MBSR, in comparison with health education, for people with insomnia after solid-organ transplant, an effect that was maintained up to 1 year after treatment. Other studies have either found the active controls to be similar or even slightly better than MBSR in improving sleep. Schmidt et al. (2011) found that, although MBSR ($d = 0.38$) and progressive muscle relaxation ($d = 0.27$) both yielded significant improvements in subjective sleep quality in patients with fibromyalgia, neither group was significantly better than the waitlist condition. Nakamura et al. (2013) found that all three active treatments provided benefit for cancer survivors ($p < 0.05$), with both Mind Body Bridging (MBB), a 3-week awareness training program, and 3 weeks of training in MM outperforming a sleep hygiene education condition. However, MBB ($d = 1.06$) was consistently superior to MM ($d = 0.70$) for self-reported sleep problems (effect size of treatment comparison, $d = 0.12$) and increasing mindfulness scores, even though MBB contained no meditation.

Gross et al. (2011) compared MBSR to sleep medication (eszopiclone) in a small RCT, and they found no statistically significant differences between treatments for any sleep-related variable. The medication group showed immediate and lasting effects on sleep diary-measured TST and SE ($d = 0.74$-1.29), and MBSR showed immediate effects for SL, but medium to large treatment effects for TST, SL, and SE after 5 months ($d = 0.69$-0.80). On actigraphy, the MBSR group showed a 9-min decrease in SL ($d = 0.31$, $p = 0.04$). The medication group exhibited significant increases in TST by more than 30 min ($d = 0.63$, $p < 0.05$), and TST in the MBSR decreased slightly. Because of the small sample size, the authors warn that the lack of statistical differences between groups "does not confirm equivalence" or noninferiority (Gross et al., 2011).

Garland et al. (2014) compared MBSR to cognitive behavioral therapy for insomnia (CBT-I) in a trial powered to test noninferiority among patients with cancer. This study, which directly compares MSBR to the gold standard nonpharmacological intervention for sleep, CBT-I, represents the most rigorous test to date of the ability of MBSR to influence sleep outcomes. Immediately after the 8-week program, MBSR was inferior to CBT-I for improving insomnia severity ($p=0.35$), but it demonstrated noninferiority at the 3-month follow-up ($p=0.02$). This suggests that the two programs may target different mechanisms, with CBT-I exerting direct effects on sleep and MBSR indirectly improving sleep via a reduction in mood disturbance and symptoms of stress, but this conclusion must be empirically tested. With regard to specific sleep outcomes, at the final assessment, both programs experienced similar reductions in diary-measured time spent awake during sleep (~40 min) and SOL (~18 min). Individuals in CBT-I and MBSR also significantly increased their TST, with an additional 0.60 h for CBT-I and 0.75 h for MBSR. These similarities in outcomes are particularly impressive considering that the MBSR program is not designed as a sleep-specific intervention.

Given the mixed findings when MBIs are compared to different control conditions, the research has not yet shown whether teaching mindfulness principles or practicing MM directly leads to improvements in sleep. Nonspecific factors, such as group dynamics or attention, could account for the observed changes in sleep. Considering the heterogeneity of the reviewed studies with regards to sample characteristics and outcome measures, we cannot easily address important theoretical issues and clinically significant questions, namely: (1) How do MBIs impact sleep? (2) For whom do MBIs improve sleep? (3) What aspects of sleep are improved? Moreover, researchers have yet to examine potential moderators, such as baseline levels of sleep disturbance/arousal or treatment preference, and knowledge of these moderators could clarify some of the mixed findings. As a whole, MBIs appear to benefit sleep, but firm conclusions cannot be drawn from the current literature.

THEORETICAL MODELS OF THE RELATIONSHIP OF MINDFULNESS TO SLEEP AND AFFECT?

Researchers have not yet developed a comprehensive model that describes how MBIs are specifically connected with the regulation of sleep and affect. However, theoretical models from cognitive psychology, neuroscience, and psychophysiology provide some insights into possible mechanisms underlying the relationship between mindfulness, sleep, and affect. Affect is most often examined in the context of wakefulness, but some models of emotion include sleep on the continuum of activation/deactivation. For example, the circumplex of emotion uses sleep as an anchor for deactivation with neutral or low emotional valence (Russell & Barrett, 1999; Yik, Russell, & Barett, 1999). However, the notion of sleep as a passive state is inconsistent with the body of research examining sleep physiology. It is now widely recognized that sleep consists of active, dynamic processes arising from within the central nervous system (Dement, 2005) and that affective states impact sleep ability, quality, and quantity. As such, researchers have most often examined the relationship between mindfulness, affect, and sleep within the context of sleep disturbance and mood disorders.

Although clinicians previously considered sleep disturbance in the form of insomnia to be a symptom of an underlying condition, such as a mood disorder, contemporary perspectives have recognized that insomnia can exist as a distinct and independent disorder (American Psychiatric Association, 2013). Etiological models of chronic insomnia have generally identified elevated cognitive and physiological arousal as key factors in the development and maintenance of sleep disturbance (Bonnet & Arand, 2010). As such, models of mindfulness and sleep have examined the cognitive and neurobiological mechanisms of mindfulness as a means of improving sleep by reducing sleep-related arousal, negative affect, and emotion dysregulation.

COGNITIVE MODELS

The metacognitive or insight model of mindfulness is predominant among cognitively oriented researchers and clinicians. This model emphasizes metacognitive insight, characterized by (1) shifting relationships with thoughts rather than changing the thought itself; (2) awareness and present-focused attention as a central component of change; and (3) an emphasis on the experiential, process-oriented nature of self-examination and self-compassion. This approach involves using mindfulness practice to "see thoughts as thoughts, with a detached but inclusive awareness," and it has been referred to by a number of terms, including "distancing," "defusing," "de-centering," and "reperceiving" (Hayes, 1987; Shapiro & Carlson, 2009; Teasdale, 1999, p. 154). Through the repeated practice of de-centering, individuals can develop an ability to detach themselves from their thoughts, acknowledging those thoughts as creations of the mind that may or may not accurately represent facts or reality. This increased awareness of dysfunctional emotional states can allow for intervention prior to the development of full-blown psychological disorders, such as a major depressive episode (see the following for a discussion of the neurobiological mechanisms related to this concept).

With respect to sleep disturbance, Lundh (2005) described a cognitive model of how mindfulness could be used to help individuals with insomnia reinterpret their sleeplessness. This new schematic targets sleep-interpreting processes (i.e., thoughts, perceptions, and attributions about sleep and its consequences), which are believed to cause presleep arousal (Lundh & Broman, 2000). Ong and colleagues (2012) have proposed a metacognitive model of sleep-related arousal, consisting of two levels. Primary arousal is the cognitive activity directly related to the inability to sleep and consists of specific thoughts about sleep and sleeplessness. For example, primary arousal might include the belief that one needs to sleep 8h to feel well-rested the following day. Secondary arousal exists at a metacognitive level and is the emotional valence, degree of attachment, and interpretive value related to the primary thoughts and beliefs. When secondary arousal develops, it creates a cognitive bias in the attention to, and perception of, sleep-related thoughts at the primary level, which amplifies the negative emotional valence. Using the example above, secondary arousal might arise when one feels that 8h of sleep is needed every night, leading to increased sleep effort and/or clock monitoring throughout the night. The practice of mindfulness would not challenge the belief (i.e., primary arousal) per se, but it would instead focus on de-centering from the attachment to this outcome and the value of sleeping 8h each night. Through this practice, de-arousal occurs at the secondary level of arousal, with potential downstream effects on the primary level of arousal.

NEUROBIOLOGICAL MODELS

The process of falling asleep entails disengaging or de-arousing from wakefulness. If these processes fail, or if arousal remains elevated, sleep disturbance occurs. In this context, arousal is a complex multidimensional construct with multiple overlapping inputs, including neurophysiological, autonomic, endocrine, cognitive, and affective signals. Most relevant to sleep-impairing arousal is excessive negative affect and associated cognitive activity such as rumination and worry. The neurobiological models with relevance to both affect and sleep suggest MM may strengthen brain areas that regulate cognitive and emotional arousal and support the meta-awareness of these processes. This model can be broken into two stages.

The first stage of the neurobiological model is a "calming" stage during which more focused forms of meditation increase the inhibitory control of the prefrontal cortex (PFC) over areas of the brain involved in emotional and cognitive arousal, the limbic system and the default mode network (DMN). The limbic system is a set of brain areas, including the amygdala, that are involved in emotional responses, and our "flight or flight" stress reaction. The amygdala is strongly coupled with the endocrine and sympathetic nervous systems, and therefore, its activation is often associated with emotional arousal or the bodily expression of emotion (Davidson, Jackson, & Kalin, 2000). The DMN consists of midline brain structures that are active during rest or when the brain is not otherwise engaged, and these structures are thought to be involved in self-referential thought and mind wandering (Qin & Northoff, 2011), as well as the rumination and worry that characterizes anxiety, depression, and insomnia (Buckner & Vincent, 2007; Farb et al., 2007; Gentili et al., 2009; Hamilton et al., 2011; Lemogne, Delaveau, Freton, Guionnet, & Fossati, 2012; Segal, 1988; Sheline et al., 2009; Whitfield-Gabrieli et al., 2009; Zhao et al., 2007).

The PFC exerts inhibitory control over both the limbic system and the DMN. If the PFC is weak or hypoactive, a condition known as hypofrontality, then an individual experiences a disinhibtion of the limbic and default mode in the form of increased emotional and cognitive arousal (Clark, Chamberlain, & Sahakian, 2009; Couyoumdjian et al., 2010). In contrast, Buddhism-derived meditation practices are associated with increased activity in the PFC (Allen et al., 2012; Baerentsen, 2001; Baron Short et al., 2010; Brefczynski-Lewis, Lutz, Schaefer, Levinson, & Davidson, 2007; Farb et al., 2010, 2007; Hasenkamp & Barsalou, 2012; Ritskes, Ritskes-Hoitinga, Stodkilde-Jorgensen, Baerntsen, & Hartman, 2003) and larger frontal gray matter volumes (Holzel et al., 2008; Lazar et al., 2005; Luders, Toga, Lepore, & Gaser, 2009), essentially reversing hypofrontality. Mindfulness and other forms of meditation also seem to reverse some of the downstream consequences of hypofrontality, including amygdala activation (Brefczynski-Lewis et al., 2007; Creswell, Way, Eisenberger, & Lieberman, 2007; Desbordes et al., 2012; Farb et al., 2007; Taylor et al., 2011; Way, Creswell, Eisenberger, & Lieberman, 2010) and sympathetic hyperarousal (Barnes, Treiber, & Davis, 2001; Carlson, Speca, Faris, & Patel, 2007; MacLean et al., 1994; Ortner, Kilner, & Zelazo, 2007; Sudsuang, Chentanez, & Veluvan, 1991; Tang et al., 2007). Similarly, a number of studies have suggested a relationship between mindfulness and reduced emotional and stress reactivity, including attenuated emotional responses to threatening situations or faster recovery from transient negative affect (Arch & Craske, 2006; Brewer et al., 2009; Britton, Haynes, et al., 2012; Britton, Shahar, et al., 2012; Broderick, 2005; Campbell-Sills, Barlow, Brown, & Hofmann, 2006; Erisman & Roemer, 2010; Goldin & Gross, 2010; Kuehner, Huffziger, & Liebsch, 2009; McKee, Zvolensky, Solomon, Bernstein, & Leen-Feldner, 2007; Ortner et al., 2007; Pace et al., 2009; Proulx, 2008; Raes, Dewulf, Van Heeringen, & Williams, 2009; Tang et al., 2007; Weinstein, Brown, & Ryan, 2009).

Multiple studies have found that various forms of meditation training are associated with decreased DMN activity (Baerentsen, 2001; Baerentsen et al., 2010; Berkovich-Ohana, Glicksohn, & Goldstein, 2012; Brewer et al., 2011; Farb et al., 2010; Goldin, Ramel, & Gross, 2009; Hasenkamp & Barsalou, 2012; Taylor et al., 2011; Travis et al., 2010). Thus, sustaining and redirecting attention back to an object (like the breath) over and over again builds the attention muscle of the dorsal attention system of the PFC (Hasenkamp & Barsalou, 2012). When the PFC is strong and engaged, the limbic and default mode systems are less active, and the mind becomes calm and tranquil.

The second stage in the neurobiological model is an "insight" stage, characterized by metacognitive awareness. The focused attention practices in mindfulness-based programs may account for the calming effects on affect, cognition, and physiological arousal. For example, MM might consist of an open monitoring practice, during which all stimuli–disturbing thoughts, judgments and emotions—are possible objects of meditation and are not viewed as distractions or obstacles. On an experiential level, the meditator sees that the emotions he or she has identified with are nothing more than passing thoughts (mental images, words) and body sensations (pressure, tightness, heat). In Buddhist terms, this is insight into the transient (impermanent) nature of thoughts and emotions. Because they are fleeting and insubstantial, thoughts and emotions are neither reliable, nor able to give lasting satisfaction or threat. As stated earlier, seeing thoughts and emotions as transient events rather than as accurate views of reality has been called "metacognitive awareness" (Teasdale et al., 2002), "decentering" (Watkins, Teasdale, & Williams, 2000), and "cognitive defusion" (Masuda, Feinstein, Wendell, & Sheehan, 2010). By calming the mind and gaining metacognitive insight, MM allows one to rediscover the process of de-arousing from wakefulness and allowing the brain to sleep.

Taken together, cognitive and neurobiological models point toward the hypothesis that the effects of MM on sleep might be mediated by reducing negative affect. The cognitive models suggest that mindfulness might serve as a means of reinterpreting sleeplessness or reducing secondary arousal, thereby improving sleep in individuals with insomnia. Similarly, neurobiological models indicate that MM is linked to greater prefrontal control over subcortical brain regions associated with negative affect, such as the amygdala. This hypothesis remains to be tested, but it could serve as a starting point in developing a neurocognitive model to understand the mechanisms involved with MBIs for sleep and affect.

FUTURE DIRECTIONS AND IMPLICATIONS

The current research on mindfulness, sleep, and affect is a growing subset of a larger literature on the mechanisms and clinical applications of cultivating mindfulness. MBIs appear to benefit self-reported sleep, with more equivocal findings on objective measures of sleep, as well as measures of mood and affect. Many of the reviewed studies consisted of relatively small sample sizes, and, with only a few exceptions (Andersen et al., 2013; Garland et al., 2014; Gross et al., 2010; Schmidt et al., 2011), most intervention studies utilized wait-list controls or did not employ control groups. Moreover, few neurobiological studies have directly examined the relationship between sleep, affect, and mindfulness. Therefore, several key agenda items for future research remain.

First, the underlying mechanisms between mindfulness, affect, and sleep remain unclear. Although the studies reviewed point toward a hypothesis that the effects of mindfulness on sleep are mediated by reducing negative affect, other plausible mechanisms should be examined. Perhaps meditation can elicit other parasympathetic effects that also play a role in relaxation and improved sleep. In addition, the adoption of other frameworks for emotion regulation, such as Compas's work on coping (Andreotti et al., 2013) and Gross's work on antecedent- and response-focused emotion regulation (Gross, 1998), could provide new perspectives for understanding the links between sleep and emotion regulation. For example, mindfulness might involve emotion regulation strategies that are antecedent-focused, as opposed to response-focused strategies. Alternatively, perhaps mindfulness functions more as distraction from arousal-eliciting stimuli. Comparing the effects of mindfulness on presleep arousal and sleep outcomes with distraction and cognitive reappraisal may yield interesting results for further understanding its potential clinical utility. Moreover, integrating mindfulness with these broader models in order to capitalize on rapidly emerging work on these models (e.g., Webb, Miles, & Sheeran, 2012) may prove advantageous.

One issue that emerges from our review is that MBIs may be more efficacious for people with elevated levels of negative affect because MBIs may primarily help sleep via reducing negative affect. Examining potential moderators could reveal who is likely to benefit from MBIs with regard to sleep. Putative moderators might include pretreatment levels of anxiety and depression or preference for complementary and alternative approaches to healing. Identifying moderators could also guide treatment recommendations with respect to medication and other nonpharmacological approaches for sleep disturbances. Other possible moderators may be interesting to consider as well (e.g., pretreatment problems with emotion regulation and previous experience with or exposure to mindfulness).

A third item for future research is the measurement of affect across a continuum of sleep and wakefulness. Thus far, the literature has focused almost entirely on insomnia or sleep disturbance in the context of another medical condition such as cancer. However, disorders involving chronic sleepiness, such as narcolepsy, could provide an interesting model for investigating the relationship between sleep, sleepiness, and affect. Moreover, the concept of healthy sleep could be examined within the context of positive affect. Buysse (2014) proposed a framework for understanding sleep health and possible connections with physical and psychological functioning, which could be incorporated to examine connections between sleep health and positive emotions.

In terms of clinical and health care implications, one issue to resolve is how MBIs would fit in or compete with CBT-I, the current gold standard nonpharmacological treatment for insomnia. MBIs have a different philosophy (acceptance-based), approach (targeting metacognitions), and technique (meditations) than CBT-I does, and thus MBIs could be viewed as alternatives to CBT-I. Should further empirical evidence support the use of MBIs for insomnia, it could open up potential opportunities for patients to receive empirically supported alternative treatments, because hundreds of hospitals, medical centers, and private clinics around the world already deliver MBIs. Another potential clinical issue is whether or not MBIs should be delivered in a group format, as typically conducted in MBSR and MBCT, or if an individual format could be equally effective. This consideration merits further investigation with respect to treatment implementation. Finally, researchers could also consider the importance of the meditation practice to the efficacy of MBIs. Although MBSR is the most widely researched MBI and places a large emphasis on meditation practice, other MBIs vary in terms of the emphasis placed on meditation. Furthermore, it is unclear whether quiet meditations, such as sitting meditations, or movement meditations, such as yoga, have relative benefits for sleep.

The existing evidence indicates that MM could serve as a viable stand-alone, or adjunctive, intervention for regulating emotion and improving sleep. Questions remain as to how to best conceptualize the role of mindfulness in sleep interventions, what mechanisms may explain the salutary sleep effects of MBIs, how individual characteristics might moderate the effectiveness of MBIs, and how MBIs for sleep compare to or complement established interventions. As technology continues to advance, new methods could be employed to answer some of these questions. Modern technology might prove to be useful in demonstrating the mechanisms and health outcomes of this ancient approach.

References

Allen, M., Dietz, M., Blair, K. S., van Beek, M., Rees, G., Vestergaard-Poulsen, P., et al. (2012). Cognitive-affective neural plasticity following active-controlled mindfulness intervention. *The Journal of Neuroscience: The Official Journal of the Society for Neuroscience, 32*(44), 15601–15610. http://dx.doi.org/10.1523/JNEUROSCI.2957-12.2012.

American Psychiatric Association. (2013). *Diagnostic and statistical manual of mental disorders* (5th ed.). Arlington, VA: American Psychiatric Publishing.

Andersen, S. R., Wurtzen, H., Steding-Jessen, M., Christensen, J., Andersen, K. K., Flyger, H., et al. (2013). Effect of mindfulness-based stress reduction on sleep quality: Results of a randomized trial among danish breast cancer patients. *Acta Oncologica, 52*(2), 336–344. http://dx.doi.org/10.3109/0284186X.2012.745948.

Andreotti, C., Thigpen, J. E., Dunn, M. J., Watson, K., Potts, J., Reising, M. M., et al. (2013). Cognitive reappraisal and secondary control coping: Associations with working memory, positive and negative affect, and symptoms of anxiety/depression. *Anxiety, Stress, and Coping, 26*(1), 20–35. http://dx.doi.org/10.1080/10615806.2011.631526.

REFERENCES

Arch, J. J., & Craske, M. G. (2006). Mechanisms of mindfulness: Emotion regulation following a focused breathing induction. *Behaviour Research and Therapy, 44*(12), 1849–1858. http://dx.doi.org/10.1016/j.brat.2005.12.007.

Baerentsen, K. B. (2001). Onset of meditation explored with fMRI. *NeuroImage, 13*, S297.

Baerentsen, K. B., Stodkilde-Jorgensen, H., Sommerlund, B., Hartmann, T., Damsgaard-Madsen, J., Fosnaes, M., et al. (2010). An investigation of brain processes supporting meditation. *Cognitive Processing, 11*(1), 57–84. http://dx.doi.org/10.1007/s10339-009-0342-3.

Barnes, V. A., Treiber, F. A., & Davis, H. (2001). Impact of transcendental meditation on cardiovascular function at rest and during acute stress in adolescents with high normal blood pressure. *Journal of Psychosomatic Research, 51*(4), 597–605.

Baron Short, E., Kose, S., Mu, Q., Borckardt, J., Newberg, A., George, M. S., et al. (2010). Regional brain activation during meditation shows time and practice effects: An exploratory FMRI study. *Evidence-Based Complementary and Alternative Medicine: eCAM, 7*(1), 121–127. http://dx.doi.org/10.1093/ecam/nem163.

Bei, B., Byrne, M. L., Ivens, C., Waloszek, J., Woods, M. J., Dudgeon, P., et al. (2013). Pilot study of a mindfulness-based, multi-component, in-school group sleep intervention in adolescent girls. *Early Intervention in Psychiatry, 7*(2), 213–220. http://dx.doi.org/10.1111/j.1751-7893.2012.00382.x.

Berkovich-Ohana, A., Glicksohn, J., & Goldstein, A. (2012). Mindfulness-induced changes in gamma band activity—Implications for the default mode network, self-reference and attention. *Clinical Neurophysiology, 123*(4), 700–710. http://dx.doi.org/10.1016/j.clinph.2011.07.048.

Bohlmeijer, E., Prenger, R., Taal, E., & Cuijpers, P. (2010). The effects of mindfulness-based stress reduction therapy on mental health of adults with a chronic medical disease: A meta-analysis. *Journal of Psychosomatic Research, 68*(6), 539–544. http://dx.doi.org/10.1016/j.jpsychores.2009.10.005.

Bondolfi, G., Jermann, F., der Linden, M. V., Gex-Fabry, M., Bizzini, L., Rouget, B. W., et al. (2010). Depression relapse prophylaxis with mindfulness-based cognitive therapy: Replication and extension in the swiss health care system. *Journal of Affective Disorders, 122*(3), 224–231. http://dx.doi.org/10.1016/j.jad.2009.07.007.

Bonnet, M. H., & Arand, D. L. (2010). Hyperarousal and insomnia: State of the science. *Sleep Medicine Reviews, 14*(1), 9–15. http://dx.doi.org/10.1016/j.smrv.2009.05.002.

Brefczynski-Lewis, J. A., Lutz, A., Schaefer, H. S., Levinson, D. B., & Davidson, R. J. (2007). Neural correlates of attentional expertise in long-term meditation practitioners. *Proceedings of the National Academy of Sciences of the United States of America, 104*(27), 11483–11488. http://dx.doi.org/10.1073/pnas.0606552104.

Brewer, J. A., Sinha, R., Chen, J. A., Michalsen, R. N., Babuscio, T. A., Nich, C., et al. (2009). Mindfulness training and stress reactivity in substance abuse: Results from a randomized, controlled stage I pilot study. *Substance Abuse, 30*(4), 306–317. http://dx.doi.org/10.1080/08897070903250241.

Brewer, J. A., Worhunsky, P. D., Gray, J. R., Tang, Y. Y., Weber, J., & Kober, H. (2011). Meditation experience is associated with differences in default mode network activity and connectivity. *Proceedings of the National Academy of Sciences of the United States of America, 108*(50), 20254–20259. http://dx.doi.org/10.1073/pnas.1112029108.

Britton, W. B., Bootzin, R. R., Cousins, J. C., Hasler, B. P., Peck, T., & Shapiro, S. L. (2010). The contribution of mindfulness practice to a multicomponent behavioral sleep intervention following substance abuse treatment in adolescents: A treatment-development study. *Substance Abuse, 31*(2), 86–97. http://dx.doi.org/10.1080/08897071003641297.

Britton, W. B., Haynes, P. L., Fridel, K. W., & Bootzin, R. R. (2010). Polysomnographic and subjective profiles of sleep continuity before and after mindfulness-based cognitive therapy in partially remitted depression. *Psychosomatic Medicine, 72*(6), 539–548. http://dx.doi.org/10.1097/PSY.0b013e3181dc1bad.

Britton, W. B., Haynes, P. L., Fridel, K. W., & Bootzin, R. R. (2012). Mindfulness-based cognitive therapy improves polysomnographic and subjective sleep profiles in antidepressant users with sleep complaints. *Psychotherapy and Psychosomatics, 81*(5), 296–304. http://dx.doi.org/10.1159/000332755.

Britton, W. B., Lindahl, J. R., Cahn, B. R., Davis, J. H., & Goldman, R. E. (2014). Awakening is not a metaphor: The effects of Buddhist meditation practices on basic wakefulness. *Annals of the New York Academy of Sciences, 1307,* 64–81. http://dx.doi.org/10.1111/nyas.12279.

Britton, W. B., Shahar, B., Szepsenwol, O., & Jacobs, W. J. (2012). Mindfulness-based cognitive therapy improves emotional reactivity to social stress: Results from a randomized controlled trial. *Behavior Therapy, 43*(2), 365–380. http://dx.doi.org/10.1016/j.beth.2011.08.006.

Broderick, P. (2005). Mindfulness and coping with dysphoric mood: Contrasts with rumination and distraction. *Cognitive Therapy and Research, 29,* 501–510.

Buckner, R. L., & Vincent, J. L. (2007). Unrest at rest: Default activity and spontaneous network correlations. *NeuroImage, 37*(4), 1091–1096. http://dx.doi.org/10.1016/j.neuroimage.2007.01.010, discussion 1097–9.

Buysse, D. J. (2014). Sleep health: Can we define it? does it matter? *Sleep, 37*(1), 9–17. http://dx.doi.org/10.5665/sleep.3298.

Campbell-Sills, L., Barlow, D. H., Brown, T. A., & Hofmann, S. G. (2006). Effects of suppression and acceptance on emotional responses of individuals with anxiety and mood disorders. *Behaviour Research and Therapy, 44*(9), 1251–1263. http://dx.doi.org/10.1016/j.brat.2005.10.001.

Carlson, L. E. (2012). Mindfulness-based interventions for physical conditions: A narrative review evaluating levels of evidence. *ISRN Psychiatry, 2012,* 651583. http://dx.doi.org/10.5402/2012/651583.

Carlson, L. E., Doll, R., Stephen, J., Faris, P., Tamagawa, R., Drysdale, E., et al. (2013). Randomized controlled trial of mindfulness-based cancer recovery versus supportive expressive group therapy for distressed survivors of breast cancer. *Journal of Clinical Oncology: Official Journal of the American Society of Clinical Oncology, 31*(25), 3119–3126. http://dx.doi.org/10.1200/JCO.2012.47.5210.

Carlson, L. E., & Garland, S. N. (2005). Impact of mindfulness-based stress reduction (MBSR) on sleep, mood, stress and fatigue symptoms in cancer outpatients. *International Journal of Behavioral Medicine, 12*(4), 278–285. http://dx.doi.org/10.1207/s15327558ijbm1204_9.

Carlson, L. E., & Speca, M. (2010). *Mindfulness-based cancer recovery.* Oakland, CA: New Harbinger.

Carlson, L. E., Speca, M., Faris, P., & Patel, K. D. (2007). One year pre-post intervention follow-up of psychological, immune, endocrine and blood pressure outcomes of mindfulness-based stress reduction (MBSR) in breast and prostate cancer outpatients. *Brain, Behavior, and Immunity, 21*(8), 1038–1049. http://dx.doi.org/10.1016/j.bbi.2007.04.002.

Carmody, J., Crawford, S., & Churchill, L. (2006). A pilot study of mindfulness-based stress reduction for hot flashes. *Menopause, 13*(5), 760–769. http://dx.doi.org/10.1097/01.gme.0000227402.98933.d0.

Carmody, J. F., Crawford, S., Salmoirago-Blotcher, E., Leung, K., Churchill, L., & Olendzki, N. (2011). Mindfulness training for coping with hot flashes: Results of a randomized trial. *Menopause, 18*(6), 611–620. http://dx.doi.org/10.1097/gme.0b013e318204a05c.

Carney, C. E., Buysse, D. J., Ancoli-Israel, S., Edinger, J. D., Krystal, A. D., Lichstein, K. L., et al. (2012). The consensus sleep diary: Standardizing prospective sleep self-monitoring. *Sleep, 35*(2), 287–302. http://dx.doi.org/10.5665/sleep.1642.

Chiesa, A., & Serretti, A. (2009). Mindfulness-based stress reduction for stress management in healthy people: A review and meta-analysis. *Journal of Alternative and Complementary Medicine, 15*(5), 593–600. http://dx.doi.org/10.1089/acm.2008.0495.

Clark, L., Chamberlain, S. R., & Sahakian, B. J. (2009). Neurocognitive mechanisms in depression: Implications for treatment. *Annual Review of Neuroscience, 32*, 57–74. http://dx.doi.org/10.1146/annurev.neuro.31.060407.125618.

Couyoumdjian, A., Sdoia, S., Tempesta, D., Curcio, G., Rastellini, E., D.E. Gennaro, L., et al. (2010). The effects of sleep and sleep deprivation on task-switching performance. *Journal of Sleep Research, 19*(1 Pt 1), 64–70. http://dx.doi.org/10.1111/j.1365-2869.2009.00774.x.

Creswell, J. D., Way, B. M., Eisenberger, N. I., & Lieberman, M. D. (2007). Neural correlates of dispositional mindfulness during affect labeling. *Psychosomatic Medicine, 69*(6), 560–565. http://dx.doi.org/10.1097/PSY.0b013e3180f6171f.

Davidson, R. J., Jackson, D. C., & Kalin, N. H. (2000). Emotion, plasticity, context and regulation: Perspectives from affective neuroscience. *Psychological Bulletin, 126*, 890–906.

Dement, W. C. (2005). History of sleep physiology and medicine. *Principles and practices of sleep medicine* (4th ed.). Philadelphia, PA: W.B. Saunders.

Desbordes, G., Negi, L. T., Pace, T. W., Wallace, B. A., Raison, C. L., & Schwartz, E. L. (2012). Effects of mindful-attention and compassion meditation training on amygdala response to emotional stimuli in an ordinary, non-meditative state. *Frontiers in Human Neuroscience, 6*, 292. http://dx.doi.org/10.3389/fnhum.2012.00292.

Edinger, J. D., Fins, A. I., Glenn, D. M., Sullivan, R. J., Jr., Bastian, L. A., Marsh, G. R., et al. (2000). Insomnia and the eye of the beholder: Are there clinical markers of objective sleep disturbances among adults with and without insomnia complaints? *Journal of Consulting and Clinical Psychology, 68*, 586–593.

Erisman, S. M., & Roemer, L. (2010). A preliminary investigation of the effects of experimentally induced mindfulness on emotional responding to film clips. *Emotion, 10*(1), 72–82. http://dx.doi.org/10.1037/a0017162.

Evans, S., Ferrando, S., Findler, M., Stowell, C., Smart, C., & Haglin, D. (2008). Mindfulness-based cognitive therapy for generalized anxiety disorder. *Journal of Anxiety Disorders, 22*(4), 716–721. http://dx.doi.org/10.1016/j.janxdis.2007.07.005.

Farb, N. A., Anderson, A. K., Mayberg, H., Bean, J., McKeon, D., & Segal, Z. V. (2010). Minding one's emotions: Mindfulness training alters the neural expression of sadness. *Emotion, 10*(1), 25–33. http://dx.doi.org/10.1037/a0017151.

Farb, N. A., Segal, Z. V., Mayberg, H., Bean, J., McKeon, D., Fatima, Z., et al. (2007). Attending to the present: Mindfulness meditation reveals distinct neural modes of self-reference. *Social Cognitive and Affective Neuroscience, 2*(4), 313–322. http://dx.doi.org/10.1093/scan/nsm030.

Garland, S. N., Carlson, L. E., Stephens, A. J., Antle, M. C., Samuels, C., & Campbell, T. S. (2014). Mindfulness-based stress reduction compared with cognitive behavioral therapy for the treatment of insomnia comorbid with cancer: A randomized, partially blinded, noninferiority trial. *Journal of Clinical Oncology: Official Journal of the American Society of Clinical Oncology, 32*(5), 449–457. http://dx.doi.org/10.1200/JCO.2012.47.7265.

Gentili, C., Ricciardi, E., Gobbini, M. I., Santarelli, M. F., Haxby, J. V., Pietrini, P., et al. (2009). Beyond amygdala: Default mode network activity differs between patients with social phobia and healthy controls. *Brain Research Bulletin, 79*(6), 409–413. http://dx.doi.org/10.1016/j.brainresbull.2009.02.002.

Germer, C. K., & Neff, K. D. (2013). Self-compassion in clinical practice. *Journal of Clinical Psychology, 69*(8), 856–867. http://dx.doi.org/10.1002/jclp.22021.

Goldenberg, D., Kaplan, K., Nadeau, M., Brodeur, C., Smith, S., & Schmid, C. (1994). A controlled trial of a stress-reduction, cognitive behavioral treatment program in fibromyalgia. *Journal of Musculoskeletal Pain, 2*, 53–56.

Goldin, P. R., & Gross, J. J. (2010). Effects of mindfulness-based stress reduction (MBSR) on emotion regulation in social anxiety disorder. *Emotion, 10*(1), 83–91. http://dx.doi.org/10.1037/a0018441.

Goldin, P., Ramel, W., & Gross, J. (2009). Mindfulness meditation training and self-referential processing in social anxiety disorder: Behavioral and neural effects. *Journal of Cognitive Psychotherapy: An International Quarterly, 23*, 242–257.

Gonzalez-Garcia, M., Ferrer, M. J., Borras, X., Munoz-Moreno, J. A., Miranda, C., Puig, J., et al. (2014). Effectiveness of mindfulness-based cognitive therapy on the quality of life, emotional status, and CD4 cell count of patients aging with HIV infection. *AIDS and Behavior, 18*, 676–685. http://dx.doi.org/10.1007/s10461-013-0612-z.

Gross, J. J. (1998). Antecedent- and response-focused emotion regulation: Divergent consequences for experience, expression, and physiology. *Journal of Personality and Social Psychology, 74*(1), 224–237.

Gross, C. R., Kreitzer, M. J., Reilly-Spong, M., Wall, M., Winbush, N. Y., Patterson, R., et al. (2011). Mindfulness-based stress reduction versus pharmacotherapy for chronic primary insomnia: A randomized controlled clinical trial. *Explore, 7*(2), 76–87. http://dx.doi.org/10.1016/j.explore.2010.12.003.

Gross, C. R., Kreitzer, M. J., Russas, V., Treesak, C., Frazier, P. A., & Hertz, M. I. (2004). Mindfulness meditation to reduce symptoms after organ transplant: A pilot study. *Advances in Mind-Body Medicine, 20*(2), 20–29.

Gross, C. R., Kreitzer, M. J., Thomas, W., Reilly-Spong, M., Cramer-Bornemann, M., Nyman, J. A., et al. (2010). Mindfulness-based stress reduction for solid organ transplant recipients: A randomized controlled trial. *Alternative Therapies in Health and Medicine, 16*(5), 30–38.

Hamilton, J. P., Furman, D. J., Chang, C., Thomason, M. E., Dennis, E., & Gotlib, I. H. (2011). Default-mode and task-positive network activity in major depressive disorder: Implications for adaptive and maladaptive rumination. *Biological Psychiatry, 70*(4), 327–333. http://dx.doi.org/10.1016/j.biopsych.2011.02.003.

Hasenkamp, W., & Barsalou, L. W. (2012). Effects of meditation experience on functional connectivity of distributed brain networks. *Frontiers in Human Neuroscience, 6*, 38. http://dx.doi.org/10.3389/fnhum.2012.00038.

Hayes, S. C. (1987). A contextual approach to therapeutic change. In N. Jacobson (Ed.), *Psychotherapists in clinical practice: Cognitive and behavioral perspectives* (pp. 327–387). New York: Guilford.

Heidenreich, T., Tuin, I., Pflug, B., Michal, M., & Michalak, J. (2006). Mindfulness-based cognitive therapy for persistent insomnia: A pilot study. *Psychotherapy and Psychosomatics, 75*(3), 188–189. http://dx.doi.org/10.1159/000091778.

Holzel, B. K., Ott, U., Gard, T., Hempel, H., Weygandt, M., Morgen, K., et al. (2008). Investigation of mindfulness meditation practitioners with voxel-based morphometry. *Social Cognitive and Affective Neuroscience, 3*(1), 55–61. http://dx.doi.org/10.1093/scan/nsm038.

Hsu, S. H., Grow, J., & Marlatt, G. A. (2008). Mindfulness and addiction. *Recent Developments in Alcoholism, 18*, 229–250.

Hughes, J. W., Fresco, D. M., Myerscough, R., van Dulmen, M. H., Carlson, L. E., & Josephson, R. (2013). Randomized controlled trial of mindfulness-based stress reduction for prehypertension. *Psychosomatic Medicine, 75*(8), 721–728. http://dx.doi.org/10.1097/PSY.0b013e3182a3e4e5.

Johansson, B., Bjuhr, H., & Ronnback, L. (2012). Mindfulness-based stress reduction (MBSR) improves long-term mental fatigue after stroke or traumatic brain injury. *Brain Injury, 26*(13–14), 1621–1628. http://dx.doi.org/10.3109/02699052.2012.700082.

Johnson, D. P., Penn, D. L., Fredrickson, B. L., Kring, A. M., Meyer, P. S., Catalino, L. I., et al. (2011). A pilot study of loving-kindness meditation for the negative symptoms of schizophrenia. *Schizophrenia Research, 129*(2–3), 137–140. http://dx.doi.org/10.1016/j.schres.2011.02.015.

Kabat-Zinn, J. (1990). *Full catastrophe living: Using the wisdom of your body and mind to face, stress, pain, and illness.* New York: Delacourt.

Kim, Y. W., Lee, S. H., Choi, T. K., Suh, S. Y., Kim, B., Kim, C. M., et al. (2009). Effectiveness of mindfulness-based cognitive therapy as an adjuvant to pharmacotherapy in patients with panic disorder or generalized anxiety disorder. *Depression and Anxiety, 26*(7), 601–606. http://dx.doi.org/10.1002/da.20552.

Klatt, M. D., Buckworth, J., & Malarkey, W. B. (2009). Effects of low-dose mindfulness-based stress reduction (MBSR-ld) on working adults. *Health Education & Behavior, 36*(3), 601–614. http://dx.doi.org/10.1177/1090198108317627.

Kreitzer, M. J., Gross, C. R., Ye, X., Russas, V., & Treesak, C. (2005). Longitudinal impact of mindfulness meditation on illness burden in solid-organ transplant recipients. *Progress in Transplantation, 15*(2), 166–172.

Kuehner, C., Huffziger, S., & Liebsch, K. (2009). Rumination, distraction and mindful self-focus: Effects on mood, dysfunctional attitudes and cortisol stress response. *Psychological Medicine, 39*(2), 219–228. http://dx.doi.org/10.1017/S0033291708003553.

Kuyken, W., Byford, S., Taylor, R. S., Watkins, E., Holden, E., White, K., et al. (2008). Mindfulness-based cognitive therapy to prevent relapse in recurrent depression. *Journal of Consulting and Clinical Psychology, 76*(6), 966–978. http://dx.doi.org/10.1037/a0013786.

Lazar, S. W., Kerr, C. E., Wasserman, R. H., Gray, J. R., Greve, D. N., Treadway, M. T., et al. (2005). Meditation experience is associated with increased cortical thickness. *Neuroreport, 16*(17), 1893–1897.

Lemogne, C., Delaveau, P., Freton, M., Guionnet, S., & Fossati, P. (2012). Medial prefrontal cortex and the self in major depression. *Journal of Affective Disorders, 136*(1–2), e1–e11. http://dx.doi.org/10.1016/j.jad.2010.11.034.

Lengacher, C. A., Reich, R. R., Post-White, J., Moscoso, M., Shelton, M. M., Barta, M., et al. (2012). Mindfulness based stress reduction in post-treatment breast cancer patients: An examination of symptoms and symptom clusters. *Journal of Behavioral Medicine, 35*(1), 86–94. http://dx.doi.org/10.1007/s10865-011-9346-4.

Luders, E., Toga, A. W., Lepore, N., & Gaser, C. (2009). The underlying anatomical correlates of long-term meditation: Larger hippocampal and frontal volumes of gray matter. *NeuroImage, 45*(3), 672–678.

Lundh, L. (2005). The role of acceptance and mindfulness in the treatment of insomnia. *Journal of Cognitive Psychotherapy: An International Quarterly, 19*(1), 29–39.

Lundh, L., & Broman, J. E. (2000). Insomnia as an interaction between sleep-interfering and sleep-interpreting processes. *Journal of Psychosomatic Research, 49*(5), 299–310.

Ma, S. H., & Teasdale, J. D. (2004). Mindfulness-based cognitive therapy for depression: Replication and exploration of differential relapse prevention effects. *Journal of Consulting and Clinical Psychology, 72*(1), 31–40. http://dx.doi.org/10.1037/0022-006X.72.1.31.

MacLean, C. R., Walton, K. G., Wenneberg, S. R., Levitsky, D. K., Mandarino, J. V., Waziri, R., et al. (1994). Altered responses of cortisol, GH, TSH and testosterone to acute stress after four months' practice of transcendental meditation (TM). *Annals of the New York Academy of Sciences, 746*, 381–384.

Marchetti, I., Koster, E. H., Sonuga-Barke, E. J., & De Raedt, R. (2012). The default mode network and recurrent depression: A neurobiological model of cognitive risk factors. *Neuropsychology Review, 22*(3), 229–251. http://dx.doi.org/10.1007/s11065-012-9199-9.

Masuda, A., Feinstein, A. B., Wendell, J. W., & Sheehan, S. T. (2010). Cognitive defusion versus thought distraction: A clinical rationale, training, and experiential exercise in altering psychological impacts of negative self-referential thoughts. *Behavior Modification, 34*(6), 520–538. http://dx.doi.org/10.1177/0145445510379632.

McCarney, R., Warner, J., Iliffe, S., van Haselen, R., Griffin, M., & Fisher, P. (2007). The hawthorne effect: A randomised, controlled trial. *BMC Medical Research Methodology, 7*, 30. http://dx.doi.org/10.1186/1471-2288-7-30.

McKee, L., Zvolensky, M. J., Solomon, S. E., Bernstein, A., & Leen-Feldner, E. (2007). Emotional-vulnerability and mindfulness: A preliminary test of associations among negative affectivity, anxiety sensitivity, and mindfulness skills. *Cognitive Behaviour Therapy, 36*(2), 91–101. http://dx.doi.org/10.1080/16506070601119314.

Morgenthaler, T., Alessi, C., Friedman, L., Owens, J., Kapur, V., Boehlecke, B., et al. (2007). Practice parameters for the use of actigraphy in the assessment of sleep and sleep disorders: An update for 2007. *Sleep, 30*(4), 519–529.

Nakamura, Y., Lipschitz, D. L., Kuhn, R., Kinney, A. Y., & Donaldson, G. W. (2013). Investigating efficacy of two brief mind-body intervention programs for managing sleep disturbance in cancer survivors: A pilot randomized controlled trial. *Journal of Cancer Survivorship: Research and Practice, 7*(2), 165–182. http://dx.doi.org/10.1007/s11764-012-0252-8.

Newman Taylor, K., Harper, S., & Chadwick, P. (2009). Impact of mindfulness on cognition and affect in voice hearing: Evidence from two case studies. *Behavioural and Cognitive Psychotherapy, 37*(4), 397–402. http://dx.doi.org/10.1017/S135246580999018X.

Omidi, A., Mohammadi, A., Zargar, F., & Akbari, H. (2013). Efficacy of mindfulness-based stress reduction on mood states of veterans with post-traumatic stress disorder. *Archives of Trauma Research, 1*(4), 151–154. http://dx.doi.org/10.5812/atr.8226.

Ong, J. C., Shapiro, S. L., & Manber, R. (2008). Combining mindfulness meditation with cognitive-behavior therapy for insomnia: A treatment-development study. *Behavior Therapy, 39*(2), 171–182. http://dx.doi.org/10.1016/j.beth.2007.07.002.

Ong, J. C., Shapiro, S. L., & Manber, R. (2009). Mindfulness meditation and cognitive behavioral therapy for insomnia: A naturalistic 12-month follow-up. *Explore, 5*(1), 30–36. http://dx.doi.org/10.1016/j.explore.2008.10.004.

Ong, J. C., Ulmer, C. S., & Manber, R. (2012). Improving sleep with mindfulness and acceptance: A metacognitive model of insomnia. *Behaviour Research and Therapy, 50*(11), 651–660. http://dx.doi.org/10.1016/j.brat.2012.08.001.

Ortner, C. M. N., Kilner, S., & Zelazo, P. D. (2007). Mindfulness meditation and emotional interference in a simple cognitive task. *Motivation and Emotion, 31*(271), 283.

Pace, T. W., Negi, L. T., Adame, D. D., Cole, S. P., Sivilli, T. I., Brown, T. D., et al. (2009). Effect of compassion meditation on neuroendocrine, innate immune and behavioral responses to psychosocial stress. *Psychoneuroendocrinology, 34*(1), 87–98. http://dx.doi.org/10.1016/j.psyneuen.2008.08.011.

Parswani, M. J., Sharma, M. P., & Iyengar, S. (2013). Mindfulness-based stress reduction program in coronary heart disease: A randomized control trial. *International Journal of Yoga, 6*(2), 111–117. http://dx.doi.org/10.4103/0973-6131.113405.

Piet, J., & Hougaard, E. (2011). The effect of mindfulness-based cognitive therapy for prevention of relapse in recurrent major depressive disorder: A systematic review and meta-analysis. *Clinical Psychology Review, 31*(6), 1032–1040. http://dx.doi.org/10.1016/j.cpr.2011.05.002.

Post, R. M. (2007). Kindling and sensitization as models for affective episode recurrence, cyclicity, and tolerance phenomena. *Neuroscience and Biobehavioral Reviews, 31*(6), 858–873. http://dx.doi.org/10.1016/j.neubiorev.2007.04.003.

Proulx, K. (2008). Experiences of women with bulimia nervosa in a mindfulness-based eating disorder treatment group. *Eating Disorders, 16*(1), 52–72. http://dx.doi.org/10.1080/10640260701773496.

Qin, P., & Northoff, G. (2011). How is our self related to midline regions and the default-mode network? *NeuroImage, 57*(3), 1221–1233. http://dx.doi.org/10.1016/j.neuroimage.2011.05.028.

Raes, F., Dewulf, D., Van Heeringen, C., & Williams, J. M. (2009). Mindfulness and reduced cognitive reactivity to sad mood: Evidence from a correlational study and a non-randomized waiting list controlled study. *Behaviour Research and Therapy, 47*(7), 623–627. http://dx.doi.org/10.1016/j.brat.2009.03.007.

Rechtschaffen, A. & Kales, A. (Eds.), (1969). *A manual of standardized terminology, techniques and scoring system of sleep stages in human subjects.* Los Angeles: Brain Information Service/Brain Research Institute, University of California.

Ritskes, R., Ritskes-Hoitinga, M., Stodkilde-Jorgensen, H., Baerntsen, K., & Hartman, T. (2003). MRI scanning during zen meditation: The picture of enlightenment? *Constructivism in Human Sciences, 8*, 85–90.

Roth, B., & Robbins, D. (2004). Mindfulness-based stress reduction and health-related quality of life: Findings from a bilingual inner-city patient population. *Psychosomatic Medicine, 66*(1), 113–123.

Russell, J. A., & Barrett, L. F. (1999). Core affect, prototypical emotional episodes, and other things called emotion: Dissecting the elephant. *Journal of Personality and Social Psychology, 76*(5), 805–819.

Salmoirago-Blotcher, E., Crawford, S., Carmody, J., Rosenthal, L., & Ockene, I. (2011). Characteristics of dispositional mindfulness in patients with severe cardiac disease. *Journal of Evidence Based Complementary & Alternative Medicine, 16*(3), 218–225. http://dx.doi.org/10.1177/2156587211405525.

Schmidt, S., Grossman, P., Schwarzer, B., Jena, S., Naumann, J., & Walach, H. (2011). Treating fibromyalgia with mindfulness-based stress reduction: Results from a 3-armed randomized controlled trial. *Pain, 152*(2), 361–369. http://dx.doi.org/10.1016/j.pain.2010.10.043.

Segal, Z. V. (1988). Appraisal of the self-schema construct in cognitive models of depression. *Psychological Bulletin, 103*(2), 147–162.

Shapiro, S. L., Bootzin, R. R., Figueredo, A. J., Lopez, A. M., & Schwartz, G. E. (2003). The efficacy of mindfulness-based stress reduction in the treatment of sleep disturbance in women with breast cancer: An exploratory study. *Journal of Psychosomatic Research, 54*(1), 85–91.

Shapiro, S. L., & Carlson, L. E. (2009). *The art and science of mindfulness: Integrating mindfulness into psychology and the helping professions.* Washington, D.C.: American Psychological Association Publications.

Shapiro, S. L., Carlson, L. E., Astin, J. A., & Freedman, B. (2006). Mechanisms of mindfulness. *Journal of Clinical Psychology, 62*(3), 373–386. http://dx.doi.org/10.1002/jclp.20237.

Sheline, Y. I., Barch, D. M., Price, J. L., Rundle, M. M., Vaishnavi, S. N., Snyder, A. Z., et al. (2009). The default mode network and self-referential processes in depression. *Proceedings of the National Academy of Sciences of the United States of America, 106*(6), 1942–1947. http://dx.doi.org/10.1073/pnas.0812686106.

Silva, G. E., Goodwin, J. L., Sherrill, D. L., Arnold, J. L., Bootzin, R. R., Smith, T., et al. (2007). Relationship between reported and measured sleep times: The sleep heart health study (SHHS). *Journal of Clinical Sleep Medicine, 3*, 622–630.

Soler, J., Valdeperez, A., Feliu-Soler, A., Pascual, J. C., Portella, M. J., Martin-Blanco, A., et al. (2012). Effects of the dialectical behavioral therapy-mindfulness module on attention in patients with borderline personality disorder. *Behaviour Research and Therapy, 50*(2), 150–157. http://dx.doi.org/10.1016/j.brat.2011.12.002.

Spek, A. A., van Ham, N. C., & Nyklicek, I. (2013). Mindfulness-based therapy in adults with an autism spectrum disorder: A randomized controlled trial. *Research in Developmental Disabilities, 34*(1), 246–253. http://dx.doi.org/10.1016/j.ridd.2012.08.009.

Sudsuang, R., Chentanez, V., & Veluvan, K. (1991). Effect of buddhist meditation on serum cortisol and total protein levels, blood pressure, pulse rate, lung volume and reaction time. *Physiology & Behavior, 50*(3), 543–548.

Tang, Y. Y., Ma, Y., Wang, J., Fan, Y., Feng, S., Lu, Q., et al. (2007). Short-term meditation training improves attention and self-regulation. *Proceedings of the National Academy of Sciences of the United States of America, 104*(43), 17152–17156. http://dx.doi.org/10.1073/pnas.0707678104.

Taylor, V. A., Grant, J., Daneault, V., Scavone, G., Breton, E., Roffe-Vidal, S., et al. (2011). Impact of mindfulness on the neural responses to emotional pictures in experienced and beginner meditators. *NeuroImage, 57*(4), 1524–1533. http://dx.doi.org/10.1016/j.neuroimage.2011.06.001.

Teasdale, J. D. (1999). Metacognition, mindfulness and the modification of mood disorders. *Clinical Psychology & Psychotherapy, 6*, 146.

Teasdale, J. D., Moore, R. G., Hayhurst, H., Pope, M., Williams, S., & Segal, Z. V. (2002). Metacognitive awareness and prevention of relapse in depression: Empirical evidence. *Journal of Consulting and Clinical Psychology, 70*(2), 275–287.

Teasdale, J. D., Segal, Z. V., Williams, J. M., Ridgeway, V. A., Soulsby, J. M., & Lau, M. A. (2000). Prevention of relapse/recurrence in major depression by mindfulness-based cognitive therapy. *Journal of Consulting and Clinical Psychology, 68*(4), 615–623.

Travis, F., Haaga, D. A., Hagelin, J., Tanner, M., Arenander, A., Nidich, S., et al. (2010). A self-referential default brain state: Patterns of coherence, power, and eLORETA sources during eyes-closed rest and transcendental meditation practice. *Cognitive Processing, 11*(1), 21–30. http://dx.doi.org/10.1007/s10339-009-0343-2.

van Son, J., Nyklicek, I., Pop, V. J., Blonk, M. C., Erdtsieck, R. J., Spooren, P. F., et al. (2013). The effects of a mindfulness-based intervention on emotional distress, quality of life, and HbA(1c) in outpatients with diabetes (DiaMind): A randomized controlled trial. *Diabetes Care, 36*(4), 823–830. http://dx.doi.org/10.2337/dc12-1477.

Wanden-Berghe, R. G., Sanz-Valero, J., & Wanden-Berghe, C. (2011). The application of mindfulness to eating disorders treatment: A systematic review. *Eating Disorders, 19*(1), 34–48. http://dx.doi.org/10.1080/10640266.2011.533604.

Watkins, E., Teasdale, J. D., & Williams, R. M. (2000). Decentring and distraction reduce overgeneral autobiographical memory in depression. *Psychological Medicine, 30*(4), 911–920.

Way, B. M., Creswell, J. D., Eisenberger, N. I., & Lieberman, M. D. (2010). Dispositional mindfulness and depressive symptomatology: Correlations with limbic and self-referential neural activity during rest. *Emotion, 10*(1), 12–24. http://dx.doi.org/10.1037/a0018312.

Webb, T. L., Miles, E., & Sheeran, P. (2012). Dealing with feeling: A meta-analysis of the effectiveness of strategies derived from the process model of emotion regulation. *Psychological Bulletin, 138*(4), 775–808. http://dx.doi.org/10.1037/a0027600.

Weinstein, N., Brown, K., & Ryan, R. M. (2009). A multi-method examination of the effects if mindfulness on stress attribution, coping and emotional well-being. *Journal of Research in Personality, 43*, 374–385.

Whitfield-Gabrieli, S., Thermenos, H. W., Milanovic, S., Tsuang, M. T., Faraone, S. V., McCarley, R. W., et al. (2009). Hyperactivity and hyperconnectivity of the default network in schizophrenia and in first-degree relatives of persons with schizophrenia. *Proceedings of the National Academy of Sciences of the United States of America, 106*(4), 1279–1284. http://dx.doi.org/10.1073/pnas.0809141106.

Williams, J. M., Alatiq, Y., Crane, C., Barnhofer, T., Fennell, M. J., Duggan, D. S., et al. (2008). Mindfulness-based cognitive therapy (MBCT) in bipolar disorder: Preliminary evaluation of immediate effects on between-episode functioning. *Journal of Affective Disorders, 107*(1–3), 275–279. http://dx.doi.org/10.1016/j.jad.2007.08.022.

Williams, J. M., Duggan, D. S., Crane, C., & Fennell, M. J. (2006). Mindfulness-based cognitive therapy for prevention of recurrence of suicidal behavior. *Journal of Clinical Psychology, 62*(2), 201–210. http://dx.doi.org/10.1002/jclp.20223.

Yik, M. S. M., Russell, J. A., & Barett, L. F. (1999). Structure of self-reported current affect: Integration and beyond. *Journal of Personality and Social Psychology, 77*, 600.

Yook, K., Lee, S. H., Ryu, M., Kim, K. H., Choi, T. K., Suh, S. Y., et al. (2008). Usefulness of mindfulness-based cognitive therapy for treating insomnia in patients with anxiety disorders: A pilot study. *The Journal of Nervous and Mental Disease, 196*(6), 501–503. http://dx.doi.org/10.1097/NMD.0b013e31817762ac.

Zainal, N. Z., Booth, S., & Huppert, F. A. (2013). The efficacy of mindfulness-based stress reduction on mental health of breast cancer patients: A meta-analysis. *Psycho-Oncology, 22*(7), 1457–1465. http://dx.doi.org/10.1002/pon.3171.

Zernicke, K. A., Campbell, T. S., Blustein, P. K., Fung, T. S., Johnson, J. A., Bacon, S. L., et al. (2013). Mindfulness-based stress reduction for the treatment of irritable bowel syndrome symptoms: A randomized wait-list controlled trial. *International Journal of Behavioral Medicine, 20*(3), 385–396. http://dx.doi.org/10.1007/s12529-012-9241-6.

Zernicke, K. A., Campbell, T. S., Speca, M., McCabe-Ruff, K., Flowers, S., Dirkse, D. A., et al. (2013). The eCALM trial-eTherapy for cancer appLying mindfulness: Online mindfulness-based cancer recovery program for underserved individuals living with cancer in alberta: Protocol development for a randomized wait-list controlled clinical trial. *BMC Complementary and Alternative Medicine, 13*, 34. http://dx.doi.org/10.1186/1472-6882-13-34.

Zhang, L., & Zhao, Z. X. (2007). Objective and subjective measures for sleep disorders. *Neuroscience Bulletin, 23*, 236–240.

Zhao, X. H., Wang, P. J., Li, C. B., Hu, Z. H., Xi, Q., Wu, W. Y., et al. (2007). Altered default mode network activity in patient with anxiety disorders: An fMRI study. *European Journal of Radiology, 63*(3), 373–378. http://dx.doi.org/10.1016/j.ejrad.2007.02.006.

SECTION 3

EVIDENCE REGARDING SLEEP AND AFFECT AMONG SPECIAL POPULATIONS

CHAPTER 17

Pain and Sleep

Timothy Roehrs and Thomas Roth

Sleep Disorders and Research Center, Henry Ford Hospital, Detroit, Michigan, USA

Department of Psychiatry and Behavioral Neuroscience, School of Medicine, Wayne State University, Detroit, Michigan, USA

PAIN AND SLEEP

Sleep and pain are vital functions that interact in complex ways to ultimately compromise the biological and behavioral capacity of the individual. Whereas the function of sleep is not yet firmly established, sleep is clearly a biological imperative. When sleep time is reduced or its continuity is disrupted, the drive to sleep is enhanced, producing rapid sleep onset. If sleep drive reaches its maximum due to persistent sleep loss, sleep intrudes into wakefulness in the form of brief microsleeps that can result in attention problems, memory and cognitive problems, automobile or workplace accidents, and decreased productivity and quality of life (Roehrs, Carskadon, Dement, & Roth, 2010). Sleepiness or fatigue can be thought of as an organism's signal that functioning is impaired and the sleep deficit must be corrected.

Pain is a perception of unpleasant or noxious sensation detected by receptors called nociceptors. Pain plays a vital role because it has protective functions. As one investigator has stated, "pain is a perceived threat or damage to one's biological integrity" (Chapman & Gavrin, 1999). Nociceptive information signals to the organism the need for defensive responses to avoid further damage. Thus, sleep and pain are restorative and protective in function, and importantly, the emerging information also shows that sleep and pain are interactive. This relationship has been described as bidirectional; that is, sleep loss enhances pain, and enhanced pain disrupts sleep. Therefore, the implication that improving sleep attenuates pain is of clinical importance.

In this chapter we describe the literature on the nature of the sleep disturbances reported in people experiencing acute pain and people with primary chronic pain disorders. We limit our presentation to primary pain disorders, although some patients with other medical disorders also experience pain and sleep disturbance (i.e., multiple sclerosis, various cancers). Emerging literature addresses the impact of experimental sleep manipulations on pain threshold and superthreshold pain stimulation, which are also reviewed. Stress and mood are known to impact both sleep and pain, and we discuss the potential modulatory effects of these factors on the sleep-pain nexus. We also describe the hypothesized neurobiological mechanisms mediating the sleep-pain nexus. Finally, we consider the impact of pharmacological and behavioral treatments for sleep and pain.

METHODOLOGICAL ISSUES

Several important methodological issues regarding the measurement of sleep and pain must be discussed. Much of the sleep-pain literature relies on self-report of both sleep and pain. First, in measuring sleep and pain by self-report, there is a potential for response bias; that is, reporting one symptom increases the likelihood of reporting another. Also, participants might expect that nocturnal pain disturbs sleep, which then tends to enhance self-report of both sleep disturbance and pain disturbance. Thus, in evaluating the literature, we place more weight on studies that use objective assessments of at least one of the two variables, and preferably both, with the ideal being the objective measurement and self-report of both variables.

Many studies have used reliable pain questionnaires, most typically in chronic pain conditions. The most broadly used scales are the McGill Pain Inventory and the Brief Pain Inventory. A variety of disease-specific pain inventories are also used. For studies with healthy volunteers, investigators typically use behavioral measures of pain threshold or sensitivity that require verbal report of pain or behavioral withdrawal responses to the presentation of hot, cold, or mechanical stimuli. Researchers have also employed brain imaging and electrophysiological studies to assess neural responses to evoked experimental pain. A complete discussion of the complexities of pain assessment is beyond the scope of this chapter.

In the sleep field, standardized continuous recording of multiple physiological signals, including electroencephalography (EEG), electromyography, and electrooculography, has been used since the 1950s to document sleep, and this methodology is termed polysomnography (PSG). PSG documents sleep versus wakefulness and the various sleep stages, which is important because pain thresholds vary as a function of sleep stage and the loss of specific sleep stages may be critical to pain sensitivity

(Keenan & Hirshkowitz, 2010). Depending on the pain condition or the nature of the sleep disturbance, self-report of sleep can either overestimate or underestimate the duration of sleep and awakenings relative to a PSG.

A less expensive, intrusive, and laboratory-bound methodology, actigraphy, has recently been employed in sleep-pain studies. Actigraphy measures movement via a wrist-worn device, and this methodology, supported by validated analytic software, can document time to sleep onset and the amount of sleep versus wake time. However, actigraphy has limitations because it may overestimate sleep time in some populations (i.e., elderly) and, in contrast, underestimate sleep time in populations with primary sleep disorders (i.e., periodic leg movements, sleep apnea) in which sleep is briefly fragmented. Actigraphy also does not yield information regarding sleep stages, which may be important in understanding the nature of the sleep disturbance in acute and chronic pain conditions, as the review below indicates (see also Chapter 4 for a detailed description of methods for sleep assessment).

ASSOCIATION OF DISTURBED SLEEP AND PAIN

Acute Pain Conditions

An increasing number of studies have used PSG to assess sleep in acute clinical pain situations. Given that acute pain is unpredictable and short-lived, making it difficult to arrange for PSG, investigators have studied the one clinical situation lending itself to PSG assessment, which is pain experienced during postoperative recovery from elective surgery. This model of acute pain is marked by a number of confounding factors, however (Chouchou, Khoury, Chauny, Denis, & Lavigne, 2013; Roehrs & Roth, 2005; Rosenberg-Adamsen, Kehlet, Dodds, & Rosenberg, 1996). The hospital environment, specifically the intensive care unit where most of these studies are conducted, is not conducive to sleep, with noise, temperature, light, and frequent vital sign assessments all disrupting slumber. The magnitude and duration of the surgical stress are certainly also critical factors potentially confounding the relationship between acute pain and sleep disturbance. The surgical stress also stimulates a cascade of hormonal, biochemical, and inflammatory responses. Finally, the medications used during the postoperative recovery, particularly the opiates, are themselves known to disrupt sleep. These confounding factors make it difficult to assess the true interactions between acute pain and sleep disturbance.

In most patients, regardless of the type of surgery (e.g., major abdominal, orthopedic, gynecologic, open-heart, and coronary artery bypass), PSG-defined sleep time is shortened and sleep is disrupted with frequent brief arousals and awakenings during the first one to two recovery nights

(Chouchou et al., 2013; Roehrs & Roth, 2005). Generally the studies also show that slow-wave and rapid eye movement (REM) sleep are reduced and in many cases totally abolished. As recovery occurs and pain diminishes, slow-wave and REM sleep reappear and normalize. A major methodological problem with many of these studies is that only nighttime sleep is recorded. One study that did PSG during the daytime period as well reported enhanced REM sleep in daytime naps (Gogenur, Wildschiotz, & Rosenberg, 2008). Thus, a shift may occur in the circadian timing of sleep and sleep stages or daytime sleep compensate for nighttime loss of a particular sleep stage.

Chronic Pain Conditions

Headache

Headache entities include a variety of disorders with various symptom expressions and etiologies. The association between sleep complaints and headaches has been established through clinical experience and numerous epidemiological studies. A representative population-based study found 25% of respondents complained of insomnia and headache (Lamberg, 1999). The PSG literature on sleep, sleep disorders, and various headache disorders has recently been reviewed (Ong & Park, 2012; Rains & Poceta, 2010), and Brennan and Charles (2009) have provided an excellent discussion of the neurobiology linking sleep and headache. Sleep apnea headache is a specific diagnostic entity in the *International Classification of Headache Disorders, 2nd ed.* (ICHD-II; Headache Classification Subcommittee of the International Headache Society, 2004), which describes a common symptom of morning headache when awakening that is associated with the sleep disorder obstructive sleep apnea syndrome (OSA). The etiology of headache in OSA is disputed and could relate to the typical hypoxemia or hypercapnea associated with each apnea event. PSGs done on patients with cluster headaches or tension-type headaches have reported an elevated incidence of OSA relative to that found in the general population (Graft-Radford & Newman, 2004; Mitsikostas, Vikelis, & Viskow, 2007; Nobre, Leal, & Filho, 2005; Sand, Hagen, & Schrader, 2003). A second specific sleep-related diagnosis in the ICHD-II is hypnic headache, which is a headache that occurs in the latter portion of the night (Headache Classification Subcommittee of the International Headache Society, 2004). PSG has shown that hypnic headache is not specific to a given sleep stage. One study suggested it is associated with arousal from slow-wave sleep (Arjona, Jimenez-Jimenez, Vela-Bueno, & Talon-Barranco, 2000), whereas hypnic headache was associated with oxygen desaturations during REM sleep in another study (Dodick, 2000).

In patients with migraine, the PSG is essentially normal outside of the migraine attack (Jennum & Jensen, 2002). During migraine attacks, PSG studies show that sleep efficiency and slow-wave sleep are reduced and awakenings are increased (Ong & Park, 2012). Some studies have reported an increase in slow-wave sleep before an attack, suggesting possible prior sleep loss. Increased slow-wave sleep is a known response to sleep loss, and sleep loss is a well known trigger for migraine attacks. Investigators have suggested that migraine is also associated with an extended nighttime sleep period, a condition referred to as weekend migraine (Ong & Park, 2012). Notably, extended sleep on weekends is often a response to insufficient sleep during weekdays due to a reduced nighttime sleep period. Weekend oversleeping is one of the diagnostic features of insufficient sleep syndrome, a disorder listed in the *International Classification of Sleep Disorders* (American Academy of Sleep Medicine, 2005). Thus, accumulated sleep loss during weekdays may trigger the attack of weekend migraine. Increased slow-wave sleep as reported in weekend migraine is also consistent with a chronic insufficient sleep etiology.

The literature on sleep and cluster and tension-type headaches also suggests the important role of sleep schedule and duration. As with migraine studies, the studies of these headaches do not document sleep schedule and duration prior to the laboratory PSG and totally neglect consideration of daytime napping. Depending on the severity and chronicity of headache, sleep efficiency is reduced and awakening increased (Ong & Park, 2012). The evidence is equivocal as to whether cluster headaches and tension-type headaches are associated with REM sleep or the transition from REM to non rapid eye movement (NREM) sleep, or they occur without sleep-stage specificity (Jennum & Jensen, 2002; Zaremba et al., 2012). In reviewing the literature on sleep in chronic headache conditions, Ong and Park (2012) have offered a complete biobehavioral model that considers the role of sleep behaviors (i.e., sleep schedule and duration), compensatory behaviors (i.e., daytime napping, caffeine use), and mood and cognitive variables as being important factors in understanding the sleep and chronic headache nexus.

Musculoskeletal Pain

Rheumatoid Arthritis

Rheumatoid arthritis is a chronic autoimmune inflammatory disorder primarily affecting the joints. Patients with rheumatoid arthritis experience sleep disturbance and pain (Drewes, 1999), and the disturbance is characterized by fragmented sleep (e.g., sleep disrupted by 3-15 s EEG arousals) and increased wakefulness (Hirsch et al., 1994; Lavie, Nahir, Lorber, & Scharf, 1991). Patients also show frequent sleep-stage shifts and a high occurrence of periodic leg movements (e.g., slow rhythmic knee or ankle flexions) during sleep (Drewes, Bjerregard & Taaghold, 1998; Drewes, Svedsen, et al., 1998). A recent study of rheumatoid arthritis with comorbid insomnia, in which other primary sleep disorders were ruled out, reported reduced sleep efficiency and increased wakefulness during sleep rather than fragmentation of sleep (Roehrs et al., 2013).

A major symptom of rheumatoid arthritis is daytime fatigue, which may relate to the primary disease process (i.e., inflammatory cytokine activation) and/or to the disrupted sleep of rheumatoid arthritis. Clinicians and patients often do *not* differentiate fatigue from sleepiness. Sleepiness can be measured objectively using the multiple sleep latency test (MSLT) in which time to PSG sleep onset is assessed at 2-hour intervals throughout the day (Roehrs & Roth, 2005). When sleep is deprived or restricted, MSLT latency to sleep is reduced. When fatigue is measured using a standard fatigue scale concurrent with the MSLT, rheumatoid arthritis patients report high levels of fatigue while showing normal MSLT levels of sleepiness (Roehrs et al., 2013). Self-reported fatigue was correlated with reports of pain severity, but not with daytime sleepiness.

Fibromyalgia

Another prevalent musculoskeletal chronic pain disorder is fibromyalgia in which the pain is widespread and not specifically localized to joints. Its etiology is unknown. Fibromyalgia is the earliest and most completely PSG-studied chronic pain disorder. The first PSG study of fibromyalgia dates to 1975, and the study found that the amounts of slow-wave sleep, REM sleep, and total sleep time were reduced relative to age-matched controls (Moldofsky, Scarisbrick, England, & Smythe, 1975). The findings of Moldofsky et al. (1975) have been replicated in numerous studies since then (Mahowald & Mahowald, 2000). This early PSG study of fibromyalgia also reported intrusions of EEG α-activity (8-12 Hz) in NREM sleep and particularly as an admixture of α-activity with slow-wave δ-activity (0.5-2.5 Hz), which correlated positively with pain scores. This anomaly first reported by Hauri and Hawkins (1973) in depressed patients was termed α-delta sleep. These early reports led to the speculation that α-EEG is a marker of the musculoskeletal pain and mood symptoms of fibromyalgia. However, the clinical significance of α-intrusions in fibromyalgia is disputed because healthy individuals and other chronic pain patients also show the abnormality (Mahowald & Mahowald, 2000). Furthermore, multisite EEG monitoring (PSG typically records only from central EEG sites) during sleep suggests that α-EEG, rather than being a marker of arousal, is actually sleep protective, a sign of sleep-maintaining processes (Pivik & Harman, 1995).

As with rheumatoid arthritis, an additional symptom of fibromyalgia is daytime sleepiness and fatigue. Compared to rheumatoid arthritis patients, fibromyalgia patients have exhibited greater self-rated fatigue and sleepiness (Roehrs et al., 2013). Yet, despite a comparable degree of disturbed nocturnal sleep, fibromyalgia patients had unusually high MSLT scores, even compared to healthy volunteers. Elevated MSLT scores coupled with low nocturnal sleep efficiencies have been reported in patients with primary insomnia and are considered a sign of hyperarousal in insomnia (Roehrs, Randall, Harris, Maan, & Roth, 2011). This apparent hyperarousal in fibromyalgia, as reflected in the MSLT, is parallel to the hypothesized central hypersensitization of sensory processing in fibromyalgia (Ynus, 2007). These MSLT findings suggest the sensitization of fibromyalgia may extend beyond sensory processing to general hyperarousal.

Neuropathic Pain

Neuropathic pain is a maladaptive response of the nervous system to damage, triggered by lesions to the somatosensory nervous system. Peripheral neuropathies, specifically diabetic peripheral neuropathy and postherpetic neuralgia, are associated with reports of persistent sleep interruption. We are unaware of sleep studies that have compared patients with neuropathic pain to patients with other chronic pain conditions or to age-matched control populations. Pharmacotherapeutic studies have assessed self-reported "sleep interference" in neuropathic pain and showed improvement relative to placebo in "sleep interference" ratings associated with the agent being studied (Roehrs & Roth, 2005). Two PSG studies were found that have assessed sleep and pain in patients with neuropathic pain. Patients with diabetic peripheral neuropathy or post-herpetic neuralgia ($N=35$) had a 68% sleep efficiency, with a sleep latency of 34 minutes and wakefulness after sleep onset of 108 minutes (Hsu, Roth, Lamoreaux, Martin, & Hotary, 2004). Compared to historical controls, this represents severe sleep disturbance and is similar to that level seen in primary insomnia (Randall, Roehrs, & Roth, 2012). In addition, 10 of the 35 patients had sleep apnea (apnea index > 10 [# apnea events per hour sleep]). A PSG study of 70 patients with diabetic peripheral neuropathy who were randomized to three different pharmacologic treatments included a placebo condition in each treatment arm (Boyle et al., 2012). In the placebo condition, the sleep efficiency averaged between 77% and 79% among the treatment arms. Other sleep and sleep-stage measures were not provided, but apnea indices ranged from 3 to 6, and periodic leg movement indices ranged from 16 to 20. These results must be viewed with caution, however, because the low sleep efficiency of the treatment groups on placebo may be due to the primary sleep disorders and their associated sleep fragmentation rather than pain per se. Yet, these provocative PSG findings do highlight the need for more thorough study of the relationship between sleep disturbance and sleep disorders in neuropathic pain.

EXPERIMENTAL SLEEP MANIPULATIONS AND PAIN SENSITIVITY

Nonspecific Sleep Deprivation

Total Deprivation

A majority of studies have clearly shown total sleep deprivation in healthy pain-free volunteers enhances pain sensitivity or reduces pain thresholds (Finan, Goodin, & Smith, 2013; Roehrs & Roth, 2005). For example, one night of total sleep deprivation in healthy volunteers increased pain sensitivity by 27% over a range of radiant heat intensities (Roehrs, Hyde, Blaisdell, Greenwald, & Roth, 2006). In another study total sleep deprivation for 40 hours reduced pain threshold to a pressure stimulus by 8% (Onen, Alloui, Gross, Eschallier, & Dubray, 2001). Kundermann, Spernal, Huber, Krieg, and Lautenbacher (2004) showed pain thresholds were enhanced with total sleep deprivation, but touch detection thresholds were not affected (Kundermann et al., 2004). The Kundermann et al. (2004) study included an important control showing that the sleep deprivation effect is specific to nociception (e.g., sensation of pain) rather than a nonspecific activation of all somatosensation (e.g., sensation of touch, vibration, or temperature). Actually, Nathanial Kleitman, who is considered the father of modern sleep research, reported this in 1939 (Kleitman, 1939).

The effects of total sleep deprivation on pain in clinical patient populations has also been assessed. In patients with major depressive disorder, total sleep deprivation reduced heat pain threshold (Kundermann, Hemmeter-Spernal, Huber, Krieg, & Lautenbacher, 2008). As often reported in the depression literature, sleep deprivation improved mood. A study of sleep deprivation in somatoform patients found an increase in self-reported pain, but pain thresholds to heat and mechanical stimulation remained unchanged (Busch et al., 2012). The clinical significance of the discrepancy between subjective and objective measures of pain remains to be explored. It is not unlike the previously mentioned discrepancy between objective and subjective fatigue or sleepiness in fibromyalgia patients (Roehrs et al., 2013).

Reduced Nighttime Sleep Period

In everyday life, total sleep deprivation is rare. Typically, people reduce the nighttime sleep period for various reasons, and chronic pain patients show reduced total sleep times. In the Roehrs et al. (2006) study of healthy volunteers, reducing the nighttime sleep period to 4 hours increased pain sensitivity by 13.5%, half the effect of total sleep deprivation in that study. In another healthy volunteer study, 4 hours of sleep compared to habitual sleep increased pain ratings accompanying a radiant heat stimulus (generated by a laser) by 30% (Tiede et al., 2010). Evoked potentials to the laser stimulation (laser-evoked potentials, LEPs) were also collected, and the N1 amplitude was *reduced* by approximately 30% after the sleep restriction, suggesting that the hyperalgesia of sleep restriction is due to perceptual amplification rather than sensory amplification, in that LEPs were reduced not increased. One of the few studies reducing sleep time in a patient population, compared fatigue, pain, and depression ratings after 8 hours versus 4 hours of nighttime sleeping in rheumatoid arthritis patients and healthy pain-free volunteers (Irwin et al., 2012). The sleep restriction increased Profile of Mood States (POMS) scores on the fatigue, depression, and anxiety subscales and increased McGill pain ratings relative to 8 hours of nighttime sleeping and relative to the sleep-restricted ratings of the healthy volunteers.

Sleep Fragmentation

The primary effect of sleep disorders such as periodic leg movements during sleep or sleep apnea syndrome is sleep fragmentation with frequent brief arousals (3-15s) following each event. These arousals can occur from 10 to 60 times per hour of sleep, and typically, the patient is unaware of the frequent awakenings because of their brevity. Earlier we noted that the prevalence of sleep disorders such as sleep apnea and periodic leg movements is heightened in some chronic pain conditions. We are aware of only one study that has assessed the impact of sleep fragmentation on pain (Smith, Edwards, McCann, & Haythornthwaite, 2007). Sleep was interrupted eight times during the night, and relative to continuous sleep, pain inhibition was diminished. The importance of sleep fragmentation to pain chronicity is also evident in a study in which pain sensitivity was measured in patients with sleep apnea syndrome before and after apnea treatment using continuous positive airway pressure (CPAP). Treatment reduced arousal frequency from 50 events per hour to 2 per hour and pain sensitivity was reduced by 29% (Khalid, Roehrs, Hudgel, & Roth, 2011).

Sleep-Stage Deprivation

Slow-Wave-Sleep Deprivation

Owing to the early description of α-delta sleep in patients with fibromyalgia, a number of studies have focused on the selective deprivation of slow-wave sleep. The results of these studies have been equivocal (Roehrs & Roth, 2005). Discrepancy among study results may relate to the extent to which the deprivation procedure also reduced total sleep time. We have already pointed out that reduced sleep time has a hyperalgesic effect. Another possibly important factor may be the age of the study participants. Hyperalgesic effects were seen in middle-aged study participants, but not younger subjects (Roehrs & Roth, 2005). As one ages, intolerance to sleep reduction appears to develop.

REM-Sleep Deprivation

The other sleep stage that may be important in nociceptive signaling is REM sleep. Brennan and Charles (2009) noted the overlap between REM and nociceptive neurobiology, and several human and animal studies have shown that REM-sleep deprivation enhances nociception (Lautenbacher, Kundermann, & Krieg, 2006; Roehrs & Roth, 2005). In healthy women the percentage of REM sleep was correlated with ratings of suprathreshold thermal pain and frequency of unsolicited pain reports (Smith, Edwards, Stonerock, & McCann, 2005). In a REM-sleep deprivation study using healthy volunteers, pain sensitivity to a radiant heat stimulus was increased relative to a yoked control NREM awakening condition with a similar number and distribution of awakenings as in the REM condition (Roehrs et al., 2006). The one negative human REM-sleep deprivation study we found measured pain *tolerance*, not pain threshold or sensitivity (Onen et al., 2001). Pain tolerance studies differ from threshold studies in the response instructions given to the subject (e.g., respond when you can no longer tolerate the pain versus when you first experience pain). In healthy subjects REM-sleep interruption relative to baseline sleep did not significantly reduce the mechanical pain *tolerance* threshold. Notably, this study employed mechanical pressure, rather than thermal stimulation, as the pain-evocative stimulus.

MECHANISMS UNDERLYING THE SLEEP-PAIN NEXUS

A number of reviews have described the overlap between various central neurotransmitter systems that control sleep and pain (Brennan & Charles, 2009; Finan et al., 2013; Roehrs & Roth, 2005). We do not review this literature because it is primarily based on nonclinical, animal studies. The neurobiology of pain and sleep is discussed in detail by Roehrs and Roth (2005). Very briefly, adenosine and acetylcholine have roles in both sleep and analgesia, and the opioid system also has sleep effects. In human chronic pain studies, the role of cytokines has received much attention, as described next.

Sleep Loss or Sleepiness and Cytokines

One line of evidence suggests that the sleep-pain nexus is mediated by proinflammatory cytokine (PIC) activation. Present both peripherally and centrally, cytokines are a large, diverse group of small proteins that have cellular signaling activity, with some promoting and others inhibiting inflammatory processes. The earlier literature was equivocal regarding sleep loss and PIC activation (Dinges, Douglas, Hamarman, Zaugg, & Kapoor, 1995), but recent studies have shown total sleep deprivation and sleep restriction to 4-6 hours nightly produces elevated levels of interleukin 6 (IL-6) and less consistently tumor necrosis factor alpha (TNF-α) (Haack, Sanchez, & Mullington, 2007; Irwin, Wang, Campomayor, Collado-Hidalgo, & Cole, 2006; Shearer et al., 2001; Vgontzas et al., 2007; Vgontzas, Zoumakis, Bixler, et al., 2004). Increased IL-6 levels and sleepiness due to sleep restriction are reversed by a midday nap (Vgontzas et al., 2007), and sleepiness and elevated TNF-α levels in sleepy patients are reversed by the TNF-α antagonist, etanercept (Vgontzas, Zoumakis, Lin, et al., 2004). TNF-α levels are also elevated in sleepy children with apnea due to enlarged tonsils. Tonsillectomy reverses the apnea and sleepiness, and most importantly, the procedure reduces TNF-α levels (Gozal et al., 2010).

Cytokine Activation and Pain

A final important linkage exists between PIC activation and pain. PICs in both the peripheral and central nervous system (CNS) are known to play a key role in acute and chronic pain conditions (Watkins, Milligan, & Maier, 2003). In the periphery macrophages release cytokines in response to tissue injury. The local hyperalgesic effects of cytokine release were demonstrated in animal studies showing that local application of TNF-α or IL-1 increased pain sensitivity (Cunha, Poole, Lorenzetti, & Ferreira, 1992). Also, the administration of cytokine-antagonists reversed the hyperalgesia (Cunha, Cunha, Poole, & Ferreira, 2000). In human chronic pain conditions, both IL-6 and TNF-α levels are elevated peripherally, and their levels are correlated with the experienced pain and hyperalgesia (Sommer & Kress, 2004).

PICs released peripherally are also recognized as having a signaling function in the CNS, which then produces de novo central synthesis and release of the cytokines from glia cells (Watkins & Maier, 2005). Thus, nociceptive transmission is facilitated downward from central to spinal pathways by the central cytokine release. Studies have also shown that PIC activation enhances dorsal root ganglion cell excitability, as well as A-delta and C nerve fiber excitability in the periphery (Watkins & Maier, 2005). PIC activation is thus one likely mechanism by which sleep loss is hyperalgesic.

MODULATORS OF THE SLEEP-PAIN NEXUS: MOOD AND COGNITIVE PROCESSES

Mood and cognitive factors may play a potential modulatory role in the sleep-pain relationship. Human brain imaging has identified neural systems engaging attentional and emotional brain regions distinct from primary nociception (Apkarian, Bushnell, Treede, & Zubieta, 2005). The question arises as to how reduced sleep alters emotional and cognitive modulation of pain. In the sleep field, a number of studies have shown that total sleep deprivation or restriction of sleep elevates measures of depression and anxiety in healthy participants. For example, 56 hours of total deprivation produced subclinical elevations on the depression, anxiety, and somatic pain scales of the Personality Assessment Inventory (Kahn-Green, Killgore, Mamimori, Balkin, & Killgore, 2007) and 30 hours of deprivation elevated the depression and fatigue scales of the POMS (Scott, McNaughton, & Polman, 2006). Reducing sleep time by 33% for seven nights produced increases on all the POMS scales (Dinges et al., 1997) and in-house officers reporting < 5 hours sleep over 31 hours of duty demonstrated elevations on all six POMS scales (e.g., anger, fatigue, depression, anxiety, vigor, confusion), compared to those sleeping > 5 hours (Orton & Gruzelier, 1989). However, the meaning of a nonspecific increase in all POMS scales is unclear; it may reflect a response bias as discussed earlier. A 12-day 50% sleep restriction reduced visual analog scale assessments of adjectives reflecting optimism and sociability (Haack & Mullington, 2005), and one 12-hour recovery night of sleep reversed the reduced optimism and sociability. Thus, the question is whether or not sleep loss has a hyperalgesic effect heightened by enhancing depression and anxiety.

One of the risk factors for progression of acute pain to chronic pain in patients following breast cancer surgery was greater preoperative emotional distress (Katz et al., 2005). In a large prospective, population-based study, anxiety, depression, and sleep problems were predictive of the development of new, chronic, widespread pain after 15 months (Gupta et al., 2007). These data point to the need to better understand the potential mediating role of mood in the sleep-pain relationship.

In the pain literature, a number of studies have shown that cognitive factors moderate pain. One important factor is referred to as "pain catastrophizing," and it is characterized as ruminating on pain sensations, exaggerating the threat associated with pain, and feeling unable to control the pain (Forsythe, Dunbar, Hennigar, Sullivan, & Gross, 2008). For example, a prospective study of patients with osteoarthritis of the knee followed the patients for 24 months after total knee arthroplasty (Forsythe et al., 2008). Preoperative pain catastrophizing scores were a significant predictor for the presence of pain on the McGill Pain Questionnaire over the 24-month follow-up period.

Another important cognitive factor involves the attention-distraction dimension. Perception of pain is altered by the attentional focus of the individual, such that pain is experienced as more intense when the individual is focused on it and less so when the individual is distracted. A recent study assessed whether or not sleep restriction alters the attentional modulation of nociception and pain perception (Tiede et al., 2010). LEPs and pain thresholds in healthy volunteers were assessed under two differing attentional conditions, stimulus intensity discrimination (e.g., pain focused) and mental arithmetic (e.g., pain distracted), after habitual sleep and restricted sleep (4 hours). Sleep restriction differentially altered nociception versus perception by reducing LEPs, but it enhanced pain ratings by 30%.

PHARMACOLOGICAL AND BEHAVIORAL TREATMENT OF SLEEP AND PAIN

Behavioral

Studies have described a number of different cognitive-behavioral treatments for chronic pain with a range of etiologies. These treatment studies have focused on the control of pain, with sleep sometimes being a secondary outcome. Review of this large literature is beyond the scope of this chapter (for those interested, see reviews by Keefe and Block (1982), McCracken and Turk (2002), and van Tulder et al. (2001)). In contrast, cognitive-behavioral treatment for insomnia (CBT-I) has been evaluated in chronic pain populations with comorbid insomnia, and both sleep and pain are outcomes of interest in these studies. The few and relatively small studies using CBT-I in chronic pain populations have produced equivocal results, however. As shown in studies of primary insomnia, CBT-I in chronic pain patients generally reduces sleep latency and increases sleep efficiency, and in some cases, it increases total sleep time (Currie, Wilson, Pontefract, & deLaplante, 2000; Edinger, Wohlgemuth, Krystal, & Rice, 2005; Jungquist et al., 2010). Currie et al. (2000) studied patients with musculoskeletal pain, excluding those with fibromyalgia, relative to a control population. In their study, sleep was improved, but their findings only indicated a trend in pain reduction. A study of fibromyalgia patients showed improved sleep in both a CBT-I and a sleep hygiene-only group, but only the sleep hygiene group showed improved pain (Edinger et al., 2005). In patients with chronic neck and back pain, CBT-I improved sleep and "sleep interference by pain" ratings, but it did not improve pain severity (Jungquist et al., 2010). A recent, large study ($N=150$) compared CBT-I and cognitive-behavioral treatment focused on pain only (CBT-P) in elderly patients with osteoarthritic pain (Vitiello et al., 2013). CBT-I improved insomnia severity relative to CBT-P and the control, but neither active treatment improved pain severity.

CBT-I typically has three components, cognitive therapy or sleep hygiene instructions, stimulus control, and sleep restriction. Stimulus control and sleep restriction both functionally reduce a patient's nighttime sleep period. Reducing the nighttime sleep period enhances sleep drive and thereby reduces sleep latency and increases sleep efficiency, but it may not necessarily increase sleep time. The sleep restriction studies reviewed earlier have all shown that a reduced nighttime sleep period and sleep time increase pain sensitivity, while reducing pain thresholds, which may explain the equivocal CBT-I effects on pain.

Regarding behavioral treatments for other sleep disorders, a persistently reduced nighttime sleep period produces chronic excessive daytime sleepiness as measured by the MSLT, a condition identified as insufficient sleep syndrome (American Academy of Sleep Medicine, 2005). An enforced nightly sleep period of 10 hours for 1-2 weeks in people with insufficient sleep syndrome results in normalized levels of sleepiness or alertness (Roehrs et al., 2010). A recent study of pain-free people with excessive sleepiness showed that four consecutive 10-hour nighttime sleep periods normalized sleepiness or alertness and also reduced sensitivity to a painful radiant heat stimulus relative to baseline and a control group that maintained their habitual nighttime sleep patterns (Roehrs, Harris, Randall, & Roth, 2012). This study suggests that identifying and correcting insufficient or irregular sleep behavior in chronic pain patients may improve their pain.

Earlier we noted that sleep fragmentation characteristically results from periodic leg movements during sleep and sleep apnea syndrome. The standard treatment for sleep apnea syndrome is CPAP. A study measured pain sensitivity to a radiant heat stimulus in apnea patients without concurrent pain. Pain sensitivity was assessed before and after CPAP and again after discontinuation of CPAP for two nights (Khalid et al., 2011). CPAP reduced brief arousals during sleep (e.g., restored sleep continuity) from approximately 50 per hour of sleep to 2 per hour, and with discontinuation of CPAP for two nights, the arousal rate returned to 30 per hour. CPAP reduced radiant heat pain sensitivity, but the sensitivity returned to pretreatment levels with CPAP discontinuation.

Pharmacologic

Chronic Pain of Various Etiologies

Studies have assessed drugs from a number of different classes as treatments for various acute and chronic pain disorders, and these drugs include analgesics, antidepressants, antiepileptics, and hypnotics. In many of these studies, pain is the major focus, and when sleep is measured, it is a secondary outcome typically only measured by self-report. Furthermore, the studies rarely included patients based on the presence of comorbid insomnia or sleep disturbance, and not all patients with chronic pain also reported disturbed sleep or insomnia, which diminishes a study's power to demonstrate positive sleep effects.

Choy et al. (2011) conducted a meta-analysis of randomized trials of the treatment of fibromyalgia. The authors concluded that there was "limited robust clinical data for some therapeutic classes including tricyclic antidepressants, analgesics, sedative hypnotics, and monoamine oxidase inhibitors." The review showed that the trials of pregabalin and the two serotonin norepinephrine reuptake inhibitors, milnacipran and duloxetine (each at their indicated dose), exhibited improved pain scores. But only pregabalin improved sleep-scale scores. Sleep improvement with pregabalin was confirmed in a large ($N=119$) PSG study in patients with fibromyalgia and comorbid sleep maintenance insomnia, which found pregabalin reduced the amount of wakefulness during the sleep period, reduced daily pain scores, and reduced daytime fatigue (Roth, Lankford, Bhadra, Whalen, & Resnick, 2012). Additionally, the percent of slow-wave sleep was modestly increased. The PSG sleep effects of pregabalin have also been studied in healthy participants relative to placebo. Although it did not improve sleep efficiency or sleep time, pregabalin did increase slow-wave sleep (Hindmarch, Dawson, & Stanley, 2005). Given the participants were volunteers without sleep disturbance, one would not expect improved sleep efficiency.

A study using mediation analyses attempted to determine whether the pregabalin sleep-pain effects were due to a pain effect or a sleep effect. The data of two large studies of patients with fibromyalgia showed that, relative to placebo, daily pain scores were improved with pregabalin, and self-rated sleep was improved as well (Russel et al., 2009). The mediation analyses showed that 43-80% of the sleep effects resulted from the direct effects of pregabalin on sleep and not through pain relief that produced the improved sleep.

A number of pharmacologic treatment studies have involved patients with neuropathic pain of various etiologies. A recent study randomized patients with diabetic neuropathic pain to amitriptyline, duloxetine, and pregabalin (Boyle et al., 2012). Relative to placebo, all three treatments improved pain, with no drug being superior in analgesic effects, but the treatments had differential effects on PSG-defined sleep. Pregabalin improved sleep at both doses, as did amitriptyline at the higher dose, but duloxetine increased sleep disturbance.

Studies have assessed standard hypnotics in chronic pain patients with the results being quite mixed (Roehrs, 2009). In fibromyalgia patients, zolpidem (Moldofsky, Lue, Mously, Roth-Schecter, & Reynolds, 1996), but not zopiclone (Gronbald, Nykanen, & Konttinen, 1993), improved some of the sleep measures, but neither drug improved pain. Triazolam improved sleep and pain in patients with rheumatoid arthritis (Walsh, Muehlbach, Lauter, Hilliker, & Schweitzer, 1996), whereas zopiclone only improved pain (Drewes, Bjerregard, et al., 1998). A large ($N=153$) study of eszopiclone, the S isomer of zopiclone, assessed its sleep and pain effects in patients with insomnia comorbid with rheumatoid arthritis (Roth et al., 2009). Compared to placebo, eszopiclone improved self-reported measures of sleep and most all of the pain measures. Hospitalized patients with mucositis associated with chemotherapy for hematologic malignancies were randomized to two nights of eszopiclone or placebo (Dimsdale et al., 2011). Pain scores throughout the day were improved with eszopiclone, and patients reported increased sleep time and fewer awakenings.

SUMMARY

Disturbed sleep is a prominent symptom in people experiencing acute and chronic pain, and studies using PSG have objectively documented sleep disturbance in any number of chronic pain disorders. The sleep-pain nexus has been modeled in healthy pain-free volunteers, showing that sleep deprivation, restriction, or fragmentation are hyperalgesic, and these studies suggest the sleep-pain relation is bidirectional. It is also known that cognitive and mood factors can modulate the sleep-pain nexus. Emerging information identifies one possible mechanism linking the sleep-pain nexus, the PICs. Some of the behavioral and pharmacological treatment studies that have shown sleep improvement have also produced improved pain control. Future research should strive for a more complete understanding of the normal physiology and pathophysiology of acute and chronic pain and sleep, which will likely enhance clinicians' ability to manage pain and sleep.

References

American Academy of Sleep Medicine. (2005). *The international classification of sleep disorders* (2nd ed.). Weschester, IL: American Academy of Sleep Medicine.

Apkarian, A. V., Bushnell, M. C., Treede, R. D., & Zubieta, J. K. (2005). Human brain mechanisms of pain perception and regulation in health and disease. *European Journal of Pain, 9*, 463–484.

Arjona, J. A., Jimenez-Jimenez, F. J., Vela-Bueno, A., & Talon-Barranco, A. (2000). Hypnic headache associated with stage 3 sleep wave sleep. *Headache, 40*, 753–754.

Boyle, J., Erriksson, M. V., Gouni, R., Johnsen, S., Coppini, D. V., & Kerr, D. (2012). Randomized, placebo-controlled comparison of amitriptyline, duloxetine, and pregabalin in patients with chronic diabetic peripheral neuropathic pain. *Diabetes Care, 35*, 2451–2458.

Brennan, K. C., & Charles, A. (2009). Sleep and headache. *Seminars in Neurology, 29*, 406–418.

Busch, V., Haas, J., Cronlein, T., Pieh, C., Geisler, P., Hajak, G., et al. (2012). Sleep deprivation in chronic somatoform pain—Effects on mood and pain regulation. *Psychiatry Research, 195*, 134–143.

Chapman, C. R., & Gavrin, J. (1999). Suffering: The contribution of persistent pain. *Lancet, 353*, 2233–2237.

Chouchou, F., Khoury, S., Chauny, J. M., Denis, R., & Lavigne, G. J. (2013). Postoperative sleep disruptions: A potential catalyst of acute pain? *Sleep Medicine Reviews, 17*, 1–10.

Choy, E., Marshall, D., Gabriel, Z. L., Mitchell, S. A., Gylee, E., & Dakin, A. A. (2011). A systematic review and mixed treatment comparison of the efficacy of pharmacological treatments for fibromyalgia. *Seminars in Arthritis and Rheumatism, 41*, 335–345.

Cunha, J. M., Cunha, S., Poole, S., & Ferreira, S. H. (2000). Cytokine-mediated inflammatory hyperalgesia limited by interleukin-1 receptor antagonist. *British Journal of Pharmacology, 130*, 1418–1424.

Cunha, F., Poole, S., Lorenzetti, B., & Ferreira, S. (1992). The pivotal role of tumor necrosis factor alpha in the development of inflammatory hyperalgesia. *British Journal of Pharmacology, 107*, 660–664.

Currie, S. R., Wilson, K. G., Pontefract, A. J., & deLaplante, L. (2000). Cognitive-behavioral treatment of insomnia secondary to chronic pain. *Journal of Consulting and Clinical Psychology, 68*, 407–416.

Dimsdale, J. E., Ball, E. D., Carrier, E., Wallace, M., Holman, P., Mulroney, C., et al. (2011). Effect of eszopiclone on sleep, fatigue, and pain in patients with mucositis associated with hematologic malignancies. *Support Care Cancer, 19*, 2015–2020.

Dinges, D. F., Douglas, S. D., Hamarman, S., Zaugg, L., & Kapoor, S. (1995). Sleep deprivation and human immune function. *Advances in Neuroimmunology, 5*, 97–110.

Dinges, D. F., Pack, F., Williams, K., Gillen, K. A., Powell, J. W., Ott, G. E., et al. (1997). Cumulative sleepiness, mood disturbance, and psychomotor vigilance performance decrements during a week of sleep restricted to 4-5 hours per night. *Sleep, 20*, 267–277.

Dodick, D. W. (2000). Polysomnography in hypnic headache syndrome. *Headache, 40*, 748–752.

Drewes, A. M. (1999). Pain and sleep disturbances with special reference to fibromyalgia and rheumatoid arthritis. *Rheumatology, 38*, 1035–1044.

Drewes, A. M., Bjerregard, K., & Taaghold, S. J. (1998). Zopiclone as night medication in rheumatoid arthritis. *Scandinavian Journal of Rheumatology, 27*, 180–187.

Drewes, A. M., Svedsen, L., Taagholt, J., Bjerregård, K., Neilsen, K. D., & Hansen, B. (1998). Sleep in rheumatoid arthritis: A comparison with healthy subjects and studies of sleep/wake interactions. *British Journal of Rheumatology, 37*, 71–81.

Edinger, J. D., Wohlgemuth, W. K., Krystal, A. D., & Rice, J. R. (2005). Behavioral insomnia therapy for fibromyalgia patients: A randomized clinical trial. *Archives of Internal Medicine, 165*, 2527–2535.

Finan, P. H., Goodin, B. R., & Smith, M. T. (2013). The association of sleep and pain: An update and a path forward. *The Journal of Pain: Official Journal of the American Pain Society, 14*, 1539–1552.

Forsythe, M. E., Dunbar, M. J., Hennigar, A. W., Sullivan, M. J., & Gross, M. (2008). Prospective relation between catastrophizing and residual pain following knee arthroplasty: Two-year follow-up. *Pain Research & Management: The Journal of the Canadian Pain Society, 13*, 335–341.

Gogenur, I., Wildschiotz, G., & Rosenberg, J. (2008). Circadian distribution of sleep phases after major abdominal surgery. *British Journal of Anaesthesia, 100*, 45–49.

Gozal, D., Serperio, L. D., Kheirandish-Gozal, L., Capdevila, O. C., Khalyfa, A., & Tauman, R. (2010). Sleep measures and morning plasma TNF-α levels in children with sleep-disordered breathing. *Sleep, 33*, 319–326.

Graft-Radford, S. B., & Newman, A. (2004). Obstructive sleep apnea and cluster headache. *Headache, 44*, 607–610.

Gronbald, M., Nykanen, J., & Konttinen, Y. (1993). Effect of zopiclone on sleep quality, morning stiffness, widespread tenderness and pain and general discomfort in primary fibromyalgia patients. A double-blind randomized trial. *Clinical Rheumatology, 12*, 186–191.

Gupta, A., Silman, A. J., Ray, D., Morriss, R., Dickens, C., MacFarlane, G. J., et al. (2007). The role of psychosocial factors in prediction the onset of chronic widespread pain: Results from a prospective population-based study. *Rheumatology, 46*, 666–671.

Haack, M., & Mullington, J. M. (2005). Sustained sleep restriction reduces emotional and physical well-being. *Pain, 119*, 56–64.

Haack, M., Sanchez, E., & Mullington, J. M. (2007). Elevated inflammatory markers in response to prolonged sleep restriction are associated with increased pain experience in healthy volunteers. *Sleep, 30*, 1145–1152.

Hauri, P., & Hawkins, D. R. (1973). Alpha-delta sleep. *Electroencephalography and Clinical Neurophysiology, 34*, 233–237.

Headache Classification Subcommittee of the International Headache Society. (2004). The international classification of headache disorders: 2nd edition. *Cephalalgia, 24*(Suppl. 1), 1–151.

Hindmarch, I., Dawson, J., & Stanley, N. (2005). A double-blind study in healthy volunteers to assess the effects on sleep of pregabalin compared with alprazolam and placebo. *Sleep, 28*, 187–193.

Hirsch, M., Carlander, B., Verge, M., Tafti, M., Anaya, J., Billiard, M., et al. (1994). Objective and subjective sleep disturbances in patients with rheumatoid arthritis. *Arthritis and Rheumatism, 37*, 41–49.

Hsu, T., Roth, T., Lamoreaux, L., Martin, S., & Hotary, L. (2004). Polysomnographic profile of patients with neuropathic pain and self-reported sleep disturbance. *Sleep, 27*, A333.

Irwin, M. R., Olmstead, R., Carrillo, C., Sadeghi, N., FitzGerald, J. D., Ranganath, V. K., et al. (2012). Sleep loss exacerbates fatigue, depression, and pain in rheumatoid arthritis. *Sleep, 35*, 537–543.

Irwin, M. R., Wang, M., Campomayor, C. O., Collado-Hidalgo, A., & Cole, S. (2006). Sleep deprivation and activation of morning levels of cellular and genomic markers of inflammation. *Archives of Internal Medicine, 166*, 1756–1762.

Jennum, P., & Jensen, R. (2002). Sleep and headache. *Sleep Medicine, 6*, 471–479.

Jungquist, C. R., O'Brien, C., Matteson-Rusby, S., Smith, M. T., Pigeon, W. R., Xia, Y., et al. (2010). The efficacy of cognitive-behavioral therapy for insomnia in patients with chronic pain. *Sleep Medicine, 11*, 302–309.

Kahn-Green, E. T., Killgore, D. B., Mamimori, G. H., Balkin, T. J., & Killgore, W. D. S. (2007). The effects of sleep deprivation on symptoms of psychopathology in healthy adults. *Sleep Medicine, 8*, 215–221.

Katz, J., Poleshuck, E. L., Andrus, C. H., Hogan, L. A., Jung, B. F., Kulick, D. I., et al. (2005). Risk factors for acute pain and its persistence following breast cancer surgery. *Pain, 119*, 16–25.

Keefe, F. J., & Block, A. R. (1982). Behavioral treatment of pain. In R. S. Surwit, Williams, R. B.Jr., A. Steptoe, & R. Biersner (Eds.), *NATO conference series: Vol. 19. Behavioral treatment of disease* (pp. 371–387).

Keenan, S., & Hirshkowitz, M. (2010). Monitoring and staging human sleep. In M. H. Kryger, T. Roth, & W. C. Dement (Eds.), *Principles and practice of sleep medicine* (5th ed., pp. 1602–1609). Philadelphia, PA: Elsevier.

Khalid, I., Roehrs, T. A., Hudgel, D. W., & Roth, T. (2011). Continuous positive airway pressure in severe obstructive sleep apnea reduces pain sensitivity. *Sleep, 34*, 1687–1691.

Kleitman, N. (1939). *Sleep and wakefulness*. Chicago, IL: University of Chicago Press.

Kundermann, B., Hemmeter-Spernal, J., Huber, M. T., Krieg, J. C., & Lautenbacher, S. (2008). Effects of total sleep deprivation in major depression: Overnight improvement of mood is accompanied by increased pain sensitivity and augmented pain complaints. *Psychosomatic Medicine, 70*, 92–101.

Kundermann, B., Spernal, J., Huber, M. T., Krieg, J. C., & Lautenbacher, S. (2004). Sleep deprivation affects thermal pain thresholds but not somatosensory threshold in healthy volunteers. *Psychosomatic Medicine, 66*, 932–937.

Lamberg, L. (1999). Chronic pain linked with poor sleep; exploration of causes and treatment. *Journal of the American Medical Association, 281*, 691–692.

Lautenbacher, S., Kundermann, B., & Krieg, J. (2006). Sleep deprivation and pain perception. *Sleep Medicine Reviews, 10*, 357–369.

Lavie, P., Nahir, M., Lorber, M., & Scharf, Y. (1991). Nonsteroidal anti-inflammatory drug therapy in rheumatoid arthritis patients: Lack of association between clinical improvement and effects on sleep. *Arthritis and Rheumatism, 34*, 655–659.

Mahowald, M. L., & Mahowald, M. W. (2000). Nighttime sleep and daytime functioning (sleepiness and fatigue) in less well-defined chronic rheumatic diseases with particular reference to the 'alpha-delta NREM sleep anomaly'. *Sleep Medicine, 1*, 77–83.

McCracken, L. M., & Turk, D. C. (2002). Behavioral and cognitive-behavioral treatment for chronic pain: Outcome, predictors of outcome and treatment process. *Spine, 27*, 2564–2573.

Mitsikostas, D. D., Vikelis, M., & Viskow, A. (2007). Refractory chronic headache associated with obstructive sleep apnoea syndrome. *Cephalalgia, 28*, 139–143.

Moldofsky, H., Lue, F. A., Mously, C., Roth-Schechter, B., & Reynolds, W. J. (1996). The effect of zolpidem in patients with fibromyalgia: A dose ranging, double blind, placebo controlled, modified crossover study. *Journal of Rheumatology, 23*, 529–533.

Moldofsky, H., Scarisbrick, P., England, R., & Smythe, H. (1975). Musculoskeletal symptoms and non-REM sleep disturbances in patients with "fibrositis syndrome" and healthy subjects. *Psychosomatic Medicine, 37*, 341–351.

Nobre, M. E., Leal, A. J., & Filho, P. M. (2005). Investigation into sleep disturbance of patients suffering from cluster headache. *Cephalalgia, 25*, 488–492.

Onen, S. H., Alloui, A., Gross, A., Eschallier, A., & Dubray, C. (2001). The effects of total sleep deprivation, selective sleep interruption and sleep recovery on pain tolerance thresholds in healthy subjects. *Journal of Sleep Research, 10*, 35–42.

Ong, J. C., & Park, M. (2012). Chronic headaches and insomnia: Working toward a biobehavioral model. *Cephalalgia, 32*, 1059–1070.

Orton, D. I., & Gruzelier, J. H. (1989). Adverse changes in mood and cognitive performance of house officers after night duty. *British Medical Journal, 298*, 21–23.

Pivik, R. T., & Harman, K. (1995). A reconceptualization of EEG alpha activity as an index of arousal during sleep: All alpha activity is not equal. *Journal of Sleep Research, 4*, 131–137.

Rains, J. C., & Poceta, J. S. (2010). Sleep and headache. *Current Treatment Options in Neurology, 12*, 1–15.

Randall, S., Roehrs, T. A., & Roth, T. (2012). Efficacy of eight months of nightly zolpidem: A placebo controlled study. *Sleep, 35*, 1551–1557.

Roehrs, T. A. (2009). Does effective management of sleep disorders improve pain symptoms. *Drugs, 69*(Suppl. 2), 5–11.

Roehrs, T., Carskadon, M. A., Dement, W. C., & Roth, T. (2010). Daytime sleepiness and alertness. In M. H. Kryger, T. Roth, & W. C. Dement (Eds.), *Principles and practice of sleep medicine* (5th ed., pp. 42–53). Philadelphia, PA: Elsevier.

Roehrs, T., Diererichs, C., Gillis, M., Burger, A. J., Stout, R. A., Lumley, M. A., et al. (2013). Nocturnal sleep, daytime sleepiness and fatigue in fibromyalgia patients compared to rheumatoid arthritis patients and healthy controls. *Sleep Medicine, 14*, 109–115.

Roehrs, T. A., Harris, E., Randall, S., & Roth, T. (2012). Pain sensitivity and recovery from mild chronic sleep loss. *Sleep, 35*, 1602–1608.

Roehrs, T. A., Hyde, M., Blaisdell, B., Greenwald, M., & Roth, T. (2006). Sleep loss and REM sleep loss are hyperalgesic. *Sleep, 29*, 145–151.

Roehrs, T. A., Randall, S., Harris, E., Maan, R., & Roth, T. (2011). MSLT in primary insomnia: Reliability and relation to nocturnal sleep. *Sleep, 34*, 1647–1652.

Roehrs, T., & Roth, T. (2005). Sleep and pain: Interaction of two vital functions. *Seminars in Neurology, 25*, 106–116.

Rosenberg-Adamsen, S., Kehlet, H., Dodds, C., & Rosenberg, J. (1996). Postoperative sleep disturbances: Mechanisms and clinical implications. *British Journal of Anaesthesia, 76*, 552–559.

Roth, T., Lankford, A., Bhadra, P., Whalen, E., & Resnick, E. M. (2012). Effect of pregabalin on sleep in patients with fibromyalgia and sleep maintenance disturbance: A randomized, placebo-controlled, 2-way crossover polysomnography study. *Arthritis Care & Research, 62,* 597–606.

Roth, T., Price, J. M., Amato, D. A., Rubens, R. P., Roach, J. M., & Schnitzer, T. J. (2009). The effect of eszopiclone in patients with insomnia and coexisting rheumatoid arthritis: A pilot study. *Primary Care Companion to the Journal of Clinical Psychiatry, 11,* 292–301.

Russel, I. J., Crofford, L. J., Leon, T., Cappelleri, J. C., Bushmakin, A. G., Whalen, E., et al. (2009). The effects of pregabalin on sleep disturbance symptoms among individuals with fibromyalgia syndrome. *Sleep Medicine, 10,* 604–610.

Sand, T., Hagen, K., & Schrader, H. (2003). Sleep apnoea and chronic headache. *Cephalalgia, 23,* 90–95.

Scott, J. P. R., McNaughton, L. R., & Polman, R. C. J. (2006). Effects of sleep deprivation and exercise on cognitive, motor performance, and mood. *Physiology & Behavior, 87,* 396–408.

Shearer, W. T., Reuben, J. M., Mullington, J. M., Price, N. J., Lee, B. N., Smith, E. O., et al. (2001). Soluble TNFα receptor 1 and IL-6 plasma levels in humans subjected to the sleep deprivation model of spaceflight. *The Journal of Allergy and Clinical Immunology, 107,* 165–170.

Smith, M. T., Edwards, R. R., McCann, U. D., & Haythornthwaite, J. A. (2007). The effects of sleep deprivation on pain inhibition and spontaneous pain in women. *Sleep, 30,* 494–505.

Smith, M. T., Edwards, R. R., Stonerock, G. L., & McCann, U. D. (2005). Individual variation in rapid eye movement sleep is associated with pain perception in healthy women: Preliminary data. *Sleep, 28,* 809–812.

Sommer, C., & Kress, M. (2004). Recent findings on how proinflammatory cytokines cause pain: Peripheral mechanisms in inflammatory and neuropathic hyperalgesia. *Neuroscience Letters, 361,* 184–187.

Tiede, W., Mageri, W., Baumgartner, U., Durrer, B., Ehlert, U., & Treede, R. (2010). Sleep restriction attenuates amplitudes and attentional modulation of pain-evoked potentials, but augments pain rations in healthy volunteers. *Pain, 148,* 36–42.

van Tulder, M. W., Ostelo, R., Vlaeyen, J. W. S., Linton, S. J., Morley, S. J., & Assendelft, W. J. J. (2001). Behavioral treatment for chronic low back pain: A systematic review within the framework of the Cochrane Back Review Group. *Spine, 26,* 270–281.

Vgontzas, A. N., Pejovic, S., Zoumakis, E., Lin, H. M., Bixler, E. O., Basta, M., et al. (2007). Daytime napping after a night of sleep loss decreases sleepiness, improves performance, and causes beneficial changes in cortisol and interleukin-6 secretion. *American Journal of Physiology Endocrinology and Metabolism, 292,* E253–E261.

Vgontzas, A. N., Zoumakis, E., Bixler, E. O., Lin, H. M., Follett, H., Kales, A., et al. (2004). Adverse effects of modest sleep restriction on sleepiness, performance and inflammatory cytokines. *The Journal of Clinical Endocrinology and Metabolism, 89,* 2119–2126.

Vgontzas, A. N., Zoumakis, E., Lin, H. M., Bixler, E. O., Trakada, G., & Chroussos, G. P. (2004). Marked decrease in sleepiness in patients with sleep apnea by Etanercept, a tumor necrosis factor-α antagonist. *The Journal of Clinical Endocrinology and Metabolism, 89,* 4409–4413.

Vitiello, M. V., McCurry, S. M., Shortreed, S. M., Balderson, B. H., Baker, L. D., Keefe, F. J., et al. (2013). Cognitive-behavioral treatment for comorbid insomnia and osteoarthritis pain in primary care: The lifestyles randomized controlled trial. *Journal of the American Geriatrics Society, 61,* 947–956.

Walsh, J. K., Muehlbach, M. J., Lauter, S. A., Hilliker, A., & Schweitzer, P. K. (1996). Effects of triazolam on sleep, daytime sleepiness, and morning stiffness in patients with rheumatoid arthritis. *Journal of Rheumatology, 23,* 245–252.

Watkins, L. R., & Maier, S. F. (2005). Immune regulation of central nervous system functions: From sickness responses to pathological pain. *Journal of Internal Medicine, 257,* 139–155.

Watkins, L. R., Milligan, E. D., & Maier, S. F. (2003). Glial proinflammatory cytokines mediate exaggerated pain states: Implications for clinical pain. *Advances in Experimental Medicine and Biology, 521,* 1–21.

Ynus, M. B. (2007). Fibromyalgia and overlapping disorders: The unifying concept of central sensitivity syndromes. *Seminars in Arthritis and Rheumatism, 36,* 339–356.

Zaremba, S., Holle, D., Wessendorf, T. E., Diener, H. C., Katsarava, Z., & Obermann, M. (2012). Cluster headache shows no association with rapid eye movement sleep. *Cephalalgia, 32,* 289–296.

CHAPTER

18

The Impact of Sleep on Emotion in Typically Developing Children

Reut Gruber[*,†], *Soukaina Paquin*[†], *Jamie Cassoff*[†,‡], *and Merrill S. Wise*[§]

[*]Department of Psychiatry, McGill University, Montreal, Quebec, Canada
[†]Attention Behavior and Sleep Lab, Douglas Mental Health University Institute, Montreal, Quebec, Canada
[‡]Department of Psychology, McGill University, Montreal, Quebec, Canada
[§]Methodist Healthcare Sleep Disorders Center, Memphis, Tennessee, USA

Emotion plays a critical role in the lives of children, affecting their social development, personal and academic success, and overall adjustment. Emotional processes also shape children's developmental trajectories, preparing them for future lifetime opportunities and challenges. Although research has identified a myriad of factors relevant to healthy emotional development in children, it has largely ignored the role of sleep in this process. Recent studies have begun to shape our understanding of how sleep critically affects emotional functioning, however, and this chapter explores the impact of sleep on emotion and mood in typically developing children. To achieve this goal, we first identify key aspects of sleep and emotion, systematically examining the associations between them. Next, we review empirical evidence that supports this framework during infancy, toddlerhood, school age, and adolescence. We then propose developmental considerations related to the proposed framework, discuss the implications of this framework, and suggest future directions for this course of work. Due to the scarcity of literature pertaining to sleep regulation and affect in pediatric populations, we occasionally present information drawn from research with adults when the child equivalent is lacking.

ASSOCIATIONS AMONG KEY ASPECTS OF SLEEP REGULATION, EMOTION, AND MOOD

Sleep Regulation

Sleep is a highly complex behavior arising from the interaction of multiple neural circuits, neurotransmitters, and hormones. Two important aspects of sleep regulation, namely sleep duration (how much sleep) and sleep timing (when sleep occurs), are regulated by two distinct endogenous physiological processes (Borbely, 1982; Czeisler et al., 1986; Moore, 1999): the homeostatic and circadian processes.

The homeostatic process primarily regulates the length and depth of sleep, and it is related to the accumulation of the neurotransmitter adenosine during prolonged periods of wakefulness. Sleep pressure accumulates as the length of awake time increases and dissipates during a sleep episode. Such pressure depends on the quality and quantity of prior sleep and individual and developmental differences in sleep needs. Insufficient sleep and nonrestorative sleep lead to homeostatic dysregulation, or a sleep debt. The above-described homeostatic processes interact with *the circadian sleep process*, which governs predictable patterns of maximum sleepiness (circadian troughs) and maximum alertness (circadian peak) throughout the 24-h day (Allada, White, So, Hall, & Rosbash, 1998; Blau & Young, 1999; Ebadi & Govitrapong, 1986) and regulates sleep onset, awakening, and the duration of the daily sleep-wake cycle. Because the human circadian clock is slightly longer than 24h, intrinsic circadian rhythms must be synchronized or "entrained" to the 24-h cycle by environmental cues called *zeitgebers*. The major environmental zeitgeber ("time giver") is light (Borbely, 1982; Czeisler et al., 1986), but circadian rhythms are also synchronized by other external time cues related to social timing (e.g., school start time), family routines, and meal times. Together, these environmental inputs realign the internal clock (circadian pacemaker) with the light-dark cycle each day (Allada et al., 1998; Blau & Young, 1999; Ebadi & Govitrapong, 1986; Moore, 1999).

In addition, sleep-related neural activity in humans has been divided into basic states: waking, rapid eye movement (REM) sleep, and non-rapid eye movement (NREM) sleep. These states are interrelated but distinct in terms of their generating systems, regulation, behavioral features, and roles in maintaining the body's physiology and the brain's cognitive functions. In the waking state, the excitatory neurotransmitters (noradrenaline, serotonin, histamine, acetylcholine, and orexin) are released from their respective neurons in the brainstem, midbrain, and basal forebrain structures, whereas the release of inhibitory neurotransmitters (GABA [γ-aminobutyric acid] and galanin) from the ventrolateral preoptic nucleus (VLPO) is suppressed (Wulff, Gatti, Wettstein, & Foster, 2010).

During sleep, the NREM and REM states arise from differential patterns of neurotransmitter release. During NREM sleep, the aminergic neurotransmitters, cholinergic neurotransmitters, and orexin are all inhibited through the VLPO-mediated release of GABA and galanin, which decreases arousal. Once NREM sleep begins, thalamocortical rhythms that include θ oscillations, δ oscillations, and sleep spindles can be detected by electroencephalography (EEG). During REM sleep, the release of aminergic neurotransmitters of the brainstem is inhibited, but acetylcholine is released from neurons in the brainstem, midbrain, and basal forebrain. Orexin is also released, and the brainstem and VLPO release GABA and galanin to inhibit the aminergic brainstem neurons (Wulff et al., 2010).

Homeostatic Sleep Process and Emotional Processes

Emotion refers to any short-term evaluative, affective, intentional, psychological state, including happiness, sadness, disgust, and other inner feelings (Colman, 2009). Affect refers to subjectively experienced feeling, such as happiness, sadness, fear, or anger. We use the terms emotion and affect interchangeably. Theorists studying social and emotional development have viewed negative emotionality and emotion-related regulation as processes that are critical for healthy adjustment (Eisenberg et al., 2005). In the next section, we define these terms and examine the impact of the homeostatic sleep process on emotional reactivity and emotional regulation, based on empirical studies. To this end, we have conducted an online search of the medical, psychological, and academic literature to identify research articles on the relationship between sleep and affect in normally developing children. Studies with healthy children were included, but reports on children with medical conditions such as primary sleep disorders or psychiatric conditions were not evaluated.

Homeostatic Sleep Process and Emotional Reactivity

A child must be able to control extreme states of arousal or reactivity in order to participate in the mutual, reciprocal social interactions that are essential for healthy socioemotional development. In studies involving young children and infants, temperament measures (Bates, 1989; Goldsmith & Campos, 1982) often include items on negative emotionality, or the intensity of emotional responses and reactivity (Strelau, 1983; Thomas & Chess, 1977). In older children, the research has focused on the intensity of the experienced positive and negative emotions (the affective intensity). Developmentally, children with high levels of negative emotionality (e.g., anger) are more likely to develop both internalizing and externalizing problems in childhood and adolescence (Bates, Pettit,

Dodge, & Ridge, 1998; Eisenberg et al., 1995; Eisenberg et al., 2005; Kim & Deater-Deckard, 2011).

A variety of correlational and experimental studies provide empirical data on the association between emotional reactivity and dysregulation of the homeostatic sleep process. Correlational studies of emotional reactivity in infants, toddlers, school-age children, and adolescents associate shorter and poorer sleep with increased negative emotion expression. In infants, Shinohara and Kodama (2012) investigated the relationship between sleep quality and crying episodes in 4-16-week-old infants. Each participating mother recorded her observations of her child's crying behavior, and infant sleep was recorded using actigraphy. Results indicated that, for 14-16-week-old infants, the duration of active sleep was negatively correlated with the amount of time that the child cried in a 24-h period. Another study (Kelmanson, 2004) found that poor sleep quality in 8-week-old infants was associated with more negative emotions as reported by the mother. Toddlers who demonstrated persistent sleep problems over 4 years were more likely to demonstrate negative emotional expressions, such as aggression and anxious or depressed mood, as compared to children without persistent sleep problems. In another study evaluating the relationship between sleep and affect in toddlers, Ward, Gay, Alkon, Anders, and Lee (2008) found that children who had difficulty napping scored high in negative affect on a parental report temperament scale.

Several studies have examined the relationship between poor sleep and negative emotional expression and processing in school-aged children. Bruni et al. (2006) investigated the association between sleep and emotional reactivity in school-aged children. Analyses revealed that poor sleep quality was related to increased emotional reactivity and decreased ability to stay focused on the task at hand. O'Brien et al. (2011) reported that students who demonstrated snoring symptoms suggestive of sleep-disordered breathing were also more likely to have conduct problems such as aggression. Oginska and Pokorski (2006) examined the relationship between insufficient sleep and mood outcomes in adolescents. Results showed a significant relationship between unmet sleep needs and negative emotions. Collectively, these studies indicate that dysregulation of the homeostatic sleep process is associated with increased emotional reactivity in infants, toddlers, school-age children, and adolescents. Although these findings are important, the existing studies are correlative in nature, and the direction of the sleep-affect relationship cannot be determined. Several experimental studies have directly examined whether sleep duration directly impacts emotional reactivity, however. One study (Berger, Miller, Seifer, Cares, & Lebourgeois, 2012) experimentally restricted daytime naps in toddlers and assessed the impact of the restriction on their negative emotional expression. Ten toddlers participated in a sleep restriction protocol whereby they first followed a strict daytime nap schedule and then were assigned to the nap or no-nap conditions, which were counterbalanced over nonconsecutive study days. An emotion-elicitation protocol (faces with different emotions were presented to each child) and a challenge protocol (participants completed a solvable puzzle to elicit positive affect and an unsolvable puzzle to elicit negative affect) were administered in each child's home following both conditions. Children in the no-nap condition showed more negative displays in reaction to neutral and negative pictures. Furthermore, when completing the unsolvable puzzle task, children in the no-nap condition showed fewer positive displays than children in the nap condition. Thus, experimentally restricting daytime naps resulted in toddlers displaying enhanced negative affect in response to neutral and negative stimuli and reduced positive affect following mastery experiences.

Gruber, Cassoff, Frenette, Wiebe, and Carrier (2012) and Vriend et al. (2013) have experimentally restricted and extended sleep in school-aged children and evaluated the consequences on emotional reactivity. Gruber et al. (2012) implemented a protocol whereby participants tried to maintain their typical sleep schedules for 1 week, before being randomly assigned to either restrict (go to bed 1 h later than usual) or extend (go to bed 1 h earlier than usual) their sleep for 1 week. Teachers who were blind to the participants' conditions assessed the emotional lability and restless-impulsive behavior of each participating student. Emotional lability and restless impulsive scores improved from baseline in the sleep-extension group, and they deteriorated from baseline in the sleep-restriction group. Vriend et al. (2013) implemented a 3-week protocol in which participants were asked to sleep as they typically would during week 1 before being assigned in a counterbalanced fashion to either the sleep-extension condition (bedtime was 1 h earlier than usual) or the sleep-restriction condition (bedtime was 1 h later than usual) for weeks 2 and 3 of the study. The investigators measured affect through parent and child reports, as well as an affective response task during which the investigators presented the children with pictures and then rated how they felt following the viewing of each picture. Results indicated that the children showed less positive affect in the sleep-restricted condition versus the sleep-extended condition. McGlinchey et al. (2011) restricted the sleep of adolescents and adults for two nights following a week of typically scheduled sleep. On the first night of the sleep restriction protocol, participants had to restrict their sleep to 6.5 h. On the second night, participants slept in the lab and were only allowed the opportunity to sleep for 2 h between 3AM and 5AM. The effect of the sleep restriction on emotional reactivity was assessed after baseline and after each night of sleep restriction. Results indicated that, following sleep deprivation, adolescents displayed less positive and more negative emotion. Thus, these experimental studies show that reducing sleep duration results in increases in emotional reactivity and negative emotion among youth.

What are the underlying mechanisms? Although the above-described studies empirically demonstrated clear associations between dysregulation of the homeostatic sleep process and emotional reactivity, questions remain regarding the mechanisms underlying these associations. The limbic system is a brain system that plays an important role in the generation, integration, and control of emotions; it is tightly linked to the autonomic nervous system and regulates the responses of the hypothalamic-pituitary-adrenal (HPA) axis via the hypothalamus (Herman, Ostrander, Mueller, & Figueiredo, 2005). The amygdala, which is part of the limbic system, is connected to the prefrontal cortex (PFC), hippocampus, septum, and dorsomedial thalamus, and it plays an important role in emotional reactivity and emotional control (Braun, 2011). Using functional magnetic resonance imaging in adults, Yoo, Gujar, Hu, Jolesz, and Walker (2007) demonstrated that one night of sleep deprivation amplified the reactivity of the amygdala in response to viewing of negative emotional stimuli. These findings suggest that sleep deprivation (i.e., dysregulation of the homeostatic sleep process) amplifies emotional reactivity by exaggerating subcortical limbic and striatal reactivities to affective stimuli (Gujar, Yoo, Hu, & Walker, 2011; Yoo et al., 2007). No such studies have been published in children to date. Together, these results show that negative expressions of affect increase as the homeostatic sleep pressure increases across the day or across multiple nights of insufficient sleep.

Homeostatic Sleep Process and Emotional Regulation

Emotional regulation refers to the processes that children use to modify the type, intensity, duration, or expression of emotion, thereby fostering an optimal level of engagement with their environment (Cicchetti, Ganiban, & Barnett, 1991; Koole, 2009). Emotional regulation is critical to initiating, motivating, and organizing adaptive behavior, as well as preventing stressful levels of negative emotions and maladaptive behavior (Cicchetti, Ackerman, & Izard, 1995). An impaired ability to regulate negative emotionality in a context-appropriate fashion is viewed as a diathesis of psychopathology (Kovacs, Joormann, & Gotlib, 2008). The literature on infants and young children discusses the capacities involved in emotional regulation within the context of temperament. Temperamental self-regulatory capacities are often labeled as "effortful control" and include the abilities to shift and focus attention as needed (e.g., shifting from threatening stimuli/thoughts to neutral/positive ones), to inhibit inappropriate behavior (i.e., inhibitory control), and to activate or perform an action when there is a strong tendency to avoid it (i.e., activation control) (Evans & Rothbart, 2007; Rothbart, Chew, & Gartstein, 2001).

In older children, studies have investigated these same skills in connection with the development of executive functioning skills. Executive function refers to the cognitive processes of attention shifting, working memory, and inhibitory control, which are utilized in planning, problem solving, and goal-directed activity (Miyake et al., 2000). Thus, the development of the executive regulatory system plays a central role in a child's emotional regulation capacity.

Multiple studies have shown that sleep loss impairs performance on measures of those executive functions that form the basis for emotional regulation. The executive functions integrate the interrelated processes of emotion ("hot" executive functions) and cognition ("cold" executive functions) that are executed by the same underlying brain mechanisms, which contribute to the regulation of attention, emotion, and cognition (Davidson, Putnam, & Larson, 2000; Davis, Bruce, & Gunner, 2002). Hot and cold executive functions most likely have a reciprocal (two-way) relationship (Davidson et al., 2000; Davis et al., 2002), and both types of executive function are directly affected by sleep. Specifically, sleep deprivation fundamentally influences various cold executive functions, including abstract thinking, creativity, integration, planning, supervisory control (Nilsson et al., 2005), problem-solving, divergent thinking capacity (Horne, 1988; Linde & Bergstrom, 1992), working memory (Jens et al., 2005), flexibility (Alhola & Polo-Kantola, 2007), attention, vigilance (Durmer & Dinges, 2005), and cognitive set shifting (Wimmer, Hoffmann, Bonato, & Moffitt, 1992). Sleep loss also affects certain hot executive functions, including behavioral inhibition (Harrison & Horne, 1998), decision-making, and impulsivity (Killgore, Balkin, & Wesensten, 2006). These sleep-related changes may combine to impair the regulation of relevant behaviors and emotions.

What are the underlying mechanisms? The sensitivity of these processes to sleep deprivation is related to the neural areas governing emotional regulation and executive functions. Executive-functioning capacities include structures in the dorsolateral prefrontal, anterior cingulate, and parietal cortices that are particularly sensitive to sleep loss (Collette, Hogge, Salmon, & Van der Linden, 2006). Numerous studies have demonstrated the impacts of sleep on the brain circuits that underlie executive functions (Drummond et al., 1999; Fischer, Barkley, Smallish, & Fletcher, 2005; Harrison & Horne, 1998, 2000; Horne, 1988; Mesulam, 1990; Wu et al., 1991). For example, one study looked at how individuals performed on serial subtraction and verbal learning tasks related to cold executive functions, under conditions of normal wakefulness and sleep deprivation. This study revealed a deprivation-associated decrease in blood-oxygen-level-dependent signaling in the prefrontal anterior cingulate gyrus, the lateral posterior parietal lobules, the pulvinar thalamus, the temporal lobes, and the visual cortices (Mesulam, 1990). Furthermore, sleep deprivation appeared to decrease the capacity of individuals to regulate strong emotional arousal (hot executive functions) upon viewing intense negative (i.e., disgusting and disturbing) photographs (Yoo et al., 2007). Compared to the non-sleep-deprived control group, sleep-deprived individuals showed a significant loss of functional connectivity between the amygdala and the medial PFC. In the nondeprived group, the prefrontal lobe was strongly connected to the amygdala, regulating and exerting inhibitory top-down control. Conversely, in sleep-deprived participants, the mPFC connection decreased, potentially negating top-down control and resulting in hyperactivity of an amygdala region known to have strong inhibitory projections (i.e., modulatory impacts) on the amygdala (Sotres-Bayon, Bush, & LeDoux, 2004; Yoo et al., 2007). Therefore, a night of sleep may reset the brain to next-day emotional challenges by maintaining the functional integrity of the PFC-amygdala circuit, thus governing appropriate behavioral repertoires (e.g., optimal social judgments and rational decisions). Collectively, these findings suggest that sleep disruption impairs PFC function, resulting in so-called executive dysfunction, which can have direct and significant impacts on an individual's ability to regulate emotions.

Circadian Sleep Process and Mood

Mood refers to a temporary but relatively sustained and pervasive affective state (Colman, 2009). In healthy individuals, mood follows a circadian rhythm and varies with the time of day; it is generally low in the morning and best toward evening, and it declines with extended wakefulness (Wirz-Justice, 2008). This observation has led to the proposal that mood is partly regulated by circadian process. In addition, some have argued that mood instability arises when the circadian system receives conflicting timing cues from the endogenous clock and the environment. For example, individuals suffering from jet lag or performing shift-work often exhibit mood disruptions when their endogenous rhythms become out-of-phase with the external environment (Srinivasan et al., 2010). Researchers have also recognized an association between circadian rhythm disruption and mood disorders, but the potential mechanistic links remain poorly understood (Grandin, Alloy, & Abramson, 2006).

Very little literature examines the association between mood and circadian processes in typically developing children and adolescents. In a recent study, Owens, Belon, and Moss (2010) investigated whether delaying school start times by 30 min would entrain significant changes in the mood, behavior, and sleep of adolescents. In a sample of 201 adolescents ranging from grades 9 to 12, the authors found that delaying school start times from 8:00AM to 8:30AM resulted in significant increases in sleep duration, as well as reduced daytime sleepiness and depressed mood (Owens et al., 2010). These results were recently replicated in a sample of 197 boarding school students who underwent a 25-min school start-time delay (Boergers, Gable, & Owens, 2014). In this study, the delay in school start time was associated with increases in sleep duration, reduced daytime sleepiness, and reduced depressed mood.

CIRCADIAN TENDENCIES AND SOCIOEMOTIONAL ADJUSTMENT

Another aspect related to the interplay between affect and the circadian rhythm is the association observed between circadian tendencies and socioemotional adjustment. The concept of morningness or eveningness refers to self-described preferences for morning or evening activities (Mongrain, Lavoie, Selmaoui, Paquet, & Dumont, 2004; Smith, Reilly, & Midkiff, 1989), and this concept reflects individual differences in circadian rhythms (Chelminski, Ferraro, Petros, & Plaud, 1997; Horne & Ostberg, 1976). Differences in circadian orientation have been demonstrated in childhood (Kim, Dueker, Hasher, & Goldstein, 2002; Park, Matsummoto, Seo, & Shinkoda, 1999), even at preschool age (Werner, LeBourgeois, Geiger, & Jenni, 2009; Wickersham, 2006). From a very young age, biological, genetic, and social factors determine circadian orientation (Jenni & O'Connor, 2005; Roenneberg et al., 2007; Roenneberg, Wirz-Justice, & Merrow, 2003; Wada et al., 2009).

Eveningness has been associated with emotional dysregulation in both adults and adolescents (Jenni & O'Connor, 2005; Takeuchi et al., 2001). Furthermore, eveningness-oriented infants and children, from a few months to 8 years of age, reported "feeling angry" significantly more often than other children. Studies have also associated affective disorders and a circadian phase shift towards eveningness (Drennan, Klauber, Kripke, & Goyette, 1991; Gaspar-Barba et al., 2009; Hidalgo et al., 2009) Conversely, the evidence suggests that morningness could play a protective role, because it has been associated with better emotional regulation, decreased emotional reactivity, decreased risk for aggressive behavior (Gau et al., 2007; Goldstein, Hahn, Hasher, Wiprzycka, & Zelazo, 2007), as well as decreased delinquent behavior (Gau et al., 2007), decreased impulsivity (Adan et al., 2010), and increased persistence (Goldstein et al., 2007). In summary, sleep or circadian patterns appear to be tangibly linked to affective processes. Unfortunately, this concept has not yet been thoroughly examined in children. In addition, some of the more robust associations between circadian rhythm disruptions and affect have been observed in clinical presentations and reports pertaining to psychiatric disease, putting those findings outside the scope of this chapter.

NEURAL SLEEP REGULATION AND ITS ASSOCIATIONS WITH AFFECT

Based on empirical data from several studies conducted in adults, researchers have hypothesized that REM sleep represents a brain state that is particularly amenable to emotional memory consolidation (Hu, Stylos-Allan, & Walker, 2006; Paré, Collins, & Pelletier, 2002). Investigators have two conflicting views on the role and impact of REM sleep on emotional memory. Based on the marked reduction in aminergic neurochemistry and reactivated limbic anatomy during REM sleep (Hobson & Pace-Schott, 2002), one view suggests that REM sleep may offer an ideal neurobiological medium within which to depotentiate prior negative experiences and restore optimal postsleep affective reactivity (Levin & Nielsen, 2009; Walker, 2009a, 2009b; Walker & van Der Helm, 2009). This hypothesis supports the view that sleep (specifically REM neurophysiology) may critically govern the optimal homeostasis of emotional brain regulation. However, an alternative viewpoint suggests that REM sleep may facilitate the acquisition and recall of negative emotional memories, thereby playing a central role in producing the neurocognitive dysfunctions of mood disorders. According to this latter view, REM sleep physiology significantly contributes to the production of cognitive distortions associated with negative self-appraisal, the production of negative affect, and the selective consolidation of negative emotional memories (Hu et al., 2006; Rauchs et al., 2004; Walker & Stickgold, 2006). This hypothesis appears to be supported by studies in which REM-related indices (e.g., REM density) were strongly correlated with depression-related neurocognitive distortions, such as self-attacks, suicidal ideations, ruminations, and concentration difficulties (Agargun & Cartwright, 2003; Ellman, Spielman, Luck, Steiner, & Halperin, 1978; Giles, Roffwarg, Schlesser, & Rush, 1986). To date, however, no study has addressed specific REM-dependent emotional processing in children. In summary, we have thus far reviewed the sleep processes and physiologies that affect emotional processes central to socioemotional development and emotional memory. We now discuss developmental considerations related to this proposed framework, examining the framework's educational, research, and clinical implications and suggesting directions for future research.

DEVELOPMENTAL CONSIDERATIONS RELATED TO THE INTERPLAY BETWEEN SLEEP AND AFFECT

Developmental Changes Related to Sleep and Affect

Multiple systems are involved in the interplay between sleep and emotional processes, and each of these systems changes dramatically during development. For example, the capacity for emotional regulation and the mechanisms underlying such regulation change dramatically across infancy, toddlerhood, school age, and adolescence, with development continuing during young adulthood. In parallel, sleep processes and sleep physiology change dramatically throughout development. Researchers do not yet know if, how, and to what extent the developmental changes in sleep affect (or are potentially interrelated with) the developmental changes in emotional reactivity, emotional processes, and emotional regulation. Many open questions related to the way in which changes in sleep can affect emotional development. For example, the frequency and duration of REM sleep changes dramatically during development. In the first months of life, an infant's sleep is divided evenly (50:50) between NREM sleep and REM sleep. The proportion of REM sleep decreases throughout early childhood, eventually reaching the adolescent/adult level of about 20-25% of nocturnal sleep. When young infants fall asleep, the initial sleep episode typically consists of REM sleep; this is called a sleep-onset REM period. After 3 months of age, sleep-onset REM periods are replaced by sleep-onset NREM periods (the adult pattern; Anders, 2006). Given the role of REM sleep in emotional processing, we might ask: Are the above-mentioned patterns functional, with sleep-onset REM reflecting the need for significant emotional processing and thus allowing healthy development at the beginning of life? What does the age-related reduction in REM-sleep time say about the nature and extent of emotional processing during different developmental periods? How does having more (or less) REM sleep during the earliest stages of life affect emotional development and adjustment? Researchers might also explore if and how sleep changes affect different aspects of emotion or circadian preference changes, which then lead to a tendency toward eveningness in puberty. Despite the demonstrated associations between these sleep processes, mood, and adjustment, we lack information on the interplay among these systems during development. Does the extent of the change (delay) in the circadian phase during puberty lead to improved or worsened emotional regulation strategies? Is it associated with more or less intense moods?

Disentangling the relationships between these processes will provide critical insights into the role of sleep in emotional development. This will improve our understanding of normative development, provide insights into the emotional problems common to the multiple pathologies that involve both sleep disturbances and emotions (e.g., depression and anxiety), and offer a new understanding of how we can achieve success in treating the domains of sleep and emotional regulation. For example, if dysregulation of a sleep system is the primary source of emotional dysregulation among children. So, preventing this sleep dysregulation or deprivation should help the affected children improve their emotional regulation.

Overlapping Brain Mechanisms

Studies of adults have shown that overlapping brain mechanisms may be implicated in emotional regulation and sleep. In children, the differential maturation rates of the amygdala, septal nuclei, cingulate gyrus, and orbital frontal lobes seem to correspond to different phases of socioemotional development (Joseph, 1999). The maturational sequence of the prefrontolimbic circuits may also be connected to the ability to express and experience various emotions, including anger, joy, and fear of strangers (Joseph, 1999). All of these processes and brain circuits are known to be affected by sleep deprivation in adults. However, we do not yet have information regarding the impact of sleep changes on the development of these brain areas and related emotional domains in children. Another relevant issue is the timing of such an impact. There appear to be critical periods in the development of emotion and related brain mechanisms, when these systems are more susceptible (either positively or negatively) to influences (Soffer-Dudek, Sadeh, Dahl, & Rosenblat-Stein, 2011). The ongoing development and heightened plasticity of the neurocognitive system renders these developmental periods vulnerable to positive and negative influences. Thus, children might benefit more from sufficient restorative sleep and/or be more susceptible to the impact of insufficient or poor sleep during these periods. Future research should examine the impact of developmental sleep changes on the brain mechanisms underlying emotions. In particular, we should examine the impact of sleep on brain development during these critical developmental windows, with an eye toward stimulating important discussions with implications for developmental and clinical research.

Implications for Abnormal Development

In the field of child development, emotional regulation and its component skills are viewed as basic capacities that can foster typical, positive development or atypical development with pathological outcomes, depending on their manifestation and other social, dispositional, and biological resources available to the child. In the first part of this chapter, we present evidence supporting the significant impact of sleep on emotional reactivity, as well as the brain mechanisms and psychological processes that form the basis for emotional regulation. This information has important conceptual and practical implications for our understanding of normal development and suggests potential strategies that could be used to enhance positive development.

On the conceptual side, we propose that knowledge about how sleep impacts emotion-relevant regulation could significantly facilitate our understanding of adaptive and maladaptive developmental outcomes. Given that sleep could have a significant impact on self-regulation, knowledge of the sleep-affect relationship should be integrated into developmental models. Specifically, optimal arousal is needed to maintain the internal homeostasis that allows children to maintain self-regulation. Self-regulation is critical for a child's emerging emotional, behavioral, and social development, and it serves as an important marker of later psychosocial risk (Cole, Michel, & Teti, 1994; Lawson & Ruff, 2004). Children with dysregulated sleep may be vulnerable to regulatory difficulties in early childhood due to increased *emotional reactivity* or decreased *emotional regulation*.

Experimental studies conducted in toddlers, school-age children, and adolescents collectively show that children with dysregulated sleep are more excitable or irritable and have more difficulty regulating their arousal and negative affect, which in turn could challenge their ability to regulate emotions at home and in school. Conversely, the few studies involving sleep extension (either via experimental sleep extension or as the result of a delayed school-start time that allowed a longer sleep duration) show that extended sleep results in improvements in positive affect and mood and decreases in negative emotion intensity and emotional reactivity. In terms of emotional regulation, Vygotsky (1981) posits that higher-order mental skills reflect voluntary rather than involuntary (i.e., automatic) reactions. Such processes are effortful, particularly as children acquire these skills and work to develop them. Thus, when a child's sleep is dysregulated, he or she might lack the mental resources needed for the effort, resulting in more reactive or impulsive regulation. However, no known study has systematically examined the relationship between the pace or level of high-order mental skill development and a child's proneness to lability or negativity following sleep deprivation or sleep dysregulation. Such studies will aid our understanding of normal and abnormal development, because children can present unstable and immature patterns of development due to fatigue. Thus, future research should explore the conditions under which normal emotional development is compromised by insufficient sleep, as well as how emotional risk factors (e.g., emotional reactivity) can be exacerbated or buffered by restricted or improved sleep, making them likely to yield atypical or typical developmental trajectories. In addition, investigators should examine how these two emotional phenomena manifest in the same individual during early and middle childhood, under conditions of optimal sleep versus sleep dysregulation or restriction. This is especially crucial because fatigue-related emotional and social difficulties may persist across middle childhood and into adolescence (Lahey et al., 1995), and they could evolve into maladjustment in adolescence.

At the applied level, information regarding the impact of sleep on emotions could be used to enhance positive development and prevent negative developmental outcomes. Given the critical nature of the domains that are adversely affected by sleep restriction, the appropriate dissemination and use of information on sleep and optimal emotional development, along with tools for optimizing sleep, could have significant positive impacts on the emotional health and social adjustment of young people. Yet, despite the wealth of evidence demonstrating the critical importance of sleep to emotional processes, as well as the adverse impacts of sleep deprivation on emotions, children and families do not yet have easy access to this knowledge, and no current programs seek to use this knowledge for the benefit of typically developing children. The existence of problems that could be avoided through healthy sleep education and the current difficulties in addressing such problems represent a translation gap. This translation gap is critical because a considerable proportion of adolescents obtain less sleep than they need, making them chronically sleep-deprived (Carskadon, 2011). Furthermore, noted declines in sleep time and increasingly delayed bedtimes suggest the emergence of sleep restriction in preadolescents and even younger children. In recent decades, the weeknight sleep time for adolescents has gradually decreased by about 1 h, with weeknight bedtimes getting later and weekend sleep times increasing (Iglowstein, Jenni, Molinari, & Largo, 2003). Yet, society tends to underestimate the impact of chronic sleep insufficiency on the emotional health of children and adolescents. The multiple negative impacts of sleep deprivation emphasize the need to provide children and their parents with education on healthy sleep and tools that can assist them in achieving such sleep. We must therefore develop means for applying research findings on sleep and its positive impact on the emotional development of young persons. Of the initial approaches for addressing this problem, knowledge transfer should be a major educational priority, with efforts directed toward the educational system. One proposed approach is to develop, implement, and assess a school-based intervention collectively designed and implemented by sleep researchers, classroom teachers, students, and parents.

DIRECTIONALITY AND THE CONTRIBUTIONS OF ADDITIONAL FACTORS

Although we have focused on the aspects of sleep that affect emotions, we emphasize that this association is bidirectional (Cousins et al., 2011; Kahn, Sheppes, & Sadeh, 2013). Emotions affect sleep, which, in turn, may further affect emotions. For example, nighttime fears, which are commonly reported in typically developing children, can directly impact a child's sleep. Furthermore, major processes during child development can influence sleep behavior, including the development of cognitive processes, attachment, and autonomy. In addition to these bidirectional associations between sleep and emotions, other factors may affect both sleep and emotion, further shaping the interplay between these systems. For example, parenting practices, such as setting limits and creating consistent routines, shape the physical and emotional home environment, thus affecting sleep and emotions. In addition, cultural factors pertaining to emotional expression, emotional regulation, and sleep habit norms may have different effects on sleep, emotions, and their interactions. Therefore, more research is needed to further explore how the associations among sleep, emotion, behavior, and circumstances (e.g., lifestyle choices, socioeconomic status, health, employment, school) interact with intrinsic processes such as the homeostatic system (sleep pressure) and the circadian timing system, how these factors combine to shape emotions and sleep, and how this interplay changes with age.

SUMMARY

In summary, we propose a conceptual framework for associating circadian processes, homeostatic sleep processes, and REM sleep with emotional reactivity, emotional regulation, and emotional processing. The available empirical evidence strongly supports the association between homeostatic sleep dysregulation and emotional reactivity in children, as well as the improvement of negative emotions following restoration of the homeostatic sleep balance. Some lines of empirical evidence support an association between mood and circadian regulation in children. However, very few empirical studies have examined the role of REM sleep in the emotional information processing of children, and no data addresses the impact of developmental changes on these processes. Empirical evidence from studies conducted mostly with adults also support associations between cognitive (i.e., executive functions) and brain mechanisms associated with emotional regulation and sleep (e.g., the PFC). Given the empirical evidence, we assert that sleep regulation and processes should be integrated into conceptual models regarding normal and abnormal development. The multiple gaps that exist in the literature point to a great need for future research that uses both cross-sectional and longitudinal designs to examine the associations between developmental changes in key sleep and emotional processes. Finally, a significant effort is needed to translate the available empirical evidence into effective action by developing intervention and prevention strategies for optimizing sleep as a means for improving emotional wellness and adjustment and generally achieving positive developmental outcomes.

References

Adan, A., Natale, V., Caci, H., & Prat, G. (2010). Relationship between circadian typology and functional and dysfunctional impulsivity. *Chronobiology International, 27*(3), 606–619.

Agargun, M. Y., & Cartwright, R. (2003). REM sleep, dream variables and suicidality in depressed patients. *Psychiatry Research, 119*(1), 33–39.

Alhola, P., & Polo-Kantola, P. (2007). Sleep deprivation: Impact on cognitive performance. *Neuropsychiatric Disease and Treatment, 3*(5), 553–567.

Allada, R., White, N. E., So, W. V., Hall, J. C., & Rosbash, M. (1998). A mutant Drosophila homolog of mammalian Clock disrupts circadian rhythms and transcription of period and timeless. *Cell, 93*(5), 791–804.

Anders, T. F. (2006). Neurophysiological studies of sleep in infants and children. *Journal of Child Psychology and Psychiatry, 23*(1), 75–83. http://dx.doi.org/10.1111/j.1469-7610.1982.tb00051.x.

Bates, J. E. (1989). Applications of temperament concepts. In G. A. Kohnstamm, M. K. Rothbart, & J. E. Bates (Eds.), *Temperament in childhood* (pp. 321–355). New York: Wiley.

Bates, J. E., Pettit, G. S., Dodge, K. A., & Ridge, B. (1998). Interaction of temperamental resistance to control and restrictive parenting in the development of externalizing behavior. *Developmental Psychology, 34*(5), 982.

Berger, R. H., Miller, A. L., Seifer, R., Cares, S. R., & Lebourgeois, M. K. (2012). Acute sleep restriction effects on emotion responses in 30- to 36-month-old children. *Journal of Sleep Research, 21*(3), 235–246.

Blau, J., & Young, M. W. (1999). Cycling vrille expression is required for a functional Drosophila clock. *Cell, 99*(6), 661–671.

Boergers, J., Gable, C. J., & Owens, J. A. (2014). Later school start time is associated with improved sleep and daytime functioning in adolescents. *Journal of Developmental & Behavioral Pediatrics, 35*(1), 11–17. http://dx.doi.org/10.1097/DBP.0000000000000018.

Borbely, A. A. (1982). A two process model of sleep regulation. *Human Neurobiology, 1*, 195–204.

Braun, K. (2011). The prefrontal-limbic system: Development, neuroanatomy, function, and implications for socioemotional development. *Clinics in Perinatology, 38*(4), 685–702.

Bruni, O., Ferini-Strambi, L., Russo, P. M., Antignani, M., Innocenzi, M., Ottaviano, P., et al. (2006). Sleep disturbances and teacher ratings of school achievement and temperament in children. *Sleep Medicine, 7*(1), 43–48.

Carskadon, M. A. (2011). Sleep in adolescents: The perfect storm. *Pediatric Clinics of North America, 58*(3), 637.

Chelminski, I., Ferraro, F. R., Petros, T., & Plaud, J. J. (1997). Horne and Ostberg questionnaire: A score distribution in a large sample of young adults. *Personality and Individual Differences, 23*(4), 647–652. http://dx.doi.org/10.1016/S0191-8869(97)00073-1.

Cicchetti, D., Ackerman, B. P., & Izard, C. E. (1995). Emotions and emotion regulation in developmental psychopathology. *Development and Psychopathology, 7*(1), 1–10. http://dx.doi.org/10.1017/S0954579400006301.

Cicchetti, D., Ganiban, J., & Barnett, D. (1991). Contributions from the study of high-risk populations to understanding the development of emotion regulation. In J. Garber & K. A. Dodge (Eds.), *The development of emotion regulation and dysregulation*. New York: Cambridge University Press.

Cole, P. M., Michel, M. K., & Teti, L. O. D. (1994). The development of emotion regulation and dysregulation: A clinical perspective. *Monographs of the Society for Research in Child Development, 59*(2–3), 73–102.

Collette, F., Hogge, M., Salmon, E., & Van der Linden, M. (2006). Exploration of the neural substrates of executive functioning by functional neuroimaging. *Neuroscience, 139*(1), 209–221. http://dx.doi.org/10.1016/j.neuroscience.2005.05.035.

Colman, A. M. (2009). *A dictionary of psychology*. Oxford University Press.

Cousins, J. C., Whalen, D. J., Dahl, R. E., Forbes, E. E., Olino, T. M., Ryan, N. D., et al. (2011). The bidirectional association between daytime affect and nighttime sleep in youth with anxiety and depression. *Journal of Pediatric Psychology, 36*(9), 969–979.

Czeisler, C. A., Allan, J. S., Strogatz, S. H., Ronda, J. M., Sanchez, R., Rios, C. D., et al. (1986). Bright light resets the human circadian pacemaker independent of the timing of the sleep-wake cycle. *Science, 233*(4764), 667–671.

Davidson, R. J., Putnam, K. M., & Larson, C. L. (2000). Dysfunction in the neural circuitry of emotion regulation—A possible prelude to violence. *Science, 289*(5479), 591–594. http://dx.doi.org/10.1126/science.289.5479.591.

Davis, E. P., Bruce, J., & Gunner, M. R. (2002). The anterior attention network: Associations with temperament and neuroendocrine activity in 6-year-old children. *Developmental Psychobiology, 40*(1), 43–56. http://dx.doi.org/10.1002/dev.10012.

Drennan, M. D., Klauber, M. R., Kripke, D. F., & Goyette, L. M. (1991). The effects of depression and age on the Horne-Ostberg morningness-eveningness score. *Journal of Affective Disorders, 23*(2), 93–98.

Drummond, S., Brown, G., Stricker, J., Buxton, R., Wong, E., & Gillin, J. (1999). Sleep deprivation-induced reduction in cortical functional response to serial subtraction. *Neuroreport, 10*, 3745–3748.

Durmer, J. S., & Dinges, D. F. (2005). Neurocognitive consequences of sleep deprivation. *Seminars in Neurology, 25*(1), 117–129. http://dx.doi.org/10.1055/s-2005-867080.

Ebadi, M., & Govitrapong, P. (1986). Neural pathways and neurotransmitters affecting melatonin synthesis. *Journal of Neural Transmission, Supplement, 21*, 125–155.

Eisenberg, N., Fabes, R. A., Murphy, B., Maszk, P., Smith, M., & Karbon, M. (1995). The role of emotionality and regulation in children's social functioning: A longitudinal study. *Child Development, 66*(5), 1360–1384.

Eisenberg, N., Sadovsky, A., Spinrad, T. L., Fabes, R. A., Losoya, S. H., Valiente, C., et al. (2005). The relations of problem behavior status to children's negative emotionality, effortful control, and impulsivity: Concurrent relations and prediction of change. *Developmental Psychology, 41*(1), 193.

Eisenberg, N., Zhou, Q., Spinrad, T. L., Valiente, C., Fabes, R. A., & Liew, J. (2005). Relations among positive parenting, children's effortful control, and externalizing problems: A three-wave longitudinal study. *Child Development, 76*(5), 1055–1071.

Ellman, S., Spielman, A., Luck, D., Steiner, S., & Halperin, R. (1978). REM deprivation: A review. In *The mind in sleep* (pp. 419–458). Hillsdale, NJ: Lawrence Erlbaum.

Evans, D. E., & Rothbart, M. K. (2007). Developing a model for adult temperament. *Journal of Research in Personality, 41*(4), 868–888. http://dx.doi.org/10.1016/j.jrp.2006.11.002.

Fischer, M., Barkley, R., Smallish, L., & Fletcher, K. (2005). Executive functioning in hyperactive children as young adults: Attention, inhibition, response perseveration, and the impact of comorbidity. *Developmental Neuropsychology, 27*, 107–133.

Gaspar-Barba, E., Calati, R., Cruz-Fuentes, C. S., Ontiveros-Uribe, M. P., Natale, V., De Ronchi, D., et al. (2009). Depressive symptomatology is influenced by chronotypes. *Journal of Affective Disorders, 119*(1), 100–106.

Gau, S.S.-F., Shang, C.-Y., Merikangas, K. R., Chiu, Y.-N., Soong, W.-T., & Cheng, A.T.-A. (2007). Association between morningness-eveningness and behavioral/emotional problems among adolescents. *Journal of Biological Rhythms, 22*(3), 268–274.

Giles, D. E., Roffwarg, H. P., Schlesser, M. A., & Rush, A. J. (1986). Which endogenous depressive symptoms relate to REM latency reduction? *Biological Psychiatry, 21*(5), 473–482.

Goldsmith, H. H., & Campos, J. J. (1982). Toward a theory of infant temperament. In *The development of attachment and affiliative systems* (pp. 161–193). Springer.

Goldstein, D., Hahn, C. S., Hasher, L., Wiprzycka, U. J., & Zelazo, P. D. (2007). Time of day, intellectual performance, and behavioral problems in Morning versus Evening type adolescents: Is there a synchrony effect? *Personality and Individual Differences, 42*(3), 431–440.

Grandin, L. D., Alloy, L. B., & Abramson, L. Y. (2006). The social zeitgeber theory, circadian rhythms, and mood disorders: Review and evaluation. *Clinical Psychology Review*, 26(6), 679–694.

Gruber, R., Cassoff, J., Frenette, S., Wiebe, S., & Carrier, J. (2012). Impact of sleep extension and restriction on children's emotional lability and impulsivity. *Pediatrics*, 130(5), e1155–e1161.

Gujar, N., Yoo, S. S., Hu, P., & Walker, M. P. (2011). Sleep deprivation amplifies reactivity of brain reward networks, biasing the appraisal of positive emotional experiences. *Journal of Neuroscience*, 31(12), 4466–4474. http://dx.doi.org/10.1523/JNEUROSCI.3220-10.2011.

Harrison, Y., & Horne, J. A. (1998). Sleep loss impairs short and novel language tasks having a prefrontal focus. *Journal of Sleep Research*, 7(2), 95–100.

Harrison, Y., & Horne, J. (2000). The impact of sleep deprivation on decision making: A review. *Journal of Experimental Psychology*, 6, 236–249.

Herman, J. P., Ostrander, M. M., Mueller, N. K., & Figueiredo, H. (2005). Limbic system mechanisms of stress regulation: Hypothalamo-pituitary-adrenocortical axis. *Progress in Neuro-Psychopharmacology and Biological Psychiatry*, 29(8), 1201–1213.

Hidalgo, M. P., Caumo, W., Posser, M., Coccaro, S. B., Camozzato, A. L., & Chaves, M. L. F. (2009). Relationship between depressive mood and chronotype in healthy subjects. *Psychiatry and Clinical Neurosciences*, 63(3), 283–290.

Hobson, J. A., & Pace-Schott, E. F. (2002). The cognitive neuroscience of sleep: Neuronal systems, consciousness and learning. *Nature Reviews Neuroscience*, 3(9), 679–693.

Horne, J. A. (1988). Sleep loss and divergent thinking ability. *Sleep*, 11, 528–536.

Horne, J. A., & Ostberg, O. (1976). A self-assessment questionnaire to determine morningness-eveningness in human circadian rhythms. *International Journal of Chronobiology*, 4(2), 97–110.

Hu, P., Stylos-Allan, M., & Walker, M. P. (2006). Sleep facilitates consolidation of emotional declarative memory. *Psychological Science*, 17(10), 891–898.

Iglowstein, I., Jenni, O. G., Molinari, L., & Largo, R. H. (2003). Sleep duration from infancy to adolescence: Reference values and generational trends. *Pediatrics*, 111(2), 302–307.

Jenni, O. G., & O'Connor, B. B. (2005). Children's sleep: An interplay between culture and biology. *Pediatrics*, 115(Suppl. 1), 204–216.

Jens, P. N., Marie, S., Andreas, U. K., Mats, L., Torbjörn, P. N., Nina Erixon, L., et al. (2005). Less effective executive functioning after one night's sleep deprivation. *Journal of Sleep Research*, 14(1), 1–6.

Joseph, R. (1999). Environmental influences on neural plasticity, the limbic system, emotional development and attachment: A review. *Child Psychiatry and Human Development*, 29(3), 189–208.

Kahn, M., Sheppes, G., & Sadeh, A. (2013). Sleep and emotions: Bidirectional links and underlying mechanisms. *International Journal of Psychophysiology*, 89(2), 218–228.

Kelmanson, I. A. (2004). Temperament and sleep characteristics in two-month-old infants. *Sleep and Hypnosis*, 6, 78–84.

Killgore, W. D., Balkin, T. J., & Wesensten, N. J. (2006). Impaired decision making following 49 h of sleep deprivation. *Journal of Sleep Research*, 15(1), 7–13. http://dx.doi.org/10.1111/j.1365-2869.2006.00487.x, JSR487 [pii].

Kim, J., & Deater-Deckard, K. (2011). Dynamic changes in anger, externalizing and internalizing problems: Attention and regulation. *Journal of Child Psychology and Psychiatry*, 52(2), 156–166.

Kim, S., Dueker, G. L., Hasher, L., & Goldstein, D. (2002). Children's time of day preference: Age, gender and ethnic differences. *Personality and Individual Differences*, 33(7), 1083–1090.

Koole, S. L. (2009). The psychology of emotion regulation: An integrative review. *Cognition & Emotion*, 23(1), 4–41. http://dx.doi.org/10.1080/02699930802619031.

Kovacs, M., Joormann, J., & Gotlib, I. H. (2008). Emotion (dys)regulation and links to depressive disorders. *Child Development Perspectives*, 2(3), 149–155. http://dx.doi.org/10.1111/j.1750-8606.2008.00057.x.

Lahey, B. B., Loeber, R., Hart, E. L., Frick, P. J., Applegate, B., Zhang, Q., et al. (1995). Four-year longitudinal study of conduct disorder in boys: Patterns and predictors of persistence. *Journal of Abnormal Psychology, 104*(1), 83.

Lawson, K. R., & Ruff, H. A. (2004). Early attention and negative emotionality predict later cognitive and behavioural function. *International Journal of Behavioral Development, 28*(2), 157–165.

Levin, R., & Nielsen, T. (2009). Nightmares, bad dreams, and emotion dysregulation: A review and new neurocognitive model of dreaming. *Current Directions in Psychological Science, 18*(2), 84–88.

Linde, L., & Bergstrom, M. (1992). The effect of one night without sleep on problem-solving and immediate recall. *Psychological Research, 54*(2), 127–136.

McGlinchey, E. L., Talbot, L. S., Chang, K.-h., Kaplan, K. A., Dahl, R. E., & Harvey, A. G. (2011). The effect of sleep deprivation on vocal expression of emotion in adolescents and adults. *Sleep, 34*(9), 1233.

Mesulam, M. (1990). Large-scale neurocognitive networks and distributed processing for attention, language, and memory. *Annals of Neurology, 28*, 597–613.

Miyake, A., Friedman, N. P., Emerson, M. J., Witzki, A. H., Howerter, A., & Wager, T. D. (2000). The unity and diversity of executive functions and their contributions to complex "Frontal Lobe" tasks: A latent variable analysis. *Cognitive Psychology, 41*(1), 49–100. http://dx.doi.org/10.1006/cogp.1999.0734.

Mongrain, V., Lavoie, S., Selmaoui, B., Paquet, J., & Dumont, M. (2004). Phase relationships between sleep-wake cycle and underlying circadian rhythms in Morningness-Eveningness. *Journal of Biological Rhythms, 19*(3), 248–257. http://dx.doi.org/10.1177/0748730404264365.

Moore, R. Y. (1999). A clock for the ages. *Science, 284*(5423), 2102–2103.

Nilsson, J. P., Soderstrom, M., Karlsson, A. U., Lekander, M., Akerstedt, T., Lindroth, N. E., et al. (2005). Less effective executive functioning after one night's sleep deprivation. *Journal of Sleep Research, 14*(1), 1–6. http://dx.doi.org/10.1111/j.1365-2869.2005.00442.x.

O'Brien, L. M., Lucas, N. H., Felt, B. T., Hoban, T. F., Ruzicka, D. L., Jordan, R., et al. (2011). Aggressive behavior, bullying, snoring, and sleepiness in schoolchildren. *Sleep Medicine, 12*(7), 652–658.

Oginska, H., & Pokorski, J. (2006). Fatigue and mood correlates of sleep length in three age-social groups: School children, students, and employees. *Chronobiology International, 23*(6), 1317–1328.

Owens, J. A., Belon, K., & Moss, P. (2010). Impact of delaying school start time on adolescent sleep, mood, and behavior. *Archives of Pediatrics and Adolescent Medicine, 164*(7), 608–614. http://dx.doi.org/10.1001/archpediatrics.2010.96.

Paré, D., Collins, D. R., & Pelletier, J. G. (2002). Amygdala oscillations and the consolidation of emotional memories. *Trends in Cognitive Sciences, 6*(7), 306–314.

Park, Y., Matsummoto, K., Seo, Y., & Shinkoda, H. (1999). Sleep and chronotype for children in Japan. *Perceptual and Motor Skills, 88*(3c), 1315–1329.

Rauchs, G., Bertran, F., Guillery-Girard, B., Desgranges, B., Kerrouche, N., Denise, P., et al. (2004). Consolidation of strictly episodic memories mainly requires rapid eye movement sleep. *Sleep, 27*(3), 395–401.

Roenneberg, T., Kuehnle, T., Juda, M., Kantermann, T., Allebrandt, K., Gordijn, M., et al. (2007). Epidemiology of the human circadian clock. *Sleep Medicine Reviews, 11*(6), 429–438.

Roenneberg, T., Wirz-Justice, A., & Merrow, M. (2003). Life between clocks: Daily temporal patterns of human chronotypes. *Journal of Biological Rhythms, 18*(1), 80–90.

Rothbart, M. K., Chew, K., & Gartstein, M. A. (2001). Assessment of temperament in early development. In C. A. Nelson & M. Luciana (Eds.), *Biobehavioral assessment of the infant* (pp. 190–208). New York: Guilford.

Shinohara, H., & Kodama, H. (2012). Relationship between duration of crying/fussy behavior and actigraphic sleep measures in early infancy. *Early Human Development, 88*(11), 847–852.

Smith, C. S., Reilly, C., & Midkiff, K. (1989). Evaluation of three circadian rhythm questionnaires with suggestions for an improved measure of morningness. *Journal of Applied Psychology*, 74(5), 728–738. http://dx.doi.org/10.1037/0021-9010.74.5.728.

Soffer-Dudek, N., Sadeh, A., Dahl, R. E., & Rosenblat-Stein, S. (2011). Poor sleep quality predicts deficient emotion information processing over time in early adolescence. *Sleep*, 34(11), 1499.

Sotres-Bayon, F., Bush, D. E. A., & LeDoux, J. E. (2004). Emotional perseveration: An update on prefrontal-amygdala interactions in fear extinction. *Learning & Memory*, 11(5), 525–535. http://dx.doi.org/10.1101/lm.79504.

Srinivasan, V., Singh, J., Pandi-Perumal, S. R., Brown, G. M., Spence, D. W., & Cardinali, D. P. (2010). Jet lag, circadian rhythm sleep disturbances, and depression: The role of melatonin and its analogs. *Advances in Therapy*, 27(11), 796–813.

Strelau, J. (1983). *Temperament personality activity*. London: Academic Press.

Takeuchi, H., Inoue, M., Watanabe, N., Yamashita, Y., Hamada, M., Kadota, G., et al. (2001). Parental enforcement of bedtime during childhood modulates preference of Japanese junior high school students for eveningness chronotype. *Chronobiology International*, 18(5), 823–829.

Thomas, A., & Chess, S. (1977). *Temperament and development*. New York: Brunner/Mazel.

Vriend, J. L., Davidson, F. D., Corkum, P. V., Rusak, B., Chambers, C. T., & McLaughlin, E. N. (2013). Manipulating sleep duration alters emotional functioning and cognitive performance in children. *Journal of Pediatric Psychology*, 38(10), 1058–1069.

Vygotsky, L. S. (1981). The genesis of higher mental functions. In *The concept of activity in Soviet psychology* (pp. 144–188).

Wada, K., Krejci, M., Ohira, Y., Nakade, M., Takeuchi, H., & Harada, T. (2009). Comparative study on circadian typology and sleep habits of Japanese and Czech infants aged 0-8 years. *Sleep and Biological Rhythms*, 7(3), 218–221.

Walker, M. P. 2009a. REM, dreams and emotional brain homeostasis. *Frontiers in Neuroscience*, 3, 442–443.

Walker, M. P. 2009b. The role of sleep in cognition and emotion. *Annals of the New York Academy of Sciences*, 1156(1), 168–197.

Walker, M. P., & Stickgold, R. (2006). Sleep, memory, and plasticity. *Annual Review of Psychology*, 57, 139–166.

Walker, M. P., & van Der Helm, E. (2009). Overnight therapy? The role of sleep in emotional brain processing. *Psychological Bulletin*, 135(5), 731.

Ward, T. M., Gay, C., Alkon, A., Anders, T. F., & Lee, K. A. (2008). Nocturnal sleep and daytime nap behaviors in relation to salivary cortisol levels and temperament in preschool-age children attending child care. *Biological Research for Nursing*, 9(3), 244–253.

Werner, H., LeBourgeois, M. K., Geiger, A., & Jenni, O. G. (2009). Assessment of chronotype in four-to eleven-year-old children: Reliability and validity of the Children's Chronotype Questionnaire (CCTQ). *Chronobiology International*, 26(5), 992–1014.

Wickersham, L. (2006). Time-of-day preference for preschool-aged children. *Chrestomathy: Annual Review of Undergraduate Research*, 5, 259–268.

Wimmer, F., Hoffmann, R. F., Bonato, R. A., & Moffitt, A. R. (1992). The effects of sleep deprivation on divergent thinking and attention processes. *Journal of Sleep Research*, 1(4), 223–230.

Wirz-Justice, A. (2008). Diurnal variation of depressive symptoms. *Dialogues in Clinical Neuroscience*, 10(3), 337.

Wu, J., Gillin, J., Buchsbaum, M., Hershey, T., Hazlett, E., Sicotte, N., et al. (1991). The effect of sleep deprivation on cerebral glucose metabolic rate in normal humans assessed with positron emission tomography. *Sleep*, 14(2), 155–162.

Wulff, K., Gatti, S., Wettstein, J. G., & Foster, R. G. (2010). Sleep and circadian rhythm disruption in psychiatric and neurodegenerative disease. *Nature Reviews Neuroscience, 11*(8), 589–599.

Yoo, S. S., Gujar, N., Hu, P., Jolesz, F. A., & Walker, M. P. (2007). The human emotional brain without sleep—A prefrontal amygdala disconnect. *Current Biology, 17*(20), R877–R878. http://dx.doi.org/10.1016/j.cub.2007.08.007.

CHAPTER 19

Sleep and Adolescents

Eleanor L. McGlinchey

Columbia University Medical Center, New York State Psychiatric Institute, New York, New York, USA

SLEEP AND ADOLESCENTS

Adolescents may be the best population in which to study the interaction between sleep and affect. Adolescence is the gradual period of transition from childhood to adulthood, both in terms of physical maturation and the accompanying psychological and emotional changes (Spear, 2000a, 2000b). The epidemic of sleep deprivation among adolescents (Carskadon, Mindell, & Drake, 2006) means that every day provides a small experiment on the adverse consequences of sleep deprivation. Studies have clearly demonstrated the negative consequences of sleep deprivation on many behavioral, cognitive, emotional, and physical functions among adults (e.g., Dinges et al., 1997). However, for most adults, sleep behaviors and preferences originate in adolescence. Similarly, most affective disorders begin during the adolescent years (Kessler et al., 2005), and there appears to be a critical window during adolescence when the sleep-affect relationship can go particularly wrong. Even in the healthiest scenarios, many of the sleep habits that define adults are ingrained during adolescence. Indeed, genes appear to have little influence on sleep duration among adolescents (Barclay, Eley, Buysse, Archer, & Gregory, 2010), suggesting that a teen's sleep duration is more a mixture of normative biological changes and environmental influences. With this in mind, this chapter discusses normative changes in sleep during adolescence, conceptual models that guide current research on sleep and affect among adolescents, and the current state of research and future directions for research on sleep and affect in teens.

CHANGES IN NORMAL SLEEP

Every stage of development is accompanied by a change in sleep (Carskadon & Dement, 2011). For the adolescent period, this change is particularly pronounced. Studying adolescent sleep is often easier because they have not yet developed a bias toward desiring better sleep (a dream of most adults). However, given this lack of desire, adolescents are more vulnerable to the negative effects of sleep deprivation when their maladaptive behaviors set in and parents have less control over bedtime. Thus, the interaction between biological changes and behavioral changes in adolescent sleep become difficult to tease apart and even more difficult to address clinically.

In terms of biological changes, sleep conforms to the 90-minute cycle of adults by mid-adolescence with rapid eye movement (REM) sleep remaining constant. However, slow-wave sleep continues to decrease across adolescence (Carskadon & Dement, 2011). The two-process model of sleep becomes particularly important in adolescence. To summarize, the two-process model consists of two independent regulatory processes that interact to control the timing, intensity, and duration of sleep: a homeostatic sleep process and a circadian sleep process (Borbély, 1982). Process S is the first of these regulatory mechanisms, and it represents a homeostatic component of sleep that gradually increases over the wake period and decreases over the subsequent sleep period. Process C is the interacting process to Process S and represents the circadian component, arising from the endogenous pacemaker in the suprachiasmatic nuclei (SCN). At the molecular level, intrinsically rhythmic cells within the SCN regulate the sleep-wake cycle to a 24-hour period, along with the help of zeitgebers ("time givers" in German) such as sunlight and meal times. Notably, both of these processes are readily open to influence from the environment and do not solely depend on the biology of a person. Both processes are particularly sensitive to the daily alteration of light and dark, but the SCN is also responsive to nonphotic cues such as arousal or locomotor activity, social cues, feeding, sleep deprivation, and temperature (Mistlberger, Antle, Glass, & Miller, 2000). Hence, although biology clearly influences sleep in adolescence, behavioral changes related to the timing of exposure to light and regularizing activities, meal times, and other factors can form the basis of healthy versus maladaptive sleep habits.

Unfortunately, any discussion of "normal" sleep in adolescence involves the topic of sleep deprivation. Given sleep's established contributions to healthy functioning, it is of great public health concern that nearly half of America's teenagers report regular insufficient sleep and excessive daytime sleepiness—particularly during school days. Late-night bedtimes and early school start times contribute to what is increasingly regarded as an epidemic of sleep deprivation in adolescents. A large study of sleep habits ($n = 1602$ students in the seventh to twelfth grades) found that 45% of adolescents reported insufficient sleep on school nights and 28% complained they often felt "irritable and cranky" as a result of getting too little sleep (Carskadon et al., 2006).

At least four sets of normal maturational changes in adolescence create increased vulnerability to sleep loss and sleep problems. First, adolescence is associated with a marked increase in social and psychological influences on daytime schedules in ways that often lead to very late bedtimes. Key social factors include less parental control over bedtime, and increased social interactions, homework, sports, hobbies, part-time employment, and use of electronic media at night, including TV, movies, video games, internet, music, cell-phones, and text-messaging (Carskadon, 2005; Van den Bulck, 2004).

The second set of changes is the lightening of nighttime sleep (less slow-wave sleep than in childhood) with increased proneness to external disruptions (Carskadon & Acebo, 2002). Moreover, daytime sleepiness increases during puberty, probably reflecting an increased need for sleep during this period of rapid physical growth, cognitive development, and emotional changes. Research on the developing brain shows that sleep need increases during periods of brain maturation (Dahl & Lewin, 2002). The consensus appears to be that human adolescents need approximately 9 hours of sleep every night (Carskadon & Dement, 2011). However, starting around puberty, adolescents report high increases in daytime sleepiness, even when keeping their total time in bed each night constant (Carskadon, 1990). Unfortunately, even with this increased sleep need relative to childhood, adolescents are decreasing their time in bed. The average time spent in bed goes from 8.3 hours in the sixth grade to 6.9 hours in the twelfth grade (Carskadon et al., 2006). Reported bedtime is delayed and sleep duration decreases with increasing age.

The third set of maturational changes associated with adolescence is biological changes in the circadian system at puberty that shift sleep-timing preferences in the direction of a delayed sleep phase (Jenni, Achermann, & Carskadon, 2005; Wolfson & Carskadon, 1998). Unfortunately, these biological changes interact with early school start times, an increase in paid work responsibilities, and increased amounts of time devoted to socializing in a way that can spiral quickly into a pattern of extreme sleep deprivation (Carskadon & Acebo, 2002; Jenni et al., 2005). Most "catch-up" sleep occurs on weekends and holidays on an extremely phase-delayed schedule (Carskadon & Acebo, 2002; Dahl & Lewin, 2002). The human circadian system adapts easily to phase delays because endogenous rhythms of body temperature and neuroendocrine function are able to quickly reset to later bed and wake times. However, the circadian system has more difficulty accommodating phase advances (earlier sleep schedules). This creates an effect that is like "jet-lag" on Monday morning; students try to adjust to a sudden phase advance in which they need to get up for school at a time that is often at least 3-4 hours earlier than their biological clock "expects" to wake up (Wolfson & Carskadon, 2003).

Finally, the prefrontal cortex (PFC), shown to be impacted by sleep deprivation in adults (Muzur, Pace-Schott, & Hobson, 2002), continues to mature slowly throughout adolescence (Giedd et al., 1999). Indeed, the the PFC decreases in relative size and is remodeled during the adolescent period (Spear, 2000a, 2000b). These changes in the PFC raise the possibility that the adolescent brain may be dramatically underprepared for the challenge sleep deprivation may impose. This is particularly relevant when discussing behavior regulation and cognitive control functions (PFC) and the brain regions controlling emotional processing and behaviors associated with reward and punishment (ventral-medial areas of the PFC) that have bidirectional connections with the amygdala and the hypothalamus (Giedd et al., 1999; Quevedo, Benning, Gunnar, & Dahl, 2009; Silk et al., 2009; Steinberg, 2010). Importantly, these same prefrontal and interrelated limbic regions also appear to be the networks with the greatest sensitivity to insufficient sleep (Yoo, Gujar, Hu, Jolesz, & Walker, 2007).

In summary, large numbers of adolescents are struggling with the burdens of sleep deprivation in the context of social influences, increased sleep need, daytime sleepiness, a natural tendency toward a delayed sleep phase, repeated circadian shifts, and a slowly maturing PFC. Hence, this epidemic of sleep and circadian problems in adolescence can be best understood within a developmental perspective that considers interactions between social and biological changes at the onset of puberty. Taken together, this research supports the hypothesis that the added burden of sleep deprivation may render adolescents more vulnerable than adults toward dysregulation of their emotions.

CONCEPTUAL MODELS

Sleep for Restoration

Many guiding models for the human need for sleep conceptualize sleep as a restorative process, whereby energy is conserved and brain plasticity functions occur (Siegel, 2009; Tononi & Cirelli, 2006). Part of the homeostatic process of sleep involves metabolic, immune, thermoregulatory, cardiovascular, and respiratory recuperation. In addition, extensive animal and human research has shown that sleep plays a critical role in many cognitive and psychological processes, including learning and memory (Diekelmann & Born, 2010), creativity (Cai, Mednick, Harrison, Kanady, & Mednick, 2009), and emotion processing (van der Helm, Gujar, & Walker, 2010). Of particular importance, even in children as young as 5 years old, sleep influences stress and coping strategies (Mikoteit et al., 2012). Taken together, mounting evidence suggests the restorative effects of sleep for physical, cognitive, *and* emotional health (for more review on this topic, see Cirelli & Tononi, 2008).

During a developmental period when physical, cognitive, and emotional health should be at their peak, adolescents must have adequate time to reap the benefits of sleep in order to potentially avoid psychiatric and physical illness. Adolescence is characterized by hormonal changes associated with the onset of puberty (Pinyerd & Zipf, 2005). In addition, dramatic changes occur in brain structure and function, accompanied by subsequent behavioral changes (Casey, Getz, & Galvan, 2008; Spear, 2000a, 2000b). These behaviors tend to be the ones by which adolescents are defined, including the development of the social brain (Blakemore, 2010), emotional processing (Romeo, 2010), and higher-order executive control functions (Uhlhaas et al., 2009). These characteristic changes often manifest as poor decisions and actions; risk taking and sensation-seeking behaviors; emotional lability; and high reactivity in relation to social, work, and school demands. Interestingly, these hormonal, neuronal, behavioral, and psychological characteristics of adolescence coincide with the changes in the sleep-wake cycle described earlier, suggesting the critical role of sleep in the proper development of these changes (O'Brien & Mindell, 2005).

Given the crucial role of sleep in many restorative processes, combined with the knowledge that adolescence is a critical period for proper cognitive, physical, and emotional growth and development, this model is a prevailing guide for most research on sleep in adolescence across health domains.

Link Between Arousal and Sleep

Considering the evolutionary background of sleep, one is only able to sleep well in safe conditions. Even animals sleep only in caves or nests that are far from predators or other dangerous conditions. Sleep inhibition in the presence of danger has obvious advantages for survival, but it also comes at a cost. Given that sleep requires "turning off" awareness and responsiveness to the environment, vigilance is naturally opposed to sleep. Therefore, the states of arousal and anxiety are antithetical to sleep. For children and adolescents, nighttime can represent a potentially dangerous situation in which fear and other sources of cognitive or emotional arousal might prohibit a sense of safety and thus contribute to difficulty falling and staying asleep. Particularly for adolescents (as well as many adults), mentally reviewing or replaying stressful events of the day or ruminating on negative thoughts can become a bedtime habit with deleterious effects (Harvey, 2002). Many teens who tend to rehash recent stressors may have been able to distract themselves from the anxiety during the day, but in the absence of any distractors in bed, arousal related to the urge to rehash can further contribute to difficulty sleeping.

Sleep Difficulties—The Three-P Model

In most health clinics, the assessment of sleep difficulties has been heavily influenced by the three-P model (3P; Spielman, Caruso, & Glovinsky, 1987), also known as the three-factor or the Spielman model. This model considers three broad areas that may be contributing to trouble sleeping, utilizing diathesis-stress theories as a guide. The 3P model assumes that there exist predisposing, precipitating, and perpetuating factors that contribute to difficulty sleeping. Predisposing factors can take on many forms and are often the target of preventative efforts aimed at reducing sleep difficulties among adolescents. These include biological vulnerabilities such as changes in melatonin associated with puberty, psychological factors such as elevated trait anxiety, or social factors such as early school start times that are misaligned with circadian preference. Precipitating factors are the ones that provoke the initial difficulty sleeping and can include factors as difficult as a traumatic event or as small as the transition from the summer sleep schedule to the school schedule. The final part of the model involves perpetuating factors and includes the "bad habits" that may have initially been perceived as a coping mechanism for poor sleep, but eventually contribute to the sleeping problem becoming a nightly difficulty. Some examples that commonly occur in adolescents include napping in the afternoon after school (hence, having difficulty falling asleep at an appropriate time later that night), sleeping in excessively late on weekends (perpetuating a constant "jet lagged" feeling on Monday morning), or getting on the computer or smartphone when they can't fall asleep. Using this model, we see that the predisposing factors remain a constant vulnerability for difficulty sleeping, whereas the precipitating factors might give anyone a few nights of poor sleep, but their effect quickly decreases. The perpetuating factors are typically the ones targeted in treatment because they tend to sustain the sleep difficulty.

CURRENT STATE OF THE RESEARCH

Bidirectional Associations Between Sleep and Mood or Emotion

Most of the research to date on the link between sleep and affect has been conducted in adults. However, given the robust evidence found in adults, the evidence for a sleep-affect link in adolescents has seen rapid growth over the last several years. Experimental investigations of sleep deprivation in adolescents have begun to find increased vulnerability to emotion dysregulation in adolescents when compared to adults (Leotta, Carskadon, Acebo, Seifer, & Quinn, 1997). For example, Talbot and colleagues found increased worry and anxiety in young adolescents compared to adults when they were reduced to 2 hours of sleep versus when they had a full night of sleep (Talbot, McGlinchey, Kaplan, Dahl, & Harvey, 2010). Also, compared to adults, adolescents experienced blunted positive emotional responding (happiness, excitement) when in a sleep-deprived state relative to when they were rested (McGlinchey et al., 2011). Along the same lines, adverse affective consequences may be partially remediated as a result of good quality sleep. In a prospective study conducted on children at high risk for major depressive disorder, Silk et al. (2007) found that the odds of developing depression in adolescence and young adulthood were reduced as a function of increased percentage of slow-wave sleep during childhood. In population-wide studies, a similar pattern of results has emerged. In a large longitudinal study of adolescents tracked from sixth through eighth grade, sleep loss and decreases in sleep duration during middle school were associated with poor self-esteem and increased risk for depression (Fredriksen, Rhodes, Reddy, & Way, 2004).

Bidirectional Associations Among Sleep, Affect, and Behavior

The link between poor sleep and behavioral problems has emerged from multiple studies conducted in samples as young as toddlers and all the way up through adolescents (e.g., Bates, Viken, Alexander, Beyers, & Stockton, 2003; Sadeh, Gruber, & Raviv, 2002). Indeed, in a meta-analysis conducted by Astill, Van der Heijden, Van IJzendoorn, and Van Someren (2012), shorter sleep duration was associated with more internalizing and externalizing behavioral problems.

Among adolescents, the link between poor sleep and substance use problems has received particular attention. Substance abuse researchers have speculated that teens often seek out drugs and alcohol for their sedation and/or emotional regulatory effects or for their rewarding and/or stimulating effects. Sleep disturbance could also represent a direct effect of certain forms of substance use. For example, in a large Finnish sample, adolescents who abused substances were also more likely to have irregular sleep patterns and excessive daytime sleepiness (Tynjala, Kannas, & Levalahti, 1997). Similarly, in another large sample of adolescents who reported having trouble sleeping, the odds of abusing alcohol, marijuana, and cigarettes were over twice as likely (Johnson, Breslau, Roehrs, & Roth, 1999).

In considering one of the most serious and fatal behaviors associated with poor sleep, suicidal ideation and suicide completion have known links with insomnia and hypersomnia. In terms of suicidal ideation, insomnia, hypersomnia, and nightmares have all been independently associated with thoughts of ending ones life (Choquet, Kovess, & Poutignat, 1993; Choquet & Menke, 1989). Research on suicide attempts has found even more robust evidence for the link between suicide and sleep. For example, Gasquet and Choquet (1994) found that 81% of teens presenting to an emergency room following a suicide attempt reported difficulty falling asleep or early morning waking immediately preceding the attempt, additionally noting that insomnia predicts inpatient hospitalization for suicide attempts. Moreover, Liu and Buysse (2006) found that nightmares were associated with increased risk for suicide attempts, and adolescents who slept less than 8 hours were more than three times as likely to attempt suicide. Unfortunately, poor sleep has also been associated with suicide completion. Using a psychological autopsy study design in which parents, siblings, and friends were interviewed between 4 and 6 months after the suicide, Goldstein, Bridge, and Brent (2008) compared adolescent suicide completers with an age/sex-matched control group. Suicide completers had higher rates of overall sleep disturbance, insomnia, and hypersomnia within the week before completing suicide, as compared to the matched control group.

Bidirectional Associations Among Sleep, Affect, and Cognition (Including Learning and Memory Consolidation)

Most of the research on the relationship between cognitive functioning and sleep in adolescents has focused on the poor academic outcomes associated with lack of sleep (for meta-analysis see Dewald et al., 2010). This research has been supported by studies on specific cognitive functions related to experimental investigations of sleep deprivation or sleep enhancement in both children and adults. In particular, the studies on memory encoding relate to the subject of sleep and affect. Sleep-deprived adults exhibit a 40% reduction in the ability to form new memories under conditions of sleep deprivation (Walker & Stickgold, 2006). In terms of sleep enhancement, a daytime nap restores the normal deterioration in learning capacity that increases during the day (Mander, Santhanam, Saletin, & Walker, 2011). Along the same lines, sleep deprivation after learning prevents the consolidation of new memories (both emotional and nonemotional), and experimentally enhancing the quality of sleep by increasing slow-wave sleep enhances consolidation and hence long-term retention of (nonemotional) memories (Diekelmann & Born, 2010; Payne et al., 2009; Stickgold & Walker, 2005).

This compelling evidence drawn from adults has prompted research on developmental implications. Consistent with adult studies, an observational study of 135 healthy school children demonstrated clear adverse consequences of inadequate sleep on cognitive functioning

(Sadeh et al., 2002). In a follow-up study, Sadeh, Gruber, and Raviv (2003) compared cognitive functioning in children when they restricted their sleep by 1 hour or extended their sleep by 1 hour, and they found a similar pattern of poor cognitive functioning when sleep was restricted. In addition to acute sleep loss, chronic sleep loss may also predict long-term cognitive impairments. In one prospective study, parent-reported sleep problems during childhood significantly predicted neuropsychological functioning in early adolescence (Gregory, Caspi, Moffitt, & Poulton, 2009). In another meta-analysis of healthy school-age children (5-12 years old) across 86 studies and 35,936 children, sleep duration was clearly associated with cognitive performance, particularly executive functioning, on tasks that addressed multiple cognitive domains, as well as with school performance, although not with intelligence, sustained attention, or memory (Astill et al., 2012). This finding has obvious implications for education and functioning in adolescents. Executive functioning and learning are both functions necessary for healthy emotion regulation as well as for academic achievement.

Bidirectional Associations Among Sleep, Affect, and Health

Sleep deprivation increases appetite, weight gain, and insulin tolerance (Spiegel, Tasali, Penev, & Van Cauter, 2004). Indeed, in a meta-analysis involving 12 studies done on children and adolescents, the investigators observed an association between short sleep and obesity across the age range. Specifically, children who were obese had 60-80% increased odds of being a short sleeper (Cappuccio et al., 2008). Furthermore, experimentally induced acute sleep deprivation in normal weight college-age research participants has been associated with disturbed appetite hormones (grehlin, leptin) so that the participants also craved more food that was less healthy (Spiegel et al., 2004). A recent experimental study also found that going to bed 2 hours earlier for 2 weeks resulted in weight loss and decreased food intake in a sample of children (Hart et al., 2013). Such research points to shifting bedtimes earlier as a possible intervention in weight management.

Technology Use

Technology use in adolescence has increased at an unprecedented rate over the past decade, and this increase appears to have consequences for adolescent sleep (National Sleep Foundation, 2011). As discussed previously, the process by which the human circadian clock is set to a 24-hour period and kept in appropriate phase with seasonally shifting day length occurs via the daily alteration of *light* and *dark*. The light-entrainable brain system synchronizes networks of other subordinate circadian oscillators controlling fluctuations in other brain regions. In particular, it synchronizes the neural structure supporting reward seeking, which is centered on the ventral striatum (Venkatraman, Chuah, Huettel, & Chee, 2007), and the neural structure supporting emotion processing, which is centered on the amygdala and orbitomedial prefrontal cortex (Yoo et al., 2007), and these two structures are highly relevant to risk taking among teens (Blakemore, Burnett, & Dahl, 2010; Giedd et al., 1999). Hence, any intervention aimed at improving sleep incorporates timed light exposure and inevitably needs to address exposure to sources of light in the hours before bed via the use of cell phones, laptops, and televisions. The National Sleep Foundation (NSF) Poll focused on technology use in 2011. In the 13-19-year-old age group, well over 90% were using some kind of technology the hour before bed: 54% watched television, 72% used their cell phones, 60% used their laptops, 64% listened to their music maker, and 23% played video games every night or almost every night within the hour before going to sleep. These behaviors have multiple potential adverse impacts on sleep. The artificial light exposure suppresses the release of the sleepiness-promoting hormone melatonin, enhances alertness, and shifts circadian rhythms to a later hour, something already discussed as a naturally occurring problem in teens at the onset of puberty. Moreover, technology is also highly arousing and engaging, often facilitating much-loved contact with peers. These influences are so potent that many teens have great difficulty switching off media devices at night, delaying bedtimes still further (NSF, 2011). In addition to delaying sleep, cell phones are too often a source of sleep disturbance during the middle of the night. About one in ten of 13-19 year olds in the NSF poll reported being awakened after they went to bed every night or almost every night by a phone call, text message, or email. Clearly, these are modifiable sources of sleep disturbance, but they can be difficult to address in teenagers.

Clinical Studies of Disordered Sleep in Adolescents

As discussed previously, evidence suggests that the onset of puberty triggers a distinct evening preference in approximately 40% of teens (Carskadon, Acebo, Richardson, Tate, & Seifer, 1997; Roenneberg et al., 2004), which is then exacerbated by various social (e.g., importance of peers, parents less involved in decisions regarding bed and wake times), psychological (e.g., increased pressure at school) and behavioral factors (e.g., use of social media in bed and during the night). Together, these factors can often contribute to an increased tendency toward an evening preference or, in the extreme cases, delayed sleep phase syndrome (DSPS). DSPS is a circadian sleep disorder characterized by late bedtimes and arising times, early night insomnia, and poor morning alertness (Regestein

& Monk, 1995). Problems with delayed sleep in children and adolescents can be further compounded by large differences between weekday and weekend schedules, so-called social jetlag.

Sleep problems among children, as reported by parents, range from 20% to 30% (e.g., Anders & Eiben, 1997; Lozoff, Wolf, & Davis, 1985; Mindell, 1996; Mindell, Kuhn, Lewin, Meltzer, & Sadeh, 2006; Stores, 2006). Reports of sleep problems can differ between parents and older children in both directions. Most of the epidemiological work in this area draws from samples collected in the United States or Europe. Interestingly, one study compared children in grades 1-4 in the United States and in China. Chinese children went to bed approximately half an hour later (9:02 vs. 8:27 PM) and woke up half an hour earlier (6:28 vs. 6:55 AM), resulting in an average sleep duration that was 1 hour less (9.25 vs. 10.15 hours; Liu, Liu, Owens, & Kaplan, 2005). Cultural sleep differences represent an understudied area, particularly as these differences relate to adolescents.

After the onset of puberty and throughout the adolescent period, the most prevalent sleep complaints tend to center on difficulty waking up for school and associated difficulties with daytime sleepiness, tiredness, and irritability. However, a broad continuum of severity exists among adolescents. This continuum ranges from normative or mild difficulties that appear to affect up to half of all high school students in the US and result in a few related symptoms (National Sleep Foundation, 2006) to severe difficulties with sleep and sleep timing that result in significant impairments (such as failures in school) for a much smaller subset of adolescents. In the extreme cases, adolescents often meet full criteria for DSPS. Currently, there is a great need to generate empirical data to help delineate when to diagnose an adolescent as having DSPS disorder from the much larger set of youth with mild to moderate problems with erratic and late sleep schedules.

Aside from the most common sleep disorders for adolescents, sleep difficulties often occur among adolescents with psychiatric disorders. In adolescents diagnosed with generalized anxiety disorder, rates of sleep disturbance range from 42% to 66% (Kendall & Pimentel, 2003; Pina, Silverman, Fuentes, Kurtines, & Weems, 2003). In one study of clinically impaired anxious adolescents with generalized anxiety disorder, social phobia, and/or separation anxiety disorder, 88% reported having at least one sleep problem, and over half the sample reported experiencing more than three sleep-related difficulties, including insomnia, nightmares, refusal to sleep alone, excessive sleepiness, refusal to sleep anywhere but home, decreased sleep, and sleep walking or talking (Alfano, Ginsburg, & Kingery, 2007). This study also found that the sleep difficulties were associated with increased anxiety severity. Unfortunately, most of the research in the area of sleep and anxiety disorders in adolescents has produced post hoc or observational data. It remains to be studied how sleep problems are related to daily functioning in teens and whether sleep problems affect treatment outcomes.

The first episode of depression typically occurs during adolescence and is associated with multiple functional and health risks later in adulthood. Episodes that occur early in adolescence are also associated with a more severe course of depression, psychosocial difficulties, poor interpersonal relationships, low education, unemployment or underemployment, and risk for suicide. Large longitudinal population-wide studies have found evidence that sleep disturbances in adolescence increase the risk for depression later in life (Breslau, Roth, Rosenthal, & Andreski, 1996; Buysse et al., 2008). Depressed adolescents experience more insomnia relative to healthy adolescents, and their sleep architecture is also disturbed (as happens with depressed adults), with depressed adolescents experiencing short REM latency, increased REM, and less slow-wave sleep (Kutcher, Williamson, Marton, & Szalai, 1992; Rao & Poland, 2008). Unsurprisingly, these disruptions in sleep architecture are associated with increased cortisol and elevated hypothalamus-pituitary-adrenal (HPA) axis activity. Hence, a depressed adolescent tends to experience less physical, cognitive, and psychological benefits from his or her sleep, compared with a healthy adolescent. Future work in this area will most likely focus on intervention and prevention of adolescent depression by managing sleep problems.

DISCUSSION OF THE IMPACT OF SLEEP ON ADOLESCENT WELL-BEING AND PREVENTION

Clinical Implications

The research to date indicates that, for optimal well-being during adolescence and into young adulthood, the duration, architecture, and timing of sleep are critically important for proper physical, cognitive, and psychological functioning. As discussed throughout this chapter, the adverse consequences of poor sleep in adolescents include obesity, poor academic achievement, and development of serious psychiatric symptoms. Related to all of these outcomes, poor sleep in adolescence also appears to directly drive poor decision-making, emotion dysregulation, impulsivity, poor social functioning, and increased substance use. Although many of these factors are common features of adolescence, sleep difficulties seem to exacerbate the vulnerabilities already present. Moreover, given that sleep disruption can deteriorate synaptic plasticity, particularly during development (Jan et al., 2010), the long-term consequences of chronic poor sleep may be a mechanism by which psychiatric disorders develop in adolescents. The relationship is not as simple as one might initially think, however, and the relationship between disturbed sleep and poor health-related outcomes appears to be bidirectional, possibly presenting a vicious cycle of increased risk for morbidity. As an example, sleep disturbances and sleep deprivation among teens have been shown to lead to increased use of alcohol and other illicit drugs (Wong, Brower, Nigg, & Zucker, 2010). In parallel, the use of alcohol and drugs can lead to changes in sleep architecture such as reduction of slow-wave sleep and REM, which may perpetuate more difficulty sleeping (Shibley, Malcolm, & Veatch, 2008). Thus, in the example of a vicious cycle such as this one, the adolescent might benefit most from having both the substance use and sleep difficulties addressed in a single treatment. Similarly, in a large epidemiological study of adolescents, less sleep was associated with poor mental health status (Kaneita et al., 2009). In the same sample 2 years later, this association remained stable over time, suggesting that reduced sleep and poor mental health are involved in a vicious cycle of increased vulnerability to psychological symptoms.

Public Health Implications

In the past few years, the field has become increasingly aware of the consequences of poor sleep among adolescents; however, the epidemic of sleep deprivation among children and adolescents remains due to the lack of increased intervention or prevention efforts. Parent education may be one promising tool, given evidence that teaching new parents about normative sleep across the first year of life has helped reduce many common sleep problems among infants and young children (Mindell et al., 2006). Parents of school-age children and teenagers do not appear to have used this approach, however. One way that sleep problems in later childhood and adolescence could be prevented is by educating parents about developmental sleep needs, healthy sleep habits, setting clear limits around technology use at bedtime, and *modeling* healthy sleep to their children. At the same time, teachers, coaches, and health care providers could enhance prevention efforts by learning to recognize the common "red flags" of poor sleep among children and teens (e.g., falling asleep in class, snoring). One study has piloted a screening tool for assessing sleep problems in pediatric primary care clinics (Owens & Dalzell, 2005). A potential next step may be to introduce prevention programs in schools and primary care clinics for the purpose of encouraging healthy sleep and providing extra support for children and teens who show early signs of sleep disruption (e.g., excessive sleepiness in class). Encouraging regular physical exercise appears to be one very promising tool. For example, adolescent males who played football three times a week had better sleep efficiency, more slow-wave sleep, and less REM sleep relative to an age-matched control group, and this effect appeared to persist over time (Brand et al., 2010). This type of work has the potential to provide the most economical and time-efficient approach to the treatment of pediatric sleep problems.

Successful prevention efforts for pediatric sleep problems must also involve removing societal and environmental barriers to healthy sleep. One of the biggest barriers to healthy sleep among adolescents may be early school start time. Recent evidence indicates that a 30-minute delay in school start time (from 8:00 to 8:30 AM) is associated with significant improvements in measures of adolescent alertness, mood, and health (Owens, Belon, & Moss, 2010). Some counties have begun to implement changes in school start time in the United States, but the progress has been slow. The biggest factor in changing legislation on school start time has been the unification of communities of parents, teachers, and policy makers who recognize the detrimental effects of requiring teens to wake for school that starts before 8:00 AM. However, adults are not always convinced when lack of information and awareness prevails.

Future Directions for Research

Although researchers have made considerable progress toward understanding the functions of sleep and consequences of poor sleep in adolescence, several areas require future research. First, the risk factors associated with the development of sleep problems need to be better identified and defined. Critically, better identification could lead to even more powerful intervention and prevention efforts. Furthermore, the impact of potential moderating variables (e.g., child temperament, household socioeconomic status, etc.) on poor sleep and treatment outcomes needs to be examined. Related to treatment outcomes, more research is needed to understand the potential long-term impacts that behavioral interventions may have on the persistence of sleep problems into adulthood, as well as the long-term consequences of sleep disturbance in adolescence. Researchers must also evaluate behavioral treatments for sleep difficulties in children and adolescents with special needs, including those with autism spectrum disorders, mental retardation, neurodevelopmental disabilities, and chronic medical and psychiatric conditions.

References

Alfano, C. A., Ginsburg, G. S., & Kingery, J. N. (2007). Sleep-related problems among children and adolescents with anxiety disorders. *Journal of the American Academy of Child and Adolescent Psychiatry*, 46(2), 224–232. http://dx.doi.org/10.1097/01.chi.0000242233.06011.8e.

Anders, T. F., & Eiben, L. A. (1997). Pediatric sleep disorders: A review of the past 10 years. *Journal of the American Academy of Child & Adolescent Psychiatry*, 36(1), 9–20.

REFERENCES

Astill, R. G., Van der Heijden, K. B., Van IJzendoorn, M. H., & Van Someren, E. J. W. (2012). Sleep, cognition, and behavioral problems in school-age children: A century of research meta-analyzed. *Psychological Bulletin, 138,* 1109–1138.

Barclay, N. L., Eley, T. C., Buysse, D. J., Archer, S. N., & Gregory, A. M. (2010). Diurnal preference and sleep quality: Same genes? A study of young adult twins. *Chronobiology International, 27*(2), 278–296. http://dx.doi.org/10.3109/07420521003663801.

Bates, J. E., Viken, R. J., Alexander, D. B., Beyers, J., & Stockton, L. (2003). Sleep and adjustment in preschool children: Sleep diary reports by mothers relate to behavior reports by teachers. *Child Development, 73*(1), 62–75.

Blakemore, S. J. (2010). The developing social brain: Implications for education. *Neuron, 65*(6), 744–747. http://dx.doi.org/10.1016/j.neuron.2010.03.004.

Blakemore, S. J., Burnett, S., & Dahl, R. E. (2010). The role of puberty in the developing adolescent brain. *Human Brain Mapping, 31,* 926–933. http://dx.doi.org/10.1002/hbm.21052.

Borbély, A. A. (1982). A two process model of sleep regulation. *Human Neurobiology, 1*(3), 195–204.

Brand, S., Gerber, M., Beck, J., Hatzinger, M., Pühse, U., & Holsboer-Trachsler, E. (2010). High exercise levels are related to favorable sleep patterns and psychological functioning in adolescents: A comparison of athletes and controls. *Journal of Adolescent Health, 46*(2), 133–141.

Breslau, N., Roth, T., Rosenthal, L., & Andreski, P. (1996). Sleep disturbance and psychiatric disorders: A longitudinal epidemiological study of young adults. *Biological Psychiatry, 39*(6), 411–418.

Buysse, D. J., Angst, J., Gamma, A., Ajdacic, V., Eich, D., & Rössler, W. (2008). Prevalence, course, and comorbidity of insomnia and depression in young adults. *Sleep, 31*(4), 473.

Cai, D. J., Mednick, S. A., Harrison, E. M., Kanady, J. C., & Mednick, S. C. (2009). REM, not incubation, improves creativity by priming associative networks. *Proceedings of the National Academy of Sciences of the United States of America, 106*(25), 10130–10134.

Cappuccio, F. P., Taggart, F. M., Kandala, N. B., Currie, A., Peile, E., Stranges, S., et al. (2008). Meta-analysis of short sleep duration and obesity in children and adults. *Sleep, 31,* 619–626.

Carskadon, M. A. (1990). Patterns of sleep and sleepiness in adolescents. *Pediatrician, 17*(1), 5–12.

Carskadon, M. A. (2005). Sleep and circadian rhythms in children and adolescents: Relevance for athletic performance of young people. *Clinics in Sports Medicine, 24*(2), 319–328. http://dx.doi.org/10.1016/j.csm.2004.12.001.

Carskadon, M. A., & Acebo, C. (2002). Regulation of sleepiness in adolescents: Update, insights, and speculation. *Sleep, 25*(6), 606–614.

Carskadon, M. A., Acebo, C., Richardson, G. S., Tate, B. A., & Seifer, R. (1997). An approach to studying circadian rhythms of adolescent humans. *Journal of Biological Rhythms, 12,* 278–289.

Carskadon, M. A., & Dement, W. C. (2011). Normal human sleep: An overview. In M. H. Kryger, T. Roth, & W. C. Dement (Eds.), *5th principles and practice of sleep medicine* (pp. 16–26). Philadelphia: Elsevier.

Carskadon, M. A., Mindell, J. A., & Drake, D. (2006). *The National Sleep Foundation: Sleep in America poll*. Retrieved December 30, 2013, from, http://www.sleepfoundation.org/sites/default/files/2006_summary_of_findings.pdf.

Casey, B. J., Getz, S., & Galvan, A. (2008). The adolescent brain. *Developmental Review, 28*(1), 62–77. http://dx.doi.org/10.1016/j.dr.2007.08.003.

Choquet, M., Kovess, V., & Poutignat, N. (1993). Suicidal thoughts among adolescents: An intercultural approach. *Adolescence, 28,* 649–659.

Choquet, M., & Menke, H. (1989). Suicidal thoughts during early adolescence: Prevalence, associated troubles, and help-seeking behavior. *Acta Psychiatrica Scandinavia (Suppl.), 81,* 170–177.

Cirelli, C., & Tononi, G. (2008). Is sleep essential? *PLoS Biology, 6*(8), e216. http://dx.doi.org/10.1371/journal.pbio.0060216.

Dahl, R. E., & Lewin, D. S. (2002). Pathways to adolescent health sleep regulation and behavior. *Journal of Adolescent Health, 31*(6), 175–184.

Dewald, J. F., Meijer, A. M., Oort, F. J., Gerard, A., Kerkhof, G. A., & Bogels, S. M. (2010). The influence of sleep quality, sleep duration and sleepiness on school performance in children and adolescents: A meta-analytic review. *Sleep Medicine Reviews, 14*, 179–189.

Diekelmann, S., & Born, J. (2010). The memory function of sleep. *Nature Reviews Neuroscience, 11*, 114–126.

Dinges, D. F., Pack, F., Williams, K., Gillen, K. A., Powell, J. W., Ott, G. E., et al. (1997). Cumulative sleepiness, mood disturbance and psychomotor vigilance performance decrements during aweek of sleep restricted to 4-5 hours per night. *Sleep, 20*, 267–277.

Foundation, National Sleep. (2006). *Adolescent sleep needs and patterns: Research report and resource guide.* Washington, DC: National Sleep Foundation.

Foundation, National Sleep. (2011). *2011 sleep in America poll: Communications technology and sleep.* Washingtone, DC: National Sleep Foundation.

Fredriksen, K., Rhodes, J., Reddy, R., & Way, N. (2004). Sleepless in Chicago: Tracking the effects of adolescent sleep loss during the middle school years. *Child Development, 75*(1), 84–95.

Gasquet, I., & Choquet, M. (1994). Hospitalization in a pediatric ward of adolescent suicide attempters admitted to general hospitals. *Journal of Adolescent Health, 15*, 416–422.

Giedd, J. N., Blumenthal, J., Jeffries, N. O., Castellanos, F. X., Liu, H., Zijdenbos, A., et al. (1999). Brain development during childhood and adolescence: A longitudinal MRI study [letter]. *Nature Neuroscience, 2*, 861–863.

Goldstein, T. R., Bridge, J. A., & Brent, D. A. (2008). Sleep disturbance preceding completed suicide in adolescents. *Journal of Consulting and Clinical Psychology, 76*, 84–91.

Gregory, A. M., Caspi, A., Moffitt, T. E., & Poulton, R. (2009). Sleep problems in childhood predict neuropsychological functioning in adolescence. *Pediatrics, 123*, 1171.

Hart, C. N., Carskadon, M. A., Considine, R. V., Fava, J. L., Lawton, J., Raynor, H. A., et al. (2013). Changes in children's sleep duration on food intake, weight, and leptin. *Pediatrics, 132*(6), e1473–e1480. http://dx.doi.org/10.1542/peds.2013-1274.

Harvey, Allison G. (2002). A cognitive model of insomnia. *Behaviour Research and Therapy, 40*(8), 869–893.

Jan, J. E., Reiter, R. J., Bax, M. C. O., Ribary, U., Freeman, R. D., & Wasdell, M. B. (2010). Long-term sleep disturbances in children: A cause of neuronal loss. *European Journal of Paediatric Neurology, 14*(5), 380–390.

Jenni, O. G., Achermann, P., & Carskadon, M. A. (2005). Homeostatic sleep regulation in adolescents. *Sleep, 28*(11), 1446–1454.

Johnson, E. O., Breslau, N., Roehrs, T., & Roth, T. (1999). Insomnia in adolescence: Epidemiology and associate problems. *Sleep, 22*, s22.

Kaneita, Y., Yokoyama, E., Harano, S., Tamaki, T., Suzuki, H., Munezawa, T., et al. (2009). Associations between sleep disturbance and mental health status: A longitudinal study of Japanese junior high school students. *Sleep Medicine, 10*(7), 780–786.

Kendall, P. C., & Pimentel, S. S. (2003). On the physiological symptom constellation in youth with Generalized Anxiety Disorder (GAD). *Journal of Anxiety Disorders, 17*(2), 211–221.

Kessler, R. C., Berglund, P., Demler, O., Jin, R., Merikangas, K. R., & Walters, E. E. (2005). Lifetime prevalence and age-of-onset distributions of DSM-IV disorders in the National Comorbidity Survey Replication. *Archives of General Psychiatry, 62*(6), 593–602.

Kutcher, S., Williamson, P., Marton, P., & Szalai, J. (1992). REM latency in endogenously depressed adolescents. *The British Journal of Psychiatry, 161*(3), 399–402.

Leotta, C., Carskadon, M. A., Acebo, C., Seifer, R., & Quinn, B. (1997). Effects of acute sleep restriction on affective response in adolescents: Preliminary results. *Sleep Research, 26*, 201.

Liu, X., & Buysse, D. J. (2006). Sleep and youth suicidal behavior: A neglected field. *Current Opinion in Psychiatry, 19*, 288–293.

Liu, X., Liu, L., Owens, J. A., & Kaplan, D. L. (2005). Sleep patterns and sleep problems among schoolchildren in the United States and China. *Pediatrics, 115*(1), 241–249. http://dx.doi.org/10.1542/peds.2004-0815F.

Lozoff, B., Wolf, A. W., & Davis, N. S. (1985). Sleep problems seen in pediatric practice. *Pediatrics, 75*(3), 477–483.

Mander, B. A., Santhanam, S., Saletin, J. M., & Walker, M. P. (2011). Wake deterioration and sleep restoration of human learning. *Current Biology, 21*, R183–R184.

McGlinchey, E. L., Talbot, L. S., Chang, K. H., Kaplan, K. A., Dahl, R. E., & Harvey, A. G. (2011). The effect of sleep deprivation on vocal expression of emotion in adolescents and adults. *Sleep, 34*, 1233–1241. http://dx.doi.org/10.5665/SLEEP.1246.

Mikoteit, T., Brand, S., Beck, J., Perren, S., von Wyl, A., von Klitzing, K., et al. (2012). Visually detected NREM Stage 2 sleep spindles in kindergarten children are associated with stress challenge and coping strategies. *The World Journal of Biological Psychiatry, 13*(4), 259–268. http://dx.doi.org/10.3109/15622975.2011.562241.

Mindell, J. A. (1996). Treatment of child and adolescent sleep disorders. In R. Dahl (Ed.), *Child and adolescent psychiatric clinics of North America: Sleep disorder* (pp. 741–752). Philadelphia: W.B. Saunders.

Mindell, J. A., Kuhn, B., Lewin, D. S., Meltzer, L. J., & Sadeh, A. (2006). Behavioral treatment of bedtime problems and night wakings in infants and young children. *Sleep, 29*, 1263–1276.

Mistlberger, R. E., Antle, M. C., Glass, J. D., & Miller, J. D. (2000). Behavioral and serotonergic regulation of circadian rhythms. *Biological Rhythm Research, 31*, 240–283.

Muzur, A., Pace-Schott, E. F., & Hobson, J. A. (2002). The prefrontal cortex in sleep. *Trends in Cognitive Sciences, 6*(11), 475–481.

O'Brien, E. M., & Mindell, J. A. (2005). Sleep and risk-taking behavior in adolescents. *Behavioral Sleep Medicine, 3*(3), 113–133. http://dx.doi.org/10.1207/s15402010bsm0303_1.

Owens, J. A., Belon, K., & Moss, P. (2010). Impact of delaying school start time on adolescent sleep, mood, and behavior. *Archives of Pediatric and Adolescent Medicine, 164*, 608–614. http://dx.doi.org/10.1001/archpediatrics.2010.96.

Owens, J. A., & Dalzell, V. (2005). Use of the 'BEARS' sleep screening tool in a pediatric residents' continuity clinic: A pilot study. *Sleep Medicine, 6*, 63–69.

Payne, J. D., Schacter, D. L., Propper, R. E., Huang, L. W., Wamsley, E. J., Tucker, M. A., et al. (2009). The role of sleep in false memory formation. *Neurobiology of Learning and Memory, 92*, 327–334.

Pina, A. A., Silverman, W. K., Fuentes, R. M., Kurtines, W. M., & Weems, C. F. (2003). Exposure-based cognitive-behavioral treatment for phobic and anxiety disorders: Treatment effects and maintenance for Hispanic/Latino relative to European-American youths. *Journal of the American Academy of Child & Adolescent Psychiatry, 42*(10), 1179–1187.

Pinyerd, B., & Zipf, W. B. (2005). Puberty—Timing is everything!. *Journal of Pediatric Nursing, 20*(2), 75–82.

Quevedo, K. M., Benning, S. D., Gunnar, M. R., & Dahl, R. E. (2009). The onset of puberty: Effects on the psychophysiology of defensive and appetitive motivation. *Development and Psychopathology, 21*, 27–45. http://dx.doi.org/10.1017/S0954579409000030.

Rao, U., & Poland, R. E. (2008). Electroencephalographic sleep and hypothalamic–pituitary–adrenal changes from episode to recovery in depressed adolescents. *Journal of Child and Adolescent Psychopharmacology, 18*(6), 607–613.

Regestein, Q. R., & Monk, T. H. (1995). Delayed sleep phase syndrome: A review of its clinical aspects. *The American Journal of Psychiatry, 152*, 602–608.

Roenneberg, T., Kuehnle, T., Pramstaller, P. P., Ricken, J., Havel, M., Guth, A., et al. (2004). A marker for the end of adolescence. *Current Biology, 14*, R1038.

Romeo, R. D. (2010). Adolescence: A central event in shaping stress reactivity. *Developmental Psychobiology, 52*(3), 244–253.

Sadeh, A., Gruber, R., & Raviv, A. (2002). Sleep, neurobehavioral functioning, and behavior problems in school-age children. *Child Development, 73,* 405–417.

Sadeh, A., Gruber, R., & Raviv, A. (2003). The effects of sleep restriction and extension on school-age children: What a difference an hour makes. *Child Development, 74,* 444–455.

Shibley, H. L., Malcolm, R. J., & Veatch, L. M. (2008). Adolescents with insomnia and substance abuse: Consequences and comorbidities. *Journal of Psychiatric Practice, 14*(3), 146–153.

Siegel, J. M. (2009). Sleep viewed as a state of adaptive inactivity. *Nature Reviews Neuroscience, 10*(10), 747–753. http://dx.doi.org/10.1038/nrn2697.

Silk, J. S., Siegle, G. J., Whalen, D. J., Ostapenko, L. J., Ladouceur, C. D., & Dahl, R. E. (2009). Pubertal changes in emotional information processing: Pupillary, behavioral, and subjective evidence during emotional word identification. *Development and Psychopathology, 21,* 87–97.

Silk, J. S., Vanderbilt-Adriance, E., Shaw, D. S., Forbes, E. E., Whalen, D. J., Ryan, N. D., et al. (2007). Resilience among children and adolescents at risk for depression: Mediation and moderation across social and neurobiological contexts. *Development and Psychopathology, 19,* 841–865.

Spear, L. (2000a). Modeling adolescent development and alcohol use in animals. *Alcohol Research & Health, 24*(2), 115–123.

Spear, L. P. (2000b). The adolescent brain and age-related behavioral manifestations. *Neuroscience & Biobehavioral Reviews, 24*(4), 417–463.

Spiegel, K., Tasali, E., Penev, P., & Van Cauter, E. (2004). Brief communication: Sleep curtailment in healthy young men is associated with decreased leptin levels, elevated ghrelin levels, and increased hunger and appetite. *Annals of Internal Medicine, 141,* 846–850.

Spielman, A. J., Caruso, L. S., & Glovinsky, P. B. (1987). A behavioral perspective on insomnia treatment. *Psychiatric Clinics of North America, 10,* 541–553.

Steinberg, L. (2010). A behavioral scientist looks at the science of adolescent brain development. *Brain and Cognition, 72,* 160–164. http://dx.doi.org/10.1016/j.bandc.2009.11.003.

Stickgold, R., & Walker, M. P. (2005). Memory consolidation and reconsolidation: What is the role of sleep? *Trends in Neuroscience, 28,* 408–415.

Stores, G. (2006). Practitioner review: Assessment and treatment of sleep disorders in children and adolescents. *Journal of Child Psychology and Psychiatry, 37*(8), 907–925.

Talbot, L. S., McGlinchey, E. L., Kaplan, K. A., Dahl, R. E., & Harvey, A. G. (2010). Sleep deprivation in adolescents and adults: Changes in affect. *Emotion, 10*(6), 831–841. http://dx.doi.org/10.1037/a0020138.

Tononi, G., & Cirelli, C. (2006). Sleep function and synaptic homeostasis. *Sleep Medicine Reviews, 10*(1), 49–62. http://dx.doi.org/10.1016/j.smrv.2005.05.002.

Tynjala, J., Kannas, L., & Levalahti, E. (1997). Perceived tiredness among adolescents and its association with sleep habits and use of psychoactive substances. *Journal of Sleep Research, 6,* 189–198.

Uhlhaas, P. J., Roux, F., Singer, W., Haenschel, C., Sireteanu, R., & Rodriguez, E. (2009). The development of neural synchrony reflects late maturation and restructuring of functional networks in humans. *Proceedings of the National Academy of Sciences of the United States of America, 106*(24), 9866–9871. http://dx.doi.org/10.1073/pnas.0900390106.

Van den Bulck, J. (2004). Television viewing, computer game playing, and Internet use and self-reported time to bed and time out of bed in secondary-school children. *Sleep, 27*(1), 101–104.

van der Helm, E., Gujar, N., & Walker, M. P. (2010). Sleep deprivation impairs the accurate recognition of human emotions. *Sleep, 33*(3), 335.

Venkatraman, V., Chuah, Y. M., Huettel, S. A., & Chee, M. W. (2007). Sleep deprivation elevates expectation of gains and attenuates response to losses following risky decisions. *Sleep, 30,* 603–609.

Walker, M. P., & Stickgold, R. (2006). Sleep, memory, and plasticity. *Annual Review of Psychology, 57,* 139–166.

Wolfson, A. R., & Carskadon, M. A. (1998). Sleep schedules and daytime functioning in adolescents. *Child Development, 69,* 875–887.

Wolfson, A. R., & Carskadon, M. A. (2003). Understanding adolescents' sleep patterns and school performance: A critical appraisal. *Sleep Medicine Reviews, 7*(6), 491–506.

Wong, M. M., Brower, K. J., Nigg, J. T., & Zucker, R. A. (2010). Childhood sleep problems, response inhibition, and alcohol and drug-related problems in adolescence and young adulthood. *Alcoholism, Clinical and Experimental Research, 34,* 1033–1044.

Yoo, S. S., Gujar, N., Hu, P., Jolesz, F. A., & Walker, M. P. (2007). The human emotional brain without sleep–A prefrontal amygdala disconnect. *Current Biology, 17,* R877–R878. http://dx.doi.org/10.1016/j.cub.2007.08.007.

CHAPTER
20

The Relationship Between Sleep and Emotion Among the Elderly

Pascal Hot[*], Isabella Zsoldos[*], and Julie Carrier[†,‡,§]*

[*]Université de Savoie, Laboratoire de Psychologie et Neurocognition (CNRS UMR-5105), Chambéry Cedex, France
[†]Functional Neuroimaging Unit, University of Montreal Geriatric Institute, Montreal, Quebec, Canada
[‡]Center for Advanced Research in Sleep Medicine (CARSM), Hôpital du Sacré-Cœur de Montréal, Montréal, Quebec, Canada
[§]Department of Psychology, University of Montreal, Montreal, Quebec, Canada

Aging is undoubtedly associated with an important increase in sleep-wake cycle complaints. Compared to young adults, the elderly (65 years and older) experience sleep that is shorter, shallower, and more fragmented (Landolt & Borbely, 2001). These age-related changes in sleep may impact functional vigilance and have cognitive and health consequences for the elderly. In parallel with sleep, our subjective experience and expression of emotions also change across the life span. Overall, findings suggest that emotional processing is different in younger and older individuals. Given that research has demonstrated that sleep is a key factor in emotion regulation among young adults, one may ask whether age-related sleep modifications are associated with emotion regulation in the elderly. In the first part of this chapter, we discuss research findings that demonstrate both preserved and impaired emotional processing during aging, and we describe the theories explaining age-related features related to well-being, emotion recognition, and emotion regulation. In the second part of this chapter, we present sleep modifications that appear with increasing age. Finally, we discuss available data on the link between sleep and emotion regulation in the elderly, and we propose future lines of research.

EMOTION REGULATION IN AGING: THE "PARADOX OF WELL-BEING"

Well-being reflects the global affective state of an individual, which is typically defined by the ratio of high positive affect to low negative affect. The subjective well-being of healthy older adults (HOA) has been relatively well documented over the past few years. HOA largely maintain well-being until the eighth decade (Barrick, Hutchinson, & Deckers, 1989; Blanchard-Fields, 1998; Gross, Carstensen, Pasupathi, Tsai, Götestam Skorpen, & Hsu, 1997; Kunzmann, Little, & Smith, 2000; Lawton, Kleban, Rajagopal, & Dean, 1992; Mroczek & Kolarz, 1998; Palmore & Cleveland, 1976; Stacey & Gatz, 1991). Indeed, research has suggested that well-being is often greater among the elderly in comparison with younger adults (Carstensen et al., 2011; Mroczek & Kolarz, 1998; Stone, Schwartz, Broderick, & Deaton, 2010). This phenomenon is referred to as the "paradox of well-being" (Baltes & Baltes, 1990; Brandtstädter & Greve, 1994; Diener & Suh, 1998) because emotional processing appears to be preserved, despite the age-related changes in the brain substrates associated with emotion, as well as the onset of health problems typically associated with lower well-being. Along with well-being, positive affect seems to remain fairly stable or to increase until very old age (Barrick et al., 1989; Biss & Hasher, 2012; Carstensen, Pasupathi, Mayr, & Nesselroade, 2000; Carstensen et al., 2011; Charles, Reynolds, & Gatz, 2001; Kessler & Staudinger, 2009; Mroczek & Kolarz, 1998; Stone et al., 2010; Teachman, Siedlecki, & Magee, 2007), when a slight decline is sometimes detected (after 80 years of age; Carstensen, Fung, & Charles, 2003; Diener & Suh, 1998; Stacey & Gatz, 1991). Researchers have also shown that people who are confronted with more functional health limitations tend to be more likely to show decreases in positive affect over the following years (Kunzmann et al., 2000). This suggests that health problems frequently associated with aging could be the cause of the declining positive affect encountered in people at very old ages (between 80 and 100 years old), rather than age *per se*. Evidence also indicates that the elderly engage in a bias in emotional processing known as the positivity bias. This refers to a bias in information processing that appears with increasing age and reflects the fact that older adults primarily focus on positive information (Charles, Mather & Carstensen, 2003; Mather & Carstensen, 2003; Werheid, Gruno, Kathmann, Fischer, Almkvist, & Winblad, 2010). The existence of this positivity bias could contribute to increased well-being by making elderly individuals focus on positive elements that can contribute to improved mood.

In parallel, negative affect declines across the life span from the late teens to approximately the mid-sixties (Barrick et al., 1989; Biss & Hasher, 2012; Carstensen et al., 2000; Charles et al., 2001; Diener & Suh, 1998; Small, Hertzog, Hultsch, & Dixon, 2003; Srivastava, John, Gosling, & Potter, 2003;

Stacey & Gatz, 1991; Teachman et al., 2007; Viken, Rose, Kaprio, & Koskenvuo, 1994), after which age negative affect remains relatively stable (Carstensen et al., 2000). Higher negative affect is sometimes reported by "very old" individuals (80 years old and more), as compared to "young old" individuals (between 60 and 80 years old; Diener & Suh, 1998; Gatz, Johansson, Pedersen, Berg, & Reynolds, 1993; Kessler, Foster, Webster, & House, 1992; Smith & Baltes, 1993; Teachman, 2006). Studies typically define negative affect based on anxiety and depressive symptoms, and findings about negative affect in old ages seem contradictory. Indeed, some studies report more frequent depression and anxiety states in HOA compared to younger adults (Eagles & Whalley, 1985; Stenback, 1980), but other studies present more frequent depression and anxiety states in younger age groups compared to HOA (Berkman et al., 1986; Clark, Aneshensel, Frerichs, & Morgan, 1981; Frerichs, Aneshensel, & Clark, 1981; Hertzog, Van Alstine, Usala, Hultsch, & Dixon, 1990; Himmelfarb, 1984; Newmann, 1989; Regier et al., 1988; Weissman, Leaf, Bruce, & Florio, 1988). Researchers have suggested that this discrepancy could be explained by the fact that typical affective states are associated with certain phases of later life. Specifically, when the older adults are split into different age groups, young adults (20-39 years) endorse the highest level of negative affect, those in middle age (40-54 years) and late middle age (55-69 years) report relatively less negative affect, and depression symptoms tend to increase after age 70 years (Gatz & Hurwicz, 1990). Other studies tend to confirm this nonlinear relationship between age and negative affect across the lifespan (Gatz et al., 1993; Kessler et al., 1992; Teachman, 2006).

The dominant theoretical model for explaining emotional profiles of HOA has been developed by Carstensen and colleagues (Carstensen, 1995, 2006; Carstensen et al., 2003; Carstensen, Isaacowitz, & Charles, 1999), and it is referred to as the socioemotional selectivity theory. As they age, older adults seem to perceive their time as more limited, and thus, they are more motivated to direct their attention to emotionally meaningful goals by focusing on the present and prioritizing emotionally gratifying experiences. According to this theory, young adults might be willing to endure unpleasant emotions in order to achieve long-term goals, but older adults might be less willing to tolerate negative emotions for the sake of future achievement, because their future is becoming limited (Carstensen, 1995; Lang & Carstensen, 2002; Tamir, 2009).

Moreover, lowering expectations and focusing on areas that are more manageable with increasing age make sense given that personal resources diminish progressively. According to Higgins (1999), the size of the discrepancy between a reference value (personal goals, aspirations in life) and the individual's current situation results in an increase in affect. Therefore, subjective well-being could be maintained in older age by regulating the extent of these discrepancies, especially in areas important to self.

In agreement with this hypothesis and with the socioemotional selectivity theory, Cheng (2004) showed that subjective well-being is maintained in old age through the decrease of the discrepancy between the person's current state and his or her aspiration for some goals that have larger effects than others (e.g., maintaining the quality of relationships), hence making positive aging possible by concentrating on these areas. On the contrary, goal discrepancies related to health seem to increase (Cheng, 2004).

To maintain high levels of positive affect, adults learn to regulate their emotions and become more skilled at regulating emotions across the life span, as evidenced by several studies showing that HOA report having greater control over their emotions (Blanchard-Fields, Mienaltowski, & Seay, 2007; Carstensen et al., 2000; Gross et al., 1997; Lawton et al., 1992) and experiencing less negative emotions than their younger counterparts (Blanchard-Fields & Coats, 2008; Carstensen et al., 2000; Charles et al., 2001; Gross et al., 1997). The tendency to generate (upregulate) positive affect and reduce (downregulate) negative affect once it is elicited is considered to be an efficient strategy of affect regulation in response to stressors (Kuhl, 2000; Kuhl & Beckmann, 1994). Consistent with claims that HOA are more effective at regulating emotions, findings indicate that intentional downregulation of negative emotions may be less costly for older adults than it is for their younger counterparts for some negative emotions such as disgust (Scheibe & Blanchard-Fields, 2009). For example, the long-term experience and practice of emotion regulation could render emotion-regulatory processes less effortful for HOA. Along with the theories related to emotion regulation, research has shown that older adults are equally or even more effective than younger adults at modulating facial expressions or inner experience of emotions (Kunzmann, Kupperbusch, & Levenson, 2005; Magai, Consedine, Krivoshekova, Kudajie-Gyamfi, & McPherson, 2006; Phillips, Henry, Hosie, & Milne, 2008). They also report higher levels of success in controlling external signs of emotions and maintaining a neutral state (Gross et al., 1997; Labouvie-Vief, DeVoe, & Bulka, 1989; Lawton et al., 1992). This can explain why older adults become more difficult to decode by others, compared to younger adults (Malatesta, Izard, Culver, & Nicholich, 1987). Furthermore, inhibiting emotions results in diminished emotional reactions in older, but not in young and middle-aged adults (Magai et al., 2006).

DECLINE IN EMOTIONAL PROCESSING WITH AGING

Our increasing comprehension of the processes that generate and maintain emotion yield a more nuanced picture of the dynamic changes in emotion abilities occurring with aging. Indeed, specific age-related differences in emotional processing exist. In particular, older adults have difficulties recognizing negative emotional facial expressions (EFEs), such as those related to fear, anger, and sadness, compared to their younger counterparts (Calder et al., 2003; Isaacowitz et al., 2007; MacPherson, Phillips, & Della Sala, 2006; Phillips, Maclean, & Allen, 2002; Sullivan & Ruffman, 2004; Sullivan, Ruffman, & Hutton, 2007; Suzuki, Hoshino, Shigemasu, & Kawamura, 2007). They also report less interest and excitement, with reduced intensity of subjective emotional experience (Lawton et al., 1992) and diminished physiological responsiveness to emotion (Gavazzeni, Wiens, & Fischer, 2008; Levenson, Carstensen, Friesen, & Ekman, 1991).

Some of these age-related differences may be linked to early cerebral changes and a decline in the cognitive processes sustaining emotions. Studies have reported changes in the limbic system, especially in the hippocampus and the amygdala. With age, the amygdala's volume decreases linearly (Allen, Bruss, Brown, & Damasio, 2005), and functional connectivity between the amygdala and posterior structures becomes impaired (St Jacques, Dolcos, & Cabeza, 2010). The amygdala's activity has been repeatedly associated with the processing of facial expressions of fear (see meta-analyses of Phan, Wager, Taylor, and Liberzon (2002) or Vytal and Hamann (2010)), and more generally, it appears involved in the processing of fear itself (expression, perception, interpretation, subjective experience; Tranel, Gullickson, Koch, & Adolphs, 2006; Vytal & Hamann (2010). The amygdala is also activated for other negative facial expressions, such as those linked to sadness and anger (Fischer, Sandblom, Gavazzeni, Fransson, Wright, & Bäckman, 2005; Larson, Schaefer, Siegle, Jackson, Anderle, & Davidson, 2006; Morris et al., 1998; Wang, McCarthy, Song, & Labar, 2005). Research has documented diminished activity of the amygdala during processing of EFE among the elderly (Gunning-Dixon et al., 2003). Damages inside and around the amygdala could explain the difficulties encountered by HOA in identifying fear, anger, and sadness.

The efforts of Gross (1998) to identify emotion regulation processes have also helped identify specific strategies that may be differentially affected by aging (Reuter-Lorenz et al., 2000; Shiota & Levenson, 2009). Most research has shown that HOA use deliberate conscious strategies to avoid negative events (Phillips et al., 2008). For example, Isaacowitz, Toner, Goren, and Wilson (2008) found that HOA focused their attention on positive elements during a negative emotional induction, which is contrary to the behaviors observed among young adults. Cumulative evidence suggests that this avoidance strategy is developed during the course of aging to compensate for the decline of cognitive function involved in emotion regulation. The early executive impairment in the elderly (Craik & Salthouse, 2008; Mayr, Spieler, & Kliegl, 2001) supports this hypothesis. Numerous studies have demonstrated that cognitive control of emotion involves interactions between regions of the prefrontal cortex that implement control processes (in particular executive functions) and subcortical regions that encode and represent emotion (Gyurak, Goodkind, Kramer, Miller, & Levenson, 2012; Gyurak, Goodkind, Madan, Kramer, Miller, & Levenson, 2009; Miller & Cohen, 2001; Ochsner, Bunge, Gross, & Gabrieli, 2002; Ochsner & Gross, 2005). These frontal areas are some of the first regions affected by brain aging (Bherer, Belleville, & Hudon, 2004; Grieve, Clark, Williams, Peduto, & Gordon, 2005; Lamar & Resnick, 2004; Rajah & D'Esposito, 2005; Raz, 2000; Raz & Rodrigue, 2006; Salthouse, Atkinson, & Berish, 2003; Tisserand et al., 2002).

In summary, studies converge to show that well-being is preserved and even improved with aging (Barrick et al., 1989; Blanchard-Fields, 1998; Carstensen et al., 2011; Gross et al., 1997; Kunzmann et al., 2000; Lawton et al., 1992; Mroczek & Kolarz, 1998; Palmore & Cleveland, 1976; Stacey & Gatz, 1991; Stone et al., 2010). However, the early atrophy of brain structures involved in emotion regulation suggests that HOA have to deal with significant changes in their abilities to manage their emotions in order to preserve this well-being. Research has suggested that older adults are not better at regulating emotions, but instead, they try to avoid unpleasant situations (Birditt & Fingerman, 2005; Coats & Blanchard-Fields, 2008). The tendency to select positive information and avoid negative information (the positivity bias) has been repeatedly observed, and it could affect several stages of emotional processing such as attentional and memory processing (Mather & Carstensen, 2005). When avoidance is not possible, a secondary strategy is to disengage from the situation and redirect attention toward nonemotional information (Isaacowitz et al., 2008; Isaacowitz, Wadlinger, Goren, & Wilson, 2006; Opitz, Gross, & Urry, 2012; Phillips et al., 2008). These findings agree with the recent theory suggesting that aging is accompanied by both strength and vulnerability in emotion regulation (Charles, 2010). Although HOA would have greater difficulty performing regulation strategies associated with high cognitive cost, attention redeployment would remain preserved.

AGE-RELATED SLEEP MODIFICATIONS

The commonality of sleep problems among older adults is now well established (Ancoli-Israel, 2005; Ancoli-Israel, Ayalon, & Salzman, 2008). According to a survey by the American National Sleep Foundation, 44% of elderly adults (65 years and older) complain about their sleep at least a few nights per week (Foley, Ancoli-Israel, Britz, & Walsh, 2004). Age-related decrements in self-reported sleep quality are observed in Western, Eastern, and African countries (Luo et al., 2013). Changes in sleep occur as early as the middle years of life (40 years and older), with over one-fourth of the population in their forties to sixties reporting sleep difficulties (Phillips & Mannino, 2005; Polo-Kantola et al., 2014). Multiple factors, including medical problems, side effects of medications, and an increase in specific sleep disorders (obstructive sleep apnea syndrome, period leg movement syndrome), account for this age-related increase in sleep difficulties. Notable modifications of the sleep-wake cycle are also observed in optimal aging (i.e., in people who do not suffer from medical, psychiatric, or specific sleep disorders). These age-related changes occur quite early and may have important repercussions for older individuals, especially when the sleep-wake system faces challenges such as those related to stress and anxiety, jet lag, and shift-work.

Aging is associated with earlier bedtime and wake time, less time asleep, more frequent awakenings of longer duration, shallower sleep, and increased rate of napping, indicating an increase in daytime sleepiness (Buysse, Browman, Monk, Reynolds, Fasiczka, & Kupfer, 1992; Carrier, Land, Buysse, Kupfer, & Monk, 2001; Carrier, Monk, Buysse, & Kupfer, 1997; Hoch et al., 1994; Landolt & Borbely, 2001; Landolt, Dijk, Achermann, & Borbely, 1996). Non-rapid-eye movement (NREM) sleep is characterized by different degrees of cortical neural synchronization, from lower synchronization in lighter sleep stages (N1 and N2) to higher synchronization in deeper stages (SWS). NREM sleep changes drastically with aging, with a substantial reduction in slow-wave sleep (SWS), and an increase in lighter NREM sleep stages (Carrier et al., 1997, 2001; Gaudreau, Carrier, & Montplaisir, 2001; Landolt & Borbely, 2001). Recent results have demonstrated that spontaneous awakenings in older adults are mainly related to a reduction in the consolidation of NREM sleep (Dijk, Duffy, & Czeisler, 2001; Klerman, Wang, Duffy, Dijk, Czeisler, & Kronauer, 2013; Salzarulo et al., 1999). During human REM sleep, the electroencephalogram (EEG) is characterized by low-frequency, high-amplitude waves (slow waves; SW <4 Hz and >75 µV) and sleep spindles (12-15 Hz). SWS is characterized at the cellular level by a hyperpolarization phase (surface EEG SW negative phase), during which cortical neurons are silent (OFF period), and a depolarization phase (EEG SW positive phase), during which cortical neurons fire intensively (ON period) (Csercsa et al., 2010; Steriade, 2006). Spindles originate from the cyclic inhibition of thalamocortical neurons. Between inhibition phases, thalamocortical neurons show rebound firing, which entrains cortical populations in spindle oscillations (Steriade, 2006). NREM-sleep oscillations between hyperpolarized and depolarized phases are believed to play a crucial role in sleep protection, brain plasticity, and memory (Steriade, 2006).

Studies report considerable changes in NREM-sleep synchronization with aging. Compared to the young, older subjects show lower SWA (spectral power between 0.5 and 4.5 Hz) and sigma (13-14 Hz) activity during NREM sleep (Carrier et al., 2001; Darchia, Campbell, Tan, & Feinberg, 2007; Landolt et al., 1996; Robillard, Massicotte-Marquez, Kawinska, Paquet, Frenette, & Carrier, 2010). Results also indicate lower spindle density, amplitude, and duration with aging (Crowley, Trinder, Kim, Carrington, & Colrain, 2002; Martin et al., 2013; Nicolas, Petit, Rompre, & Montplaisir, 2001), as well as lower SW density and amplitude (Carrier et al., 2011). In addition, studies suggest that, in young subjects, cortical neurons synchronously enter the SW hyperpolarization and depolarization phases, whereas this process takes longer in older subjects, leading to lower slope and longer SW positive and negative phases. Age-related effects on spindle or SW density and amplitude are more prominent in anterior brain areas (Carrier et al., 2011; Martin et al., 2013).

Compared to NREM sleep, the effects of aging on REM sleep are more controversial because studies have produced mixed findings. Some studies have reported a reduction in REM-sleep latency, less REM sleep during the night, and more REM sleep in the first part of the sleep episode, but other studies find no age-related changes in these variables (Carrier et al., 1997; Feinberg, 1974; Reynolds, Hoch, Buysse, Monk, Houck, & Kupfer, 1993). Two meta-analyses suggest a small but significant decrease of REM sleep across adulthood (Floyd, Janisse, Jenuwine, & Ager, 2007; Ohayon, Carskadon, Guilleminault, & Vitiello, 2004).

Finally, older adults not only show sleep changes under habitual conditions, but they also appear to be more sensitive to challenges to the sleep-wake cycle (e.g., stress, caffeine, recovery sleep at an abnormal time of day as might occur with jet lag and shift-work; Carrier, Monk, Buysse, & Kupfer, 1996; Carrier et al., 2009; Dijk et al., 2001; Moline et al., 1992; Monk, Buysse, Reynolds, & Kupfer, 1995; Vgontzas et al., 2003). For instance, some authors have proposed that the increased prevalence of insomnia in older populations may be caused by a lower sleep tolerance for the arousal-producing effects of stress (Vgontzas, Bixler, Lin, et al., 2001). To our knowledge, only one study has tested this hypothesis pharmacologically by comparing the effects of ovine CRH on the sleep of young and middle-aged subjects (Vgontzas, Bixler, Wittman, 2001). Compared to the young, middle-aged men showed a higher increase in wakefulness and more suppression of SWS than the young adults, despite similar elevations of ACTH and cortisol. The authors concluded that older adults may be at higher risk of developing insomnia when faced with equivalent stressors.

ARE AGE-RELATED SLEEP MODIFICATIONS LINKED TO EMOTION REGULATION IN AGING?

A collection of studies using young subjects systematically assessed the link between emotion regulation and sleep (Cartwright, Young, Mercer, & Bears, 1998; Lara-Carrasco, Nielsen, Solomonova, Levrier, & Popova, 2009; Pace-Schott, Milad, Orr, Rauch, Stickgold, & Pitman, 2009; Pace-Schott, Nave, Morgan, & Spencer, 2012; Pace-Schott et al., 2011;

van der Helm, Yao, Dutt, Rao, Saletin, & Walker, 2011; Walker & van der Helm, 2009). Neuroimaging findings demonstrate that sleep deprivation impairs emotion regulation by degrading the top-down inhibitory control of the medial prefrontal cortex on the amygdala (Rosales-Lagarde, Armony, Del Rio-Portilla, Trejo-Martinez, Conde, & Corsi-Cabrera, 2012; Yoo, Gujar, Hu, Jolesz, & Walker, 2007). To our knowledge, no studies have investigated the hypothesis that sleep-dependent emotion regulation may differ with aging.

When considering the paradox of well-being, age-related decreases in sleep intensity, quality, and continuity do not seem to negatively impact global affective states. Thus, emotion regulation may be less sensitive to sleep loss as we get older. Age-related sleep reduction and fragmentation may underlie a reduced need for sleep to regulate specific functions such as emotions (i.e., with aging, less sleep would be needed to maintain optimal levels). Most studies support this hypothesis for vigilance regulation (Bonnet, 1989; Bonnet & Rosa, 1987; Brendel et al., 1990; Carskadon & Dement, 1985; Duffy, Willson, Wang, & Czeisler, 2009; Landolt, Retey, & Adam, 2012; Smulders, Kenemans, Jonkman, & Kok, 1997; Vojtechovsky, Brezinova, Simane, & Hort, 1969). Overall, compared to young adults, older adults show similar or even smaller vigilance deterioration during sleep loss, and they need less sleep for their vigilance to recover to baseline levels after sleep deprivation. Interestingly, findings from a recent study by Ready, Marquez, and Akerstedt (2009) show that, compared to the young, older adults reported less negativity in association with poor sleep and less benefit of longer sleep duration on negative mood (Ready et al., 2009). These results suggest that emotion regulation is less sensitive to sleep in aging. Indeed, as emotion regulation capacities increase, moods may be less swayed by external events, including sleep disturbance.

Another explanation can be found in the fact that emotion regulation seems to be sleep stage-dependent. In particular, the main theoretical model specifically associates REM sleep with emotion regulation processes during sleep (van der Helm et al., 2011). Together, these results demonstrate that offline time containing REM sleep may offer a neurobiological state that is especially well suited to the preferential processing of emotional memory. At the same time, the affective tone of an emotional event would be progressively dissociated from the memory of the event during the iterations of REM sleep (across one night or after several nights). Because sleep changes in the elderly affect NREM sleep first, preserved processes of emotion regulation could be an expected result. However, recent studies have allowed researchers to more precisely define how the different sleep stages could be involved in emotion processing, as well as how elderly individuals react to major sleep disturbances. First, cumulative findings suggest that NREM sleep may play a crucial role in the control of emotion (Lara-Carrasco et al., 2009; Pace-Schott et al., 2011; see also for a review: Deliens, Gilson, & Peigneux, 2014). For instance, recent findings suggest that fear conditioning, recognition of negative or high-arousal memories, and emotional attenuation during the sleep period are associated with NREM sleep (Hellman & Abel, 2007; Kaestner, Wixted, & Mednick, 2013; Pace-Schott et al., 2011; Talamini, Bringmann, de Boer, & Hofman, 2013). Given that NREM sleep changes drastically with aging, emotional tasks more closely associated with NREM sleep, compared to REM sleep, may also be more sensitive to age-related differences.

Large-cohort studies suggest a link between sleep loss or disturbances and emotion dysregulation in elderly individuals. For instance, researchers have associated insomnia with depressed mood in elderly individuals (Foley, Monjan, Brown, Simonsick, Wallace, & Blazer, 1999; Sukegawa et al., 2003), and anxiety symptoms have been correlated with sleep disturbances in older women (Spira, Stone, Beaudreau, Ancoli-Israel, & Yaffe, 2009). Future studies should evaluate whether sleep deprivation affects young and older adults differently, and these studies should estimate the amount of sleep needed by both age groups for emotion regulation to recover after sleep loss.

In conclusion, very few studies have compared young and older adults on emotional tasks sensitive to either REM or NREM sleep. In addition, researchers must further investigate sleep-dependent emotion regulation, taking into account the different subprocesses implicated in emotion regulation (Gross, 1998). To date, the emotion regulation protocols that researchers employ to assess sleep influences mainly deal with basic subprocesses, such as extinction. Ochsner and Gross described a wide range of subprocesses and strategies of emotion regulation that can be placed on a continuum from automatic, bottom-up processes to voluntary, top-down processes. Such a distinction may allow researchers to unravel the paradox reported in sleep-dependent emotion regulation associated with aging. Whereas sleep studies have been largely focused on automatic regulation, research on the effects of aging on emotional processing has mainly investigated voluntary regulation strategies, such as suppression strategy, corresponding to the inhibition of ongoing emotion-expressive behavior or top-down reappraisal strategies.

Future research on age-related sleep changes would benefit from the use of carefully designed experimental tasks that include the evaluation of emotion regulation strategy. Emotional tasks sensitive to the effects of aging, such as negative EFE recognition, subjective emotional experience, and physiological responsiveness to emotion, should also be investigated (Gavazzeni et al., 2008; Levenson et al., 1991). Because it is accompanied by typical changes in both sleep and affect, aging constitutes a relevant model for unraveling the way these processes are connected. Future studies should further investigate the interactions between sleep and emotion among the elderly, with a particular focus on the causal and bidirectional links between sleep alterations and emotional disturbances.

References

Allen, J., Bruss, J., Brown, C., & Damasio, H. (2005). Normal neuroanatomical variation due to age: The major lobes and a parcellation of the temporal region. *Neurobiology of Aging, 26*, 1245–1260. http://dx.doi.org/10.1016/j.neurobiolaging.2005.05.023.

Ancoli-Israel, S. (2005). Sleep and aging: Prevalence of disturbed sleep and treatment considerations in older adults. *Journal of Clinical Psychiatry, 66*(Suppl. 9), 24–30.

Ancoli-Israel, S., Ayalon, L., & Salzman, C. (2008). Sleep in the elderly: Normal variations and common sleep disorders. *Harvard Review of Psychiatry, 16*(5), 279–286. http://dx.doi.org/10.1080/10673220802432210.

Baltes, P. B., & Baltes, M. M. (1990). *Psychological perspectives on successful aging: The model of selective optimization with compensation.* New York, NY: Cambridge University Press.

Barrick, A. L., Hutchinson, R. L., & Deckers, L. H. (1989). Age effects on positive and negative emotions. *Journal of Social Behavior and Personality, 4*, 421–429.

Berkman, L. E., Berkman, C. S., Kasl, S., Freeman, D. H., Jr., Leo, L., Ostfeld, A. M., et al. (1986). Depressive symptoms in relation to physical health and function in the elderly. *American Journal of Epidemiology, 124*, 372–388.

Bherer, L., Belleville, S., & Hudon, C. (2004). Le déclin des fonctions exécutives au cours du vieillissement normal, dans la maladie d'Alzheimer et dans la démence fronto-temporale. *Psychologie & NeuroPsychiatrie du Vieillissement, 2*, 181–189.

Birditt, K. S., & Fingerman, K. L. (2005). Do we get better at picking our battles? Age group differences in descriptions of behavioral reactions to interpersonal tensions. *The Journals of Gerontology. Series B, Psychological Sciences and Social Sciences, 60*(3), 121–128. http://dx.doi.org/10.1093/geronb/60.3.P121.

Biss, R. K., & Hasher, L. (2012). Happy as a lark: Morning-type younger and older adults are higher in positive affect. *Emotion, 12*(3), 437–441. http://dx.doi.org/10.1037/a0027071.

Blanchard-Fields, F. (1998). The role of emotion in social cognition across the adult life span. *Annual Review of Gerontology and Geriatrics, 17*, 238–265.

Blanchard-Fields, F., & Coats, A. H. (2008). The experience of anger and sadness in everyday problems impacts age differences in emotion regulation. *Developmental Psychology, 44*, 1547–1556. http://dx.doi.org/10.1037/a0013915.

Blanchard-Fields, F., Mienaltowski, A., & Seay, R. B. (2007). Age differences in everyday problem-solving effectiveness: Older adults select more effective strategies for interpersonal problems. *The Journals of Gerontology. Series B, Psychological Sciences and Social Sciences, 62*, 61–64. http://dx.doi.org/10.1093/geronb/62.1.P61.

Bonnet, M. H. (1989). The effect of sleep fragmentation on sleep and performance in younger and older subjects. *Neurobiology of Aging, 10*, 21–25. http://dx.doi.org/10.1016/S0197-4580(89)80006-5.

Bonnet, M. H., & Rosa, R. R. (1987). Sleep and performance in young adults and older normals and insomniacs during acute sleep loss and recovery. *Biological Psychology, 25*, 153–172. http://dx.doi.org/10.1016/0301-0511(87)90035-4.

Brandtstädter, J., & Greve, W. (1994). The aging self: Stabilizing and protective processes. *Developmental Review, 14*, 52–80. http://dx.doi.org/10.1006/drev.1994.1003.

Brendel, D. H., Reynolds, C. F., Jennings, J. R., Hoch, C. C., Monk, T. H., Berman, S. R., et al. (1990). Sleep stage physiology, mood, and vigilance responses to total sleep deprivation in healthy 80-year-olds and 20-year-olds. *Psychophysiology, 27*, 677–685. http://dx.doi.org/10.1111/j.1469-8986.1990.tb03193.x.

Buysse, D. J., Browman, K. E., Monk, T. H., Reynolds, C. F., 3rd, Fasiczka, A. L., & Kupfer, D. J. (1992). Napping and 24-hour sleep/wake patterns in healthy elderly and young adults. *Journal of American Geriatrics Society, 40*(8), 779–786.

Calder, A. J., Keane, J., Manly, T., Sprengelmeyer, R., Scott, S., Nimmo-Smith, I., et al. (2003). Facial expression recognition across the adult life span. *Neuropsychologia, 41*(2), 195–202. http://dx.doi.org/10.1016/S0028-3932(02)00149-5.

Carrier, J., Land, S., Buysse, D. J., Kupfer, D. J., & Monk, T. H. (2001). The effects of age and gender on sleep EEG power spectral density in the middle years of life (ages 20–60 years old). *Psychophysiology, 38*(2), 232–242. http://dx.doi.org/10.1111/1469-8986.3820232.

Carrier, J., Monk, T. H., Buysse, D. J., & Kupfer, D. J. (1996). Inducing a 6-hour phase advance in the elderly: Effects on sleep and temperature rhythms. *Journal of Sleep Research, 5*(2), 99–105. http://dx.doi.org/10.1046/j.1365-2869.1996.00015.x.

Carrier, J., Monk, T. H., Buysse, D. J., & Kupfer, D. J. (1997). Sleep and morningness–eveningness in the 'middle' years of life (20–59 y). *Journal of Sleep Research, 6*(4), 230–237. http://dx.doi.org/10.1111/j.1365-2869.1997.00230.x.

Carrier, J., Paquet, J., Fernandez-Bolanos, M., Girouard, L., Roy, J., Selmaoui, B., et al. (2009). Effects of caffeine on daytime recovery sleep: A double challenge to the sleep–wake cycle in aging. *Sleep Medicine, 10*(9), 1016–1024. http://dx.doi.org/10.1016/j.sleep.2009.01.001.

Carrier, J., Viens, I., Poirier, G., Robillard, R., Lafortune, M., Vandewalle, G., et al. (2011). Sleep slow wave changes during the middle years of life. *European Journal of Neuroscience, 33*(4), 758–766. http://dx.doi.org/10.1111/j.1460-9568.2010.07543.x.

Carskadon, M. A., & Dement, W. C. (1985). Sleep loss in elderly volunteers. *Sleep, 8*, 207–221.

Carstensen, L. L. (1995). Evidence for a life-span theory of socioemotional selectivity. *Current Directions in Psychological Science, 4*, 151–156. http://dx.doi.org/10.1111/1467-8721.ep11512261.

Carstensen, L. L. (2006). The influence of a sense of time on human development. *Science, 312*, 1913–1915. http://dx.doi.org/10.1126/science.1127488.

Carstensen, L. L., Fung, H. H., & Charles, S. T. (2003). Socioemotional selectivity theory and the regulation of emotion in the second half of life. *Motivation and Emotion, 27*(2), 103–123. http://dx.doi.org/10.1023/A:1024569803230.

Carstensen, L. L., Isaacowitz, D. M., & Charles, S. T. (1999). Taking time seriously: A theory of socioemotional selectivity. *American Psychologist, 54*, 165–181. http://dx.doi.org/10.1037/0003-066X.54.3.165.

Carstensen, L. L., Pasupathi, M., Mayr, U., & Nesselroade, J. R. (2000). Emotional experience in everyday life across the adult life span. *Journal of Personality and Social Psychology, 79*, 644–655. http://dx.doi.org/10.1037/0022-3514.79.4.644.

Carstensen, L. L., Turan, B., Scheibe, S., Ram, N., Ersner-Hershfield, H., Samanez-Larkin, G. R., et al. (2011). Emotional experience improves with age: Evidence based on over 10 years of experience sampling. *Psychology and Aging, 26*, 21–33. http://dx.doi.org/10.1037/a0021285.

Cartwright, R., Young, M. A., Mercer, P., & Bears, M. (1998). Role of REM sleep and dream variables in the prediction of remission from depression. *Psychiatry Research, 80*(3), 249–255. http://dx.doi.org/10.1016/S0165-1781(98)00071-7.

Charles, S. T. (2010). Strength and vulnerability integration: A model of emotional well-being across adulthood. *Psychological Bulletin, 136*(6), 1068–1091. http://dx.doi.org/10.1037/a0021232.

Charles, S. T., Mather, M., & Carstensen, L. L. (2003). Aging and emotional memory: The forgettable nature of negative images for older adults. *Journal of Experimental Psychology: Human Perception and Performance, 132*, 310–324. http://dx.doi.org/10.1037/0096-3445.132.2.310.

Charles, S. T., Reynolds, C. A., & Gatz, M. (2001). Age-related differences and change in positive and negative affect over 23 years. *Journal of Personality and Social Psychology, 80*, 136–151. http://dx.doi.org/10.1037/0022-3514.80.1.136.

Cheng, S. T. (2004). Age and subjective well-being revisited: A discrepancy perspective. *Psychology and Aging, 19*(3), 409–415. http://dx.doi.org/10.1037/0882-7974.19.3.409.

Clark, V. A., Aneshensel, C. S., Frerichs, R. R., & Morgan, T. M. (1981). Analysis of effects of sex and age in response to items on the CES-D scale. *Psychiatry Research, 5*, 171–181. http://dx.doi.org/10.1016/0165-1781(81)90047-0.

Coats, A. H., & Blanchard-Fields, F. (2008). Emotion regulation in interpersonal problems: The role of cognitive-emotional complexity, emotion regulation goals, and expressivity. *Psychology and Aging, 23*(1), 39–51. http://dx.doi.org/10.1037/0882-7974.23.1.39.

Craik, F. I. M., & Salthouse, T. A. (2008). *The handbook of aging and cognition* (3rd ed.). New York: Psychology Press.

Crowley, K., Trinder, J., Kim, Y., Carrington, M., & Colrain, I. M. (2002). The effects of normal aging on sleep spindle and K-complex production. *Clinical Neurophysiology, 113*(10), 1615–1622. http://dx.doi.org/10.1016/S1388-2457(02)00237-7.

Csercsa, R., Dombovari, B., Fabo, D., Wittner, L., Eross, L., Entz, L., et al. (2010). Laminar analysis of slow wave activity in humans. *Brain, 133*(9), 2814–2829. http://dx.doi.org/10.1093/brain/awq169.

Darchia, N., Campbell, I. G., Tan, X., & Feinberg, I. (2007). Kinetics of NREM delta EEG power density across NREM periods depend on age and on delta-band designation. *Sleep, 30*(1), 71–79.

Deliens, G., Gilson, M., & Peigneux, P. (2014). Sleep and the processing of emotions. *Experimental Brain Research,*. http://dx.doi.org/10.1007/s00221-014-3832-1.

Diener, E., & Suh, E. M. (1998). Subjective well-being and age: An international analysis. *Annual Review of Gerontology and Geriatrics: Focus on Emotion and Adult Developmental Psychology, 17*, 304–324.

Dijk, D. J., Duffy, J. F., & Czeisler, C. A. (2001). Age-related increase in awakenings: Impaired consolidation of nonREM sleep at all circadian phases. *Sleep, 24*(5), 565–577.

Duffy, J. F., Willson, H. J., Wang, W., & Czeisler, C. A. (2009). Healthy older adults better tolerate sleep deprivation than young adults. *Journal of American Geriatrics Society, 57*(7), 1245–1251. http://dx.doi.org/10.1111/j.1532-5415.2009.02303.x.

Eagles, J. M., & Whalley, L. J. (1985). Ageing and affective disorders: The age at first onset of affective disorders in Scotland, 1969–1978. *British Journal of Psychiatry, 147*, 180–187. http://dx.doi.org/10.1192/bjp.147.2.180.

Feinberg, I. (1974). Changes in sleep cycle patterns with age. *Journal of Psychiatric Research, 10*(3–4), 283–306. http://dx.doi.org/10.1016/0022-3956(74)90011-9.

Fischer, H., Sandblom, J., Gavazzeni, J., Fransson, P., Wright, C. I., & Bäckman, L. (2005). Age-differential patterns of brain activation during perception of angry faces. *Neuroscience Letters, 386*, 99–104. http://dx.doi.org/10.1016/j.neulet.2005.06.002.

Floyd, J. A., Janisse, J. J., Jenuwine, E. S., & Ager, J. W. (2007). Changes in REM-sleep percentage over the adult lifespan. *Sleep, 30*(7), 829–836.

Foley, D., Ancoli-Israel, S., Britz, P., & Walsh, J. (2004). Sleep disturbances and chronic disease in older adults: Results of the 2003 National Sleep Foundation Sleep in America Survey. *Journal of Psychosomatic Research, 56*(5), 497–502. http://dx.doi.org/10.1016/j.jpsychores.2004.02.010.

Foley, D. J., Monjan, A. A., Brown, S. L., Simonsick, E. M., Wallace, R. B., & Blazer, D. G. (1999). Sleep complaints among elderly persons: An epidemiologic study of three communities. *Sleep, 18*, 425–432.

Frerichs, R. R., Aneshensel, C. S., & Clark, V. A. (1981). Prevalence of depression in Los Angeles County. *American Journal of Epidemiology, 113*, 691–699.

Gatz, M., & Hurwicz, M. L. (1990). Are old people more depressed? Cross-sectional data on Center for Epidemiological Studies Depression Scale factors. *Psychology and Aging, 5*(2), 284–290. http://dx.doi.org/10.1037/0882-7974.5.2.284.

Gatz, M., Johansson, B., Pedersen, N., Berg, S., & Reynolds, C. (1993). A cross-national self-report measure of depressive symptomatology. *International Psychogeriatrics, 5*, 147–156. http://dx.doi.org/10.1017/S1041610293001486.

Gaudreau, H., Carrier, J., & Montplaisir, J. (2001). Age-related modifications of NREM sleep EEG: From childhood to middle age. *Journal of Sleep Research, 10*(3), 165–172. http://dx.doi.org/10.1046/j.1365-2869.2001.00252.x.

Gavazzeni, J., Wiens, S., & Fischer, H. (2008). Age effects to negative arousal differ for self-report and electrodermal activity. *Psychophysiology*, *45*(1), 148–151. http://dx.doi.org/10.1111/j.1469-8986.2007.00596.x.

Grieve, S. M., Clark, C. R., Williams, L. M., Peduto, A. J., & Gordon, E. (2005). Preservation of limbic and paralimbic structures in aging. *Human Brain Mapping*, *25*, 391–401. http://dx.doi.org/10.1002/hbm.20115.

Gross, J. J. (1998). The emerging field of emotion regulation: An integrative review. *Review of General Psychology*, *2*, 271–299. http://dx.doi.org/10.1037/1089-2680.2.3.271.

Gross, J. J., Carstensen, L. L., Pasupathi, M., Tsai, J., Götestam Skorpen, C., & Hsu, A. Y. C. (1997). Emotion and aging: Experience, expression, and control. *Psychology and Aging*, *12*, 590–599. http://dx.doi.org/10.1037/0882-7974.12.4.590.

Gunning-Dixon, F. M., Gur, R. C., Perkins, A. C., Schroeder, L., Turner, T., Turetski, B. I., et al. (2003). Age-related differences in brain activation during emotional face processing. *Neurobiology of Aging*, *24*, 285–295. http://dx.doi.org/10.1016/S0197-4580(02)00099-4.

Gyurak, A., Goodkind, M. S., Kramer, J. H., Miller, B. L., & Levenson, R. W. (2012). Executive functions and the down-regulation and up-regulation of emotion. *Cognition & Emotion*, *26*(1), 103–118. http://dx.doi.org/10.1080/02699931.2011.557291.

Gyurak, A., Goodkind, M. S., Madan, A., Kramer, J. H., Miller, B. L., & Levenson, R. W. (2009). Do tests of executive functioning predict ability to downregulate emotions spontaneously and when instructed to suppress? *Cognitive, Affective, & Behavioral Neuroscience*, *9*(2), 144–152. http://dx.doi.org/10.3758/CABN.9.2.144.

Hellman, K., & Abel, T. (2007). Fear conditioning increases NREM sleep. *Behavioral Neuroscience*, *121*(2), 310–323. http://dx.doi.org/10.1037/0735-7044.121.2.310.

Hertzog, C., Van Alstine, J., Usala, P. D., Hultsch, D. F., & Dixon, R. (1990). Measurement properties of the Center for Epidemiological Studies Depression Scale (CES-D) in older populations. *Psychological Assessment*, *2*, 64–72. http://dx.doi.org/10.1037/1040-3590.2.1.64.

Higgins, E. T. (1999). When do self-discrepancies have specific relations to emotions? The second-generation question of Tangney, Niedenthal, Covert, and Barlow (1998). *Journal of Personal and Social Psychology*, *77*(6), 1313–1317. http://dx.doi.org/10.1037/0022-3514.77.6.1313.

Himmelfarb, S. (1984). Age and sex differences in the mental health of older persons. *Journal of Consulting and Clinical Psychology*, *52*, 844–856. http://dx.doi.org/10.1037/0022-006X.52.5.844.

Hoch, C. C., Dew, M. A., Reynolds, C. F., 3rd, Monk, T. H., Buysse, D. J., Houck, P. R., et al. (1994). A longitudinal study of laboratory- and diary-based sleep measures in healthy "old old" and "young old" volunteers. *Sleep*, *17*(6), 489–496.

Isaacowitz, D. M., Löckenhoff, C. E., Lane, R. D., Wright, R., Sechrest, L., Riedel, R., et al. (2007). Age differences in recognition of emotion in lexical stimuli and facial expressions. *Psychology and Aging*, *22*(1), 147–159. http://dx.doi.org/10.1037/0882-7974.22.1.147.

Isaacowitz, D. M., Toner, K., Goren, D., & Wilson, H. (2008). Looking while unhappy: Mood congruent gaze in young adults, positive gaze in older adults. *Psychological Science*, *19*, 848–853. http://dx.doi.org/10.1111/j.1467-9280.2008.02167.x.

Isaacowitz, D. M., Wadlinger, H. A., Goren, D., & Wilson, H. R. (2006). Is there an age-related positivity effect in visual attention? A comparison of two methodologies. *Emotion*, *6*, 511–516. http://dx.doi.org/10.1037/1528-3542.6.3.511.

Kaestner, E. J., Wixted, J. T., & Mednick, S. C. (2013). Pharmacologically increasing sleep spindles enhances recognition for negative and high-arousal memories. *Journal of Cognitive Neuroscience*, *25*(10), 1597–1610. http://dx.doi.org/10.1162/jocn_a_00433.

Kessler, R. C., Foster, C., Webster, P. S., & House, J. S. (1992). The relationship between age and depressive symptoms in two national surveys. *Psychology and Aging*, *7*, 119–126. http://dx.doi.org/10.1037/0882-7974.7.1.119.

Kessler, E.-M., & Staudinger, U. M. (2009). Affective experience in adulthood and old age: The role of affective arousal and perceived affect regulation. *Psychology and Aging, 24*, 349–362. http://dx.doi.org/10.1037/a0015352.

Klerman, E. B., Wang, W., Duffy, J. F., Dijk, D. J., Czeisler, C. A., & Kronauer, R. E. (2013). Survival analysis indicates that age-related decline in sleep continuity occurs exclusively during NREM sleep. *Neurobiology of Aging, 34*(1), 309–318. http://dx.doi.org/10.1016/j.neurobiolaging.2012.05.018.

Kuhl, J. (2000). A functional-design approach to motivation and selfregulation: The dynamics of personality systems interactions. In M. Boekaerts, P. Pintrich, & M. Zeidner (Eds.), *Handbook of selfregulation* (pp. 111–169). San Diego, CA: Academic Press.

Kuhl, J., & Beckmann, J. (1994). Action versus state orientation: Psychometric properties of the Action Control Scale (ACS-90). In J. K. J. Beckmann (Ed.), *Volition and personality* (pp. 47–59). Goettingen, Germany: Hogrefe.

Kunzmann, U., Kupperbusch, C. S., & Levenson, R. W. (2005). Behavioral inhibition and amplification during emotional arousal: A comparison of two age groups. *Psychology and Aging, 20*, 144–158. http://dx.doi.org/10.1037/0882-7974.20.1.144.

Kunzmann, U., Little, T. D., & Smith, J. (2000). Is age-related stability of subjective well-being a paradox? Cross-sectional and longitudinal evidence from the Berlin Aging Study. *Psychology and Aging, 15*, 511–526. http://dx.doi.org/10.1037/0882-7974.15.3.511.

Labouvie-Vief, G., DeVoe, M., & Bulka, D. (1989). Speaking about feelings: Conceptions of emotion across the life span. *Psychology and Aging, 4*, 425–437. http://dx.doi.org/10.1037/0882-7974.4.4.425.

Lamar, M., & Resnick, S. M. (2004). Aging and prefrontal functions: Dissociating orbitofrontal and dorsolateral abilities. *Neurobiology of Aging, 25*, 553–558. http://dx.doi.org/10.1016/j.neurobiolaging.2003.06.005.

Landolt, H. P., & Borbely, A. A. (2001). Age-dependent changes in sleep EEG topography. *Clinical Neurophysiology, 112*(2), 369–377. http://dx.doi.org/10.1016/S1388-2457(00)00542-3.

Landolt, H. P., Dijk, D. J., Achermann, P., & Borbely, A. A. (1996). Effect of age on the sleep EEG: Slow-wave activity and spindle frequency activity in young and middle-aged men. *Brain Research, 738*(2), 205–212. http://dx.doi.org/10.1016/S0006-8993(96)00770-6.

Landolt, H. P., Retey, J. V., & Adam, M. (2012). Reduced neurobehavioral impairment from sleep deprivation in older adults: Contribution of adenosinergic mechanisms. *Frontiers in Neurology, 3*, 62. http://dx.doi.org/10.3389/fneur.2012.00062.

Lang, F. R., & Carstensen, L. L. (2002). Time counts: Future time perspective, goals, and social relationships. *Psychology and Aging, 17*, 125–139. http://dx.doi.org/10.1037/0882-7974.17.1.125.

Lara-Carrasco, J., Nielsen, T. A., Solomonova, E., Levrier, K., & Popova, A. (2009). Overnight emotional adaptation to negative stimuli is altered by REM sleep deprivation and is correlated with intervening dream emotions. *Journal of Sleep Research, 18*(2), 178–187. http://dx.doi.org/10.1111/j.1365-2869.2008.00709.x.

Larson, C. L., Schaefer, H. S., Siegle, G. J., Jackson, C. A. B., Anderle, M. J., & Davidson, R. J. (2006). Fear is fast in phobics: Amygdala activation in response to fear-relevant stimuli. *Biological Psychiatry, 60*, 410–417. http://dx.doi.org/10.1016/j.biopsych.2006.03.079.

Lawton, M. P., Kleban, M. H., Rajagopal, D., & Dean, J. (1992). Dimensions of affective experience in three age groups. *Psychology and Aging, 7*, 171–184. http://dx.doi.org/10.1037/0882-7974.7.2.171.

Levenson, R. W., Carstensen, L. L., Friesen, W. V., & Ekman, P. (1991). Emotion, physiology, and expression in old age. *Psychology and Aging, 6*, 28–35. http://dx.doi.org/10.1037/0882-7974.6.1.28.

Luo, J., Zhu, G., Zhao, Q., Guo, Q., Meng, H., Hong, Z., et al. (2013). Prevalence and risk factors of poor sleep quality among Chinese elderly in an urban community: Results from the Shanghai aging study. *PLoS One*, *8*(11), e81261. http://dx.doi.org/10.1371/journal.pone.0081261.

MacPherson, S. E., Phillips, L. H., & Della Sala, S. (2006). Age-related differences in the ability to perceive sad facial expressions. *Aging Clinical and Experimental Research*, *18*, 418–424. http://dx.doi.org/10.1007/BF03324838.

Magai, C., Consedine, N. S., Krivoshekova, Y. S., Kudajie-Gyamfi, E., & McPherson, R. (2006). Emotion experience and expression across the adult life span: Insights from a multimodal assessment study. *Psychology and Aging*, *21*, 303–317. http://dx.doi.org/10.1037/0882-7974.21.2.303.

Malatesta, C. Z., Izard, C. E., Culver, C., & Nicholich, M. (1987). Emotion communication skills in young, middle-aged, and older women. *Psychology and Aging*, *2*, 193–203. http://dx.doi.org/10.1037/0882-7974.2.2.193.

Martin, N., Lafortune, M., Godbout, J., Barakat, M., Robillard, R., Poirier, G., et al. (2013). Topography of age-related changes in sleep spindles. *Neurobiology of Aging*, *34*(2), 468–476. http://dx.doi.org/10.1016/j.neurobiolaging.2012.05.020.

Mather, M., & Carstensen, L. L. (2003). Aging and attentional biases for emotional faces. *Psychological Science*, *14*, 409–415. http://dx.doi.org/10.1111/1467-9280.01455.

Mather, M., & Carstensen, L. L. (2005). Aging and motivated cognition: The positivity effect in attention and memory. *Trends in Cognitive Science*, *9*(10), 496–502. http://dx.doi.org/10.1016/j.tics.2005.08.005.

Mayr, U., Spieler, D. H., & Kliegl, R. (2001). *Ageing and executive control*. New York: Routledge.

Miller, E. K., & Cohen, J. D. (2001). An integrative theory of prefrontal cortex function. *Annual Review of Neuroscience*, *24*, 167–202. http://dx.doi.org/10.1146/annurev.neuro.24.1.167.

Moline, M. L., Pollak, C. P., Monk, T. H., Lester, L. S., Wagner, D. R., Zendell, S. M., et al. (1992). Age-related differences in recovery from simulated jet lag. *Sleep*, *15*(1), 28–40.

Monk, T. H., Buysse, D. J., Reynolds, C. F., 3rd, & Kupfer, D. J. (1995). Inducing jet lag in an older person: Directional asymmetry. *Experimental Gerontology*, *30*(2), 137–145. http://dx.doi.org/10.1016/0531-5565(94)00059-X.

Morris, J. S., Friston, K. J., Büchel, C., Frith, C. D., Young, A. W., Calder, A. J., et al. (1998). A neuromodulatory role for the human amygdala in processing emotional facial expressions. *Brain*, *121*(1), 47–57. http://dx.doi.org/10.1093/brain/121.1.47.

Mroczek, D. K., & Kolarz, C. M. (1998). The effect of age on positive and negative affect: A developmental perspective on happiness. *Journal of Personality and Social Psychology*, *75*, 1333–1349. http://dx.doi.org/10.1037/0022-3514.75.5.1333.

Newmann, J. P. (1989). Aging and depression. *Psychology and Aging*, *4*, 150–165. http://dx.doi.org/10.1037/0882-7974.4.2.150.

Nicolas, A., Petit, D., Rompre, S., & Montplaisir, J. (2001). Sleep spindle characteristics in healthy subjects of different age groups. *Clinical Neurophysiology*, *112*(3), 521–527. http://dx.doi.org/10.1016/S1388-2457(00)00556-3.

Ochsner, K. N., Bunge, S. A., Gross, J. J., & Gabrieli, J. D. (2002). Rethinking feelings: An FMRI study of the cognitive regulation of emotion. *Journal of Cognitive Neuroscience*, *14*(8), 1215–1229. http://dx.doi.org/10.1162/089892902760807212.

Ochsner, K. N., & Gross, J. J. (2005). The cognitive control of emotion. *Trends in Cognitive Science*, *9*(5), 242–249. http://dx.doi.org/10.1016/j.tics.2005.03.010.

Ohayon, M. M., Carskadon, M. A., Guilleminault, C., & Vitiello, M. V. (2004). Meta-analysis of quantitative sleep parameters from childhood to old age in healthy individuals: Developing normative sleep values across the human lifespan. *Sleep*, *27*(7), 1255–1273.

Opitz, P. C., Gross, J. J., & Urry, H. L. (2012). Selection, optimization, and compensation in the domain of emotion regulation: Applications to adolescence, older age, and major depressive disorder. *Social and Personality Psychology Compass*, *6*(2), 142–155. http://dx.doi.org/10.1111/j.1751-9004.2011.00413.x.

Pace-Schott, E. F., Milad, M. R., Orr, S. P., Rauch, S. L., Stickgold, R., & Pitman, R. K. (2009). Sleep promotes generalization of extinction of conditioned fear. *Sleep, 32*(1), 19–26.

Pace-Schott, E. F., Nave, G., Morgan, A., & Spencer, R. M. (2012). Sleep-dependent modulation of affectively guided decision-making. *Journal of Sleep Research, 21*(1), 30–39. http://dx.doi.org/10.1111/j.1365-2869.2011.00921.x.

Pace-Schott, E. F., Shepherd, E., Spencer, R. M., Marcello, M., Tucker, M., Propper, R. E., et al. (2011). Napping promotes inter-session habituation to emotional stimuli. *Neurobiology of Learning and Memory, 95*(1), 24–36. http://dx.doi.org/10.1016/j.nlm.2010.10.006.

Palmore, E., & Cleveland, W. (1976). Aging, terminal decline, and terminal drop. *Journal of Gerontology, 31*, 76–81. http://dx.doi.org/10.1093/geronj/31.1.76.

Phan, K. L., Wager, T., Taylor, S. F., & Liberzon, I. (2002). Functional neuroanatomy of emotion: A meta-analysis of emotion activation studies in PET and fMRI. *NeuroImage, 16*, 331–348. http://dx.doi.org/10.1006/nimg.2002.1087.

Phillips, L. H., Henry, J. D., Hosie, J. A., & Milne, A. B. (2008). Effective regulation of the experience and expression of negative affect in old age. *Journals of Gerontology Series B, Psychological Sciences and Social Sciences, 63*, 138–145. http://dx.doi.org/10.1093/geronb/63.3.P138.

Phillips, L. H., Maclean, R. D. J., & Allen, R. (2002). Age and the understanding of emotions: Neuropsychological and socio-cognitive perspectives. *Journals of Gerontology Series B, Psychological Sciences and Social Sciences, 57B*, 526–530. http://dx.doi.org/10.1093/geronb/57.6.P526.

Phillips, B., & Mannino, D. (2005). Correlates of sleep complaints in adults: The ARIC study. *Journal of Clinical Sleep Medicine, 1*(3), 277–283.

Polo-Kantola, P., Laine, A., Aromaa, M., Rautava, P., Markkula, J., Vahlberg, T., et al. (2014). A population-based survey of sleep disturbances in middle-aged women – Associations with health, health related quality of life and health behavior. *Maturitas, 77*(3), 255–262. http://dx.doi.org/10.1016/j.maturitas.2013.11.008.

Rajah, M. N., & D'Esposito, M. (2005). Region-specific changes in prefrontal function with age: A review of PET and fMRI studies on working and episodic memory. *Brain, 128*, 1964–1983. http://dx.doi.org/10.1093/brain/awh608.

Raz, N. (2000). Aging of the brain and its impact on cognitive performance: Integration of structural and functional findings. In F. I. M. Craik. & T. A. Salthouse (Eds.), *The handbook of aging and cognition* (pp. 1–90). Mahwah: Erlbaum.

Raz, N., & Rodrigue, K. M. (2006). Differential aging of the brain: Patterns, cognitive correlates and modifiers. *Neuroscience and Biobehavioral Reviews, 30*(6), 730–748. http://dx.doi.org/10.1016/j.neubiorev.2006.07.001.

Ready, R. E., Marquez, D. X., & Akerstedt, A. (2009). Emotion in younger and older adults: Retrospective and prospective associations with sleep and physical activity. *Experimental Aging Research, 35*(3), 348–368. http://dx.doi.org/10.1080/03610730902922184.

Regier, D. A., Boyd, H. J., Burke, J. D., Rae, D. S., Myers, J. K., Kramer, M., et al. (1988). One-month prevalence of mental disorders in the United States. *Archives of General Psychiatry, 45*, 977–986. http://dx.doi.org/10.1001/archpsyc.1988.01800350011002.

Reuter-Lorenz, P. A., Jonides, J., Smith, E. E., Hartley, A., Miller, A., Marshuetz, C., et al. (2000). Age differences in the frontal lateralization of verbal and spatial working memory revealed by PET. *Journal of Cognitive Neuroscience, 12*(1), 174–187. http://dx.doi.org/10.1162/089892900561814.

Reynolds, C. F., Hoch, C. C., Buysse, D. J., Monk, T. H., Houck, P. R., & Kupfer, D. J. (1993). Symposium: Normal and abnormal REM sleep regulation: REM sleep in successful, usual, and pathological aging: The Pittsburgh experience 1980–1993. *Journal of Sleep Research, 2*(4), 203–210. http://dx.doi.org/10.1111/j.1365-2869.1993.tb00091.x.

Robillard, R., Massicotte-Marquez, J., Kawinska, A., Paquet, J., Frenette, S., & Carrier, J. (2010). Topography of homeostatic sleep pressure dissipation across the night in young and middle-aged men and women. *Journal of Sleep Research, 19*(3), 455–465. http://dx.doi.org/10.1111/j.1365-2869.2010.00820.x.

Rosales-Lagarde, A., Armony, J. L., Del Rio-Portilla, Y., Trejo-Martinez, D., Conde, R., & Corsi-Cabrera, M. (2012). Enhanced emotional reactivity after selective REM sleep deprivation in humans: An fMRI study. *Frontiers in Behavioral Neuroscience, 6,* 25. http://dx.doi.org/10.3389/fnbeh.2012.00025.

Salthouse, T. A., Atkinson, T. M., & Berish, D. E. (2003). Executive functioning as a potential mediator of age-related cognitive decline in normal adults. *Journal of Experimental Psychology: General, 132*(4), 566–594. http://dx.doi.org/10.1037/0096-3445.132.4.566.

Salzarulo, P., Fagioli, I., Lombardo, P., Gori, S., Gneri, C., Chiaramonti, R., et al. (1999). Sleep stages preceding spontaneous awakenings in the elderly. *Sleep Research Online, 2*(3), 73–77.

Scheibe, S., & Blanchard-Fields, F. (2009). Effects of regulating emotions on cognitive performance: What is costly for young adults is not so costly for older adults. *Psychology and Aging, 24*(1), 217–223. http://dx.doi.org/10.1037/a0013807.

Shiota, M. N., & Levenson, R. W. (2009). Effects of aging on experimentally instructed detached reappraisal, positive reappraisal, and emotional behavior suppression. *Psychology and Aging, 24*(4), 890–900. http://dx.doi.org/10.1037/a0017896.

Small, B. J., Hertzog, C., Hultsch, D. F., & Dixon, R. A. (2003). Stability and change in adult personality over 6 years: Findings from the Victoria Longitudinal Study. *The Journals of Gerontology. Series B, Psychological Sciences and Social Sciences, 58,* 166–176. http://dx.doi.org/10.1093/geronb/58.3.P166.

Smith, J., & Baltes, P. B. (1993). Differential psychological aging: Profiles of the old and very old. *Ageing and Society, 13,* 551–587. http://dx.doi.org/10.1017/S0144686X00001367.

Smulders, F. T., Kenemans, J. L., Jonkman, L. M., & Kok, A. (1997). The effects of sleep loss on task performance and the electroencephalogram in young and elderly subjects. *Biological Psychology, 45*(1–3), 217–239. http://dx.doi.org/10.1016/S0301-0511(96)05229-5.

Spira, A. P., Stone, K., Beaudreau, S. A., Ancoli-Israel, S., & Yaffe, K. (2009). Anxiety symptoms and objectively measured sleep quality in older women. *The American Journal of Geriatric Psychiatry, 17*(2), 136–143. http://dx.doi.org/10.1097/JGP.0b013e3181871345.

Srivastava, S., John, O. P., Gosling, S. D., & Potter, J. (2003). Development of personality in early and middle adulthood: Set like plaster or persistent change? *Journal of Personality and Social Psychology, 84,* 1041–1053. http://dx.doi.org/10.1037/0022-3514.84.5.1041.

St Jacques, P., Dolcos, F., & Cabeza, R. (2010). Effects of aging on functional connectivity of the amygdala during negative evaluation: A network analysis of fMRI data. *Neurobiology of Aging, 31,* 315–327. http://dx.doi.org/10.1016/j.neurobiolaging.2008.03.012.

Stacey, C. A., & Gatz, M. (1991). Cross-sectional age differences and longitudinal change on the Bradburn Affect Balance Scale. *The Journals of Gerontology. Series B, Psychological Sciences and Social Sciences, 46,* 76–78. http://dx.doi.org/10.1093/geronj/46.2.P76.

Stenback, A. (1980). Depression and suicidal behavior in old age. *Handbook of mental health and aging* (pp. 616–652). Englewood Cliffs, NJ: Prentice-Hall.

Steriade, M. (2006). Grouping of brain rhythms in corticothalamic systems. *Neuroscience, 137*(4), 1087–1106. http://dx.doi.org/10.1016/j.neuroscience.2005.10.029.

Stone, A. A., Schwartz, J. E., Broderick, J. E., & Deaton, A. (2010). A snapshot of the age distribution of psychological well-being in the United States. *Proceedings of the National Academy of Sciences, 107,* 9985–9990. http://dx.doi.org/10.1073/pnas.100374410.

Sukegawa, T., Itoga, M., Seno, H., Miura, S., Inagaki, T., Saito, W., et al. (2003). Sleep disturbances and depression in the elderly in Japan. *Psychiatry and Clinical Neurosciences, 57*(3), 265–270. http://dx.doi.org/10.1046/j.1440-1819.2003.01115.x.

Sullivan, S., & Ruffman, T. (2004). Emotion recognition deficits in the elderly. *International Journal of Neurosciences, 114,* 403–432. http://dx.doi.org/10.1080/00207450490270901.

Sullivan, S., Ruffman, T., & Hutton, S. B. (2007). Age differences in emotion recognition skills and the visual scanning of emotion faces. *The Journals of Gerontology. Series B, Psychological Sciences and Social Sciences, B62,* 53–60. http://dx.doi.org/10.1093/geronb/62.1.P53.

Suzuki, A., Hoshino, T., Shigemasu, K., & Kawamura, M. (2007). Decline or improvement? Age-related differences in facial expression recognition. *Biological Psychology, 74,* 75–84. http://dx.doi.org/10.1016/j.biopsycho.2006.07.003.

Talamini, L. M., Bringmann, L. F., de Boer, M., & Hofman, W. F. (2013). Sleeping worries away or worrying away sleep? Physiological evidence on sleep–emotion interactions. *PLoS One, 8*(5), e62480. http://dx.doi.org/10.1371/journal.pone.0062480.

Tamir, M. (2009). What do people want to feel and why? Pleasure and utility in emotion regulation. *Current Directions in Psychological Science, 18,* 101–105. http://dx.doi.org/10.1111/j.1467-8721.2009.01617.x.

Teachman, B. A. (2006). Aging and negative affect: The rise and fall and rise of anxiety and depression symptoms. *Psychology and Aging, 21*(1), 201–207. http://dx.doi.org/10.1037/0882-7974.21.1.201.

Teachman, B. A., Siedlecki, K. L., & Magee, J. C. (2007). Aging and symptoms of anxiety and depression: Structural invariance of the tripartite model. *Psychology and Aging, 22*(1), 160. http://dx.doi.org/10.1037/0882-7974.22.1.160.

Tisserand, D. J., Pruessner, J. C., Sanz Arigita, E. J., van Boxtel, M. P., Evans, A. C., Jolles, J., et al. (2002). Regional frontal cortical volumes decrease differentially in aging: An MRI study to compare volumetric approaches and voxel-based morphometry. *NeuroImage, 17*(2), 657–669. http://dx.doi.org/10.1006/nimg.2002.1173.

Tranel, D., Gullickson, G., Koch, M., & Adolphs, R. (2006). Altered experience of emotion following bilateral amygdala damage. *Cognitive Neuropsychiatry, 11*(3), 219–232. http://dx.doi.org/10.1080/13546800444000281.

van der Helm, E., Yao, J., Dutt, S., Rao, V., Saletin, J. M., & Walker, M. P. (2011). REM sleep depotentiates amygdala activity to previous emotional experiences. *Current Biology, 21*(23), 2029–2032. http://dx.doi.org/10.1016/j.cub.2011.10.052.

Vgontzas, A. N., Bixler, E. O., Lin, H. M., Prolo, P., Mastorakos, G., Vela-Bueno, A., et al. (2001a). Chronic insomnia is associated with nyctohemeral activation of the hypothalamic–pituitary–adrenal axis: Clinical implications. *Journal of Clinical Endocrinology and Metabolism, 86*(8), 3787–3794. http://dx.doi.org/10.1210/jcem.86.8.7778.

Vgontzas, A. N., Bixler, E. O., Wittman, A. M., Zachman, K., Lin, H. M., Vela-Bueno, A., et al. (2001b). Middle-aged men show higher sensitivity of sleep to the arousing effects of corticotropin-releasing hormone than young men: Clinical implications. *Journal of Clinical Endocrinology and Metabolism, 86*(4), 1489–1495. http://dx.doi.org/10.1210/jcem.86.4.7370.

Vgontzas, A. N., Zoumakis, M., Bixler, E. O., Lin, H. M., Prolo, P., Vela-Bueno, A., et al. (2003). Impaired nighttime sleep in healthy old versus young adults is associated with elevated plasma interleukin-6 and cortisol levels: Physiologic and therapeutic implications. *Journal of Clinical Endocrinology and Metabolism, 88*(5), 2087–2095. http://dx.doi.org/10.1210/jc.2002-021176.

Viken, R. J., Rose, R. J., Kaprio, J., & Koskenvuo, M. (1994). A developmental genetic analysis of adult personality: Extraversion and neuroticism from 18 to 59 years of age. *Journal of Personality and Social Psychology and Aging, 66,* 722–730. http://dx.doi.org/10.1037/0022-3514.66.4.722.

Vojtechovsky, M., Brezinova, V., Simane, Z., & Hort, V. (1969). An experimental approach of sleep and aging. *Human Development, 12,* 64–72. http://dx.doi.org/10.1159/000270684.

Vytal, K., & Hamann, S. (2010). Neuroimaging support for discrete neural correlates of basic emotions: A voxel-based meta analysis. *Journal of Cognitive Neurosciences, 22*(12), 2864–2885. http://dx.doi.org/10.1162/jocn.2009.21366.

Walker, M. P., & van der Helm, E. (2009). Overnight therapy? The role of sleep in emotional brain processing. *Psychological Bulletin, 135*(5), 731–748. http://dx.doi.org/10.1037/a0016570.

Wang, L., McCarthy, G., Song, A. W., & Labar, K. S. (2005). Amygdala activation to sad pictures during high-field (4 tesla) functional magnetic resonance imaging. *Emotion, 5,* 12–22. http://dx.doi.org/10.1037/1528-3542.5.1.12.

Weissman, M., Leaf, P. J., Bruce, M. L., & Florio, L. P. (1988). The epidemiology of dysthymia in five communities: Rates, risks, comorbidity and treatment. *American Journal of Psychiatry, 145,* 815–819.

Werheid, K., Gruno, M., Kathmann, N., Fischer, H., Almkvist, O., & Winblad, B. (2010). Biased recognition of positive faces in aging and amnestic mild cognitive impairment. *Psychology and Aging, 25,* 1–15. http://dx.doi.org/10.1037/a0018358.

Yoo, S. S., Gujar, N., Hu, P., Jolesz, F. A., & Walker, M. P. (2007). The human emotional brain without sleep—A prefrontal amygdala disconnect. *Current Biology, 17*(20), R877–R878. http://dx.doi.org/10.1016/j.cub.2007.08.007.

PART 4

FUTURE DIRECTIONS

CHAPTER
21

Sleep and Affect: An Integrative Synthesis and Future Directions

Kimberly A. Babson[*,†] *and Matthew T. Feldner*[‡,§]

*National Center for PTSD, VA Palo Alto Health Care System, Menlo Park, California, USA
†Department of Psychiatry and Behavioral Sciences, Stanford School of Medicine, Stanford, California, USA
‡Department of Psychological Science, University of Arkansas, Fayetteville, Arkansas, USA
§Laureate Institute for Brain Research, Tulsa, Oklahoma, USA

Poor sleep among children, adolescents, and adults is reaching epidemic proportions (CDC, 2014). Mental health and substance use disorders account for over 7% of the disease burden worldwide (Whiteford et al., 2013), and sleep problems and other types of psychopathology are frequently co-occurring (Roth et al., 2006). Therefore, it is increasingly important to understand the links between sleep and affect. In fact, a growing number of researchers are investigating the interplay between sleep and both physical and psychological functioning. Their results have shed light on the interrelations between sleep and psychological disorders, but few studies have summarized the relationship between sleep and affect outside the context of a specific disorder. This is a key area of work, however, because understanding the interactions between sleep and affect from biological, cognitive, and behavioral perspectives can advance models of the etiology and maintenance of psychological disorders, thus informing both prevention and clinical intervention development. This book attempts to fill this critical gap by pulling together state-of-the-science findings on the interrelations between sleep and affect. In this chapter we provide a synthesis and extension of the findings reviewed throughout this volume, and we recommend areas for future research.

THE COMPLEX INTERPLAY OF SLEEP AND AFFECT

Methods for studying sleep and affect have never been more sophisticated and accessible to the average researcher. For example, methods for studying sleep vary from technically challenging and expensive neuroimaging to user-friendly and relatively inexpensive actigraphy methods, which can compliment self-report. The increasing availability of objective measures of sleep has allowed for greater examination of the convergence of subjective and objective measures of sleep, and, as we describe below, this examination is yielding very interesting areas of divergence in relation to affect. Akin to methods for studying sleep, methods for studying affect range from the highly technical and costly (e.g., functional magnetic resonance imaging) to the free and easily implemented (e.g., International Affective Picture System). As a result, researchers are increasingly integrating state-of-the-art methods for studying sleep and affect, documenting the complex interplay between them.

Sleep Problems Influence Affect

Studies have shown that sleep loss increases negative affect and influences positive affect in both positive and negative ways, depending on the presence of psychopathology. What follows is an overview of patterns observed in research examining how sleep may impact affect.

Negative Affect

The majority of sleep-affect research has focused on the interrelations between sleep and negative affect. Overall, poor sleep appears to increase multiple forms of negative affect, including sadness, anger, and anxiety. The majority of the work in this area has focused on the impact of sleep on anxiety, however. Sleep deprivation has been shown to increase fear and anxiety among healthy controls and individuals with disorders of anxiety (Babson, Feldner, Trainor, & Smith, 2009; Babson, Trainor, Feldner, & Blumenthal, 2010; Roy-Byrne, Uhde, & Post, 1986; Sagaspe, Sanchez-Ortuno, & Charles, 2006). In addition, among adolescents, sleep deprivation has reportedly increased worry and rumination (Talbot, McGlinchey, Kaplan, Dahl, & Harvey, 2010). Longitudinal studies have shown that sleep problems predict anxiety levels and the onset of anxiety disorders (Ford & Kamerow, 1989; Neckelmann, Mykletun, & Dahl, 2007). Researchers have also observed this relation among children and adolescents, such that sleep problems in childhood appear to potentiate the onset of anxiety in adolescence and adulthood (Gregory et al., 2005; Gregory, Eley, O'Connor, & Plomin, 2004; Gregory & O'Connor, 2002; Ong, Wickramaratne, Tang, & Weissman, 2006).

A substantial amount of work has also examined the impact of sleep disturbances on sadness and depressed mood. For example, longitudinal work combining 21 studies has demonstrated that poor sleep is associated with a two-fold increase in risk for the development of depression 1 year or more later after the sleep distrubance (Baglioni, Spiegelhalder, Lombardo, & Riemann, 2010). Furthermore, among adolescents, sleep loss appears to increase the risk for future onset of depression (Fredriksen, Rhodes, Reddy, & Way, 2004).

Relatively less work has addressed the relationship between sleep and anger and aggression, and existing studies have produced mixed results. Although some studies have shown that sleep loss is associated with an increase in anger and decrease in anger control (Kahn-Greene, Lipizzi, Conrad, Kamimori, & Killgore, 2006), other studies have suggested no impact of sleep loss on anger and aggression (Cote, McCormick, Geniole, Renn, & MacAulay, 2013; Vohs, Glass, Maddox, & Markman, 2011).

Positive Affect

Limited research has also examined the impact of sleep on positive affect. The current findings indicate that the impact of sleep on positive affect varies depending on the presence of psychological disorders, particularly depression. Among individuals without psychological disorders, sleep loss has been shown to decrease positive affect, and researchers have also observed this phenomenon among adolescents (McGlinchey et al., 2011). In comparison, among individuals with mood disorders, poor sleep appears to result in transient improvements in positive affect, with sleep loss triggering manic episodes in extreme cases (Barbini, Bertelli, Colombo, & Smeraldi, 1996).

Summary

Taken together, studies suggest that sleep disturbance is often related to an increase in anxiety, fear, worry, and/or sadness. Some research supports a relationship between sleep loss and anger and aggression, but additional research is needed to draw definitive conclusions about this connection. Sleep loss has also been shown to decrease positive affect among healthy individuals, while having the opposite effect (i.e., transient improvement in positive affect) among individuals with mood disorders.

Affect Influences Sleep

In addition to work documenting that sleep can influence affect, research has also shown that affect can impact subsequent sleep. An overview of this body of work is presented next.

Negative Affect

As with existing research on how sleep influences affect, the majority of work examining the influence of affect on sleep relates to negative affect and, more specifically, the impacts of anxiety and sadness/depression on sleep behavior. In examining the influence of negative affect on sleep, researchers have noted that the level of arousal (as opposed to hedonic valence) associated with the affect seems to be most strongly linked to sleep disturbance. For example, the degree to which sadness and depression contribute sleep problems is relatively limited. In other words, low arousal negative affect appears to have less of an impact on sleep. In comparison, high arousal affect (anxiety and anger) appears to have a greater impact on sleep. In terms of anxiety, experimental research employing lab-based anxiety inductions has demonstrated that induced stress prior to bed is associated with sleep problems (Gross & Borkovec, 1982). Longitudinal studies further support this relationship, demonstrating the onset of sleep problems after exposure to traumatic events (Vahtera et al., 2007).

Another high arousal negative affective state is anger or aggression. Indeed, anger and aggression have both been linked to poor sleep. For example, violent offenders with antisocial personality disorder have often demonstrate disrupted sleep marked by increased nighttime awakenings and decreased sleep efficiency (Lindberg et al., 2003). This finding is supported by basic research conducted on rats, which has demonstrated that an experimentally induced aggressive interaction is associated with sleep disruption the following night (Lancel, Droste, Sommer, & Reul, 2003; Meerlo, Pragt, & Daan, 1997; Meerlo & Turek, 2001).

Positive Affect

The relationship between positive affect and sleep disturbance varies as a function of the level of positive affect and the regulation or reactivity of positive affect. For example, low levels of positive affect have been associated with poor sleep. In comparison, adults who report high levels of positive affect demonstrate better sleep (Fosse, Stickgold, & Hobson, 2002; Steptoe, O'Donnell, Marmot, & Wardle, 2008), but those who experience difficulties regulating positive affect report greater sleep disturbances (Ong et al., 2013; Talbot, Hairston, Eidelman, Gruber, & Harvey, 2009). In one study, Ong et al. (2013) demonstrated that individuals with high trait positive affect reported better overall sleep quality despite lower sleep efficiency. However, individuals with high levels of positive affect reactivity to daily events had poorer sleep quality, particularly those with high trait positive affect.

Summary

Taken together, negative affect marked by high levels of arousal (e.g., anger, anxiety) is associated with poorer subsequent sleep, but negative affect characterized by low arousal (e.g., sadness/depression) appears to have less of an impact on sleep. On the other hand, low levels of positive affect and high positive affect reactivity to daily events are associated with poorer subsequent sleep.

Conclusions and Future Directions

The aggregated research suggests a complex, bidirectional interplay between negative and positive affect and sleep disturbances. Recognizing this pattern, researchers have focused on better understanding the factors that may account for the interrelations between sleep and affect, and we consider these factors in greater detail below. Additional research could help improve the assessment strategies used in the study of sleep and affect. As illustrated in this volume, both sleep and affect are complex, multidimensional constructs, and concurrent measurements of the multiple facets of these constructs often produce desynchrony. Research explaining the nature of relationships between specific factors of sleep and affect is needed. For example, researchers should explore the differences between self-reported and objectively measured sleep within the context of different affective profiles. Based on the observed differences based on measurement method, research could then integrate multimodal assessment methods in order to obtain the most informative and comprehensive data. For example, the effects of sleep deprivation on emotion regulatory capacities suggest that sleep-deprived healthy adults should exhibit greater emotional reactivity to negatively valenced and arousing stimuli, in terms of peak response and recovery time across behavioral, physiological, and self-report measures. To the best of our knowledge, researchers have yet to attempt studies with this level of specificity. Yet, such work will yield important information regarding the breadth of the impact of sleep on affect.

FACTORS IMPLICATED IN LINKS BETWEEN SLEEP AND AFFECT

Explanations of the links between sleep and affect tend to involve predisposition and causal models. In terms of the former, predisposing factors may account for the correlations observed between sleep and affect. Such factors may increase the likelihood that problems emerge in both domains (e.g., risk factors), or they may buffer against the development of problems (e.g., protective factors). In contrast, causal models focus how one factor contributes to changes in the other.

PREDISPOSING FACTORS

A factor central to both sleep and affect is physiological arousal (Bonnet & Arand, 2010). Sleep has an inverse relation with arousal, with activation of the arousal-related hypothalamic-pituitary-adrenocortical (HPA) system interrupting sleep and slow-wave sleep downregulating the activity of the HPA-axis (Steiger, 2002). Comparably, elevated arousal is a core feature of affect in dimensional models of emotion (Mehrabian, 1995), and elevated arousal can increase the likelihood of a strong emotional response (Koenigsberg, Pollak, & Sullivan, 1987). Therefore, hyperarousal may be a risk factor for problems related to both affect and sleep. Relatedly, scholars are now focusing on what may account for the elevated arousal resulting in these problems.

Researchers have recently presented a model linking biological factors to the emergence of sleep problems that may contribute to emotion dysregulation and the development of psychopathology (Harvey, Murray, Chandler, & Soehner, 2011). In considering this model, we focus on hyperarousal as a factor that may predispose people to the development of both sleep and affect-related problems, because this factor repeatedly emerges across topics in the current volume. People can come to experience elevated arousal via many pathways. Emotional arousal can be characterized as a response that unfolds across time, and it can vary across individuals, situations, and time points (e.g., peak intensity, time to reach peak intensity, time to return to baseline; Davidson, 2000). Accordingly, emotional arousal depends on a wide array of physiological mechanisms involved at different points in time. These mechanisms range from those underlying the early detection of arousal-eliciting stimuli (e.g., amygdala) to those involved in the postperceptual automatic and voluntary regulation of responses to such stimuli (e.g., dorsal regions of the anterior cingulate cortex, dorsomedial prefrontal cortex (PFC); Phillips, Ladouceur, & Drevets, 2008; Ray & Zald, 2012). The involvement of these regions in the elicitation and regulation of anxious arousal has been well documented (Cisler & Koster, 2010; Hofmann, Ellard, & Siegle, 2012). Indeed, affective neuroscience is increasingly elucidating how subcortical regions involved in emotional reactivity (e.g., amygdala) interact with cortical regions involved in regulating initial emotional reactions (e.g., dorsolateral PFC, dorsomedial PFC, dorsal anterior cingulate; Ochsner & Gross, 2005; Phillips et al., 2008). Moreover, these neurobiological structures function, in part, to regulate activity of more downstream mechanisms involved in emotional arousal, such as the activity of the HPA-axis, which also mediates arousal reactions to the environment. Accordingly, abnormalities can develop at several points in the system responsible for regulating arousal elicited by the environment, and these abnormalities can result in atypical levels of arousal that may lead to problems related to sleep and/or affect. These abnormalities likely differ across individuals, and abnormalities may result in problematic changes to other components within the system. For example, repeated stress in rats results in both increased dendritic density in the basolateral amygdala and atrophy of medial PFC neurons

(Miller & McEwen, 2006). Hyperarousal resulting from abnormalities in HPA-axis functioning (e.g., deficits in downregulation) may result in chronic stress and alter the structure and function of other substrates of emotional arousal and regulation.

The impact of repeated stress on the structure and function of key regions implicated in emotional reactivity and regulation highlights the role that environmental factors can play in the development of problems with sleep and emotion. For example, classical and operant conditioning of both internal (or interoceptive) and external cues influence functioning of the arousal system (Bouton, Mineka, & Barlow, 2001; Gorman, Kent, Sullivan, & Coplan, 2000). It is therefore likely that biological predispositions and environmental influences on biological functioning underlie the hyperarousal that may set the stage for problems with both affect and sleep.

Exercise may protect against the development of problems with both sleep and affect. Increasing exercise appears to improve sleep (Buman & King, 2010). Similarly, exercise appears to increase psychological well-being and to reduce negative affect (Stathopoulou, Powers, Berry, Smits, & Otto, 2006). Although compelling evidence indicates that exercise influences both sleep and mental health, researchers know less about the mechanisms underlying these effects. Researchers are considering hypothesized mechanisms that might influence sleep and affect specifically, while investigating whether positive effects on sleep can benefit affect or visa versa. In terms of specific effects, researchers have suggested (see Buman and Youngstedt, chapter 15) that exercise may affect adenosine and/or body temperature, which could account for associated improvements in sleep. With respect to affect, hypothesized mechanisms include effects of exercise on serotonin, betaendorphins, cognitive functioning associated with PFC functions, and improved health-related behavior, including behavioral tolerance of emotional and somatic distress (Stathopoulou et al., 2006). In addition, it has been hypothesized that exercise may improve sleep via its effect on negative affect, and exercise may improve affect via its effect on sleep (Buman & Youngstedt, this volume; Stathopoulou et al., 2006). Exercise may certainly function as a protective factor, reducing the likelihood of sleep and affect-related problems, for multiple reasons that may vary across people. Research on the mechanisms accounting for the effects of exercise on sleep and affect is an exciting area for future research, which will advance our understanding of the links between sleep and affect.

Many factors may predispose individuals to problems related to sleep and affect. Even within the specific area of hyperarousal, the arousal system is complex, and abnormalities at any point in the system could result in varying degrees of hyperarousal, setting the stage for problems related to sleep and affect. Moreover, as discussed by Fairholme and Manber (chapters 3 and 11), hedonic valence, a second core feature of affective experiences, also appears to influence sleep, but in ways that differ from the effects of arousal. As such, several predisposing factors are likely involved in sleep-affect links, and research employing broader levels of analysis, such as those examining the functioning of systems, will continue to shed light on the myriad reasons for which sleep and affect are interrelated.

CAUSAL LINKS

Research examining the factors underlying causal links between sleep and affect also continues to uncover new relations between the two. One particularly interesting line of work, which is highlighted throughout this book, indicates that sleep problems may impair a person's ability to regulate affective experiences. This emerging area of study suggests that sleep may help regulate the brain functioning associated with emotion and emotion regulation (Walker, 2009). Regions within the PFC are central to the sleep and emotion regulation systems. Research and theory has demonstrated that sleep loss weakens PFC-driven regulation of other brain regions, which can result in alterations in goal-directed behavior, a decrease in the ability to regulate emotional experience and expression, and a decrease in impulse control (Dahl, 1996). For example, sleep loss amplifies negative affective responses to daily events, while, at the same time, blunting the experience of positive affect in response to rewarding and pleasurable activities (Walker, 2009; Zohar, Tzischinsky, Epstein, & Lavie, 2005). This finding is supported by both clinical observation and empirical study. Similarly, 1 week of sleep restriction (i.e., limited to 5h of sleep per night) seems to progressively increase negative affect (Dinges et al., 1997). Both acute sleep deprivation and prolonged sleep restriction have been shown to increase amygdala reactivity and to decrease connectivity between the amygdala and the medial PFC (Motomura et al., 2013; Yoo, Gujar, Hu, Jolesz, & Walker, 2007). In concert with models of emotion regulation (e.g., Gross, 2014; Ochsner & Gross, 2005; Phillips et al., 2008), these findings suggest that sleep loss may cause problems downregulating emotional experience, which may account for the potentiation of anxious reactivity observed following sleep deprivation (e.g., Babson et al., 2009, 2010; Roy-Byrne et al., 1986; Sagaspe et al., 2006). Across more naturalistically occurring periods of prolonged sleep deprivation, the resulting elevation in anxious and fearful reactivity may lead to chronic deficits in emotion regulation and repeated anxiety- and fear-related conditioning trials, ultimately culminating in chronic, environment-elicited anxious hyperarousal and pathological levels of anxiety and fear. Moreover, sleep enhances extinction learning (Kleim et al., 2013; Pace-Schott et al., 2011, 2014; Pace-Schott, Verga, Bennett, & Spencer, 2012), and therefore, chronic sleep deprivation likely interferes with normative safety learning, allowing fear- and anxiety-related conditioning to accumulate to pathological levels.

Elevated levels of affect, particularly affect characterized by hyperarousal, may also interfere with sleep and cause the development of sleep problems. The elicitation of arousal-related negative affect interferes with sleep (Gross & Borkovec, 1982; Haynes, Adams, & Frazen, 1981). Also, disordered negative affect predicts the development of sleep problems (Johnson, Roth, & Breslau, 2006; Vahtera et al., 2007). It is possible that disordered affect causes problems sleeping via elevated arousal inherent to many types of negative affect (e.g., anxiety, fear, anger, mania). Indeed, as discussed above, arousal is central to both sleep and several types of negative affect.

Factors that Influence Links Between Sleep and Affect

Another research avenue for advancing our understanding of the links between sleep and affect is identifying factors that influence the magnitude and direction of the links between the two phenomena. Two such factors discussed in the current volume are developmental stage and diagnostic status.

Developmental Stage

Sleep, affect, and developmental stage appear to be interrelated in many different ways. Broadly, age, which can be considered a proxy for developmental stage, appears to influence the relationship between sleep and emotion regulation. For example, older adults report less negative mood in the context of poor sleep and less improvement in mood following increased sleep duration, as compared to younger adults (Ready, Marquez, & Akerstedt, 2009). Adolescence is an important developmental stage for understanding the interplay between sleep and affect. Unfortunately, shifts toward sleep deprivation in this developmental epoch are increasingly common, and they appear to have pronounced effects on affect dysregulation, even when adolescents are compared to sleep-deprived adults (McGlinchey et al., 2011; Talbot et al., 2010). Additional research, particularly among elderly people and across cohorts, is now needed to improve our currently limited understanding of why links between sleep and affect appear to vary as a function of developmental stage.

Diagnostic Status

Another particularly interesting set of findings suggests that sleep deprivation increases negative affect and depressive symptoms in healthy adults, whereas it temporarily decreases symptoms of depression among a substantial proportion of people already suffering from depression

(Campos-Morales, Valencia-Flores, & Castano-Meneses, 2005; Cutler & Cohen, 1979; Kahn-Greene, Killgore, Kamimori, Balkin, & Killgore, 2007; Naylor, King, & Lindsay, 1993; Pflug, 1976; Vogel, Vogel, McAbee, & Thurmond, 1980; Wu & Bunney, 1990; Zohar et al., 2005). A recent hypothesis for the (temporary) antidepressant effects of acute sleep deprivation, which contrast with the wide array of negative effects on healthy individuals, is that acute sleep deprivation may reset abnormal clock gene machinery (Bunney & Bunney, 2013). This hypothesis will undoubtedly stimulate novel research in the field. Indeed, researchers must address many questions related to how factors marked by diagnostic status may moderate links between sleep and affect.

Conclusions and Future Directions

Understanding the relationships between sleep and affect will continue to challenge researchers for the foreseeable future. Research that enhances our depth of understanding is certainly needed. Such researcher could further explain the factors underlying hyperarousal and how this arousal relates to problems sleeping and dysregulated affect. Additional work could also address the impact of affective valence relative to arousal. Increasingly sophisticated models of sleep and emotion regulation will allow researchers to refine their understanding of common features of problems with sleep and affect. However, research must investigate the full breadth of the sleep-affect associations in order to define the parameters of the links between sleep and affect. For example, the increasingly apparent role of arousal in both sleep and several affective states has yielded an array of sophisticated studies that have advanced our understanding of sleep-affect linkages. Work that continues to outline how normative and dysregulated affect and sleep interact will undoubtedly yield novel, testable hypotheses in this domain. Research has already linked chronic infant nighttime waking to maternal depressive symptoms (Karraker & Young, 2007). Research among adolescents and adults can usefully examine the impact of sleep problems on social networks, because such networks are involved in psychopathology related to dysregulated affect. Similarly, understanding the role of substance use may help advance our understanding of the links between sleep and affect. A large literature beyond the scope of this chapter has documented links between use of various substances and dysregulated affect. Moreover, the use of certain types of substances affects sleep (Johnson, Breslau, Roehrs, & Roth, 1999; Tynjala, Kannas, & Levalahti, 1997). Of particular relevance to the current context, research suggests people often use substances in an attempt to modulate dysregulated affect (Khantzian, 1985), thus leading to sleep problems.

CLINICAL IMPLICATIONS

Assessment of Sleep and Affect

Gold standard assessment tools have been established to measure both affect and sleep. These assessments range from self-report (e.g., retrospective, structured interview or daily diaries) to objective measures (e.g., physiological, behavioral coding). The integration of sleep and affect measures within research and clinical settings is critical in order for scholars to best understand the co-occurrence and interrelatedness of affect and sleep. Furthermore, multimodal assessment is critical because different forms of assessment can provide very different results, depending on the characteristics of the sample or individual. A prime example of these differences can be seen in comparing self-reported and objective measures of sleep among individuals with different symptoms of psychopathology.

Objective Versus Self-report

A broad literature has demonstrated mixed findings when comparing objective sleep assessment (polysomnography/actigraphy) and the self-report assessment of sleep. In addition, these mixed findings vary based on the type of affect examined (i.e., sadness, fear, anxiety). A substantial body of work has documented the presence of alterations in the objective sleep architecture of individuals with depression compared to nonpsychiatric controls (Krystal, Thakur, & Roth, 2008). Consistent differences in the sleep architecture of individuals with depression include increased time spent awake, reduction in slow-wave sleep, and abnormalities in REM sleep. Although objective abnormalities in sleep have been consistently associated with depressed mood, this association does not exist in individuals with anxiety. In fact, objective sleep patterns among individuals with anxiety have yielded mixed findings based on the form of anxiety (e.g., posttraumatic stress, generalized anxiety, panic, obsessions), suggesting a role for third variables that are currently less well understood. A complicated literature has resulted, with studies inconsistently observing objective sleep abnormalities, while consistently observing self-reported sleep disturbances. Research on sleep among individuals with posttraumatic stress disorder (PTSD) provides a prime example of these mixed findings. Although self-reported sleep disturbances are consistently reported among people with PTSD, some studies have documented no changes in sleep architecture when measured in the laboratory (Breslau et al., 2004; Lavie, Katz, Pillar, & Zinger, 1998), but other studies have noted abnormalities (Dow, Kelsoe, & Gillin, 1996; Fuller, Waters, & Scott, 1994; Germain & Nielsen, 2003). However, a recent meta-analysis conducted on the topic demonstrated that objective sleep abnormalities do exist within the context of PTSD. These abnormalities include increased stage 1 sleep, decreased slow-wave sleep, and increased REM density in those with PTSD, as compared to those without the condition. Furthermore, researchers have determined that moderating factors can account for the mixed findings. Specifically, age, sex, type of trauma (military versus civilian), time since trauma, comorbid depression, and comorbid substance use disorders all moderate the relationship between PTSD and objective sleep abnormalities (Kobayashi, Boarts, & Delahanty, 2007), potentially accounting for the mixed findings related to objectively measured sleep parameters. Although a substantial amount of work using objective sleep measures has been conducted in relation to depressed mood and anxiety, little objective sleep research has addressed sleep behavior in the context of anger and positive affect.

Interventions Targeting Sleep to Improve Affect

Pharmacological (e.g., benzodiazepine hypnotics) and behavioral interventions (cognitive behavioral therapy for insomnia [CBT-I]) have been shown to be efficacious and effective interventions for the treatment of sleep disturbances and insomnia (Morgenthaler et al., 2006; Morin, Culbert, & Schwartz, 1994; Roehrs & Roth, 2012). Studies directly comparing pharmacological and behavioral interventions for sleep disturbances have yielded comparable treatment outcomes, with each having different impacts on sleep. Pharmacological interventions seem to offer a more immediate impact on sleep disturbances, and behavioral interventions may take longer to work, but tend to have greater long-term outcomes and sustained treatment gains (McClusky, Milby, Switzer, Williams, & Wooten, 1991; Morin, Colecchi, Stone, Sood, & Brink, 1999, Morin et al., 2009). This research suggests that optimal sleep treatment outcomes occur among those who begin treatment with a combination of pharmacotherapy and behavioral interventions (specifically CBT-I) and then discontinue pharmacotherapy while maintaining CBT-I (Morin et al., 2009).

In addition to the efficacy and effectiveness of CBT-I among individuals with primary insomnia (Morgenthaler et al., 2006; Morin et al., 1994), recent work has also demonstrated that CBT-I is an effective treatment for individuals with comorbid conditions. This group includes individuals with symptoms of depression (Manber et al., 2008; Morawetz, 2003; Taylor, Lichstein, Weinstock, Sanford, & Temple, 2007), PTSD (Germain, Shear, Hall, & Buysse, 2007; Zayfert & DeViva, 2004), chronic pain (Currie, Wilson, & Curran, 2002; Edinger, Wohlgemuth, Krystal, & Rice, 2005; Jungquist et al., 2010; Vitiello, Rybarczyk, Von Korff, & Stepanski, 2009), and substance use disorders (Arnedt, Conroy, Armitage, & Brower, 2011).

Given the impact of CBT-I among individuals with affective disturbances, it is not surprising that the benefits of CBT-I extend well beyond sleep improvements. CBT-I has been shown to result in improvements in mood and quality of life among individuals with comorbid conditions. The majority of this work has been conducted on individuals with co-occurring depressive symptoms, and such studies indicate that CBT-I improves general well-being, while decreasing depressive symptom severity and suicidal ideation among individuals with elevated depressive symptoms (Manber et al., 2011). Furthermore, among individuals with insomnia and co-occurring depressed mood, CBT-I appears to be linked to improvements in self-reported sleep and decreases in depressive symptoms (Kuo, Manber, & Loewy, 2001; Manber et al., 2008). Relatively less work has been conducted on CBT-I used to treat individuals with a broad range of anxiety disorders. However, increasing work is being done in terms of CBT-I use for individuals with PTSD. For example, CBT-I has been shown to improve self-reported sleep problems and to decrease posttraumatic stress symptom severity, general anxiety, and low mood among individuals with PTSD (Germain et al., 2007; Krakow et al., 2001; Ulmer, Edinger, & Calhoun, 2011). CBT-I has also been shown to be efficacious among individuals with medical conditions. For example, among women with breast cancer, CBT-I was shown to improve self-reported sleep disturbances and to decrease trait anxiety and depression symptoms (Dirksen & Epstein, 2008). In a second study conducted among individuals with chronic pain, the administration of CBT-I resulted in a significant increase in overall self-reported sleep quality and a significant reduction in the amount that the pain interfered with daily functioning (Jungquist et al., 2010).

Interventions Targeting Affect to Improve Sleep

Although preliminary work has suggested that sleep interventions may impact mood and anxiety symptoms, pharmacological and behavioral interventions for mood and anxiety disorders may also affect sleep outcomes. However, relatively less work has been conducted in this area, and of the work that has been conducted, results have been mixed. Extensive research has demonstrated the efficacy and effectiveness of pharmacotherapy and behavioral interventions for mood and anxiety symptoms. This is an extensive literature well beyond the scope of this chapter. For this reason, we briefly discuss pharmacotherapy and two behavioral interventions. One intervention is a well-established effective treatment, and the other is a promising new behavioral approach for mood and anxiety symptoms, which has been examined within the context of co-occurring sleep disturbances. These interventions are cognitive behavioral therapy (CBT) and mindfulness-based interventions (MBIs), respectively. In terms of pharmacotherapy, a range of hypnotic sleep medications can be used to treat sleep problems (see section "Interventions Targeting Sleep to Improve Affect" above for details). However, mood and anxiety medications have also been used off-label for sleep interventions. Antidepressants (Trazadone, Amytriptyline,

Mirtazapine), atypical antipsychotics (Quetiapine, Olanzepine), and anxiolytics (Lorazepam, Cloazepam) have all been used, in low doses, to treat sleep. Very limited research has been conducted on the use of these off-label medications for sleep, however. At this point, little evidence indicates the safe and effective dose range or the short- and long-term side effects of these drugs (Roehrs & Roth, 2012). Within the context of CBT, researchers have observed mixed results, depending on the form of anxiety investigated. For example, among individuals with generalized anxiety disorder, CBT seemed to result in a reduction in worry and rumination, as well as an improvement in self-reported sleep quality (Bélanger et al., 2004). In comparison, among individuals with PTSD, clinically significant levels of insomnia remained after successful treatment of PTSD (Galvoski, Monson, Bruce, & Resick, 2009; Zayfert & DeViva, 2004). Research is only now beginning to investigate the impact of MBIs on mood and sleep, and few studies have employed control conditions, indicating a need for continued research. To date, this research has yielded mixed results based on method of sleep assessment. For example, MBI has been shown to improve self-reported sleep quality, but it did not effect or decrease objective indices of sleep (Britton, Haynes, Fridel, & Bootzin, 2010). These preliminary findings suggest that MBI may impact affect, thereby improving perception of sleep rather than structurally changing sleep. Future research is needed to better understand the role of MBI on sleep.

Emerging research is showing that exercise may benefit sleep. As a stand-alone or adjunct intervention involving aerobic or resistance training, exercise appears to reduce anxiety and depression and to improve sleep. As discussed above, researchers have hypothesized several mechanisms underlying the relationships between sleep, affect, and exercise. One hypothesis is that exercise improves affect through a reduction in arousal and depression, which may then result in improved sleep. As is the case with MBI, additional research is needed to better understand the role of exercise as a potential clinical and prevention intervention for sleep and affective disturbances.

Combined Interventions for Sleep and Affect

Recent work has started to investigate combined interventions for problems related to both sleep and affect. For example, Manber et al. (2008) have provided support for the combination of antidepressant medication and CBT-I for individuals with insomnia and comorbid depressive symptoms. Their results demonstrated that those receiving this combined treatment had greater rates of remission for depression and insomnia symptoms, as well as an improvement in self-reported and objective measures of sleep. Additional research is needed in this area to further examine the benefit of combined sleep-affect interventions.

Conclusions and Future Directions

Preliminary evidence suggests that sleep interventions (particularly CBT-I) may improve mood. However, less data supports the idea that mood interventions benefit sleep, with some research highlighting the residual nature of sleep problems following cognitive behavioral interventions for PTSD. Thus, there is a critical need to incorporate intervention components targeting improved affect in sleep treatment and vice versa in order to optimize treatment outcomes.

Researchers must attempt to uncover the role of sleep within affective interventions and the role of affect in sleep interventions. This is an up-and-coming area of work that will undoubtedly help clinicians optimize treatment outcomes. Studies are currently examining integrated sleep-affect interventions for PTSD, suicide, depression, and substance use disorders. Mobile technology also offers a promising option for future work. The development of empirically supported interventions delivered via mobile applications will allow clinicians to disseminate empirically supported interventions to individuals who would otherwise not receive treatment due to lack of resources or stigma. Furthermore, delivering an intervention via a mobile application can allow for adjunctive interventions to be added to treatment as usual, while decreasing clinician burden and allowing the client to have access to the intervention around the clock, including those times when symptoms may peak. In fact, CBT-I has been developed into a mobile application called CBT-I Coach (Hoffman et al., 2013), and clinicians treating negative affect (e.g., depression or anxiety treatment) can assign CBT-I coach to clients in order to concurrently address sleep problems. Given the infancy of this approach, additional research is needed before conclusions can be drawn about its efficacy.

GENERAL CONCLUSIONS

Our understanding of the associations between sleep and affect is rapidly growing, and this improved knowledge will enhance interventions aimed at ameliorating the extreme suffering and costs associated with problems of sleep and affect. Technological advances are making the study of these complex phenomena more sophisticated and increasingly available to researchers with various levels of expertise and experience. Although sophisticated models spanning psychological and biological levels of analysis are now being refined, an incredible amount of work remains to be done in order to fully understand why sleep and affect interact in the complex and myriad ways they do. Nonetheless, clinicians and scholars continue to develop and refine interventions with very promising effects on this constellation of problems, and the future advancement of nonclinical research will certainly produce improvements in treatment and prevention approaches.

References

Arnedt, J., Conroy, D., Armitage, R., & Brower, K. (2011). Cognitive-behavioral therapy for insomnia in alcohol-dependent patients: A randomized controlled pilot trial. *Behaviour Research and Therapy, 49,* 227–233. http://dx.doi.org/10.1016/j.brat.2011.02.003.

Babson, K. A., Feldner, M. T., Trainor, C. D., & Smith, R. C. (2009). An experimental investigation of the effects of acute sleep deprivation on panic-relevant biological challenge responding. *Behavior Therapy, 40,* 239–250. http://dx.doi.org/10.1016/j.beth.2008.06.001.

Babson, K. A., Trainor, C. D., Feldner, M. T., & Blumenthal, H. (2010). A test of the effects of acute sleep deprivation on general and specific self-reported anxiety and depressive symptoms: An experimental extension. *Journal of Behavior Therapy and Experimental Psychiatry, 41,* 297–303. http://dx.doi.org/10.1016/j.jbtep.2010.02.008.

Baglioni, C., Spiegelhalder, K., Lombardo, C., & Riemann, D. (2010). Sleep and emotions: A focus on insomnia. *Sleep Medicine Reviews, 14,* 227–238. http://dx.doi.org/10.1016/j.smrv.2009.10.007.

Barbini, B., Bertelli, S., Colombo, C., & Smeraldi, E. (1996). Sleep loss, a possible factor in augmenting manic episode. *Psychiatry Research, 65,* 121–125. http://dx.doi.org/10.1016/S0165-1781(96)02909-5.

Bélanger, L., Morin, C. M., Langlois, F., & Ladouceur, R. (2004). Insomnia and generalized anxiety disorder: Effects of cognitive behavior therapy for GAD on insomnia symptoms. *Journal of Anxiety Disorders, 18,* 561–571. http://dx.doi.org/10.1016/S0887-6185(03)00031-8.

Bonnet, M. H., & Arand, D. L. (2010). Hyperarousal and insomnia: State of the science. *Sleep Medicine Reviews, 14,* 9–15. http://dx.doi.org/10.1016/j.smrv.2009.05.002.

Bouton, M. E., Mineka, S., & Barlow, D. H. (2001). A modern learning theory perspective on the etiology of panic disorder. *Psychological Review, 108,* 4–32. http://dx.doi.org/10.1037/0033-295X.108.1.4.

Breslau, N., Roth, T., Burduvali, E., Kapke, A., Schultz, L., & Roehrs, T. (2004). Sleep in lifetime posttraumatic stress disorder: A community-based polysomnographic study. *Archives of General Psychiatry, 61,* 508–516. http://dx.doi.org/10.1001/archpsyc.61.5.508.

Britton, W. B., Haynes, P. L., Fridel, K. W., & Bootzin, R. R. (2010). Polysomnographic and subjective profiles of sleep continuity before and after mindfulness-based cognitive therapy in partially remitted depression. *Psychosomatic Medicine, 72,* 539–548. http://dx.doi.org/10.1097/PSY.0b013e3181dc1bad.

Buman, M. P., & King, A. C. (2010). Exercise as a treatment to enhance sleep. *American Journal of Lifestyle Medicine, 4,* 500–514. http://dx.doi.org/10.1177/1559827610375532.

Bunney, B. G., & Bunney, W. E. (2013). Mechanisms of rapid antidepressant effects of sleep deprivation therapy: Clock genes and circadian rhythms. *Biological Psychiatry, 73,* 1164–1171. http://dx.doi.org/10.1016/j.biopsych.2012.07.020.

Campos-Morales, R., Valencia-Flores, M., & Castano-Meneses, A. (2005). Sleepiness, performance and mood state in a group of Mexican undergraduate students. *Biological Rhythm Research, 96,* 1–8. http://dx.doi.org/10.1080/09291010400028484.

Centers for Disease Control and Prevention. (2014). *Insufficient sleep is a public health epidemic.* National Center for Chronic Disease and Prevention and Health Promotion, Division of Adult and Community Health. http://www.cdc.gov/features/dssleep/. (Accessed 01.06.14).

Cisler, J. M., & Koster, E. H. W. (2010). Mechanisms of attentional biases towards threat in anxiety disorders: An integrative review. *Clinical Psychology Review, 30,* 203–216, PubMed: 20005616.

Cote, K. A., McCormick, C. M., Geniole, S. N., Renn, R. P., & MacAulay, S. D. (2013). Sleep deprivation lowers reactive aggression and testosterone in men. *Biological Psychology, 92,* 249–256. http://dx.doi.org/10.1016/j.biopsycho.2012.09.011.

REFERENCES

Currie, S. R., Wilson, K. G., & Curran, D. (2002). Clinical significance and predictors of treatment response to cognitive-behavior therapy for insomnia secondary to chronic pain. *Journal of Behavioral Medicine, 25,* 135–153. http://dx.doi.org/10.1023/A:1014832720903.

Cutler, N., & Cohen, H. (1979). The effect of one night's sleep loss on mood and memory in normal subjects. *Comprehensive Psychiatry, 20,* 61–66. http://dx.doi.org/10.1016/0010-440X(79)90060-9.

Dahl, R. E. (1996). The regulation of sleep and arousal: Development and psychopathology. *Development and Psychopathology, 8,* 3–27. http://dx.doi.org/10.1017/S0954579400006945.

Davidson, R. J. (2000). Affective style, psychopathology, and resilience: Brain mechanisms and plasticity. *American Psychologist, 55,* 1196–1214. http://dx.doi.org/10.1037/0003-066X.55.11.1196.

Dinges, D. F., Pack, F., Williams, K., Gillen, K. A., Powell, J. W., Ott, G. E., et al. (1997). Cumulative sleepiness, mood disturbance and psychomotor vigilance performance decrements during aweek of sleep restricted to 4-5 hours per night. *Sleep: Journal of Sleep Research & Sleep Medicine, 20,* 267–277, PubMed: 9231952.

Dirksen, S. R., & Epstein, D. R. (2008). Efficacy of an insomnia intervention on fatigue, mood and quality of life in breast cancer survivors. *Journal of Advanced Nursing, 61,* 664–675. http://dx.doi.org/10.1111/j.1365-2648.2007.04560.x.

Dow, B. M., Kelsoe, J. R., Jr., & Gillin, J. C. (1996). Sleep and dreams in Vietnam PTSD and depression. *Biological Psychiatry, 39,* 42–50. http://dx.doi.org/10.1016/0006-3223(95)00103-4.

Edinger, J. D., Wohlgemuth, W. K., Krystal, A. D., & Rice, J. R. (2005). Behavioral insomnia therapy for fibromyalgia patients: A randomized clinical trial. *Archives of Internal Medicine, 165,* 2527–2535. http://dx.doi.org/10.1001/archinte.165.21.2527.

Ford, D. E., & Kamerow, D. B. (1989). Epidemiologic study of sleep disturbances and psychiatric disorders. An opportunity for prevention? *Journal of the American Medical Association, 262,* 1479–1484. http://dx.doi.org/10.1001/jama.1989.03430110069030.

Fosse, R., Stickgold, R., & Hobson, A. (2002). Emotional experience during rapid-eye-movement sleep in narcolepsy. *Sleep, 25,* 724–732, PubMed: 12405607.

Fredriksen, K., Rhodes, J., Reddy, R., & Way, N. (2004). Sleepless in Chicago: Tracking the effects of adolescent sleep loss during the middle school years. *Child Development, 75,* 84–95. http://dx.doi.org/10.1111/j.1467-8624.2004.00655.x.

Fuller, K. H., Waters, W. F., & Scott, O. (1994). An investigation of slow-wave sleep processes in chronic PTSD patients. *Journal of Anxiety Disorders, 8,* 227–236. http://dx.doi.org/10.1016/0887-6185(94)90004-3.

Galvoski, T., Monson, C., Bruce, S., & Resick, P. (2009). Does cognitive-behavioral therapy for PTSD improve perceived health and sleep impairment? *Journal of Traumatic Stress, 22,* 197–204. http://dx.doi.org/10.1002/jts.20418.

Germain, A., & Nielsen, T. A. (2003). Sleep pathophysiology in posttraumatic stress disorder and idiopathic nightmare sufferers. *Biological Psychiatry, 54,* 1092–1098. http://dx.doi.org/10.1016/S0006-3223(03)00071-4.

Germain, A., Shear, M., Hall, M., & Buysse, D. J. (2007). Effects of a brief behavioral treatment for PTSD-related sleep disturbances: A pilot study. *Behaviour Research and Therapy, 45,* 627–632. http://dx.doi.org/10.1016/j.brat.2006.04.009.

Gorman, J. M., Kent, J. M., Sullivan, G. M., & Coplan, J. D. (2000). Neuroanatomical hypothesis of panic disorder, revised. *American Journal of Psychiatry, 157,* 493–505. http://dx.doi.org/10.1176/appi.ajp.157.4.493.

Gregory, A. M., Caspi, A., Eley, T. C., Moffitt, T. E., O'Connor, T. G., & Poulton, R. (2005). Prospective longitudinal associations between persistent sleep problems in childhood and anxiety and depression disorders in adulthood. *Journal of Abnormal Child Psychology, 33,* 157–163. http://dx.doi.org/10.1007/s10802-005-1824-0.

Gregory, A. M., Eley, T. C., O'Connor, T. G., & Plomin, R. (2004). Etiologies of the associations between childhood sleep and behavioral problems in a large twin sample. *Journal of the American Academy of Child and Adolescent Psychiatry, 43,* 748–757. http://dx.doi.org/10.1097/01.chi/0000122798.47863.a5.

Gregory, A. M., & O'Connor, T. G. (2002). Sleep problems in childhood: A longitudinal study of developmental change and association with behavioral problems. *Journal of the American Academy of Child and Adolescent Psychiatry, 41,* 964–971. http://dx.doi.org/10.1097/00004583-200208000-00015.

Gross, J. J. (2014). Emotion regulation: Conceptual and empirical foundations. In J. J. Gross (Ed.), *Handbook of emotion regulation* (2nd ed., pp. 3–20).

Gross, R. T., & Borkovec, T. D. (1982). Effects of a cognitive intrusion manipulation on the sleep-onset latency of good sleepers. *Behavior Therapy, 13,* 112–116. http://dx.doi.org/10.1016/S0005-7894(82)80054-3.

Harvey, A. G., Murray, G., Chandler, R. A., & Soehner, A. (2011). Sleep disturbance as transdiagnostic: Consideration of neurobiological mechanisms. *Clinical Psychology Review, 31,* 225–235. http://dx.doi.org/10.1016/j.cpr.2010.04.003.

Haynes, S., Adams, A., & Frazen, M. (1981). The effects of presleep stress on sleep-onset insomnia. *Journal of Abnormal Psychology, 90,* 601–606. http://dx.doi.org/10.1037/0021-843X.90.6.601.

Hoffman, J., Taylor, K., Manber, R., Trockel, M., Gehrman, P., Woodward, S., et al. (2013). CBT-I coach (version 1.0) [Mobile application software]. Retrieved from: http://itunes.apple.com.

Hofmann, S. G., Ellard, K. K., & Siegle, G. J. (2012). Neurobiological correlates of cognitions in fear and anxiety: A cognitive-neurobiological information-processing model. *Cognition and Emotion, 26,* 282–299. http://dx.doi.org/10.1080/02699931.2011.579414.

Johnson, E. O., Breslau, N., Roehrs, T., & Roth, T. (1999). Insomnia in adolescence: Epidemiology and associate problems. *Sleep, 22,* s22, Retrieved via PubMed.

Johnson, E., Roth, T., & Breslau, N. (2006). The association of insomnia with anxiety disorders and depression: Exploration of the direction of risk. *Journal of Psychiatric Research, 40,* 700–708. http://dx.doi.org/10.1016/j.jpsychires.2006.07.008.

Jungquist, C. R., O'Brien, C., Matteson-Rusby, S., Smith, M. T., Pigeon, W. R., Xia, Y., et al. (2010). The efficacy of cognitive-behavioral therapy for insomnia in patients with chronic pain. *Sleep Medicine, 11,* 302–309. http://dx.doi.org/10.1016/j.sleep.2009.05.018.

Kahn-Greene, E. T., Killgore, D. B., Kamimori, G. H., Balkin, T. J., & Killgore, W. D. S. (2007). The effects of sleep deprivation on symptoms of psychopathology in healthy adults. *Sleep Medicine, 8,* 215–221. http://dx.doi.org/10.1016/j.sleep.2006.08.007.

Kahn-Greene, E. T., Lipizzi, E. L., Conrad, A. K., Kamimori, G. H., & Killgore, W. D. S. (2006). Sleep deprivation adversely affects interpersonal responses to frustration. *Personality and Individual Differences, 41,* 1433–1443. http://dx.doi.org/10.1016/j.paid.2006.06.002.

Karraker, K. H., & Young, M. (2007). Night waking in six-month-old infants and maternal depressive symptoms. *Journal of Applied Developmental Psychology, 28,* 493–498, PMID: 19050747.

Khantzian, E. (1985). The self-medication hypothesis of addictive disorders: Focus on heroin and cocaine dependence. In D. Allen (Ed.), *The cocaine crisis.* New York: Plenum.

Kleim, B., Wilhelm, F. H., Temp, L., Margraf, J., Wiederhold, B. K., & Rasch, B. (2013). Sleep enhances exposure therapy. *Psychological Medicine, 44,* 1511–1519, PubMed: 23842278.

Kobayashi, I., Boarts, J. M., & Delahanty, D. L. (2007). Polysomnographically measured sleep abnormalities in PTSD: A meta-analytic review. *Psychophysiology, 44,* 660–669. http://dx.doi.org/10.1111/j.1469-8986.2007.537.x.

Koenigsberg, H. W., Pollak, C., & Sullivan, T. (1987). The sleep lactate infusion: Arousal and the panic mechanism. *Biological Psychiatry, 22,* 789–791. http://dx.doi.org/10.1016/0006-3223(87)90215-0.

Krakow, B., Johnston, L., Melendrez, D., Hollifield, M., Warner, T. D., & Chavez-Kennedy, D. (2001). An open-label trial of evidence-based cognitve behavior therapy for nightmares and insomnia in crime victims with PTSD. *American Journal of Psychiatry, 158,* 2043–2047, PubMed: 11729023.

Krystal, A. D., Thakur, M., & Roth, T. (2008). Sleep disturbance in psychiatric disorders: Effects on function and quality of life in mood disorders, alcoholism, and schizophrenia. *Annals of Clinical Psychiatry, 20,* 39–46. http://dx.doi.org/10.1080/10401230701844661.

Kuo, T. F., Manber, R., & Loewy, D. (2001). Insomniacs with comorbid depression achieved comparable improvement in a cognitive behavioral group treatment program as insomniacs without comborbid depression. *Sleep, 14,* A62–A63, Retrieved via PubMed.

Lancel, M., Droste, S. K., Sommer, S., & Reul, J. M. (2003). Influence of regular voluntary exercise on spontaneous and social stress-affected sleep in mice. *European Journal of Neuroscience, 17,* 2171–2179, PubMed: 12786984.

Lavie, P., Katz, N., Pillar, G., & Zinger, Y. (1998). Elevated awaking thresholds during sleep: Characteristics of chronic war-related posttraumatic stress disorder patients. *Biological Psychiatry, 44,* 1060–1065. http://dx.doi.org/10.1016/S0006-3223(98)00037-7.

Lindberg, N., Tani, P., Appelberg, B., Stenberg, D., Naukkarinen, H., Rimón, R., et al. (2003). Sleep among habitually violent offenders with antisocial personality disorder. *Neuropsychobiology, 47,* 198–205, PubMed: 12824743.

Manber, R., Bernert, R. A., Suh, S., Nowakowski, S., Siebern, A. T., & Ong, J. C. (2011). CBT for insomnia in patients with high and low depressive symptom severity: Adherence and clinical outcomes. *Journal of Clinical Sleep Medicine, 7,* 645–652. http://dx.doi.org/10.5664/jcsm.1472.

Manber, R., Edinger, J., Gress, J., San Pedro-Salcedo, M., Kuo, T., & Kalista, T. (2008). Cognitive behavioral therapy for insomnia enhances depression outcome in patients with comorbid major depressive disorder and insomnia. *Sleep, 31,* 489–495, Retrieved via PubMed.

McClusky, H. Y., Milby, J. B., Switzer, P. K., Williams, V., & Wooten, V. (1991). Efficacy of behavioral versus triazolam treatment in persistent sleep-onset insomnia. *American Journal of Psychiatry, 148,* 121–126, PMID: 1888345.

McGlinchey, E. L., Talbot, L. S., Chang, K. H., Kaplan, K. A., Dahl, R. E., & Harvey, A. G. (2011). The effect of sleep deprivation on vocal expression of emotion in adolescents and adults. *Sleep, 34,* 1233–1241. http://dx.doi.org/10.5665/SLEEP.1246.

Meerlo, P., Pragt, B. J., & Daan, S. (1997). Social stress induces high intensity sleep in rats. *Neuroscience Letters, 225*(1), 41–44. http://dx.doi.org/10.1016/S0304-3940(97)00180-8.

Meerlo, P., & Turek, F. W. (2001). Effects of social stimuli on sleep in mice: Non-rapid-eye-movement (NREM) sleep is promoted by aggressive interaction but not by sexual interaction. *Brain Research, 907,* 84–92. http://dx.doi.org/10.1016/S0006-8993(01)02603-8.

Mehrabian, A. (1995). Framework for a comprehensive description and measurement of emotional states. *Genetic, Social, and General Psychology Monographs, 121,* 339–361, PubMed: 7557355.

Miller, M. M., & McEwen, B. S. (2006). Establishing an agenda for translational research on PTSD. *Annals of the New York Academy of Sciences, 1071,* 294–312. http://dx.doi.org/10.1196/annals.1364.023.

Morawetz, D. (2003). Insomnia and depression: Which comes first? *Sleep Research Online, 5,* 77–81. Retrieved from www.sro.org/2003/Morawetz/77/.

Morgenthaler, T., Kraemer, M., Alessi, C., Friedman, L., Boecklecke, B., Brown, T., et al. (2006). Practice parameters for the psychological and behavioral treatment of insomnia: An update. An AASM Report. *Sleep, 29,* 1415–1419, PubMed: 17162987.

Morin, C., Colecchi, C., Stone, J., Sood, R., & Brink, D. (1999). Behavioral and pharmacological therapies for late-life insomnia: A randomized controlled trial. *Journal of the American Medical Association, 28,* 991–999. http://dx.doi.org/10.1001/jama.281.11.991.

Morin, C., Culbert, J., & Schwartz, S. (1994). Nonpharmacological interventions for insomnia: A meta-analysis of treatment efficacy. *American Journal of Psychiatry, 151,* 1172–1180, PubMed: 8037252.

Morin, C. M., Vallières, A., Guay, B., Ivers, H., Savard, J., Mérette, C., et al. (2009). Cognitive behavioral therapy, singly and combined with medication, for persistent insomnia: A randomized controlled trial. *Journal of the American Medical Association, 301,* 2005–2015. http://dx.doi.org/10.1001/jama.2009.682.

Motomura, Y., Kitamura, S., Oba, K., Terasawa, Y., Enomoto, M., Katayose, Y., et al. (2013). Sleep debt elicits negative emotional reaction through diminished amygdala-anterior cingulate functional connectivity. *PLoS ONE, 8,* e56578. http://dx.doi.org/10.1371/journal.pone.0056578.

Naylor, M., King, C., & Lindsay, K. (1993). Sleep deprivation in depressed adolescents and psychiatric controls. *Journal of the American Academy of Child & Adolescent Psychiatry, 32,* 753–759. http://dx.doi.org/10.1097/00004583-199307000-00008.

Neckelmann, D., Mykletun, A., & Dahl, A. (2007). Chronic insomnia as a risk factor for developing anxiety and depression. *Sleep, 30,* 873–880, PubMed: 17682658.

Ochsner, K. N., & Gross, J. J. (2005). The cognitive control of emotion. *Trends in Cognitive Sciences, 9,* 242–249. http://dx.doi.org/10.1016/j.tics.2005.03.010.

Ong, A. D., Exner-Cortens, D., Riffin, C., Steptoe, A., Zautra, A. J., & Almeida, D. M. (2013). Linking stable and dynamic features of positive affect to sleep. *Annals of Behavioral Medicine, 46,* 52–61. http://dx.doi.org/10.1007/s12160-013-9484-8.

Ong, S., Wickramaratne, P., Tang, M., & Weissman, M. (2006). Early childhood sleep and eating problems as predictors of adolesncent and adult mood and anxiety disorders. *Journal of Affective Disorders, 96,* 1–8. http://dx.doi.org/10.1016/j.jad.2006.05.025.

Pace-Schott, E. F., Shepherd, E., Spencer, R. M., Marcello, M., Tucker, M., Propper, R. E., et al. (2011). Napping promotes inter-session habituation to emotional stimuli. *Neurobiology of Learning and Memory, 95,* 24–36. http://dx.doi.org/10.1016/j.nlm.2010.10.006.

Pace-Schott, E. F., Tracy, L. E., Rubin, Z., Mollica, A. G., Ellenbogen, J. M., Bianchi, M., et al. (2014). Interactions of time of day and sleep with between-session habituation and extinction memory in young adult males. *Experimental Brain Research, 232,* 1443–1458. http://dx.doi.org/10.1007/s00221-014-3829-9.

Pace-Schott, E. F., Verga, P. W., Bennett, T. S., & Spencer, R. M. (2012). Sleep promotes consolidation and generalization of extinction learning in simulated exposure therapy for spider fear. *Journal of Psychiatric Research, 46,* 1036–1044. http://dx.doi.org/10.1016/j.jpsychires.2012.04.015.

Pflug, B. (1976). The effect of sleep deprivation on depressed patients. *Acta Psychiatrica Scandinavica, 53,* 148–158. http://dx.doi.org/10.1111/j.1600-0447.1976.tb00068.x.

Phillips, M. L., Ladouceur, C. D., & Drevets, W. C. (2008). A neural model of voluntary and automatic emotion regulation: Implications for understanding the pathophysiology and neurodevelopment of bipolar disorder. *Molecular Psychiatry, 13,* 833–857. http://dx.doi.org/10.1038/mp.2008.65.

Ray, R. D., & Zald, D. H. (2012). Anatomical insights into the interaction of emotion and cognition in the prefrontal cortex. *Neuroscience and Biobehavioral Reviews, 36,* 479–501. http://dx.doi.org/10.1016/j.neubiorev.2011.08.005.

Ready, R., Marquez, D., & Akerstedt, A. (2009). Emotion in younger and older adults: Retrospective and prospective associations with sleep and physical activity. *Experimental Aging Research, 35,* 348–368. http://dx.doi.org/10.1080/03610730902922184.

Roehrs, T., & Roth, T. (2012). Insomnia pharmacotherapy. *Neurotherapeutics, 9,* 728–738. http://dx.doi.org/10.1007/s13311-012-0148-3.

Roth, T., Jaeger, S., Jin, R., Kalsekar, A., Stang, P. E., & Kessler, R. C. (2006). Sleep problems, comorbid mental disorders, and role functioning in the National Comorbidity Survey Replication. *Biological Psychiatry, 60,* 1364–1371. http://dx.doi.org/10.1016/j.biopsych.2006.05.039.

Roy-Byrne, P. P., Uhde, T. W., & Post, R. M. (1986). Effects of one night's sleep deprivation on mood and behavior in panic disorder. *Archives of General Psychiatry, 43,* 895–899. http://dx.doi.org/10.1001/archpsyc.1986.01800090085011.

Sagaspe, P., Sanchez-Ortuno, M., & Charles, A. (2006). Effects of sleep deprivation on colorword, emotional, and specific stroop interference and on self-reported anxiety. *Brain and Cognition, 60,* 76–87. http://dx.doi.org/10.1016/j.bandc.2005.10.001.

REFERENCES

Stathopoulou, G., Powers, M. B., Berry, A. C., Smits, J. A. J., & Otto, M. W. (2006). Exercise interventions for mental health: A quantitative and qualitative review. *Clinical Psychology: Science and Practice, 13,* 179–193. http://dx.doi.org/10.1111/j.1468-2850.2006.00021.x.

Steiger, A. (2002). Sleep and the hypothalamo-pituitary-adrenocortical system. *Sleep Medicine Reviews, 6,* 125–138, PubMed: 12531148.

Steptoe, A., O'Donnell, K., Marmot, M., & Wardle, J. (2008). Positive affect, psychological well-being, and good sleep. *Journal of Psychosomatic Research, 64,* 409–415. http://dx.doi.org/10.1016/j.jpsychores.2007.11.008.

Talbot, L., Hairston, I., Eidelman, P., Gruber, J., & Harvey, A. (2009). The effect of mood on sleep onset latency and REM sleep in interepisode bipolar disorder. *Journal of Abnormal Psychology, 118,* 448. http://dx.doi.org/10.1037/a0016605.

Talbot, L. S., McGlinchey, E. L., Kaplan, K. A., Dahl, R. E., & Harvey, A. G. (2010). Sleep deprivation in adolescents and adults: Changes in affect. *Emotion, 10,* 831–841. http://dx.doi.org/10.1037/a0020138.

Taylor, D. J., Lichstein, K. L., Weinstock, J., Sanford, S., & Temple, J. R. (2007). A pilot study of cognitive-behavioral therapy of insomnia in people with mild depression. *Behavior Therapy, 38,* 49–57. http://dx.doi.org/10.1016/j.beth.2006.04.002.

Tynjala, J., Kannas, L., & Levalahti, E. (1997). Perceived tiredness among adolescents and its association with sleep habits and use of psychoactive substances. *Journal of Sleep Research, 6,* 189–198. http://dx.doi.org/10.1046/j.1365-2869.1997.00048.x.

Ulmer, C., Edinger, J., & Calhoun, P. (2011). A multi-component cognitive-behavioral intervention for sleep disturbance in veterans with PTSD: A pilot study. *Journal of Clinical Sleep Medicine, 7,* 57–68, PubMed: 21344046.

Vahtera, J., Kivimaki, M., Hublin, C., Korkeila, K., Suominen, S., Paunio, T., et al. (2007). Liability to anxiety and severe life events as predictors of new-onset sleep disturbances. *Sleep, 30,* 1537–1546, PubMed: PMC2082106.

Vitiello, M. V., Rybarczyk, B., Von Korff, M., & Stepanski, E. J. (2009). Cognitive behavioral therapy for insomnia improves sleep and decreases pain in older adults with co-morbid insomnia and osteoarthritis. *Journal of Clinical Sleep Medicine, 5,* 355–362, PubMed: 19968014.

Vogel, G. W., Vogel, F., McAbee, R. S., & Thurmond, A. J. (1980). Improvement of depression by REM sleep deprivation. New findings and a theory. *Archives of General Psychiatry, 37,* 247–253. http://dx.doi.org/10.1001/archpsyc.1980.01780160017001.

Vohs, K. D., Glass, B. D., Maddox, W. T., & Markman, A. B. (2011). Ego depletion is not just fatigue: Evidence from a total sleep deprivation experiment. *Social Psychological and Personality Science, 2*(2), 166–173. http://dx.doi.org/10.1177/1948550610386123.

Walker, M. (2009). The role of sleep in cognition and emotion. *Annals of the New York Academy of Science, 1156,* 168–197. http://dx.doi.org/10.1111/j.1749-6632.2009.04416.x.

Whiteford, H. A., Degenhardt, L., Rehm, J., Baxter, A. J., Ferrari, A. J., Erskine, H. E., et al. (2013). Global burden of disease attributable to mental and substance use disorders: Findings from the Global Burden of Disease Study 2010. *The Lancet, 382,* 1575–1586, http://0-dx.doi.org.library.uark.edu/10.1016/S0140-6736(13)61611-6.

Wu, J. C., & Bunney, W. E. (1990). The biological basis of an antidepressant response to sleep deprivation and relapse: review and hypothesis. *American Journal of Psychiatry, 147,* 14–21, PubMed: 2403471.

Yoo, S., Gujar, N., Hu, P., Jolesz, F., & Walker, M. (2007). The human emotional brain without sleep-a prefrontal amygdala disconnect. *Current Biology, 17,* R877–R878. http://dx.doi.org/10.1016/j.cub.2007.08.007.

Zayfert, C., & DeViva, J. C. (2004). Residual insomnia following cognitive behavioral therapy for PTSD. *Journal of Traumatic Stress, 17,* 69–73. http://dx.doi.org/10.1023/B:-JOTS.0000014679.31799.e7.

Zohar, D., Tzischinsky, O., Epstein, R., & Lavie, P. (2005). The effects of sleep loss on medical residents' emotional reactions to work events: A cognitive-energy model. *Sleep: Journal of Sleep and Sleep Disorder Research, 28,* 47–54, PubMed: 15700720.

Index

Note: Page numbers followed by *f* indicate figures and *t* indicate tables.

A

Actigraphy
  anxious arousal, 52
  assessment of sleep, 67t, 74–76, 74f, 75f
  circadian rhythms, 14–15
  infant sleep recording, 402
  MBSR group, 357
  and sleep, 209, 231, 250–251
  sleep-pain studies, 379
  wrist, 354–355
Acute exercise, 326, 327
Acute sleep deprivation
  antidepressant effects, 471–472
  fear/anxiety, 147
  and prolonged sleep
    restriction, 470–471
ADHD. *See* Attention deficit hyperactivity
    disorder (ADHD)
Adolescent sleep, 421
  behavioral interventions, 434
  clinical implications, 432–433
  cognitive function and, 428–429
  developmental stage, 471
  DSPS in, 430–432
  and early-onset bipolar disorder, 300–301
  and health, 429
  maturational changes, 423–424
  and mood/emotion, 426–427
  National Sleep Foundation poll, 429–430
  normal sleep, 422–424
  poor sleep *vs.* behavioral problems,
    427–428
  public health implications, 433–434
  sleep deprivation, 464
  sleep disturbances, 147–149
  technology use, 429–430
  two-process model, 422
  weeknight sleep time, 411–412
Adrenocorticotropin (ACTH)
  fear/anxiety, 152–153
  heart rate, 262
  SWS, 448
Aerobic exercise, 324
Affective Norms for English Words
    (ANEW), 93, 100

Affect regulation
  experiential/analytic strategy, 54
  sleep disturbances on, 303–305
  sleep impact, 47–50, 53–54
Age-related sleep modifications, 441, 446–450
Aggression
  animal study, 258–259
  clinical implications, 264–265
  HPA axis system, 261
  human study, 257–258
  PFC, 259–260
  populations, 251–252
  at risk, 263–264
  serotonin, 263
  subtypes, 248–249
Aging
  emotional processing with, 444–446
  emotion regulation and, 442–444, 448–450
  sleep-wake cycle, 441, 447
Alpha amylase, 128
American Academy of Sleep Medicine
    (AASM)
  circadian misalignment, 15
  MSLT testing, 72–73
  sleep medicine experts, 77–78
American National Sleep Foundation,
    446–447
Amygdala
  emotional processing with aging, 445
  heuristic model, 172
  homeostatic sleep process, 404
  insufficient sleep, 49
  and medial PFC, 405–406
  neurobiological models, 360–361
  overactivation, 165
  during REM, 172
  sleep deprivation, 17
ANEW. *See* Affective Norms for English
    Words (ANEW)
Anger
  clinical implications, 264–265
  correlation, 249–250
  definition, 248–249
  inductions, 123
  irritability, 248–249

Anger *(Continued)*
  measurement, 248–249
  at risk, 263–264
Animal model
  of emotion, 26–28
  fear/anxiety, 154–155
ANS. *See* Autonomic nervous system (ANS)
Antidepressant effects
  acute sleep deprivation, 471–472
  through epigenetic processes, 231–232
Anxiety
  and depression, 282–284
  and fear (*see* Fear/anxiety)
  mood and, 475–476
Anxiety and Preoccupation about Sleep Questionnaire (APSQ), 203
ARAS. *See* Ascending reticular activating system (ARAS)
Arousal
  anxious, 52
  bedtime, 153
  cognitive, 54, 212–215
  emotional, 468–469
  negative affect, 240
  NREM sleep, 9
  physiological, 52, 54
  primary *vs.* secondary, 359–360
  region, 7–8
  vs. sleep, 425–426, 467–468
  valence and, 52–53
Ascending reticular activating system (ARAS), 6–7
  anatomy and function, 7
  bypasses the thalamus, 7
  cholinergic neurons, 7
Attentional deployment, 104
Attention deficit hyperactivity disorder (ADHD), 302–303
Auditory stimuli
  IADS, 95
  musical scores, 95–96
  script-driven imagery, 94–95
Automated computer scoring, 131
Autonomic nervous system (ANS)
  physiological assessment, 104–105
  in REM nightmares, 171, 172

B

Basal forebrain (BF), 5–6, 10–11
Bedtime arousal, 153

Behavior
  assessment, 107
  disinhibition, 16
  problems, poor sleep and, 427–428
Benzodiazepines
  hypnotics, 474
  mania symptoms, 307
  sleep-wake cycle, 297
BF. *See* Basal forebrain (BF)
Biological stimuli
  carbon dioxide-enrichment, 97
  hyperventilation, 96–97
  pharmacological agents, 98
Bipolar disorder, 284
  adolescence and early-onset, 300–301
  depression and, 284–285
  manic illness, 296, 298–299
  risks, 305–306
  symptoms, 298
Blended stimuli
  stressful film paradigm, 98–99
  virtual reality, 99
Bliss, 194
Brain
  activation and REM, 165
  imaging, 67t, 102
  MRS, 82
  overlapping mechanisms, 410
Brainstem
  nerve cells swelling, 4
  reticular formation, 5–6, 6f
  wake-promoting circuits, 5–6

C

CAR. *See* Cortisol awakening response (CAR)
Carbon dioxide-enriched air biological challenge, 97
Cardiovascular measures, 129
CBT. *See* Cognitive behavioral therapy (CBT)
CBT-I. *See* Cognitive behavioral therapy for insomnia (CBT-I)
Central nervous system (CNS)
  hyperarousal, 213–214
  signaling function, 387
Children
  emotional regulation
    abnormal development, 410–412
    affect sleep, 412–413
    definition, 404–405
    overlapping brain mechanisms, 410

homeostatic process (*see* Homeostatic process)
sleep disturbances, 147–149
Cholinergic neurons, 7
Chronic disease prevention, 330–332
Chronic pain
　cognitive-behavioral treatments, 389–390
　conditions, 380–381
　various etiologies, 390–392
Circadian dysrhythmia, 17
Circadian rhythms
　ADHD, 303
　bipolar disorder, 299, 300–301
　clock cells, 13–14
　cortisol, 83
　free-running period, 13–14
　intrinsic, 13
　and light, 14
　misalignment/dyssynchrony, 15
　OCD, 302
　pattern, mood switches, 295–296
　SCN, 66–69
　sleep/wake cycle, 3–4, 12
　time awake length, 13, 13f
Circadian sleep process, 387–389, 400, 406–407
Circadian tendencies, 407
Classical conditioning, emotion, 36–37
Clinical interview, 67t, 76, 77
Cognition
　anxiety, 205
　approaches of emotion, 26–28
　arousal, 54, 212–215
　fear/anxiety, 151
　function and adolescents sleep, 428–429
　mania, 304–305
　mindfulness, 359–360
Cognitive behavioral therapy (CBT)
　fear/anxiety, 153–154
　insomnia, 308
　PTSD and, 475–476
Cognitive behavioral therapy for insomnia (CBT-I)
　clinical interventions, 156–157
　coach, 477
　dysfunctional attitudes, 215–216
　efficacy and effectiveness, 474
　impact, 474–475
　and MBSR, 357–358
　sleep and pain, 389–390
Continuous positive airway pressure (CPAP), 385, 390

Coping style
　bedtime arousal, 153
　fear/anxiety, 154
Corticotropin-releasing hormone (CRH), 152–153
Cortisol
　ACTH and, 152–153, 262
　and alpha amylase, 128
　assessment of sleep, 67t, 83
　salivary assay, 128, 129f
　secretion, 14–15
Cortisol awakening response (CAR), 213–214
CPAP. *See* Continuous positive airway pressure (CPAP)
CRH. *See* Corticotropin-releasing hormone (CRH)
Cyclothymia, 301
Cytokines
　activation and pain, 387
　sleep loss/sleepiness and, 386–387

D

Daily positive affect, 278, 279t
Daytime Insomnia Symptom Response Scale (DISRS), 208, 211–212
Daytime insomnia symptom scale (DISS), 79–80, 277t
Default mode network (DMN), 360–362
Deficiency of sleep, 45
　on emotional functioning, 45–46
　hypothesized effects, 57f
　self-reported, 48
Delayed sleep phase syndrome (DSPS), 430–432
Depression
　anxiety and, 282–284
　and bipolar disorder, 284–285
　evidence, 228–229
　implications, 239–241
　and insomnia, 216–217, 229–230, 236–237, 282–283
　meta-analysis, 229
　and nightmares, 168–170
　severity of, 232–233
　sleep
　　abnormalities, 234–236
　　symptoms, 234
　　treatment, 233–234
Developmental stage, 471
Diagnostic Statistical Manual 5 (DSM-5), 184, 241, 293–294

Diathesis-stress model
  fear/anxiety, 150
  nightmares, 172
  positive affect, 282
Dimensional theory of emotion, 108–109
Dim light melatonin onset (DLMO) testing procedure, 83
Directed emotional expressions, 122–123
Discrete emotion theory, 108–109
Discrete positive affect, 278–281, 280t
DISRS. *See* Daytime Insomnia Symptom Response Scale (DISRS)
DISS. *See* Daytime insomnia symptom scale (DISS)
DMN. *See* Default mode network (DMN)
DSPS. *See* Delayed sleep phase syndrome (DSPS)

E

Early-onset bipolar disorder, 300–301
Ecological momentary assessment (EMA), 105–106, 204
ECQ-R. *See* Emotion Control Questionnaire-Rehearsal Scale (ECQ-R)
EFEs. *See* Emotional facial expressions (EFEs)
Electroencephalography (EEG)
  diagnostic sleep, 69
  fibromyalgia, 382
  recordings, fourier analysis, 72
  REM sleep, 447
Electromyography (EMG), 69, 70f, 130
Electrooculogram (EOG), 69
EMA. *See* Ecological momentary assessment (EMA)
EMFACS, 130
EMG. *See* Electromyography (EMG)
Emotion, 46–47
  adolescent sleep and, 426–427
  affect and, 23
  affective stimuli, 34–35
  animal models, 26–28
  appraisal theories of, 26–27, 32
  arousal, 30–31, 468–469
  basic emotions, 29, 30
  classical conditioning, 36–37
  classifications, 33
  cognitive approaches, 26–28
  definition, 24–25, 29–33
  dimensional approach, 30–31
  empirical approaches, 33
  imagery, 35–36
  introspection and observation, 25–26
  lower-tier sensory processes, 32–33
  mammalian responses, 24–25
  motivation, 24–25
  nonlinearity and parallel processing, 32–33
  observation and, 25–26
  physiological and behavioral measurements, 27–28
  pleasant/unpleasant affects, 30–31
  psychophysiology and human neuroscience, 28–29
  quasi-dualistic approach, 38
  science, 25
  self-reported affect, 39–40
Emotional facial expressions (EFEs), 444–445
Emotional functioning, 45–46, 399
Emotional reactivity
  correlational studies, 402
  homeostatic process, 401–404
Emotional regulation
  abnormal development, 410–412
  affect sleep, 412–413
  definition, 404–405
  homeostatic sleep process, 404–406
  overlapping brain mechanisms, 410
Emotional stimuli, 131–132
Emotional Stroop task, 93–94, 101
Emotion Control Questionnaire-Rehearsal Scale (ECQ-R), 207, 216
Emotion elicitation
  assessment time point selection, 103–104
  auditory stimuli
    IADS, 95
    musical scores, 95–96
    script-driven imagery, 94–95
  behavioral assessment, 107
  behavioral tasks
    anger inductions, 123
    directed emotional expressions, 122–123
    stressors, 123–125
  biological stimuli
    carbon dioxide-enrichment, 97
    hyperventilation, 96–97
    pharmacological agents, 98

blended stimuli
   stressful film paradigm, 98–99
   virtual reality, 99
developmental considerations, 99–102, 107–108
emotional stimuli, neural responses, 131–132
environment, 132–133
fMRI assessment, 106–107
impact of sleep, 109–110
matching method, 108–109
musical scores for, 100
olfactory stimuli, 94
physiological assessment, 104–105
primary perspectives, 92
psychophysiological measurement
   automated computer scoring, 131
   cardiovascular measures, 129
   electromyography, 130
   facial displays, 129–130
   human scoring, 130–131
   pupillometry, 127–128
   saliva assays, 128
research designs, 134–135
self-report assessment, 105–106
sleep manipulations, 133–134
subjective/self-report measures, 125–126
   behavioral tasks, 127
   mood states, 126–127
   visual stimuli, 126
virtual reality, 102–103
visual stimuli
   film clips, 121–122
   IAPS, 92–93
   photographs, 120–121
word stimuli
   ANEW, 93
   emotional Stroop task, 93–94
Emotion-generative process, 46–47, 57–58
Emotion regulation, 33, 37–39
   age-related sleep modifications, 441
   and aging, 442–446
   emotions and, 46–47
   protocols, 450
   sophisticated models, 472
Encephalitis lethargica, 5–7
EOG. *See* Electrooculogram (EOG)
Epworth sleepiness scale (ESS), 78–79
Escitalopram, 155–156
ESM. *See* Experience sampling methodology (ESM)

Eszopiclone, 155–156
Exercise, 322
   effects, 325–327
   and obstructive sleep apnea, 327–328
   and restless leg syndrome, 328
   vs. sleep, 322–325
   and slow-wave sleep, 326
Experience sampling methodology (ESM), 54–55
Eye-tracking methodology, 107

F

Facial action coding system (FACS), 107, 130
Facial displays, of emotion, 129–130
Facial Expression Coding System (FACES), 130–131
Fatigue
   countermeasures, 133–134
   Flinders fatigue scale, 79
   National Transportation Safety Board, 16
   rheumatoid arthritis, 382–383
Fear/anxiety
   adaptive responses and defenses, 144–145
   animal models, 154–155
   and bliss, 194
   clinical interventions, 155–157
   cognitive-behavioral model, 153–154
   cognitive model, 151
   defining, 144–145
   diathesis-stress model, 150
   neurobiological model, 152–153
   neurocognitive model, 151
   and sleep disturbances
      childhood and adolescence, 147–149
      clinical designs, 149–150
      elevations, 147–149
      epidemiological research, 149
      experimental work, 149–150
      insomnia, 145
      panic-relevant biological challenge, 147
      PSG, 145–146
   stimulus control model, 150–151
   symptoms, 145
   top-down approach, 151
Fear-conditioning models, 154–155
Fearful ISP, 182, 183*t*, 184–185
Fibromyalgia
   α-intrusions in, 382
   musculoskeletal pain, 382–383
   PSG study, 382, 391
   zolpidem and, 391–392

Film clips
  emotion elicitation, 121–122
  stressful film paradigm, 98–99
Fisher theory, nightmares, 171
Flinders fatigue scale, 79
fMRI. *See* Functional magnetic resonance imaging (fMRI)
Ford Insomnia in Response to Stress Test (FIRST), 216
FOSQ. *See* Functional outcomes of sleep questionnaire (FOSQ)
Fourier analysis, of EEG recordings, 72
Freud's original dream theory, 170
Functional magnetic resonance imaging (fMRI), 3–4
  assessment, 106–107
  brain activation, 81
  emotional stimuli, 131, 132
  emotion elicitation, 102–103
  homeostatic sleep process, 404
  negative visual stimuli, 49
Functional outcomes of sleep questionnaire (FOSQ), 79

G
Generalized anxiety disorder (GAD), 146
Glucocorticoids, fear/anxiety, 152–153

H
Hallucinations, ISP, 189, 190*t*, 191
Happiness, 95–96, 100, 278–281
Hartmann theory, nightmares, 171
Headache, 380–381
Health education, and MBSR, 356–358
Healthy older adults (HOA), 442–444
Heart rate acceleration, 104–105
Heart rate (HR), presleep worry, 213
Heart rate variability (HRV), 202, 213
HOA. *See* Healthy older adults (HOA)
Homeostatic process, 400, 401
  drive for, 66–69
  emotional reactivity, 401–404
  emotional regulation, 404–406
  executive function, 405
  functional magnetic resonance imaging, 404
  parts, 424–425
Hormones, involvement in sleep, 82
Hostility, 248–249
HPA axis. *See* Hypothalamic-pituitary-adrenal (HPA) axis
HRV. *See* Heart rate variability (HRV)
Human neuroscience, and emotion, 28–29
Human scoring, 130–131

Hyperarousal, 467–469
Hypersomnia, 241, 284
Hyperventilation, 96–97
Hypnogram, 69–70, 71*f*
Hypocretin, 7–8
Hypothalamic-pituitary-adrenal (HPA) axis
  activity, 213–214, 216–217
  conduct problems, 261–262
  fear/anxiety, 152–153
  function, 261
  hyperarousal and hyperreactivity, 261–262
  neurobiological mechanisms, 152–153
  relation with sleep, 262
  stressors, 123–125

I
IADS. *See* International Affective Digital Sounds (IADS)
IAPS. *See* International Affective Picture System (IAPS)
ICSD-2. *See* International Classification of Sleep Disorders (ICSD-2)
IES-I. *See* Impact of Events Scale-Intrusion subscale (IES-I)
Imagery of emotion, 35–36
Imagery rehearsal therapy (IRT), 173
Impact of Events Scale-Intrusion subscale (IES-I), 207–208
Insomnia
  CBT-I (*see* Cognitive behavioral therapy for insomnia (CBT-I))
  characterization, 45–46
  classifications, 144
  clinical interventions, 155–156
  cognitive behavioral therapy, 308
  definition, 282–283
  and depression, 216–217, 229–230, 236–237, 282–283
  DISS, 79–80
  MBSR, 355–356
  persistent, 321–322
  severity index, 80
  sleep disturbance and, 145, 216–217
Instrumental aggressive behavior, 248–249
Insufficient sleep
  attentional process, 57–58
  consequence, 56
  diagnostic features, 380–381
  emotional reactivity, 47–48, 49
  future depressive episode, 229–230
  imaging study, 50
  and nonrestorative sleep, 400
  syndrome, 390

International Affective Digital Sounds (IADS), 95, 101
International Affective Picture System (IAPS), 34–35, 92–93, 120–121
International Classification of Sleep Disorders (ICSD-2), 144, 184
Intrinsically photoreceptive retinal ganglion cells, 14
Introspection, 25–26, 30–31
IRT. *See* Imagery rehearsal therapy (IRT)
Isolated sleep paralysis (ISP). *See also* Sleep paralysis
  associated comorbidities, 185–186
  atonia and related sensations, 188–189
  components, 187
  cultural and historical contexts, 186–188
  definition, 184–185
  description, 181, 182
  diagnostic criteria for, 183*t*
  empirical research on, 181
  hallucinations and experiences, 189, 190*t*, 191
  phenomenology and specific features, 188–192
  prevalence rates, 185
  sensorium in, 192–193
  sleep paralysis and, 185–186
  tactile sensations, 189–190
Isotemporal substitution paradigm, 329–330

K
Klüver-Bucy syndrome, 27–28

L
Lack of sleep, 77–78, 143–144
Laser-evoked potentials (LEPs), 384–385, 388–389
Lateral hypothalamic area (LHA), 7–8, 8*f*
Likert-type scale, 207
Locus coeruleus (LC), 7, 8*f*
Long sleep, 234–235, 241, 325–326

M
Magnetic resonance imaging (MRI), 80–81
Magnetic resonance spectroscopy (MRS), 80, 82
Maintenance of wakefulness test (MWT), 67*t*, 73–74, 78–79
Mania
  attention deficit hyperactivity disorder, 302–303
  bipolar disorder, 296, 298–299
  adolescence and early-onset, 300–301
  symptoms, 298
  cognition, 304–305
  cyclothymia, 301
  decreased need for sleep, 293–295
  deregulation of energy levels, 293–295
  diagnostic criteria, 293
  intervention pathways, 306–309
  mood switches, 295–297
  obsessive-compulsive disorders, 301–302
  sedative agents, 294
  symptoms, 295–297
  therapeutic implications, 305–309
  typical cyclic nature, 295–297
MBCT. *See* Mindfulness-based cognitive therapy (MBCT)
MBIs. *See* Mindfulness-based interventions (MBIs)
MBSR. *See* Mindfulness-based stress reduction (MBSR)
Medial PFC, amygdala and, 260, 405–406, 468–469, 470–471
Medulla oblongata, 4
Melanopsin, 14
Melatonin
  assessment of sleep, 67*t*
  chronobiotic impacts, 307
  with puberty, 426
  research-focused measures, 83–84
Memory
  affected by sleep deprivation, 16
  emotional, 35–36, 174–175, 408
  fear extinction, 171–172
Mental health, 463
Mindfulness-based cognitive therapy (MBCT), 215, 341–342, 357, 364
Mindfulness-based interventions (MBIs), 340–342
  descriptive summary, 343*t*
  effects, 356
  efficacy, 342
  implications, 362–364
  investigate impacts, 475–476
  theoretical models, 358–359
  usage, 342
Mindfulness-based stress reduction (MBSR), 340–341
  CBT-I and, 357–358
  components, 341–342
  health education and, 356–358
  implications, 362–364
  insomnia disorder, 355–356
  mindfulness meditations and, 341*t*

Mindfulness-based stress reduction (MBSR) *(Continued)*
  nonspecific factors, 358
  sleep duration, 356
Mindfulness meditation (MM), 355
  calming stage, 360–361
  cognitive models, 359–360
  implications, 362–364
  insight stage, 362
  MBSR, 341*t*
  neurobiological models, 360–362
  practice, 340–341
  randomized controlled trials, 356
  theoretical models, 358–359
  time spent, 355–356
Mood. *See also* Depression
  adolescent sleep and, 426–427
  and anxiety, 475–476
  circadian sleep process and, 387–389, 406–407
  and cognitive processes, 387–389
  consistent predictor, 231–232
  definition, 406
  disorder, 282–284
  disturbances, 168–170
  as emotional climate, 91
  impacts next-day sleep, 238–239
  and nightmares (*see* Nightmares)
  nondepressed samples, 230–231
  Profile of Mood States, 126
  regulation, 163–164
  sleep and, 426–427
  subjective/self-report measures, 126–127
MRI. *See* Magnetic resonance imaging (MRI)
MRS. *See* Magnetic resonance spectroscopy (MRS)
Multiple sleep latency test (MSLT), 390
  assessment of sleep, 67*t*
  diagnostic measures, 72–73
  rheumatoid arthritis, 382
  scores, 382–383
Musculoskeletal pain
  fibromyalgia, 382–383
  neuropathic pain, 383
  rheumatoid arthritis, 381–382
Musical scores, 95–96, 100
MWT. *See* Maintenance of wakefulness test (MWT)

N

NA. *See* Negative affect (NA)
Narcolepsy, 7–8
  MSLT, 72–73
  nightmares associated with, 168
National Sleep Foundation (NSF) poll, 429–430
Negative affect (NA), 55–56, 276–278
  affect influences sleep, 466
  aging, 442–443
  basal, 48–49
  definition, 442–443
  direct effect, 56
  sleep problems influence affect, 464–465
Negative-emotion induction, 51–52
Nerve cells, swelling, 4
Neural sleep regulation, and its affect, 408
Neurobiological models
  fear/anxiety, 152–153
  mindfulness, 360–362
Neurocognitive model, fear/anxiety, 151
Neuroimaging of sleep, 80
Neurological methodology, 107–108
Neuropathic pain, 383, 391
Nightmares, 165
  associated with narcolepsy, 168
  cognitive-behavioral techniques, 172–173
  depression and, 168–170
  diathesis-stress model, 172
  disturbances of mood and, 168–170
  Fisher theory, 171
  Freud's original dream theory, 170
  Hartmann theory, 171
  imagery rehearsal therapy, 173
  and mood regulation, 163–164
  Nielsen and Levin theory on, 171–172
  N2 NREM, 168
  phenomenology, 166
  and PTSD, 170
  REM behavior disorder, 166–168
  sleep disorders with, 166–168
  syndrome of recurring, 166
  theory and treatment strategies, 172–175
Nighttime sleep period, reduction, 384–385, 389–390
Night-time Thoughts Questionnaire (NTQ), 202–203
Nociceptors, 377
Nonfearful affect, 194

Non-rapid eye movement (NREM) sleep, 3–4, 109, 447–448
  in adults, 70
  arousal structures, 9
  aspects, 174–175
  basal forebrain, 10–11
  definition, 447–448
  emotion regulation, 449–450
  flip-flop switch, 12
  and PET, 82
  presleep worry, 214–215
  role of, 9
  slow wave sleep, 4–5, 72
Normal sleep
  changes, 422–424
  controls, 49
  types, 4–5
NREM sleep. *See* Non-rapid eye movement (NREM) sleep
NSF poll. *See* National Sleep Foundation (NSF) poll
NTQ. *See* Night-time Thoughts Questionnaire (NTQ)

O

Objective diagnostic measurement
  actigraphy, 67t, 74–76, 74f, 75f
  EEG, 69
  MSLT, 72–74
  MWT, 67t, 73–74
  NREM sleep, 70
  PSA, 72
  PSG, 69–70
  REM sleep, 70–72
  SWS, 70, 71–72
Objective sleep assessment, 473–474
Obsessive-compulsive disorders (OCD), 301–302
Obstructive sleep apnea (OSA), 321–322
  exercise and, 327–328
  headache, 380
  moderate-to-severe, 72–73
Obstructive sleep apnea syndrome (OSAS), 264–265
OCD. *See* Obsessive-compulsive disorders (OCD)
Olfactory stimuli, 94
Oneiromancy, 187
Orexin. *See* Hypocretin
OSA. *See* Obstructive sleep apnea (OSA)

P

PA. *See* Positive affect (PA)
Pain, 377–378
  acute conditions, 379–380
  behavioral measures, 378
  catastrophizing, 388
  chronic conditions, 380–381
  cognitive-behavioral treatments, 389–390
  mediation analyses, 391
  methodological issues, 378–379
  mood and cognitive processes, 387–389
  musculoskeletal pain, 381–383
  nighttime sleep period, 384–385
  pharmacologic treatment, 390–392
  PIC activation and, 386–387
  tolerance, 386
  total sleep deprivation, 384
PANAS. *See* Positive and Negative Affect Scale (PANAS)
Panic disorder, 98, 186
Paradox of well-being, 442, 449
Penn-State Worry Questionnaire (PSWQ), 202, 204, 211–212, 215
Perceived sleep quality, 331–332
Persistent insomnia, 321–322
PET. *See* Positron emission tomography (PET)
PFC. *See* Prefrontal cortex (PFC)
Pharmacological agents, emotion elicitation, 98
Phasic REM sleep, 4–5, 164–165, 174
Photographs, emotion elicitation, 120–121
Physical activity. *See* Exercise
Physiological arousal groups, 52, 54
Physiological assessment, 104–105
PIC activation. *See* Proinflammatory cytokine (PIC) activation
Pittsburgh sleep quality index (PSQI), 78, 202
Point Subtraction Aggression Paradigm (PSAP), 123
Polysomnography (PSG), 54, 67t, 69–70, 205, 378–379
  acute pain conditions, 379–380
  fibromyalgia, 382, 391
  headache disorders, 380–381
  during migraine attacks, 380–381
  neuropathic pain, 383
  outcome measures for sleep, 354–356
  sleep disturbances, 145–146
  sleep effects of pregabalin, 391

POMS. *See* Profile of Mood States (POMS)
Pons, 4, 11
Pontine-geniculo-occipital (PGO) waves, 11, 164–165
Poor sleep, 283–284, 321
    and behavioral problems, 427–428
    cognitive and physiological arousal, 54
    depression treatment, 233–234
    and emotional reactivity, 48
    future depressive episode, 240
    and insufficient sleep, 56, 57–58, 229–230
    nondepressed sample, 230–231
    nonpharmacological treatments, 322
    observational and experimental studies, 239–240
    pharmacological treatments, 322
    quality, 249–250
    severity of depression, 232–233
    stress and, 51
Positive affect (PA), 275
    affect influences sleep, 466
    daily, 278, 279*t*
    definition, 276
    depression/bipolar disorder, 284–285
    discrete, 278–281, 280*t*
    dysregulation, 285
    emotion regulation in aging, 442, 444
    evidence, 275–276
    mechanisms, 281
    promotive effects, 276
    protective effects, 276, 281
    and resilience, 276
    risk and vulnerability factors, 282
    sleep problems influence affect, 465
    trait, 276–278, 277*t*, 281
Positive airway pressure (PAP), 327–328
Positive and Negative Affect Scale (PANAS), 126–127
Positivity bias, 442
Positron emission tomography (PET), 3–4
    emotional stimuli, 131
    radioactive tracer, 81, 82
Post-traumatic stress disorder (PTSD), 94–95, 146, 473–474
    behavioral interventions, 155–156
    CBT-I and, 156–157
    clinical interventions, 155–156
    hypothesis from, 48
    nightmares and, 170
    polysomnographic studies, 170
    post-trauma intervals, 147–148
Power spectral analysis (PSA), 72, 82

Prefrontal cortex (PFC)
    dorsal attention system, 361–362
    Gage incident, 259–260
    importances, 259–260
    inhibitory control, 360–361
    sleep deprivation in adults, 424
    sleep loss, 260 (*see also* Sleep)
Presleep cognitive activity, 203–204
Presleep rumination, 209–210
Primordial emotions, 23–44
Profile of Mood States (POMS), 126, 384–385
Proinflammatory cytokine (PIC) activation, 386–387
PSA. *See* Power spectral analysis (PSA)
PSAP. *See* Point Subtraction Aggression Paradigm (PSAP)
PSG. *See* Polysomnography (PSG)
PSQI. *See* Pittsburgh sleep quality index (PSQI)
PSWQ. *See* Penn-State Worry Questionnaire (PSWQ)
Psychophysiological measurement, emotion elicitation
    automated computer scoring, 131
    cardiovascular measures, 129
    electromyography, 130
    facial displays, 129–130
    human scoring, 130–131
    pupillometry, 127–128
    saliva assays, 128
Psychophysiology, and emotion, 28–29
PTSD. *See* Post-traumatic stress disorder (PTSD)
Public health implications, 433–434
Pupillometry, 127–128

Q

Questionnaire assessments, 67*t*, 78

R

Randomized controlled trials (RCTs), MM, 356
Raphe nuclei (RN), 7, 8*f*
Rapid eye movement (REM) sleep, 3–4
    in adults, 70–72
    aging effects, 448
    amygdala during, 172
    aspects, 174–175
    brain activation and, 165
    deprivation, 386
    disturbances, 300
    dream mentation, 4–5, 11
    electroencephalogram, 11, 447

emotional processing, 408–409
emotion elicitation, 109
exercise and, 326, 327
fragmentation, 284
generator location, 11
during manic episodes, 303
muscle paralysis, 4–5
nightmares and (*see* Nightmares)
NREM-REM cycle, 164
phasic component, 164–165
presleep worry, 214–215
proportion, 408–409
recent findings/current research, 174–175
regulation of emotion, 449–450
REM behavior disorder, 166–168
REM-on and REM-off neurons, 12
roles and impacts, 408
sleep latency, 51
slow-wave sleep and, 298
tonic component, 164
tonic portion, 4–5
Reactive aggressive behavior, 248–249, 261–262
Recurrent fearful ISP, 182, 183*t*, 184–185
Regulation
  affect
    experiential/analytic strategy, 54
    impact on sleep, 53–54
    sleep impact on, 47–50
  emotion (*see* Emotion regulation)
  sleep disturbances on, 303–305
REM behavior disorder (RBD), 166–168
REM sleep. *See* Rapid eye movement (REM) sleep
Repetitive thought, 201, 213–214
Research-focused measurement
  combined fMRI/PET-EEG studies, 82
  cortisol, 83
  fMRI, 81
  hormones involved in sleep, 82
  melatonin, 83–84
  MRI, 80–81
  MRS, 82
  neuroimaging of sleep, 80
  PET, 81, 82
Resilience, 276. *See also* Positive affect (PA)
Response Style Questionnaire, Rumination Scale (RRS), 207, 217–218
Restless leg syndrome (RLS), 321–322
  exercise and, 328
  symptoms, 328
Restorative process, 424–425
Reticular formation (RF), 5–7, 6*f*

Rheumatoid arthritis, 381–382
RN. *See* Raphe nuclei (RN)
Rosenzweig picture frustration study, 255, 256*f*
Rumination
  clinical implications, 216–218
  common and distinguishing features, 210–212
  phenomenology and assessment, 206–208
  self-report measures, 210–211
  and sleep, 201, 208–210
  theoretical models, 212–215
  trait measure, 53–54
  treatment implications, 215–216

S

Sadness. *See also* Depression
  arousal negative affect, 240
  definition, 227
  and depression impact, 236–238
  impacts, 228–229, 231, 234
  and sleep, 241
Saliva assays, 128
Salivary cortisol (SC), 213–214
SAMs. *See* Self-assessment manikins (SAMs)
SCN. *See* Suprachiasmatic nucleus (SCN)
SCR. *See* Skin conductance response (SCR)
Script-driven imagery, 94–95, 101
Sedentary behavior, 328–329
Self-assessment manikins (SAMs), 106, 107, 126
Self-report assessment, 105–106, 473–474
Self-reported sleep quality, 53, 475–476
Sensations, isolated sleep paralysis, 188–190
Sensorium, in ISP, 192–193
Serotonin, 263, 264
Shift work sleep disorder, 15
Skin conductance response (SCR), 95–96
Sleep
  animal studies, 255–257
  apnea headache, 380
  arousal structures, 9, 10*f*
  clinical implications, 216–218, 264–265
  correlation, 249–250
  definition, 248
  deprivation study, 254
  diary
    ESM, 55
    measures, 67*t*, 77–78
    variables, 354

Sleep *(Continued)*
  disorders, 166–168, 236–238
  duration, 48–49
  forensic psychiatric inpatients, 252–254, 253f
  fragmentation, 54, 134, 385
  homeostasis, 294–295
  human studies, 254
  impact
    low mood, 240
    next-day, mood, 238–239
    sadness, 228–229
  implications, 239–241
  loss, 252–254
    causes, 260
    hypothesized causal mechanism, 261f
    serotonergic function, 264
  measurement, 231, 248, 354–356
  regulation, 400–401
  restriction, partial, 126, 133–134
  at risk, 263–264
  self-report questionnaires, 250–251
  stages, 70f
  theoretical models, 212–215
  treatment implications, 215–216
Sleep deprivation, 15–17, 296–297
  adolescents, 464
  in adults, 428
  affective regulation, 303–304
  anxiety, 47–48
  behavioral tasks
    anger inductions, 123
    directed emotional expressions, 122–123
    stressors, 123–125
  emotional stimuli, neural responses, 131–132
  emotion elicitation, 109, 110
  environment, 132–133
  negative consequences, 421
  nonspecific, 384–385
  and prolonged sleep restriction, 470–471
  psychophysiological measurement
    automated computer scoring, 131
    cardiovascular measures, 129
    electromyography, 130
    facial displays, 129–130
    human scoring, 130–131
    pupillometry, 127–128
    saliva assays, 128
  REM sleep, 294–295, 386
  research designs, 134–135

  sleep manipulations, 133–134
  slow-wave sleep, 385–386
  subjective/self-report measures, 125–126
    behavioral tasks, 127
    mood states, 126–127
    visual stimuli, 126
  visual stimuli
    film clips, 121–122
    photographs, 120–121
Sleep disturbances
  pharmacological and behavioral interventions, 474
  rumination, 211
Sleep latency, 51
  arousal, 52–53
  cognitive and physiological arousal, 54
  negative emotions, 55, 59
  stressors, 51
Sleep onset latency (SOL)
  actiwatch data, 209
  fragmentation, 284
  prolonged, 305–306
  worry-induced, 205
Sleep paralysis. *See also* Isolated sleep paralysis (ISP)
  clinical impairment, 194–195
  cultural and historical contexts, 186–188
  and ISP, 185–186
  nonfearful affect and, 194
Sleep-wake cycle
  circadian rhythm, 3–4, 12, 297f
  disturbances, 298–299, 305
  notable modifications, 446–447
  regulation, 151
  restoration, 294
  during temporal isolation, 299
Slow-wave sleep (SWS), 4–5, 51, 70, 71–72, 294–295
  characterization, 447
  deprivation, 385–386
  exercise and, 326
  growth hormone secretion, 324–325
  and REM sleep, 298
Socioemotional adjustment, 407
Socioemotional selectivity theory, 443
SOL. *See* Sleep onset latency (SOL)
Spielman model, 426
Spinal cord, 4
State Trait Anxiety Inventory State Scale, 205
Stimulus control model, 150–151
Stressful film paradigm, 98–99

Stressors
  appraisal of, 153–154
  behavioral tasks, 123–125
Stress, sleep latency, 51
Subjective diagnostic measurement
  clinical interview, 76, 77
  DISS, 79–80
  ESS, 78–79
  Flinders fatigue scale, 79
  FOSQ, 79
  insomnia severity index, 80
  PSQI, 78
  questionnaire assessments, 78
  sleep diary measures, 77–78
Subjective/self-report measures, 125–127
Suprachiasmatic nucleus (SCN), 13–14, 422
  arousal, 14
  circadian timing, 66–69
SWS. *See* Slow-wave sleep (SWS)
Sympathoadrenomedullary system, 152–153

T

Testosterone, 128
Thalamus, 7, 8*f*
Three-P model, sleep difficulties, 426
Time course
  emotion, 91, 104–105, 109
  neural regions, 106–107
Total sleep deprivation, 16, 384
  acute, 133–134
  mood and, 231–232
Total sleep time (TST), 70–71, 77, 355–356
Trait measures
  mindfulness, 215
  rumination, 53–54, 205–207
  worry, 204, 205–206
Trait positive affect, 276–278, 277*t*, 281
Treatment(s)
  continuous positive airway pressure, 78, 79
  depression, 233–234, 322
  insomnia, 325
  mania, 307
  pain, 389–392
  restless leg syndrome, 328
  rumination, 215–216
  for sleep, 215–216, 474–476
  strategies of Nightmares, 172–175
  for worry, 215–216
Trier Social Stress Task (TSST), 124
  research designs, 135
  salivary cortisol concentrations, 129*f*
TST. *See* Total sleep time (TST)
Tumor necrosis factor alpha (TNF-α), 386–387

V

Valence. *See also* Emotion
  and arousal, 52–53
  hedonic, 30–31
  negative emotions, 51, 54, 191
  positive emotions, 238–239
  primary perspectives, 92
VASs. *See* Visual analog scales (VASs)
Ventral periaqueductal gray matter (vPAG), 7
Ventrolateral preoptic area (VLPO)
  in anterior hypothalamus, 9, 10*f*
  inhibitory projections, 12
  REM, 12
Violence, 248–249
Virtual reality, 99
Visual analog scales (VASs), 106, 107
Visual stimuli
  film clips, 121–122
  IAPS, 92–93
  photographs, 120–121
  SAM, 126
VLPO. *See* Ventrolateral preoptic area (VLPO)
Voluntary hyperventilation task, 96–97
vPAG. *See* Ventral periaqueductal gray matter (vPAG)

W

Wakefulness, 3–4, 5–9
  during episodes, 192
  lateral hypothalamic area, 7–8
Weeknight sleep time, adolescents, 411–412
Word stimuli
  ANEW, 93
  emotional Stroop task, 93–94
Worry
  clinical implications, 216–218
  common and distinguishing features, 210–212
  phenomenology and assessment, 202–203
  and sleep, 201, 203–206
  theoretical models, 212–215
  trait rumination and, 205–206
  treatment implications, 215–216
  work-related, 204

Y

Yoga, 325

Printed and bound by CPI Group (UK) Ltd, Croydon, CR0 4YY
11/06/2025
01899189-0006